Water Power
The Development and Application of the Energy of Flowing Water

by Joseph P. Frizell

with an introduction by Roger Chambers

Self Reliance Books

Get more historic titles on animal and stock breeding, gardening and old fashioned skills by visiting us at:

http://selfreliancebooks.blogspot.com/

Introduction

I am pleased to present yet another title on Homesteading and Farm Life.

This volume is entitled "Water Power: The Development and Application of the Energy of Flowing Water" and was published in 1910.

The work is in the Public Domain and is re-printed here in accordance with Federal Laws.

As with all reprinted books of this age that are intended to perfectly reproduce the original edition, considerable pains and effort had to be undertaken to correct fading and sometimes outright damage to existing proofs of this title. At times, this task is quite monumental, requiring an almost total "rebuilding" of some pages from digital proofs of multiple copies. Despite this, imperfections still sometimes exist in the final proof and may detract from the visual appearance of the text.

I hope you enjoy reading this book as much as I enjoyed making it available to readers again.

Roger Chambers

PREFACE.

WATER-POWER, which was formerly the chief reliance of mankind in industry, has been greatly overshadowed in recent times by the cheapness of coal and the development of the steam-engine. This condition, while permanent to all immediate practical intents, is nevertheless, upon a broad view of the subject, merely temporary. Two hundred and fifty years ago, when unbroken forests stretched from the Penobscot to the Wabash, the idea of a scarcity of wood and timber appeared grotesque. At the present day, the idea of the exhaustion of existing deposits of coal appears equally so. Nevertheless that time will certainly come. Our deposits of coal are finite; every ton taken from the mines leaves a ton less to be mined. Coal does not have the power of reproduction even to the extent that wood and timber have. It does not, like water, have the power of rising in the form of vapor after having developed energy by descending. Coal deposits are simply an enormous store of fuel in rapid process of destruction by fire. Water, on the other hand, will continue to stand mankind in good stead long after the reluctant earth has yielded up its last ton of coal; at least, long after coal has become too inaccessible and consequently too expensive to be used for power.

Recent remarkable developments in electricity and other modes of transmitting mechanical energy have recalled water-power to something like its former position in industrial economy. Whereas formerly it was necessary that the industry should be located at the waterfall, it is now possible to make large use of water-power for industries located many miles distant. These changed conditions have given a new interest to water-power and directed the attention of investors to sources which formerly appeared entirely outside the range of practical consideration. They have also led to some noteworthy improvements in utilization, development, and transmission of power.

These improvements are embodied in records of engineering societies, in the pages of engineering journals, in the plans of skillful engineers, in catalogues and advertisements of machinists. It has appeared to the writer no unworthy task to review these sources of information, to select what appears worthy of preservation, or necessary for the illustration of the subject and put it in an accessible and preservable form. In other words, to bring the entire subject of water-power up to date.

This is in no sense an elementary work. It does not deal in axioms, definitions, or the elements of science. It assumes the reader to be possessed of elementary notions of mechanics and mathematics. It discusses some subjects which cannot be made clear without recourse to the higher mathematics, and in these cases the writer does not hesitate to plunge boldly into these branches of analysis. He does not conceive it necessary to offer any apology for this proceeding. It is true that there are successful engineers—successful at least in the sense of obtaining and retaining lucrative employments—who affect to despise mathematical knowledge, and make a boast of their ignorance of these branches. As well might the lawyer boast his ignorance of the common law. Mathematics is the common law of engineering. It is to engineering what the common law is to the legal profession. Granting that in practice no scientific principle can be applied, the truth of which has not been avouched by experiment, the fact remains that experiments cannot be intelligently conducted until their nature and scope have been indicated by the results of mathematical inquiry; neither can their results be intelligently coördinated and understood without recourse to the same potent instrument of investigation.

The work, however, is by no means valueless to readers of limited mathematical knowledge. These will be obliged to omit certain demonstrations, and content themselves with the results of the inquiry. They will find the main part of the work entirely intelligible.

The work does not attempt to go much into constructive details. It directs the reader's attention to the principles involved in the design, and the results to be aimed at, without invading the province of the machinist, the carpenter, or the mason. The illustrations are conformable to this view, being generally no more elaborate than necessary to elucidate the principles to which they relate; and these principles are often illustrated by sketches of very little value in a constructive point of view. Photographs, except to show general situations, have been avoided. A photograph of a finished machine is generally no more instructive to the student of mechanism than the photograph of a man in an overcoat to the student of anatomy.

PREFACE TO THE THIRD EDITION.

THE passing of this book to a third edition appears to approve the author's judgment in attempting the work. In the interval a fact has transpired well calculated to awaken reflections as to the nature of the agency now chiefly relied on for mechanical energy. We have seen a dispute between coal-mine owners and their employees carry the prices of all grades of coal to more than double their ordinary level, and nearly a year elapse without witnessing a return to the same level. This fact appears to show an abnormal sensitiveness in the coal market to any dis-

turbance of the supply, and has an ominous significance as showing what the coal consumer has in future to expect from combinations of mine-owners or of mine workers.

In this edition some typographical and numerical errors
which, despite great care, had crept into the former, have been
eliminated. The subject of dams on soft bottom has been considerably extended, and a section added on dams in tidal waters.
The length of wasteways in relation to the appurtenant drainage
area has been more fully considered. The subject of tidal power
has been outlined, and extended attention has been given to
storage reservoirs. The subject of leases and rentals of water-
power and land appurtenant thereto has also been more fully
elaborated.

BOSTON, April, 1903.

CONTENTS.

WATER-POWER.

CHAPTER I.

NATURAL WATERCOURSES.

PHYSICISTS are wont to assert that the sun is the source of all power upon the earth. It creates fuel by separating carbon from the carbonic acid of the atmosphere and storing it in the substance of plants. In former times, when this constituent of the atmosphere was much more abundant than at present, it thus originated the vast accumulations of vegetable growth which now constitute the coal-mines. By rarefying parts of the atmosphere and disturbing the barometric equilibrium, it creates the winds, which involve enormous power. A current of air one mile wide, 100 feet high, and moving with the moderate velocity of 30 miles an hour represents an expenditure of more than 100 000 horse-power. The sun creates the food which nourishes animals and gives them the strength to perform labor. It also sets in motion the agencies which give rise to water-power.

Watercourses.—About three-fourths of the earth's surface is water, from which evaporation is constantly going on. Water in the form of vapor is constantly rising from the seas. It is distributed by the winds in very unequal measure throughout the land, falls in the form of rain or of snow which resumes the liquid form, and, gathering into streams, pursues its course

toward the sea.   Rivers owe their existence to this ceaseless
play of natural forces, and to the further fact that the contraction of the earth's crust has tended to form its surface in a
series of ridges and valleys.

Flow of Streams.—The agencies which sustain the flow of
streams are of extremely variable and intermittent character,
whence arises great variation in the flow of streams.   No feature
of streams is more striking than this.   Rivers as large as the
Connecticut at Hartford, with a drainage-ground of 10000
square miles, carry forty times as much water at one time as
another.   The maximum flow of the Merrimac at Lowell,
where its drainage-area is some 4000 square miles, is sixty or
seventy times the minimum.   The Kanawha River at Charles-
ton, W. Va., with a drainage-area of 9000 square miles, has
varied in the ratio of one to one hundred.   The Minnesota
River at Fort Snelling, Minn., draining some 16000 square
miles, has been known to carry 8 or 10 cubic feet per second
for every square mile of its drainage-area; at other times,
not over 1 cubic foot per second to every 30 square miles.   In
small streams draining 10 square miles and under, the variations are almost without limit.   The extent of the drainage-
area together with the rainfall of the region forms the most
reliable indication of the flow of a stream both in flood and low
water.

The Slope of a stream or fall per mile, together with the
quantity of water to be relied on, determines, its value for water-
power.   Nearly all the streams emptying into the Atlantic and
the Gulf, except the Mississippi and its lower tributaries, have
some points of resemblance in respect to slope.   From the
point where they are large enough to be regarded as rivers to
where the influence of the tide becomes apparent, they have a
fall of from 2 to 10 feet per mile.   The slope is greatest in the
upper reaches of the stream, and diminishes toward tide-water.
It is by no means uniform.   Shoals and rapids occur, often
many miles in extent, with a fall much above the average,
and stretches of quiet water with a fall much below.

Streams in Different Sections of the United States.—In other respects there is a material difference between the streams of the Northern and those of the Southern States. This difference arises from three causes, viz.: 1. The effect of glacial action, occasioning great deposits of drift over the Northern and especially the Northeastern section, which are wanting at the south. 2. The greater effects of volcanic action in the former. 3. The difference in climate. From the first of these causes it results that Northern rivers flow in beds of gravel and boulders, with valleys sloping gently upward from the stream, while Southern rivers flow mainly in canyons or trenches worn out of the original rock with nearly vertical escarpments, and often partly filled with alluvial deposits. To glacial action is also due the immense number of lakes which dot the surface of the Northern section, and exert a marked influence on the flow of streams, a feature entirely wanting at the South. To the second cause is due the prevalence of metamorphic rocks in the Northern section which by their upheaval have created so many falls and rapids, and by their great hardness defy the abrasive action of the water. The Southern river-beds usually lie in sedimentary rocks of nearly horizontal stratification and often very fragile structure. In the third place the heat is more intense in the Southern States, the hot weather of longer duration, and in the Southwestern part the rainfall is less. Little snow falls and no ice accumulates. The rains, when they do come, are more violent. These causes, together with the absence of any regulating effect of lakes or great bodies of drift, tend decidedly to increase the fluctuations of streams, diminishing the low-water flow and increasing the floods. In some streams of the extreme Southwest the variations of flow are literally infinite, their beds being dry at times and roaring torrents at others.

In regions where snow accumulates, streams flowing east or west, like the Ohio, the Missouri, or the Minnesota, are liable to severer floods than streams flowing south, like the Connecticut, the Hudson, and the Upper Mississippi. In the first, the melting takes place simultaneously along its course, and the water is all precipitated into the channel. In the latter, the melting extends slowly up the stream, the result of one day's melting being well on its way before the next day's contribution enters the channel. Streams flowing in a northerly direction, like the Red River of the North and some of the tributaries of the St. Lawrence, are subject to great floods during the breaking up of the ice, which commences at the head of the stream and moves northward, constantly encountering firmer ice. The ice accumulates in immense masses, jams and dams the water, which is all the time increasing from the melting of snow. On the Red River of the North it is no uncommon thing to see driftwood among the branches of trees 40 feet above the ground. Such conditions are very trying to any artificial structure placed in the river.

In the Northeastern section, in the latitude of Boston, the total volume of water carried annually by streams is distributed throughout the year very much as follows:

January....	10 per cent.	July.......	2 per cent.
February..	14 "	August.....	3 "
March	20 "	September..	3 "
April......	15 "	October....	5 "
May......	10 "	November..	6 "
June......	4 "	December...	8 "

In the Mississippi valley, in the same latitude, this proportion is changed by a different distribution of the rainfall. In the extreme North, the lowest water occurs in the winter, the temperature remaining below freezing, often for months in succession, during which time no water enters the streams except from springs. The extremely low stage of the Minnesota above referred to occurred in the winter, after a long period of extremely cold weather. In the South Atlantic and Gulf States there is no snow to affect the spring flow. The streams are usually high during the winter and become very low during the latter part of the summer and autumn. In Texas and New

Mexico the normal flow of the streams is maintained by springs, the result of rainfall in previous years. The drainage-areas being vastly greater than for streams of equal flow in other sections, a concurrence of rains in all the tributaries gives rise to an enormous flood, which comes down with the velocity of a bursting reservoir or a tidal bore. It subsides rapidly and the river very soon returns to its normal condition.

Nearly one-half the territory of the United States, called the Arid Region, comprising the entire area west of the 100th meridian except the narrow margin between the Pacific Ocean and the Coast Range, is a region deficient in water, the rainfall ranging from 6 inches per annum in Nevada to 30 inches in other parts. The entire State of Nevada and a part of Utah lie in a basin which has no drainage to the sea, all the water that falls upon it being removed by evaporation. In many parts of the Arid Region no agricultural production is possible without artificial irrigation, and in no part of it can agriculture be carried on with full success without this aid. The interests of irrigation on all the streams are held to be paramount to those of water-power, and the latter thus becomes of quite secondary importance throughout this part of the country.

Table 1 gives some facts of interest in regard to the flow of streams in different parts of the United States. It is taken from vol. 16 of the 10th United States Census Report, and from the 18th Annual Report of the U. S. Geological Survey.

Relation of Slope, Depth, and Velocity in Streams.—If we carefully determine, by levelling, the fall in a given length of a running stream, and divide the fall expressed in feet by the length in feet, this quotient is called the slope and is represented by the letter S. Suppose the length to be a mile, and the fall 2 feet, the slope is $\frac{2}{5280} = 0.000379$. Let Fig. 1 represent the cross-section of a river-channel, ab the surface, $acdb$ the bottom. Measure the depth at a number of equidistant points; the average of these measurements is the mean depth. Measure also the velocity at a number of equidistant

TABLE 1.—DATA REGARDING THE FLOW OF STREAMS.

Stream.	Place of Measurement.	Drainage Square Miles.	Spring.	Summer.	Autumn.	Winter.	Year.	Max.	Min.	Minimum Flow per Square Mile.	Ordinary Low-water Flow of Stream.	Ordinary Low-water Flow per Sq. Ml.
Merrimac	Lawrence, Mass.	4 599	10	11	13	9	43	96 000	1 400	0.30	2 800	0.60
"	Lowell, Mass.	4 085	10	11	13	9	43	81 000	1 275	0.31		
Concord	Framingham, Mass.	361	11	11	12	10	44	4 449	60	0.17	126	0.35
Sudbury	Hanover, N. H.	78	11	11	12	10	44	3 228	2.8	0.036	12.5	0.16
Connecticut	Hartford, Conn.	3 316	10	12	12	10	44		1 006	0.303	1 210	0.365
Croton	Croton Dam, N. Y.	154	10	12	13	10	44	205 464	5 508	0.503		
Passaic	Paterson, N. J.	739	12	14	13	10	48	25 360	60	0.178		
Delaware	Lambertville, N. J.	813	11	13	11	9	44	17 913	195	0.24		
Potomac	Cumberland, Md.	6 820	11	12	9	8	39	350 000	2 000	0.29		
"	Great Falls, Md.	920	11	12	9	8	44	175 000	35	0.002		
James	Richmond, Va.	11 476	12	12	9	10	43	175 000	1 065	0.093		
Allegheny	Roberts Run, Pa.	6 800	12	13	9	10	44		1 300	0.191		
Ohio	Near Pittsburgh, Pa.	6 000	12	12	9	10	43		2 070	0.34		
Kanawha	Charleston, W. Va.	18 723	12	13	9	10	41		2 271	0.12		
Red River of the North	National Boundary	8 900	5	8	6	3	18	118 291	1 100	0.123		
Mississippi	Grand Rapids, Minn.	39 577	7	12	7.5	3.5	30				2 800	0.07
"	Rock Island, Ill.	87 842	7	12	6	3	30.5				19 000	0.97
Illinois	Mouth	29 013	8			8	24		1 600	0.055	1 750	0.216
Kansas	Topeka, Kan.	56 354							2 000	0.035	3 000	0.06
Ark. meas.	Mouth	160 000									3 000	0.043
White	Mouth	97 925	13.5	11	10	9	43.5	17 600	250			0.019
Potomac, 1895-6	Cumberland, Md.	801						31 950	300			0.1074
James, 1896	Buchanan, Va.	2 068						52 340	480			
Cape Fear, 1896	Fayetteville, N. C.	4 493						64 200	1 000			
Yadkin, 1896	Salisbury, N. C.	3 400						63 550	1 330			
Catawba, 1896	Rock Hill, S. C.	2 087						36 000	370			
Ocmulgee, 1894-5-6.	Macon, Ga.	2 495						445 120	16 360			
Tennessee, 1890-96	Chattanooga, Tenn.	21 832						27 200	96			
Platte, 1895 Aug. 17 to Nov. 1, 1896.	Columbus, Neb.							0	0			
Solomon, 1895	Beloit, Kan.	5 530						24 000	7			
Kansas, 1895-6	Lawrence, Kan.	59 841						53 508	700			
Arkansas, 1895-6	Hutchinson, Kan.	34 000						10 600	16			
Rio Grande, 1895-6	San Marcial, N. Mex.	28 067						11 300	0			
Gila, 1889-96	Buttes, Ariz.	11 710						12 000	1			
Humboldt, 1896	Oreana, Nev.	11 840						1 508	20			

points by methods to be hereafter explained, and thus determine the mean velocity, which is designated by v.   Determine
the length *acdb* of the line representing the bottom.   This

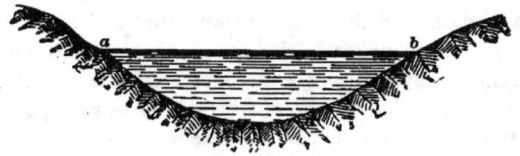

FIG. 1.

length is called the "wetted perimeter."   The mean depth
multiplied by the width *ab* is the area of the cross-section.
This area, divided by the wetted perimeter, is called the
"hydraulic mean depth" or the "mean radius," and is represented by R.   It is obvious that in a natural river-channel the
wetted perimeter is not materially different from the width,
and that no great error can result from taking the one equal to
the other.   In an artificial canal with vertical side walls,
Fig. 2, the wetted perimeter may differ very materially from

FIG. 2.

the width.   The relation between the slope, depth, and
velocity, in a stream, is ordinarily expressed by the formula.

$$v = c \sqrt{RS}. \quad . \quad . \quad . \quad .. \quad . \quad . \quad (1)$$

Though this is the best formula that modern hydraulic science
has to offer for the case in hand, it is still very imperfect.
The factor c varies between such wide limits that no general
value can be assigned to it.   It is not only different for different
streams and for different localities on the same stream, but it
varies for the same locality at different stages of the stream.

which shows that it does not truly express the law of the phenomenon. Nevertheless, in default of a better expression, we have to make such use as we can of it. The only rational use that the writer is able to make of it is by the aid of a table of data contained in the work of Ganguillet and Kutter, translated by Trautwine and Hering. This contains results of measurements by different experimenters on a great number of streams and channels from which it is usually possible to select the value of c suited to the case we have in view.

For rivers and canals with fairly regular channels in earth, with values of R less than 0.5 c may be 30;

from 0.5 to 1   "   45;

"   1   "   2   "   55;

"   2   "   3   "   65;

"   3   "   4   "   80;

"   4   "   10   "   100.

On great rivers, where R is from 30 to 75 c may run from 125 to 275.

Flowage Occasioned by Dams.—Questions as to the extent to which the surface of a stream is raised by a dam are of very frequent occurrence. Such questions as to an existing dam can be best determined by levelling. It is often required, however, to find how far the flowage of a proposed dam will extend, which can only be determined by calculation. Even in that case a survey of the river-channel would be made and numerous cross-sections taken of the stream and valley. In endeavoring to elucidate the principles of the case, with no existing data, we must proceed upon mere supposition as to the contour of the ground.

We will assume a stream 100 feet wide, a maximum depth of 6 feet, and a mean depth of 4 feet, the area of cross-section being 400 square feet. Suppose the fall to be 2 feet per mile and the mean velocity 3 feet per second, which would make $s = 0.000379$ and would imply a value of 77 for c, the relation of slope, velocity, etc., being expressed by the formula

$$v = 77 \sqrt{RS}.$$

We will assume a parabolic cross-section for the valley in which the river flows, a supposition which ordinarily represents the facts reasonably well. The equation of the parabola is $y^2 = 2Px$, where x and y are the coördinates of any point in the curve, x being, in this case, the maximum depth, and y the semi-width. We have the relation $\overline{50}^2 = 2P \times 6$, so that $2P = \dfrac{2500}{6} = 417$, and when the water rises, the relation between depth and width will be expressed by

$$(\text{semi-width})^2 = 417 \times \text{max. depth.}$$

That is to say, for purposes of computation we may put the width of the stream as raised by a dam equal to

$$2 \sqrt{417 \times \text{max. depth}} = 40.8 \sqrt{\text{max. depth.}}$$

We neglect the small error that results from assuming the width equal to the wetted perimeter, which allows us to put R equal to two-thirds the maximum depth.

Now suppose the surface of the stream to be raised 12 feet by a dam. If the water in the pond stood at an exact level, the flowage would extend 6 miles up the stream. This point where the natural surface of the stream is at the same level as the water at the dam is called the *hydrostatic limit* of the flowage, and its distance from the dam is called the *hydrostatic amplitude*. The surface of water, however, cannot be at an exact level while it is in motion. The water in the pond is moving, very slowly near the dam, more rapidly further up. Some head is required to maintain the movement, and the stream will be somewhat raised at the hydrostatic limit. The velocity increases and the depth diminishes as we go up the stream, so that no formula can apply to the entire reach under consideration. The best we can do is to divide the channel into lengths such that the velocity and depth can, without serious error, be regarded as constant in each, and compute the slope for each length. For an approximate computation, we will take lengths of a mile each, and will consider the cross-section and velocity at the middle of each mile section.

At a distance of half a mile up-stream from the dam, we should have the maximum depth $6 + 12 - 1 = 17$ feet. The mean depth we take at two thirds the maximum, viz., 11.333, and this is approximately equal to R. The width is $40.8 \sqrt{17} = 168.3$. For the increased depth we must adopt a higher value of c, say $c = 105$. The cross-section is $11.333 \times 168.3 = 1907.4$, and the velocity $= \dfrac{400 \times 3}{1907.4} = 0.629$. We have therefore for the fall in the first mile

$$5280s = 5280 \frac{(0.629)^2}{(105)^2 \times 11.333} = 0.01671 \text{ feet.}$$

For the second mile we have the depth $1\frac{1}{2}$ miles above the dam $12 + 6 - 3 = 15$, $R = 10$, width $= 40.8 \sqrt{15} = 158$, cross-section $= 1580$, velocity $= \dfrac{1200}{1580} = 0.760$. We may take $c = 100$. We have therefore for the fall in the second mile

$$5280s = 5280 \frac{(0.76)^2}{10 \times 100^2} = 0.03050.$$

For the third mile we have the max. depth $= 12 + 6 - 5 = 13$, $R = 8.67$, width $= 40.8 \sqrt{13} = 147.11$, cross-section $= 1275$. In this section the slope of the surface sensibly affects the cross-section.

We have for the fall in the first mile 0.0167
Second mile. 0.0305
Estimated in half the third mile 0.0300

    Total. 0.0772

Correction for the cross-section $0.0772 \times 147.11 = 11.36$. Making the corrected cross-section 1286.36,

$$\text{velocity} = \frac{1200}{1286.36} = 0.933. \quad \text{We take } c = 95.$$

Fall in third mile $= 5280\dfrac{(0.933)^2}{(95)^2 \times 8.67} = 0.0587.$

For the fourth mile we have max. depth at $3\frac{1}{2}$ miles

$12 + 6 - 7 =$ 11.0000

Add fall in first, second, and third miles and half of
fourth, estimated at 0.06.................... .1659

Corrected max. depth........................ 11.1659

Mean depth $= R = 7.444$, width $= 40.8\sqrt{11.1659} =$ 136.3, cross-section $= 1014.62$, velocity $= \dfrac{1200}{1014.62} = 1.183.$ We may take $c = 90$.

Fall in fourth mile $= 5280\dfrac{(1.183)^2}{(90)^2 \times 7.444} = 0.1225.$

For the fifth mile we have for the max. depth

$12 + 6 - 9 =$ 9.0000

Add fall in first, second, third, fourth, and half of fall
in fifth, estimated at 0.1450................. 0.3734

Corrected max. depth........................ 9.3734

$R = 6.249$, width $= 40.8\sqrt{9.3734} = 124.91$, cross-section $= 780.58$, $c = 86$, velocity $= \dfrac{1200}{780.58} = 1.537.$

Fall in fifth mile $= 5280\dfrac{(1.537)^2}{(86)^2 \times 6.249} = 0.2700.$

For the sixth mile max. depth $= 7.8188$, $R = 5.213$, width $= 40.8\sqrt{7.8188} = 114.09$, cross-section $= 594.73$, velocity $= \dfrac{1200}{594.73} = 2.018$, c may be taken at 81.

Fall in sixth mile $= 5280\dfrac{(2.018)^2}{(81)^2 \times 5.213} = 0.6285.$

Collecting these several quantities we have:

Fall in first mile...................... 0.0167
"   " second " 0.0305
"   " third   " 0.0587
"   " fourth " 0.1225
"   " fifth   " 0.2700
"   " sixth   " 0.6285

At the hydrostatic limit the water is raised.. 1.1269

To pursue the computations above the hydrostatic limit we adopt a method of trial and error which is in many hydraulic problems the only available method. We assume a value for the swell and execute the computations on that assumption. The result will show how the assumption should be amended. Guided by preliminary computations which need not be introduced, I adopt 0.732 as the rise of the surface at middle of the seventh mile. This makes the max. depth 6.732, mean depth 4.488, width $40.8 \sqrt{6.732} = 105.86$, cross-section 475.10, velocity = 2.5258. We here take $c = 79$.

$$\text{Fall in the seventh mile} = 5280\frac{(2.5258)^2}{(79)^2 \times 4.488} = 1.2026.$$

This result makes the surface at the middle of the
seventh mile $1.1269 + 0.6013 =$............... 1.7282
above what we may call the pond-level. Our
assumption would make it $0.732 + 1$ foot fall in
original surface........................... 1.7320
At the end of the seventh mile the surface of the
water is $1.1269 + 1.2026 =$................. 2.3295
above pond-level. The natural surface was.... 2.
Swell of stream 1 mile above the hydrostatic limit.. 0.3295

In like manner, guided by preliminary computations, we assume the swell at the middle of the eighth mile 0.189, max. depth 6.189, mean 4.126, width 101.50, cross-section = 418.79, velocity = 2.8654. Take $c = 78$.

$$\text{Fall in eighth mile} = 5280\frac{(2.8654)^2}{(78)^2 \times 4.126} = 1.7269.$$

This result makes the swell at the middle of the

eighth mile $1.1269 + 1.202 + 0.8634 - 3 = \ldots$ 0.1929

Our assumption makes it.......................... 0.1890

At the end of the eighth mile the swell is

$1.1269 + 1.2026 + 1.7269 - 4 = \ldots\ldots\ldots\ldots$ 0.0564

That is, at 2 miles above the hydrostatic limit the swell is about two-thirds of an inch.

In applying these methods to an important case, the engineer would take half-mile and quarter-mile sections near the hydrostatic limit of the flowage. He would draw the cross-sections, compute the areas, measure the wetted perimeter, and thence compute the values of R. He would also be guided by the nature of the bed and banks of the stream in finding the values of c.

It is worth while to remark that before the building of the dam the entire power of the stream was expended in giving motion to the water. After the water is raised by the dam a very slight fall suffices for the movement of the water, leaving the greater part available for power.

It is also to be remembered that the above computation of the flowage takes account only of the immediate and usually temporary effect of the dam. In the course of time other agencies come into operation which may have the effect of extending the flowage much farther up the stream. The stream, in high water, is constantly bringing down sediment, either earthy matters held in suspension, or gravel and pebbles rolling on the bottom. The heavier matters are arrested at the head of the pond where the velocity begins to slacken; the lighter are deposited farther down. Both, by diminishing the waterway, tend to extend the swell of the dam. In the course of years these deposits consolidate and harden and do not yield readily to an increased velocity of the current. Some dams are carried away with considerable regularity at every recurrence of extreme high water, an event which carries with it the removal of the accumulated deposits. In this case, the evil referred to only becomes apparent upon the construction of a permanent dam. In cases of litigation between millowners and parties claiming to be aggrieved by flowage, the dam, in order to demonstrate that it does not create the flowage complained of, is virtually removed by opening waste-gates and drawing down the pond till the fall is practically obliterated. It is urged that the flowage cannot be created by the dam, since it does not disappear upon removal of the same. This is an inconclusive test. It is necessarily made at a low stage of the stream and lasts but a few hours, during which time the current is entirely inadequate to the removal of the deposits which have accumulated in the course of years at the head of the pond. The fact remains that these deposits were occasioned by the dam and would disappear upon its permanent removal.

CHAPTER II.

DAMS FOR WATER-POWER.

IN common speech a dam for water-power is a structure designed to create a head or fall in a running stream. Strictly speaking, the fall is not created by the dam. It exists in the natural condition of the stream distributed over a greater or less portion of its length. The effect of the dam is to concentrate it at a single point and apply the power previously expended in idly chafing the channel, to the turning of wheels and giving motion to machinery. Another class of dams of distinctly different construction have important relations to water-power, viz., storage-dams designed to create reservoirs for holding the flow of abundant periods for use during periods of scarcity. We exclude this class of dams from present consideration.

Considerations as to the Form of Dams.—A dam may fail in several ways:

1. It may be overturned or shoved aside by the pressure

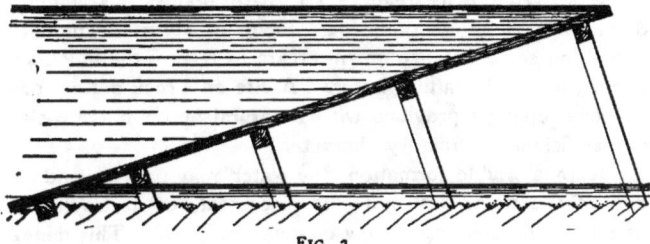

FIG. 3.

of the water, for which reason the body of the dam must have a strength and solidity sufficient to resist the utmost pressure that the water can exert against it. Often the stability of the dam is derived from the weight of the water itself (Fig. 3).

15

The dam is inclined so that the pressure acts mainly downward. A wooden dam of this kind sometimes fails by floating. A jam of ice, drift, or logs occurs in the stream, obliterating the fall. The force tending to hold the dam down then wholly disappears and it floats, or its tendency to float brings it into such disarrangement that it is destroyed when the jam breaks. A dam under such conditions should never depend for its stability wholly upon the pressure of the water.

2. A dam may leak so much as to wholly destroy or seriously impair its value. For this reason, every dam has a part specially applied with reference to stopping the water. This may be a layer of planking, of stone laid in cement mortar, of mortar alone applied to rough stone, or of a bank of earth. Examples have recently occurred in high dams of a "skin" consisting of a continuous riveted steel plate.

3. The water may find its way under the dam, and, on a soft bottom, destroy it by washing away its foundation. For this reason, the water-tight portion of the dam must go deep enough into the bottom to avoid all danger of undermining.

4. The water falling over a dam, in the case of a soft bottom, may excavate to such a depth as to engulf the dam in the cavity or so weaken its hold on the bottom as to allow it to be shoved down-stream. To guard against this action, a dam, in such a situation, must have an apron or platform extending some distance down-stream to receive the shock and commotion of the falling water. A site on a rock bottom may be considered as provided with a natural apron if the rock is of a sufficiently refractory character.

5. In a friable formation, the water may pass around the end of the dam, destroying it piecemeal, or destroying its usefulness by creating a new channel for itself. This danger is met by abutments at each end of the dam, rising above high water and connected in a water-tight manner with the shores. On rock bottoms the abutments are often furnished by nature. Very often the mill to which the dam pertains forms one of the abutments. The abutment sometimes rises above the

natural level of the bank and is joined to the high ground by an earth embankment or wall.

A dam therefore must, in general, have these several parts: (1) a body called the overflow, rollway, or spillway; (2) a water-tight skin or diaphragm; (3) a water-tight connection with the bed and banks; (4) an apron; (5) abutments.

The cross-section of the dam is governed, first, by the necessity of opposing a sufficient resistance to the pressure of the water. This is determined by a simple computation in which the principle to be observed is: That the moment of the forces tending to overturn the dam must be less than the moment of the forces tending to resist overturning. Of course these rigorous principles of statics apply but imperfectly to this case, and the results are of but limited value. The method assumes an unyielding foundation, which is true only of rock. It ignores the strength which the dam derives from its connection with abutments natural or artificial, and it assumes the impossibility of one layer of the dam yielding with reference to another. Still it is an aid to a correct comprehension of the problem.

Let Fig. 4 represent a dam sustaining the pressure of water. Consider one foot in length of such a dam. Let H represent the height of the dam, h the greatest height of water to be expected, in floods, above the top. The weight of a cubic foot of water varies somewhat with the temperature, the maximum being 62.4 pounds. For convenience of calculation we take it at 16 cubic feet to the 1000 pounds, being 62.5 pounds per cubic foot. The pressure against the face of the dam is

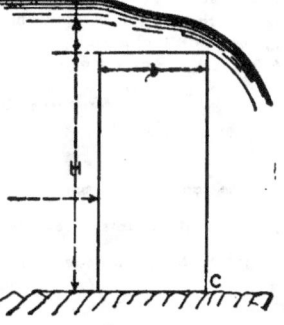

FIG. 4.

$$62.5H\left(\frac{H}{2} + h\right).$$

This force tends to overturn the dam around the point C. It is distributed in varying intensity over the face of the dam. Its effect on the dam is the same as that of an equal force acting at a single point, called the centre of pressure. When the surface of the water is at the top of the dam, the centre of pressure is at one-third the height above the bottom. The moment of pressure with reference to C is, in that case,

$$62.5\frac{H^3}{6}.$$

In the general case under consideration, it is easier to find the moment directly, without reference to the centre of pressure. This moment, as explained in the accompanying note,* is

$$62.5H^2\left(\frac{H}{6}+\frac{h}{2}\right). \quad . \quad . \quad . \quad . \quad (2)$$

The force tending to resist overturning is the weight of 1 linear foot of the dam. It acts through the centre of the dam, at a distance from c equal to one-half the breadth b.

It is often convenient to designate the weight per cubic foot of the masonry and the water by symbols.

---

* The moment of a force with reference to a given point is the product of the force (in pounds) by the perpendicular distance of the point from the line of action of the force. To find the moment of pressure on the vertical face of the dam, let x represent heights with reference to the point C. The moment of the pressure of any film of water whose height is dx is $62.5x(H + h - x)dx = 62.5(Hxdx + hxdx - x^2dx)$. The total moment is

$$62.5\int_0^H Hxdx + hxdx - x^2dx = 62.5\left(\frac{H^3}{2} + \frac{H^2h}{2} - \frac{H^3}{3}\right)$$

$$= 62.5H^2\left(\frac{H}{6}+\frac{h}{2}\right). \quad . \quad . \quad . \quad (2)$$

The centre of pressure is the point where the same moment would be produced if the total pressure $62.5\left(\frac{H^2}{2} + Hh\right)$ acted there. Let $X =$ the height of this point above the bottom, then

$$62.5\left(\frac{H^2}{2} + Hh\right)X = 62.5\,H^2\left(\frac{H}{6}+\frac{h}{2}\right),$$

whence

$$X = \frac{H^3 + 3H^2h}{3H^2 + 6Hh} = \frac{H^2 + 3Hh}{3H + 6h} = \frac{H}{3}\left(1 + \frac{h}{H + 2h}\right). \quad . \quad . \quad (2a)$$

Let w represent the weight of the masonry or material of the dam per cubic foot, y the weight of the water per cubic foot. Then $s = \dfrac{w}{y} =$ the specific gravity of the dam. We have wbH for the weight of a linear foot of the dam, and for its moment with reference to C, $wH\dfrac{b^2}{2}$. We have therefore as the condition of stability

$$wH\frac{b^2}{2} > yH^2\left(\frac{H}{6} + \frac{h}{2}\right),$$

or otherwise expressed,

$$sb^2 > H\left(\frac{H}{3} + h\right), \quad \text{or} \quad b > \sqrt{\frac{H}{s}\left(\frac{H}{3} + h\right)}.$$

In addition to the tendency of the pressure to overturn the dam, its tendency to shove the dam bodily down-stream must be considered. This is opposed indirectly by the weight of the dam on which the resistance to sliding depends. Smooth stone slides on smooth stone under a horizontal force of two-thirds its weight. To cause a stone to slide on gravel or clay requires a force nearly equal to its weight. Generally if a dam is massive enough to resist overturning it will resist sliding. In the case just considered, the horizontal pressure on the dam is $yH\left(\dfrac{H}{2} + h\right)$. The weight $= wHb$. If the dam rested on smooth rock, it would be liable to slide if

$$yH\left(\frac{H}{2} + h\right) = \frac{2}{3}wHb, \quad \text{or} \quad b = \frac{3}{4}\frac{H + 2h}{s}. \quad . \quad (3)$$

We are not here considering the diminution of weight of masonry by immersion.

A rock bottom is ordinarily very rough and irregular by nature, and the irregularities may be artificially increased, so that the chance of sliding is very remote. On a soft bottom, a high dam has been known to move down-stream, not by sliding, but by a yielding of the formation on which it rests.

Effects of Flood Height.—We have considered the increased pressure on the vertical face of a dam due to high water. High water affects the forces acting on a dam in another way. The water generally rises much higher on the down-stream side than above. This acts, in one way, in favor of the dam, by diminishing the pressure tending to overturn it. It acts, in another way, against the dam by diminishing the effective weight. Both these effects are easily calculated.

Where water goes over a rollway in an unbroken sheet, and is in contact with abutments at both ends, another force comes into action against the dam. The air under the sheet is rarefied, the atmospheric pressure against the down-stream face of the dam is diminished, while it acts in full force on the up-stream side. No accurate experiments have been made, so far as I am aware, to determine the extent to which this action takes place. That the force cannot be serious will appear from the following consideration. Its tendency is to diminish the leap of the falling stream. Its greatest possible deleterious effect would be to press the stream against the down-stream face of the dam (see Fig. 4). Any action in excess of this would be in favor of the dam, as tending to resist overturn. Its maximum effect would therefore be to annul the horizontal component of the water's movement. This is known to be the velocity due to one-third of the depth h. It appears to me, therefore, that this effect is fully allowed for by adding one-third to the value of h in computing the force acting to overturn the dam.

Dams of rectangular cross-section have thus far been considered, as exhibiting the simpler applications of the principle of moments. Dams of masonry are, however, more commonly built with an inclined face, there being some economy of material in this arrangement. The inclination of the up-stream face, Fig. 5, causes the resultant pressure to pass closer to the point C, thus diminishing the lever-arm of the moment of pressure. The inclination may be such as to cause the resultant pressure to pass through C or even below it, in which case the dam has stability without reference to the weight. Dams are often built upon this principle, as the one indicated at Fig. 3, in which the stability of the dam depends wholly on the pressure of the water.

The inclination of the down-stream face, as in Fig. 6, lengthens the lever-arm of the moment of resistance.

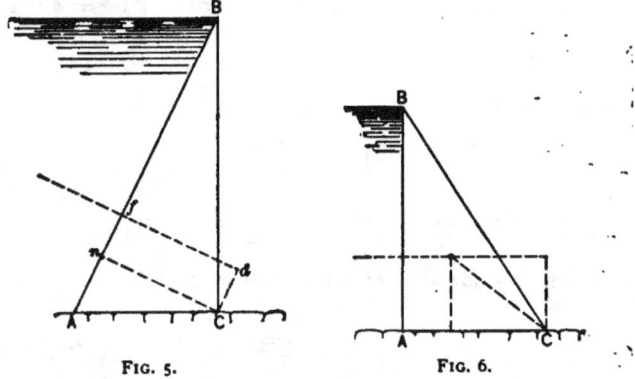

FIG. 5.   FIG. 6.

In comparing the stability of different forms of dam we do not need to take account of the surcharge h.

For the case of a triangular dam of masonry with down-stream face vertical, the water standing at the top of the dam (Fig. 5), the resultant pressure is perpendicular to AB and goes through a point f, such that $Af = \frac{1}{3}fB$. If we make the inclination of AB such that the resultant goes through C, we shall have $BC = AC\sqrt{2}$. $\therefore AC = \dfrac{H}{\sqrt{2}}$. This may be easily shown. In this case the dam has an excess of stability represented by the moment of its weight with reference to C. The weight acts vertically through the centre of gravity, and its line of direction meets the base at a distance from C equal to $\frac{1}{3}AC$. For greater convenience we will designate the sides of the triangle by the small letters a, b, c,—a being opposite the angle A, b opposite B, etc.

To determine the dimensions of such a dam, which will barely stand, under the pressure of the water, we proceed as follows:

Draw Cn perpendicular to AB, and Cd perpendicular to fd. The pressure on AB is $yc\frac{a}{2}$. Its moment with reference to C is $yc\frac{a}{2}Cd$. We have the proportion $c : b = b : An$, whence $An = \frac{b^2}{c}$ and $Cd = Af - An = \frac{1}{3}c - \frac{b^2}{c}$. The equation of moments is

$$y\frac{ca}{2}\left(\frac{c}{3} - \frac{b^2}{c}\right) = W\frac{ab}{2}\cdot\frac{b}{3} = W\frac{ab^2}{6}. \quad \cdot \quad \cdot \quad \cdot \quad (4)$$

Therefore $\frac{c^2a}{6} - \frac{ab^2}{2} = S\frac{ab^2}{6}$ or $\frac{c^2a}{6} = \left(\frac{a}{2} + \frac{Sa}{6}\right)b^2$, or, replacing c^2 by its value $a^2 + b^2$, we find, since $a = H$,

$$b = \frac{H}{\sqrt{s + 2}}. \quad \cdot \quad \cdot \quad \cdot \quad \cdot \quad \cdot \quad (5)$$

The cross-section of the dam is $\frac{bH}{2} = \frac{H^2}{2\sqrt{S + 2}}$.

For a rectangular section of equal stability the equation of moments is $wH\frac{b^2}{2} = y\frac{H^3}{6}$, whence $Sb^2 = \frac{H^2}{3}$ and $b = \frac{H}{\sqrt{3S}}$. The cross-section is $\frac{H^2}{\sqrt{3S}}$. Therefore

$$\frac{\text{cross-section of triangular dam}}{\text{cross-section of rectangular dam}} = \frac{1}{2}\sqrt{\frac{3s}{s + 2}}. \quad \cdot \quad (6)$$

If we put $s = 2.5$, we find the cross-section of the triangular dam 64.5 per cent of that of a rectangular dam of equal stability.

For a dam with up-stream face vertical and down-stream face inclined, Fig. 6, the weight acts through the centre of gravity I, at a horizontal distance from C equal to $\frac{2}{3}b$.

Moment of water-pressure $= y\dfrac{H^3}{6}$. Moment of resistance

$= w\dfrac{Hb}{2} \cdot \dfrac{2}{3}b = \dfrac{W}{3}Hb^2$, whence $b = H\sqrt{\dfrac{1}{2S}}$. Cross-section

$= b\dfrac{H}{2} = \dfrac{H^2}{2}\sqrt{\dfrac{1}{2S}}$. For a rectangular section of equal stability

we have for the equation of moments $y\dfrac{H^3}{6} = wHb\dfrac{b}{2}$, whence

$b^2 = \dfrac{1}{3}H^2\dfrac{y}{w}$ and $b = H\sqrt{\dfrac{1}{3S}}$.

Cross-section $= Hb = H^2\sqrt{\dfrac{1}{3S}}$, whence

$$\frac{\text{triangular section}}{\text{rectangular section}} = \frac{1}{2}\sqrt{\frac{3s}{2s}} = \frac{1}{2}\sqrt{\frac{3}{2}} = 0.612, \qquad (7)$$

or the triangular section is 61.2 per cent of the rectangular section.

In this case the relation of the cross-sections appears to be independent of the specific gravity. In eq. (6), if we make the specific gravity 2.25, the ratio is 0.631. If we make it 2, the ratio is 0.612, the same as in (7). It appears, therefore, to be practically immaterial, as regards stability, which side of the dam is inclined.

The mode of construction of a dam, and its connection with the bottom, often complicate the application of the principle of moments. The preceding discussion assumes that the upstream face of the dam is connected with the bottom. If we could suppose the dam connected with the bottom at C, and the bottom accessible between B and C to the water of the pond, then we should have to consider the pressure acting upward on the bottom of the dam as one of the forces tending to overturn it, and should find it necessary to greatly increase the weight of the structure.

The preceding results show an economy in inclining one

face of the dam, but the dimensions arrived at are not such as would be adopted in practice. In Figs. 5 and 6 the resultant of all the forces acting on the dam goes through the point *C*. Practically the corner would crush and the dam would fall. Aside from this, if the corner holds, the dam has no stability, being as likely to fall as stand. Moreover, a sharp crest on the top of a dam is usually inadmissible. Considerations apart from stability require the top to have a certain breadth. Therefore, to the outline of Fig. 5 or 6 a rectangular part should be added to give the required stability.

It is generally admitted among engineers that the resultant of all the forces acting on a dam should pass through the middle third of its base. This gives the dam a wider margin of safety than is usual in heavy structures. It may be adopted in the case of a dam of great importance whose failure would be attended with disastrous results. It is also convenient to remember in the frequent case where hasty and approximate estimates are required. In a case like that of Fig. 3 it finds no application. The rule may also be adopted in the case of bottom which cannot be regarded as entirely unyielding. The effect of the horizontal pressure is to throw the weight of the dam upon one side, and if the resultant comes too close to the foot of the down-stream face, the ground may not be able to sustain the weight. Often the down-stream face is curved, spreading out into a broad toe or apron, in which case the literal application of the rule would lead to inadequate dimensions.

The application of this rule to a dam with up-stream face vertical, down-stream face sloped, is indicated at Fig. 7. The forces are: the weight acting through the centre of gravity; the pressure of the water acting horizontally. *I* is the centre of gravity of the triangle *ABC*. To find it, bisect the base at *d*. Draw *Bd* and lay off thereon, from *d*, *dI* equal to one-third *dB* The centre of gravity *I'* of the rectangular part *ABef* is at the centre of figure,—mid-height and mid-width. Draw *I'I* and divide it at *i*, inversely as the weights. That is:

if the weight of ABC is represented by 7, and that of $ABef$ by 12, then $I'i$ must be $\frac{7}{12}$ of $I'I$. As before, let h represent the depth flowing over the dam. The horizontal pressure

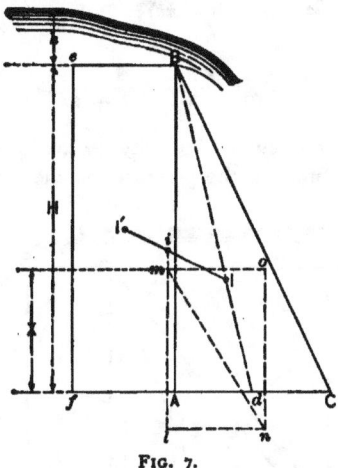

FIG. 7.

of the water acts, eq. (2a), at a height $X = \dfrac{H}{3}\left(1 + \dfrac{h}{H + 2h}\right)$ above the bottom. When $h = 0$, $X = \dfrac{H}{3}$. We can draw the two lines representing the forces acting on the dam. From m, their intersection, we lay off the vertical ml to represent the weight of the dam, and the horizontal mo to represent the pressure. Complete the rectangle $mlno$ and draw the diagonal mn. This represents the resultant of the forces acting on the dam, and must, according to the above rule, fall within the middle third of the base. If this condition is not fulfilled, we must redraw the figure with a different value of Be or a different inclination of BC.

Fig. 8 represents the construction when the up-stream face is inclined and the down-stream face vertical. In this case, putting c to represent the side AB, the pressure on the inclined face is $yc\left(\dfrac{H}{2} + h\right)$.   Moment with reference to A

$$= y(\tfrac{1}{3}c \cdot \tfrac{1}{2}Hc + \tfrac{1}{2}chc) = yc^2\left(\dfrac{H}{6} + \dfrac{h}{2}\right).$$

$$X = Af = \dfrac{c}{3}\left(1 + \dfrac{h}{H + 2h}\right). \quad \cdot \quad \cdot \quad \cdot \quad \cdot \quad (8)$$

i is the common centre of gravity, m the intersection of the two forces, mo represents the pressure, ml the weight, $mlno$ the

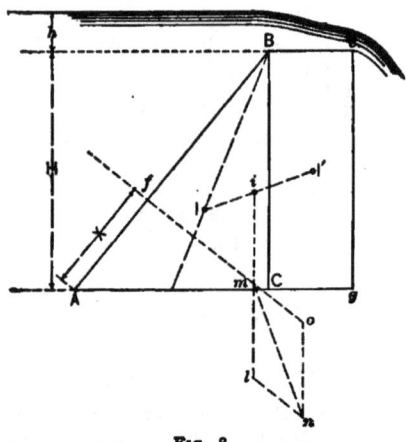

Fig. 8.

parallelogram of forces, mn the resultant.   The width Be, or the inclination of AB, must be varied till mn intersects the base at the right point.

Form with Reference to Discharge.—The cross-section is governed, secondly, by the consideration of passing the water in a manner to expose the dam to as little injury or chance of injury as possible.   To understand the conditions of this requirement, we must examine attentively the action of falling water. When a sheet of water falls over the perpendicular face of a dam, and strikes on a level surface, the stream, at certain stages, takes the form shown at Fig. 9. It changes its

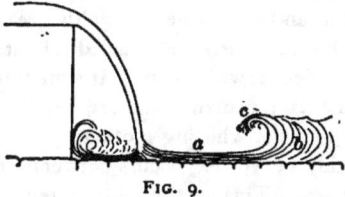

FIG. 9.

direction and flows with unabated velocity along the bottom. At *a* the stream is moving horizontally with substantially the velocity due the fall. At *b* the stream has taken its normal depth, that is, the depth determined by the quantity of water and declivity of the bed, and moves at a much slower velocity. The transition from the depth *a* to the depth *b* takes place suddenly. The elevated surface of the water moves up-stream toward *c*, and is constantly tumbling over into the swift water. *c* is the crest of a wave, which does not move like waves in deep water, but remains fixed, and is constantly breaking. It is called a standing wave. The distance of *c* from the falling stream depends upon the normal depth. As this increases, *c* approaches the stream and in a high stage of water the condition is as shown at Fig. 10. The water closes in upon the

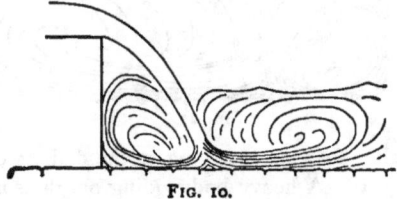

FIG. 10.

falling stream, which has the same direction as before, causing a swift current for some distance on the bottom. On the surface, the current is toward the falling stream, and a heavy body, as a log or block of ice, is carried down and along the bottom with great velocity, rises to the surface and moves up-stream, reaches the falling stream with momentum sufficient to pass through it and strike the dam with great violence. It is then seized by the current, whirled about, striking the bottom, and is carried down-stream. It sometimes repeats this cycle of movements for many minutes, till it chances to get beyond the range of the whirling motion.

A similar eddy or vortex occurs between the stream and the face of the dam. This is the more extended as the stage of the water is higher. The water in this enclosed space may rise considerably higher than the general level below the dam, owing to the exhaustion of the air as already alluded to. These commotions destroy the velocity and momentum of the water, as brakes destroy the momentum of a train. At a short distance below the dam, the water pursues its course with a velocity and depth determined by the character of its bed.

The highest flood, one whose normal depth would be equal to the height of the dam, goes over the latter in the form indicated at Fig. 11, and, provided the abutments hold, it is

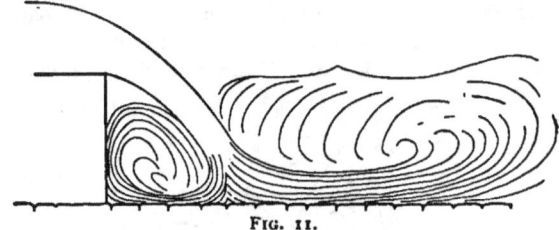

FIG. 11.

doubtful if such a flood is as trying to the dam as one of more moderate height. A heavy body riding on the surface would not, as in the case of Fig. 10, strike the dam, but the eddy under the flood-sheet takes a greater development, and bodies drawn into it are liable to beat against the dam.

It would appear, therefore, that, except for a hard rock bottom and a very solid construction of the dam, a perpendicular face down-stream is to be avoided.

The down-stream face often has the outline of Fig. 12,

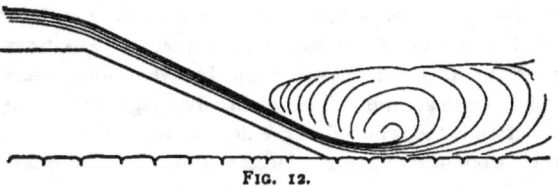

FIG. 12.

leading the water down in a long slope. In considering such dispositions, this fact must never be lost sight of: a given amount of fall imparts the same velocity to water whether the latter goes down an incline or falls freely. The friction on the incline has some slight effect of retardation, but not enough to affect the question. Fig. 12 represents the situation in high water. It differs from the case of a perpendicular face only in the absence of the eddy next the dam. In this case, as in the others, there is a strong current toward the falling stream, and heavy bodies are liable to stay in the vicinity of the dam, moving up-stream till caught by the water coming over the dam, then plunging under water and battering the inclined face, then rising to the surface and again moving up-stream. A floating body goes over the dam in a different manner from an equal mass of water, on account of its buoyancy. Ice has a buoyancy of 5 pounds per cubic foot, timber often as much as 25. A log, coming over such a dam, plunges into the swift current, going down-stream near the bottom, and would get clear of the eddy if it did not rise. Its buoyancy causes it to rise into the reverse current, which arrests its motion and carries it back up-stream. The action on the bottom is probably somewhat less severe in this case than in the case of a vertical face. The stream meets with more resistance before reaching the bottom, which diminishes its velocity. On a smooth artificially prepared bottom this consideration has force; but on a natural bottom with obstructions and irregularities which so readily change the direction of the current, the latter must be supposed to act with equal force in all directions. The direction in which the stream approaches the bottom is not so important as is ordinarily supposed. On a bottom of stratified rock, the beds horizontal or dipping down-stream, the action of an inclined stream might be even more severe than that of a vertical one. The main advantage secured by this construction is the removal of the destructive action to a greater distance from the vital parts of the dam. It is presumed, of course, that this construction would not be adopted except for a rock bottom. Many formations, however, which are classed as rock are susceptible of rapid wear. On such a bottom the effect of the form of Fig. 12 would be to scour out a hollow place at the foot of the slope. This extends and gives full development to the reverse eddy, which has the effect of extending the excavation under the slope. If the foot of the slope rests upon the bottom, it is rapidly destroyed. If it is so constructed that a considerable length of it will stand without support, the excavation extends, more slowly but still surely, till the safety of the whole structure is brought in question. The sloping part may have the effect to greatly defer this result, but cannot prevent it.

Dams with Steps.—An advantage is sometimes thought to be gained by dividing the fall into a number of small falls

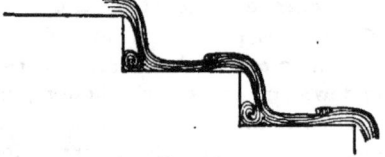

FIG. 13.

or steps separated by level "treads," as in Fig. 13. The water is assumed to lose at each step the velocity acquired in falling, and to reach the bottom with only the velocity due to

the last step. This would be a correct assumption if the steps
were far enough apart. To judge whether this result is obtainable within any reasonable constructive limits, it is necessary
to form some idea of the mode in which the water traverses
such a dam.

We have seen that water falling over a dam and striking
on the bed of the channel, supposed to be smooth, glides away
from the foot of the dam in a thin sheet with a velocity very
near that due the fall. The depth *a*, Fig. 9, for a fall of
4 feet would not in ordinary stages be more than one-fourth or
one-fifth that in the general bed of the stream, which we have
termed the normal depth. This normal depth tends to establish itself in all parts of the channel below the dam. It results
from this tendency that a mass of water is constantly rolling
back upon the escaping stream and being drawn away with it,
an action tending to the rapid extinction of velocity. The
distance below the dam at which the velocity becomes uniform
and normal is marked by the point where the water ceases to
flow up-stream. This distance is much less on a rough bottom
than on a smooth bottom. The movement is often greatly
modified by a condition which commonly exists at the foot of
a dam, viz., a deep pit or basin excavated by the water. On
a smooth, level, planked bottom it is not probable that this
distance would be less than five times the fall. We can readily
see that if, at the point *b*, Fig. 9, the water took a second
drop, the tendency to the extinction of velocity would be
greatly diminished. The tendency to the establishment of a
normal depth would be counteracted by the second fall.
There would be no mass of water flowing up-stream, and the
water would shoot over the second step with but little abatement of the velocity acquired by going over the first. Briefly
stated, the water, in going over such a dam, can undergo no
commotion tending to destroy its velocity, and therefore its
velocity is not destroyed. Of course, if the second step were
removed to a very great distance from the first, a normal depth
corresponding to the nature of the channel and volume of water

would establish itself. The preceding considerations, however, show how exceedingly unfavorable the conditions are to the extinction of velocity. These conditions become more unfavorable as the depth increases, till in high flood the water goes over the dam almost precisely as over a straight incline. In any stage above the lowest, it is evident that the steps can have little effect in moderating the velocity, without giving the dam a profile greatly in excess of that called for by considerations of stability. Should such a dam be built under the impression that water going over it would reach the bottom with a low velocity, dispositions necessary to secure the bottom from abrasion would be likely to be neglected, to the increase of the risk of disaster. This form of dam would be improved by giving the tread between two consecutive steps an inclination up-stream, forming a basin for stilling the commotion and moderating the shock of floating bodies. It is to be doubted, however, whether any form of dam has less to recommend it for adoption than this. It will appear later that no form of dam suffers more severely than this from ice and logs.

Inclination of Top of Dam.—It is usual to give the top of the dam an inclination, rising toward the down-stream side, to facilitate the passage of ice and floating bodies, the portion *ab*, Fig. 14, rising from 12 to 30 inches, according to the size of

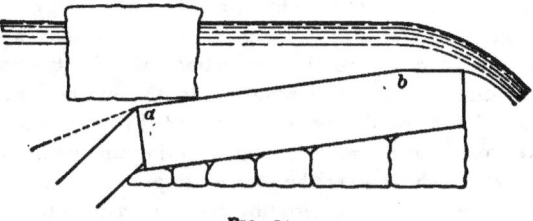

FIG. 14.

the bodies to be looked for. Blocks of ice on northern rivers may float to a depth of 40 inches, but such blocks would not move at all unless there was some depth of water going over the dam. Ice is liable to move in masses which make it formidable even at low velocities. This arrangement is to be recommended for a low dam, but the necessity for it diminishes as the height of the dam increases. It is manifest that in the case of a high dam, in any stage of water such that floating bodies could strike the dam, they must approach it with a very low velocity.

Dams of rectangular outline have sometimes been built with a sloping bank of earth or loose stone on the up-stream side. Whatever advantage may be expected from this arrangement, it certainly has one serious objection. It shoals the waterway approaching the dam, and increases the velocity with which floating bodies strike it. At a dam on a rapid stream, such a bank may be expected to form naturally in the course of time, and the crest should be shaped with reference to that contingency.

Dams Curved in Cross-section.—The execution of curved outlines in timber-work presents no insuperable difficulty, as is well understood in ship-building, but they involve methods not familiar to millwrights and are seldom attempted in dam construction. Masonry is not subject to this limitation, and of late many masonry dams have been built with curved faces, whereby it is thought that some advantage is gained.

The typical outline of such a dam is the line *mabd*, Fig. 15, viz., the curve *ma*, the inclined straight line *ab*, and the curve *bd*. In some cases the line *ab* is wanting, and *a* and *b* coincide, forming a reversed curve. The summit *m* is sometimes joined to the up-stream face by a horizontal line *mi*, sometimes by an inclined line. Down-stream from *d* the dam is constructed according to the nature of the ground. For rock of moderate hardness the surface would continue level for some little distance and terminate in a vertical face. This vertical face is usually rendered necessary by the irregular shape of the bed. If the bed were level and smooth, the down-stream face could join it by a curve having its tangent-point at *d*, but this is usually impossible. Sometimes the toe is so formed as to give an upward direction to the water, as at

Fig. 15*a*, in the expectation that the abrasive action is thereby carried farther from the dam. It may be doubted if any real advantage results from this arrangement. It gives a wider leap to the stream and tends to diminish the velocity with which it strikes the bottom. It gives greater opportunity for the development of horizontal eddies under the stream, and exposes the toe in greater measure to the impact of floating

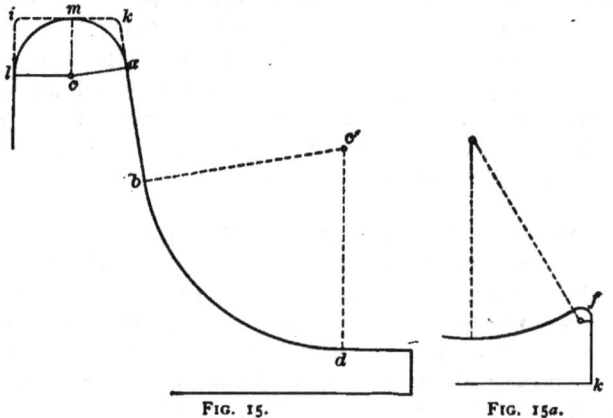

FIG. 15.   FIG. 15*a*.

bodies carried by the reverse current. In high water, when the down-stream surface comes well up on the curved face of the dam, such refinements of construction cut no figure, unless it be the following: Logs and blocks of ice, in passing the curved surface of the dam, are subject to a centrifugal force which causes them to hug the surface. This action is not necessarily injurious between *b* and *d*, where there is no danger of dislodging a stone. It is liable to act injuriously upon the exterior stones of the toe. Suppose a 30-foot dam and assume a radius of 20 feet in the curve of the toe. A heavy body would reach this curve with a velocity of something like 44 feet per second and would exert thereon, while passing, a pressure of $\frac{W}{g} \times \frac{44 \times 44}{20} = W \times 3$, i.e., a pressure of three times its normal weight. The frictional pull resulting from this pressure must have a strong tendency to dislodge the stones of the toe.

The top or upper part of the face, commencing at *m*, is sometimes formed to the curve which the overflowing stream would take if the down-stream face were perpendicular. This curve varies with the quantity of water passing the dam, but if it is correct for the highest stage of the water, the water will follow it at all other stages. The form of this curve is determined as follows:

Let v = the horizontal velocity of the stream at *m*. Then, at any short time *t* after the particle of water under consideration has passed *m*, it will have moved horizontally a distance vt and fallen vertically a distance $\frac{1}{2}gt^2$, where g represents the velocity acquired by a heavy body in falling one second. It is well known that the curve is a parabola whose equation is of the form $y^2 = 2Px$, in which $y = vt$, and $x = \frac{gt^2}{2}$. Therefore $v^2t^2 = Pgt^2$, and $P = \frac{v^2}{g}$, from which we have the equation of the curve

$$y^2 = 2\frac{v^2}{g}x. \quad . \quad . \quad . \quad . \quad . \quad . \quad (9)$$

From an investigation published by the writer in the *Engineering News* of September 29, 1892, it appears that v is the velocity due to one-third the "depth on the dam;" this expression being understood to mean the height of pond-level above the crest of the dam.

It is manifest that the form of the top might deviate materially from that of Fig. 15. It might have the outline *l i m a* without affecting the movement of the water injuriously. It might have the outline *l i m k a* with no more serious disadvantage than a tendency in the stream to leap clear of the dam and strike it again at a lower point, a matter of no importance. It must be borne in mind that where the overflowing stream is confined between abutments, the action of the atmosphere tends strongly to hold the stream to the down-stream face of the dam.

The form of the dam at the bottom is much more important than that at the top. Water falling over a dam reaches the lower level with the velocity due to the fall. At a distance down-stream which may not exceed four or five times the height of the dam, we find the water flowing with a uniform current and moderate velocity. At the former point the actual energy of the water, due to its mass and velocity, may be 50 000 horse-power; at the latter not more than two or three hundred. What has become of this enormous amount of energy? As a train of cars is brought to rest by the friction of the brakes, this mass of swift-moving water is brought to rest by the friction of its individual particles upon each other and upon the bed. The latter kind of friction tends to abrasion and injury; the former is harmless. In the case of the train as well as that of the water, the energy is transformed into heat. In the former, the energy being consumed by the brakes, these parts are heated to a very high temperature. In the case of the water, when each particle takes part in the expenditure of energy, the resultant heat is distributed through the entire mass and does not, in the ordinary case, raise the temperature enough to be detected by the most delicate thermometer. To raise the temperature of the water by one degree Fahrenheit would require a fall of 772 feet. There is another point of difference not to be lost sight of between the two cases, viz.: in the moving train the energy consumed by friction is proportional to the velocity; in the other case, to the square of the velocity. Energy, therefore, is destroyed much more rapidly in the latter case than in the former.

The toe of the dam should be formed with reference to the extinction of velocity without injury to the structure. On a bottom liable to abrasion the ideal form would be that of Fig. 16, in which the water falls into a basin of considerable extent, with smooth and rounded outlines, exhausts its energy in whirlings, tumblings, and commotions, and resumes its course in the natural bed of the stream divested of all power for mischief. Even in this arrangement the parts at the entrance to the basin are liable to heavy blows from floating

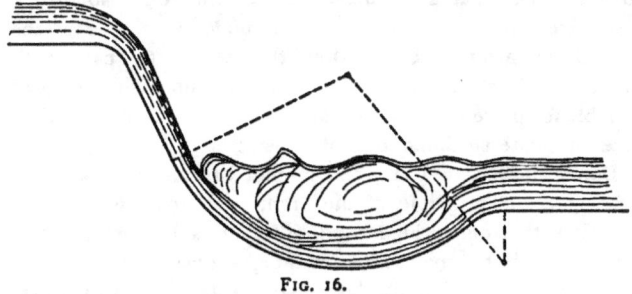

Fig. 16.

bodies carried up-stream by the reverse current, and should have the solidity to withstand such blows.

Outline of Dam in Plan.—Old dams are often met with in the form indicated by Fig. 17, *a* and *b*, obtaining a great length by running obliquely across the stream, as at *a*, or consisting

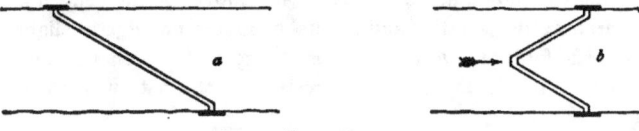

Fig. 17.

of two branches converging to a point, as at *b*. The only rational purpose that can be discovered in this arrangement is the diminution of the swell in time of flood. By giving the dam a greater length of overflow it prevents the water from rising as high as it otherwise would. In every other point of view the arrangement is bad. It is especially bad in its effects upon the banks and bed below the dam. In the arrangement *a* the current is thrown with great force upon the bank at the upper end, and it forms an eddy and reverse current at the lower end. In *b* the water concentrates in the middle of the stream, causing a very swift current there, which continues a long distance down-stream, causing eddies and reverse currents on both sides. On a soft bottom these effects are very objectionable, as greatly increasing the extent and cost of works for protection of the bed and banks. On a formation of solid rock these objections have no force. In moderate stages of the river these arrangements allow the water to pass with diminished depth upon the dam, and so diminish the head available for power. In the highest stages it is doubtful if they cause any material diminution of the swell.

Dams Curved in Plan, and intended to resist pressure of water upon the principle of the arch, are often built. From analogy with stone bridges and other arched structures, this form is thought to give some advantages over a straight dam in point of stability. This claim may be answered as follows: All material resists pressure or strain by undergoing a certain deformation. A beam sustains a load by bending. It is only in virtue of this deformation—extension of the fibres on one side and compression on the other—that the power of the beam to sustain a load is called into action. An arch resting on its centring has no strain to resist. On removal of the centring, the arch settles a little, and all its members undergo a slight change of form and volume. It is in virtue of this change that the arch acquires the power to resist the strain to which it is exposed.

An arched dam is in a radically different condition as to strain. Its necessarily close connection with the bed of the stream does not permit the lower part to act as an arch without breaking its connection with the bed, which would be attended with inconvenience and danger.

For a dam exceeding 500 feet in length, even assuming it to act as an arch, the dimensions required to resist the arch strain, and the workmanship and materials required to fit it for that strain, would make the dam more expensive than if built straight and proportioned to meet the strain acting on a straight dam.

The fact nevertheless remains that curved dams have

been built, and have stood successfully, with dimensions much lighter than would have been thought necessary for straight dams. These are mostly dams of great height, built to close narrow ravines and form reservoirs for purposes of irrigation. The Bear Valley Dam * in San Bernardino County, California, is 64 feet high and is curved to a radius of 335 feet. The lower 16 feet of it has a thickness of 22 feet. The upper 48 has a thickness of $8\frac{1}{4}$ feet at bottom and $2\frac{1}{2}$ at top.

These dams are not liable to overflow. The drainage-area commanded by them is not of great extent, and the water seldom rises to the summit of the dam. When it approaches that level, sluices are opened. These conditions—a narrow gorge bounded by nearly perpendicular walls of rock, a limited drainage-area and consequent absence of overflow—seem to give some application for curved dams. For ordinary overflow dams of moderate height and considerable length there appears to be no advantage in a curved horizontal outline.

As to its effect upon the bed and banks of the stream this form has no advantage. The escaping water concentrates in the middle of the stream, forming a very swift current there, accompanied by eddies and reverse currents on each shore. In a formation liable to abrasion, any arrangement of the dam which concentrates the current in one part of the stream is bad, as the swift water extends further down the stream than it would if the water left the dam with a uniform velocity.

A straight dam running square across the stream has a great advantage in this respect. The water leaves it with uniform velocity, encounters uniform retardation, and all comes to the normal depth at the same distance below the dam. Where liability to undermining is an important consideration, this outline should, with very rare exceptions, be adopted. On a hard rock bottom, with rock banks, such considerations are of no force, and in such situations the formation and contour often dictate a curved or broken outline for the dam.

---

* Eighteenth Annual Report U. S. Geol. Survey, Part IV, p. 683.

Wasteway or Spillway.—*Length in Relation to Drainage Area.*—Let L be the length of overflow, wasteway, or spillway of a dam which has stood long enough to warrant the belief that it has undergone the maximum depth of water. Let h represent this depth and A the appurtenant drainage area. Let L_1, A_1, h_1 be corresponding quantities at a different point on the same stream or on another stream subject to similar climatic conditions. What relation should L_1 have to L when $h_1 = h$, and what relation should h_1 have to h when $L_1 = L$?

The line L_1 must stand in the same relation to A_1 as L to A. We always find, moreover, a certain similarity of shape between two such figures, and may, for this purpose, regard them as similar figures. The lines L_1 and L are homologous dimensions of the similar areas A_1 and A. Similar areas are to each other as the squares of their homologous dimensions, and homologous dimensions of similar areas are as the square roots of the areas. Hence

$$L_1 = L \sqrt{\frac{A_1}{A}}. \quad \ldots \ldots \ldots \quad (10)$$

If, however, the depth on the second wasteway is limited to h_1 instead of h, then the length L_1 must be such as to make good the quantity of water discharged, viz., we must have

$$L_1 = L \sqrt{\frac{A_1}{A}} \frac{h}{h_1} \sqrt{\frac{h}{h_1}}. \quad \ldots \ldots \quad (11)$$

Moreover, let q be the maximum flow per second per square mile of the area A going over L, and q_1 the same from A_1 going over L_1.

We have

$$Aq : A_1 q_1 = L : L_1 = \sqrt{A} : \sqrt{A_1},$$

whence

$$q \sqrt{A} = q_1 \sqrt{A_1} \text{ and } q : q_1 = \sqrt{A_1} : \sqrt{A}, \text{ also } q_1 = q \sqrt{\frac{A}{A_1}} \quad (12)$$

Or, the maximum flow of a stream per square mile of drainage area is inversely as the square root of the said area. It is obvious that we make no error by taking L in feet and A in square miles, since, n being the number of feet in a mile, $\dfrac{An^2}{A_1 n^2} = \dfrac{A}{A_1}$.

Although equations (10) and (11) appear rationally well founded, they cannot be offered to the acceptance of the engineering profession without more convincing proof of their correctness. Eq. (10) cannot be directly compared with facts of nature, since nature does not furnish wasteways, and existing wasteways are not constructed upon infallible principles. Eq. (12), however, can be directly compared with observed facts of nature, and, being directly deduced from (10), the confirmation of the former is the confirmation of the latter.

The dam across the Connecticut River at Holyoke, Mass., has been in existence more than fifty years, with a wasteway 1017 feet long. The greatest depth observed during that time is 12.5 feet, viz., in April, 1862,[*] and this may be regarded as the maximum flow. By formula (25) this would indicate a total flow of

$$3.09 \times 12.5 \sqrt{12.5} \times 1017 = 138\,862 \text{ cubic feet per second,}$$

being at the rate of 17.35 cubic feet per second per square mile of drainage area. The formula, therefore, for the flood discharge of any drainage area A, under similar climatic conditions, will be

$$17.35 \sqrt{\frac{8006}{A}} = q. \quad \cdots \quad \cdots \quad (13)$$

It is to be borne in mind that the maximum flow of a stream stands upon a different footing from what may be called the ordinary flood. In the latter case some of the tributaries may contribute in much greater measure than others; while in the great floods that occur but once in a generation or more, all

---

[*] Clemens Herschel in Trans. Am. Soc. C. E., 1886, p. 549.

tributaries concur in making their maximum contribution, and local peculiarities disappear. The formula (13) is susceptible of a wider application than would appear at first view, for streams under diverse climatic conditions, which differ greatly in their relative aggregate flow, will differ but little in their maximum flow. Table 1a shows the application of formula (13) to a large number of observed flood discharges. For areas under 1000 square miles these results are taken from a table recently compiled by Mr. Emil Kuichling, and given in his report on water-supply for the proposed extension of the New York State canals.* Mr. Kuichling's report classes as maximum the largest flow that has ever been observed. We must, however, understand by the maximum flow from a given area the largest flow that has ever been observed from an area of that extent. We have, therefore selected examples which may be supposed to represent the flow from an area of 1 square mile, of 4 to 6 square miles, of 20 square miles, of 100, 400, and 1000 square miles.

The flood-flow of the Potomac at Great Falls, Va., is adopted as authentic, being presumably determined by the U. S. engineer officers in charge of the Washington aqueduct. Some other results, in the Tenth Census Report as well as in Mr. Kuichling's Report, are so utterly abnormal and improbable that they cannot be accepted. Several results of undoubted reliability are given on the authority of the U. S. Mississippi River Commission. On this point it may be observed that the highest absolute as well as relative flow given in the reports of the commission was at Memphis, Tenn.† This was at the rate of 1.97 cubic feet per second per square mile, while formula (13) gives 1.6. A letter to the author from Col. Amos Stickney, President of the Mississippi River Commission of August 21, 1901, states that the result obtained at Warrenton near Vicksburg, April 15, 1897, is regarded as the maximum flow of the Mississippi.

---

* State Report on Barge Canal. Albany, N. Y., 1901.
† Report of Chief of Engineers, U. S. A., 1890, p. 3220.

TABLE 1a.

Stream.	Area in Square Miles.	Max. Flow per Sq. Mile in Cu. Ft. per Sec.		Authority, etc.
		By Obser- vation.	By Formu- la 13.	
Witgendorfbach in Saxony	1.3	1116	1361	Report on Water-supply, etc.
" "	3.6	1015	818	" " " "
River Spree in Saxony	6.7	940	600	" " " "
Landwasser, in Germany......	3.8	896	796	" " " "
" " "	6.1	828	628	" " " "
" " "	20.1	366	346	" " " "
Quela River " "	116	265	144	" " " "
Kinzig " "	386	100	70	" " " "
Raritan River, N. J...........	879	59.3	52.4	
Merrimac at Lawrence	4600	19.12	22.9	10th U. S. Census Report.
Potomac at Great Falls	11476	15.2	14.5	
Colorado at Austin, Texas ,....	32000	7.8	8.7	Author's personal knowledge.
Mississippi near Vicksburg....	1116430	1.59	1.47	Miss. River Commission.
" at Arkansas City ..	1102600	1.58	1.48	" " "
" " Columbus, Ky...	895550	1.79	1.64	" " "
South Fork Dam, Pa..........	48.6	206	223	Trans Am. Soc. C. E., vol. XXIV. p. 421.

The flow of the Merrimac at Lawrence, Mass., is computed from the greatest observed depth on the Essex Company's dam, viz., 10 feet on April 23, 1852, a depth which has never since been exceeded. The flow per square mile, it will be noticed, differs materially from that computed by formula (13). The length of overflow is 900 feet. If we put, in formula (11), $L = 1017$, $A = 8006$, $h = 12.5$, $L_1 = 900$, $A_1 = 4600$, we find $h_1 = 11.27$. That is to say, the water does not rise so high by some 15 inches on the Lawrence Dam as would be expected from analogy with the Holyoke Dam. This difference in flood conditions undoubtedly results from the number of large lakes in the upper part of the Merrimac Valley, a feature almost wholly wanting in the Connecticut.

The Colorado River at Austin, Texas, probably does not carry in the aggregate a tenth part as much water per square mile of drainage area as the Connecticut at Holyoke, yet in the memorable flood of July, 1865, its flow computed from cross-section and slope must have been some 250,000 cubic feet per second.

The disastrous failure of the reservoir at Johnstown, Pa., in 1889 was due to an inadequate wasteway. In this case h_1 was limited to 10 feet and A_1 was 48.6 square miles. Our formula (11) applied to this case would give

$$L_1 = 1017 \sqrt{\frac{48.6}{8006}} \frac{12.5\sqrt{12.5}}{10\sqrt{10}} = 111 \text{ feet.}$$

A wasteway of this length, discharging freely, would have saved the dam and would have had little capacity to spare.

Had the depth in this case been limited to 5 feet, as is more commonly the fact, the length required would have been

$$L_1 = 1017 \sqrt{\frac{48.6}{8006}} \cdot \frac{12.5}{5} \sqrt{\frac{12.5}{5}} = 313.$$

A sort of thumb-rule has prevailed among engineers to the effect that works for the passage of flood-waters should be adapted to a flow equal to 6 inches in depth on the entire drainage area, discharged in 24 hours. The preceding discussion shows that this rule makes inadequate provision for very small areas, and a ridiculously extravagant one for large areas.

It will, of course, not be forgotten that the author's rule, here given provides for the greatest flow that can ever be expected, and that the probability of such a flow is very remote. In the design of such works it is always more or less an open question whether to go to the extreme limit and provide for the greatest flow that is to be expected in the course of a hundred years, or adopt more economical arrangements and take the chance of failure. In water-power dams, bridges and culverts, where failure involves loss of property only, some weight can be given to this economical consideration. In important storage reservoirs, where failure involves wide-spread devastation and loss of life, the rule should be applied except in so far as it is modified by the accumulation of water in the reservoir while rising on the wasteway, as pointed out, p. 186. It is obvious that the accumulation of water in the pond and channels appurtenant to the dam would affect the flood-flow to a larger extent on a small drainage area than on a large one.

CHAPTER III.

CONSTRUCTION OF DAMS.

HAVING set forth some of the general principles governing the form of dams, we may now proceed to consider some practical examples of their construction, commencing with the simplest case. Figs. 18, 19, and 20 represent a dam built by

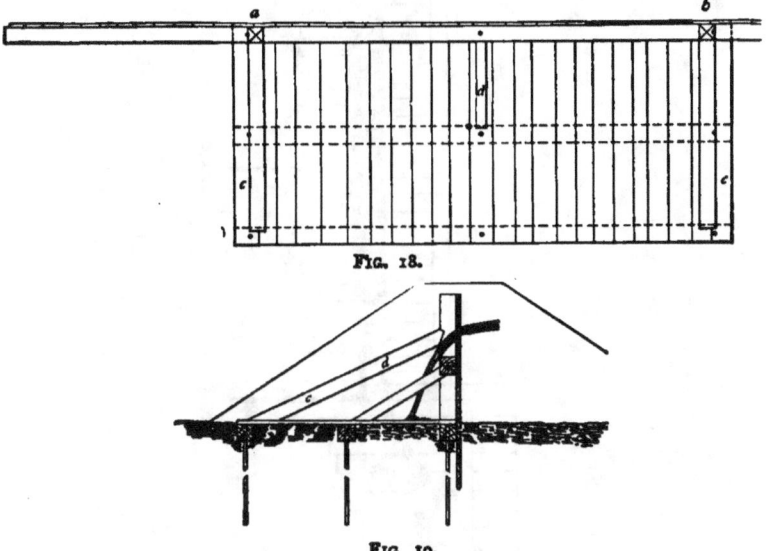

FIG. 18.

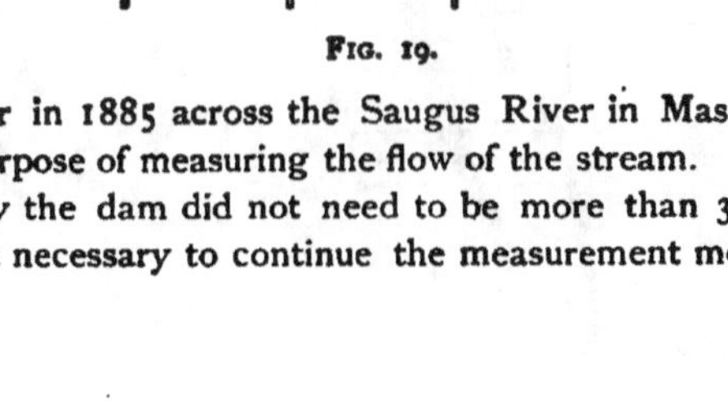

FIG. 19.

the author in 1885 across the Saugus River in Massachusetts for the purpose of measuring the flow of the stream. The head created by the dam did not need to be more than 3 feet, and it was not necessary to continue the measurement more than a

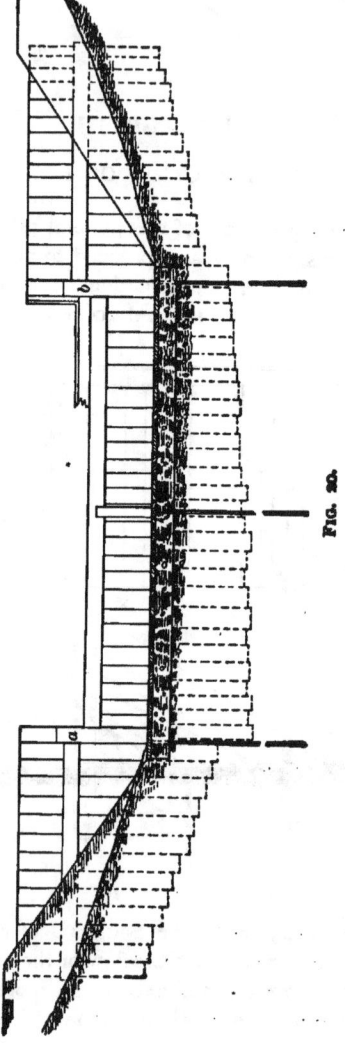

FIG. 20.

year, so that a substantial structure was not required. The bed of the stream was gravel to an indefinite depth. At the proposed location of the dam the bed was levelled for a space 12 feet up and down stream, 24 feet transverse. In this space were imbedded across the stream three 8 × 8-inch sills, each confined by three 1-inch iron rods driven through holes in the sill and 12 feet deep in the ground, the process of driving forming a head on the rod which held the timbers securely. On these timbers was laid a flooring of 2-inch plank securely spiked down, forming an apron over which the water flowed 2 or 3 inches deep until everything was ready for raising it. Along the up-stream timber, a row of spiling,* consisting of 2-inch tongued and grooved plank, was driven to a depth of 2 or 3 feet, and continued on each side to meet the embankment which formed the remainder of the dam. Along the sill this spiling was cut off 2 or 3 inches below the top of the sill, and spiked thereto, leaving a footing on the sill for the closing plank. Near the ends of the up-stream sill two upright posts *a* and *b* were inserted, confined by mortise and tenon, and connected by a light beam *ab*. Two similar beams were also inserted with their outer ends resting on the natural bank, and to these were spiked the spiling outside the weir, both the spiling and timber being partly imbedded in the embankment. The arrangements required to fit the dam for water-measuring, which need not be here described, were then added, and the posts were braced from the down-stream side as indicated at *c* and *d*. All this work was done without requiring the workmen to stand in water over their shoes. Then the short planks connecting the beam *ab* with the up-stream sill were inserted, narrowing the opening until the rush of water became formidable. Then the remaining plank were joined together in the form of a gate, and dropped into place, whereupon the water rose and flowed over the weir. The planking being dry soon swelled so as to stop all leakage.

This, although but little above the toy size, contained all

---

* The term *Spiling* is applied to sheet-piling driven by hand.

the elements of the largest dam, and could, no doubt, be applied on a larger scale. On the 13th of February, 1886, a flood in the stream went over the dam so deep that its position was only indicated by a ripple. On the subsidence of the water, both dam and embankment were found entirely un-injured, the latter owing its safety to the fact of its being solidly frozen.

Figs. 21 and 22 show a dam of timber and earth, suited to a formation of sand, gravel, or clay, with no rock or heavy boulders. It is represented as raising a head of 8 feet, and might no doubt be applied up to a head of 12. There is no rational purpose to be gained by giving such a dam a curved outline in plan, neither do the situations to which it is appli-cable ever call for a change of direction by an angle, which would introduce serious constructive difficulties. It usually runs straight across the stream, at a right angle to the general direction of the latter, as near as may be. In profile, all that is attempted in the way of moderating the action of the water upon the apron is to lead the water from the dam in a slope.

The crest of the spillway consists of a heavy timber sup-ported on round piles and joined to a line of sheet-piling. The sills for sustaining the planking also rest upon round piles. A row of sheet-piling runs along the up-stream and down-stream faces of the spillway and extends beyond the same as far as is necessary to sustain the bank which joins the spillway to the high ground on each side of the stream. Reliance for water-tightness is placed mainly upon the central line of sheet-piling. The upper row is chiefly useful in the process of con-struction, otherwise it might be dispensed with. In any case it need not be sheet-piling properly so called, but *spiling* driven by hand. For a dam not exceeding 8 feet in height it is probable that all the sheet-piling shown might be of that character. The lower row is necessary to protect the interior of the spillway against the wash of the overflow, as well as to exclude the water of the lower level during construction. Fig. 21 shows the sheet-piling spiked to the sill. This is more

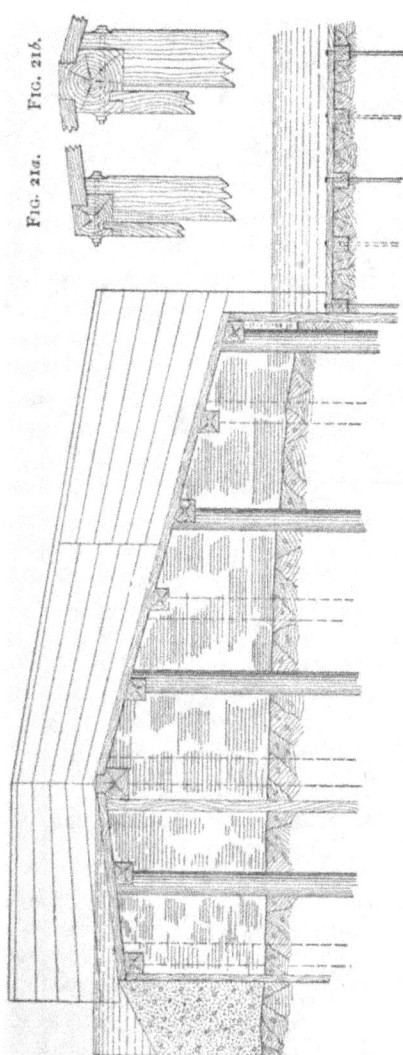

FIG. 21b.   FIG. 21a.

FIG. 21.

FIG. 23.

in accordance with modern methods, which avoid carpenter-work by the use of iron wherever possible. In old dams the methods indicated at *a* and *b* are more common. A continuous groove is cut in the sill to receive a continuous tenon cut on the heads of the sheeting-plank. This tenon is covered with a strip of canvas soaked in hot tar before putting the sill in place. The crest-sill should be 18 inches square, giving 4 inches each for the rabbets in which the planking rests, and 10 inches face showing above the planking. The other sills may be 12 × 12. The side walls of the spillway are of planking bolted to special piles, which are indicated as round piles flatted to receive the planking after they are driven. They are planked on both sides, and are fitted with a cap which is not shown. These piles are sometimes dispensed with and are replaced by uprights mortised into the sills at the bottom and into a cap at the top, which latter is sustained by iron rods running into the bank and fastened by earth anchors, a construction which facilitates repairs. The line of sheet-piling connected with the crest-sill, after passing the side walls of the spillway, reaches to the surface of the ground and extends into the bank as far as need be. This part of the work if left in contact with vegetable mould is peculiarly liable to decay; much more so than if in contact with clay or sand. On this account a construction is sometimes adopted which does not require the piling to reach the surface. The piling is cut off at the level of the crest of the spillway, where it may be assumed to be always wet, and, flush with the top, wale-pieces are bolted on to form a bed 12 inches wide. On this bed a wall of brick one or one and a half bricks thick is laid in hydraulic mortar reaching to the surface. Such a wall has, it is true, but slight stability, but it will stand well if the earth embankment remains intact. The sheet-piling which sustains the bank above and below is provided with wale-pieces, which are united by long tie-rods reaching through the bank and furnished with gaskets to prevent the water from following them.

The space under the spillway is packed with clay or binding gravel in the manner required for water-tight embankments, the material being spread in layers 6 to 7 inches thick and thoroughly rammed. The space must be kept free of water while this work is going on. The filling must not be reduced to a semi-fluid condition, neither must it be put in in too dry a state. It must be moist enough so that two lumps can be readily kneaded together. It must be brought fully up to the line of the planking and trimmed off, so that the plank will bear firmly on the filling when they are spiked down.

The apron shown for this dam is the same as the one already described (Fig. 18), being a platform of plank resting on timbers which are confined by long iron rods driven into the ground. There is no doubt that an apron can be securely confined in that manner. The cost of the rods is not serious. The spillway is represented as 200 feet long, and the fall is 8 feet. Making the width of the apron five times the fall and putting the timbers 3 feet apart and the rods 8, the work would call for some 320 rods, costing a dollar or a dollar and a quarter each—a trifling item in the cost of such a dam. In the construction of such an apron great care should be taken to make the planking tight, as the impact of the falling stream transmitted through any crevice in the planking would tend to cause a strong upward pressure.

The general method of carrying on such a work and controlling the water during construction would be this: The surface-soil is to be removed from the entire area occupied by the dam, and the portion occupied by the spillway is to be excavated to the level of low water. The ground occupied by the apron should be brought to a level such that there will be 30 inches of water over it at a low stage of the stream. This can be done in low water, and while the water is warm, by means of scrapers, the men and horses working in the water. After completion of this excavation there will be little current in the basin so formed. The apron can be put together afloat, the planks spiked down to the timbers; the holes bored for the rods and the latter inserted; then the platform weighted down to the bottom and the rods driven home. Stone or earth can be used in weighting down the apron, and left to be swept away by the water after the work is done. After the rods are driven down to the water-surface by hammers they can be driven home by slipping a pipe over each rod and using a long heavy rod, working in the pipe, for a hammer. The water flows over the apron during the construction of the dam.

While the latter is in progress, the stream is confined to as narrow a channel as it can take without giving it too high a velocity. This channel may be near one end of the spillway, or outside of the same should the nature of the ground admit of that arrangement. This latter disposition is to be preferred even should it require some excavation to provide such a channel, and the raising of the water 3 or 4 feet to turn it into the same. In any case, it is not necessary to raise the water till one embankment and some two-thirds of the spillway are complete. In raising the water, the row of sheet-piling, or spiling, along the up-stream face of the spillway comes into use. The arrangement to effect the raising of the water is precisely similar to that described for the small dam, Fig. 18. A section of the spiling can be driven below the natural surface of the water and spiked to a mud-sill, leaving a footing on the mud-sill for short plank to join it to the sill of the dam. This method does not disturb the natural flow of the stream sufficiently to cause any scour. When the time comes for raising the water, short pieces of plank are rapidly inserted bearing against the mud-sill and spiked to the sill of the dam, and a bank of earth is thrown in on the up-stream side. Where the new channel reenters the stream a fall occurs and some work of protection is necessary to prevent the scour extending back to the dam. Where brush is plentiful, as is usual on the margins of streams, this is best accomplished by covering the channel at its infall with sink-fascines, viz., loose stone enclosed in bundles of brush.

Upon completion of the spillway, the next step will be to

raise the water still further and throw it over the latter. As we are dealing with ground liable to rapid abrasion, this fact must be kept in view in all arrangements for controlling the water. To attempt to stop the flow in such a channel by a bank of loose stone, or by cribwork filled with stone, or by bags of earth, is usually futile owing to the rapid scour caused by these obstructions. Some preliminary protection of the channel is indispensable. Most naturally this protection takes the form of a thick brush mattress covering the bottom for a length equal to five times the depth of water after raising, and extending up the slopes as high as the water is to be raised. With this foundation, a bank of loose stone can be thrown in and backed by a bank of earth with good show of success. The better way, however, is to provide the channel, before admitting the water to it, with a bulkhead for readily stopping the flow. Figs. 23 and 24 give an idea of such a bulkhead. A mud-sill is imbedded in the bottom of the channel; a row of spiling driven to a sufficient depth, spiked to the mud-sill, and cut off to leave a footing for stop-plank. Uprights are mortised into the mud-sill and a cap put on which extends across the channel, and is imbedded in the earth at the sides. The cap is well braced from the bottom and sides of the channel. When the water is let into the channel, it flows gently over the mud-sill without causing any scour. When the time comes for raising the water, there will be a considerable stretch of slack water above the dam, and the stop-plank can all be put in before the water rises enough to create any serious scour.

Where the conformation is such that the water cannot be turned outside the spillway, the water is confined to a narrow channel near one end of the latter, and flows there until the larger part of the spillway is completed and planked. One of the embankments may also be fully completed, all the round piles driven, and all the sills put in place, before turning the water, but the sheet-piling can only be completed as far as the edge of the stream. A gap cannot be left in a line of sheet-

piling to be subsequently closed, as the closure cannot be
effected with such accuracy as to avoid leakage. To finish the
dam, the water must be excluded from the uncompleted end
and thrown over the finished part, and room enough must be
left to do the work in a proper manner. The best way to do

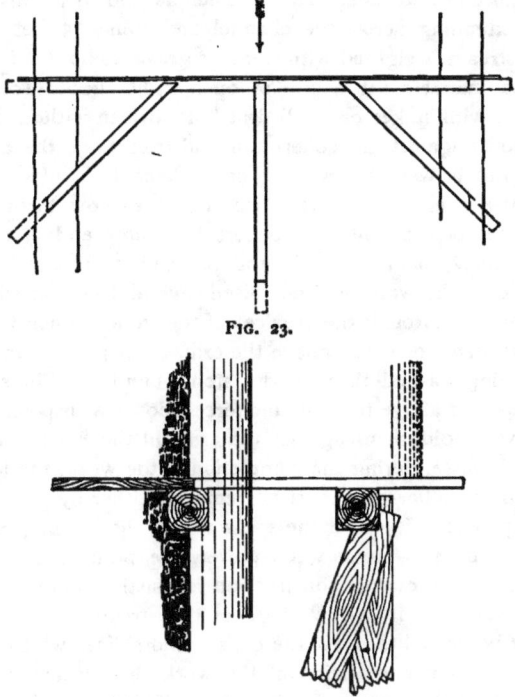

FIG. 23.

FIG. 24.

this is to construct a wale-and-tie coffer-dam, running up and
down stream, to separate the completed from the incomplete
part of the dam. It should lap on to the dam as far as the
crest and extend up stream 50 feet or more. The problem
now is to close the channel between the coffer-dam and the

shore. There are many ways of doing this: one method being the mud-sill spiling and bulkhead already described. I will detail a different method, one which can be rapidly put in execution, admitting the employment of a large number of unskilled workmen. First, dig out the channel to a tolerably level bottom and steep bank. Then lay down a brush mattress extending across the channel and some 25 feet up and down stream, weighted with stone or gravel to hold it in place. On the mattress set a crib of open light timber-work held together with spikes or small drift-bolts driven without boring. This abuts against the coffer-dam and reaches to the opposite bank, and is weighted with stone to keep it in place. This will raise the water a little, causing it to flow through the interstices of the timbers. Secure the shore end of the crib with a heavy bank of earth, and proceed to fill it with stone. This done, the water will be raised several feet and will continue to flow through the crevices of the stone. Then throw in a heavy bank of earth above the crib. To prevent the earth from being washed through the stone, bundles of brush may be thrown in above the crib and forced down with poles. The water will hold them against the crib till the earth is thrown in. Of course, during the completion of the work, the leakage through the coffer-dam must be removed either by pumping or by a siphon leading over the spillway and discharging below.

For a dam to create water-power, the problem of controlling the water during construction is usually simpler than is here assumed. In a low stage of the stream the water can usually be turned through the canals and sluices which supply the mills. On completion of the work the closing of these sluices turns the water over the dam. This arrangement does not allow coffer-dams to be dispensed with, but it admits of lower and less expensive ones. Permanent sluices are often made in the dam itself through which the water flows during construction, and which can be closed by gates on completion of the work. Such an arrangement is imperative where the necessity exists for draining the mill-pond, and it is generally advisable on account of repairs.

A pile dam similar in its general features to this, though built with less care than is here contemplated, was constructed on the Cedar River, at Waterloo in Iowa, in 1883, raising a head of some 8 feet. It is still (1897) intact, having suffered no damage from ice or floods.

The form of dam shown at Fig. 25 * has, at present, a historical rather than an engineering interest. It is the general type of dam adopted in the improvement of navigation on the Muskingum River in the State of Ohio, in a system of canals and river improvement undertaken by that State about 1830. This particular drawing has reference to the dam at Marietta, O., where the Muskingum enters the Ohio. The drawing of which this is a copy is not accompanied by any explanation and must be supplemented by the imagination of the reader. It is presumed that the timbers resting on the pile-caps in the apron form a continuous flooring, as also the top timbers of the step. As to the top of the dam the drawing is manifestly in error. It appears to show the planking running lengthwise of the dam, which would be a very unskilful arrrangement. Moreover, no millwright or carpenter would put on the upper timbers running to a point in the manner shown. It is probable that the arrangement at the top was something like that indicated at Fig. 25a. It is also to be presumed that the cribwork was filled with stone, and that there were abutments of cribwork. It is stated that a coffer-dam of brush, stone, and gravel was built preparatory to the construction of the apron, and that the cribwork was located thereon. These dams created falls of 8 to 17 feet.

Fig. 26 represents a type of dam said to be much in use on the upper Allegheny River for falls of 5 to 8 feet. This particular sketch relates to the dam at Olean, New York. The information is derived from the same source as in the preceding

* Tenth U. S. Census Report, vol. XVII. The Ohio River Basin, p. 36.

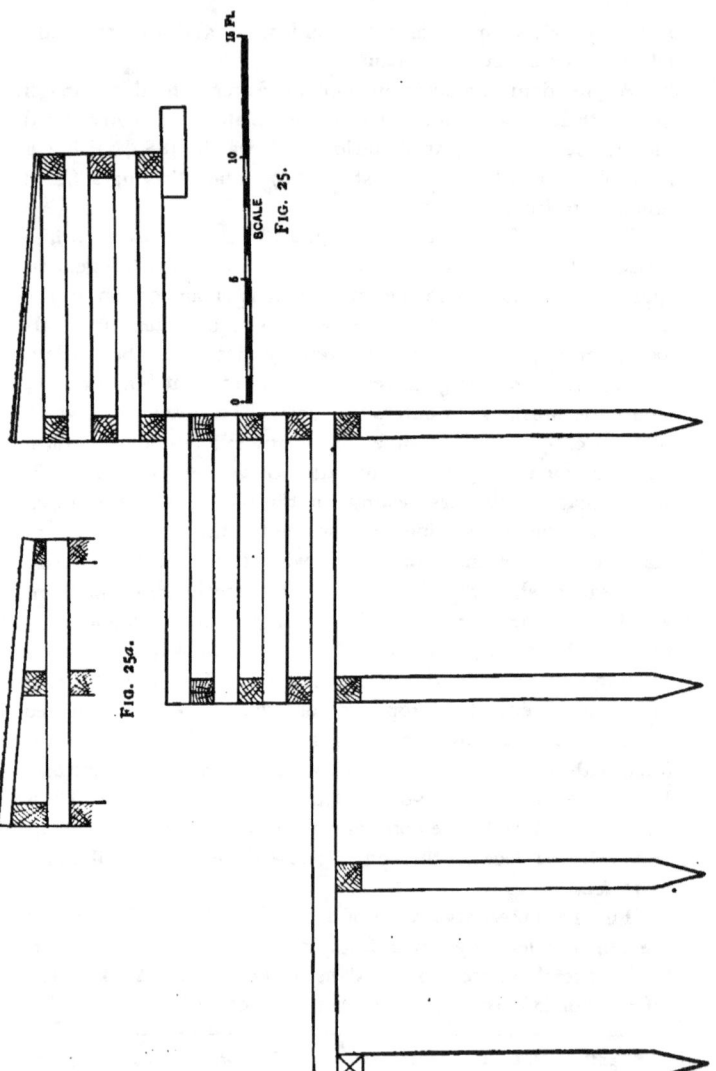

FIG. 25.

FIG. 25a.

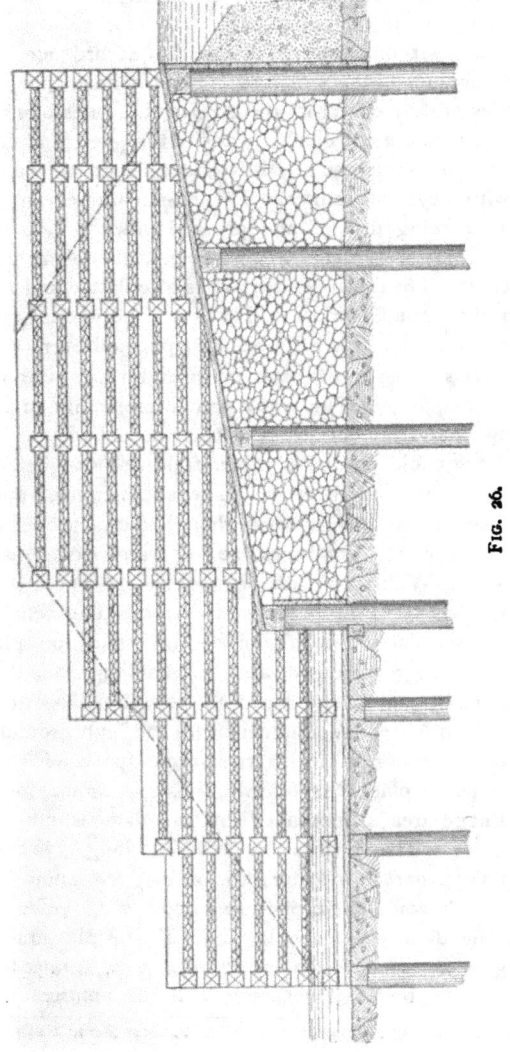

FIG. 26.

case.* Fig. 26 is compiled from the description, which is not accompanied by any drawing. The report is silent as to the apron, and an apron is added, designed in accordance with the general character of the structure. For the dam, four rows of round piles are driven across the stream, the piles being 8 feet apart. Tenons are cut on the heads of the piles, and caps are secured to the same by mortises and pins or by dovetailed tenons with keys or wedges. The up-stream row of piles is flatted after being driven, and planked down a foot or two below the natural surface of the ground, the plank running horizontally. The down-stream row is also flatted and planked down to the natural surface. The space between these rows of piling is filled with loose stone and planked over. Water-tightness is secured by a bank of earth thrown in above the up-stream planking. This bank should preferably be earth of a binding character, but where a bank is confined, as in this case, a reasonable degree of water-tightness can be secured with almost any kind of earth. The abutments of this dam are represented as of cribwork, though more properly they should be of stone, woodwork above water being liable to rapid decay. With abutments of masonry the whole structure might be considered as permanent. If the abutments are of cribwork their faces toward the spillway must be planked, although not so represented, and the planking should return along the up-stream face and extend, as spiling, into the bank of earth which unites the abutment with the high ground.

As to the order in which the several parts of this dam should be put in place, the following observation may be made. The drainage-area commanded by the dam is about 1100 square miles. The spillway is 220 feet long. We should expect an ordinary low-water flow of, say, 300 cubic feet per second, which would correspond to a depth of 6¾ inches on the crest of the dam. We should, first, dig for the abutments, throwing up a bank of earth and forming a pit from which the water could be bailed if necessary. If the abutments are to

* Tenth U. S. Census Report, vol. XVII. The Ohio River Basin, p. 11.

be of stone, the pit should be freed from water. If of crib-work, it can be dug 2 or 2½ feet deep with shovels or scrapers, the lower timbers put together afloat, and as the work progresses it will settle firmly upon the bottom. If of masonry, the up-stream portion should be laid in hydraulic mortar. The apron must be put in in like manner by men working in the water, at the lowest stage of the stream. The planking should be well below low water, not only for preservation of the timber, but for moderating the shock of floating bodies. After driving and capping the piles for the spillway, a trench is dug along the up-stream face as deep as practicable, the up-stream side of the piles flatted and the planking applied. The plank are applied in this manner. Suppose the trench to be 2 feet below the natural surface. One tier of plank is put in, forced down to the bottom, and fastened. The earth is solidly packed on the up-stream side of them. The insertion of these plank creates no scour. The second tier are then put in, and the earth packed behind them. This will create a slight fall and tend to scour the earth under the spillway. A layer of the stone filling is put in to prevent this. The down-stream planking are put in in like manner and backed with earth. When this is finished the water is pouring over the down-stream planking on to the apron. The stone filling is thrown in and brought up to the proper level, the up-stream planking being applied at the same time. A light bulkhead is thrown up along the up-stream face of the dam, excluding the overflow from half the length, while the planking of the spillway is being put on. The bulkhead is then removed and applied on the other half, etc. Another method is, confine the flow of the stream during low water to as narrow a channel as practicable near one end of the proposed dam. Complete one abutment, and the greater half of the dam and apron, pumping for the apron and abutment if necessary. Then raise the water as above, throw it over the completed portion, and finish the rest.

Other dams on the same stream are said to have two slopes; the up-stream slope descending below the natural bed of the stream and covered by a bank of earth, forming a water-tight connection with the bottom. It does not appear, however, that this form has any advantage either in construction or operation.

Tree Dams.—Figs. 27 and 28 show an example of a tree dam, built across Schoharie Creek,* in the State of New York, for the purpose of diverting the water of that stream into a feeder of the Erie Canal. It is composed of straight and regular pine-trees, 60 to 70 feet long, 18 to 22 inches diameter at the butt, divested of their branches. The butts of each tier lay in contact, and each tier is raised above the one below it by a 12-inch timber. Each 12-inch timber is drift-bolted to the tier of trees below it, and receives the drift-bolts which go through the tier above. The thickness of these longitudinal timbers combines with the taper of the tree to increase the slope of each successive layer of trees. On the upper layer are bolted four 12 × 12-inch timbers supporting a flooring of thick plank. The layer of brush and gravel which forms the water-tight backing of the dam laps on to the flooring and abuts against a longitudinal timber above the planking. The abutments of the dam are of masonry supported by piles, which show in elevation in the longitudinal section Fig. 28, and in Fig. 27a. In moderate stages of the stream the water goes over this dam in a series of small cascades. In high stages it rushes down in an unbroken stream.

The length of this dam between abutments was 436 feet, the drainage-area commanded by it being some 1500 square miles. The construction of the work must have required more than 1200 trees. It is quite within the limits of reason to say that this quantity of timber could have been combined in a manner to form half a dozen dams of the same length and height as this, and built with equal prospect of permanence. The great length of abutment and retaining-wall required in this form of dam is a serious objection. Another objection is: The up-stream slope is so slight that a sufficient backing of

* Report of the State Engineer of New York for 1865.

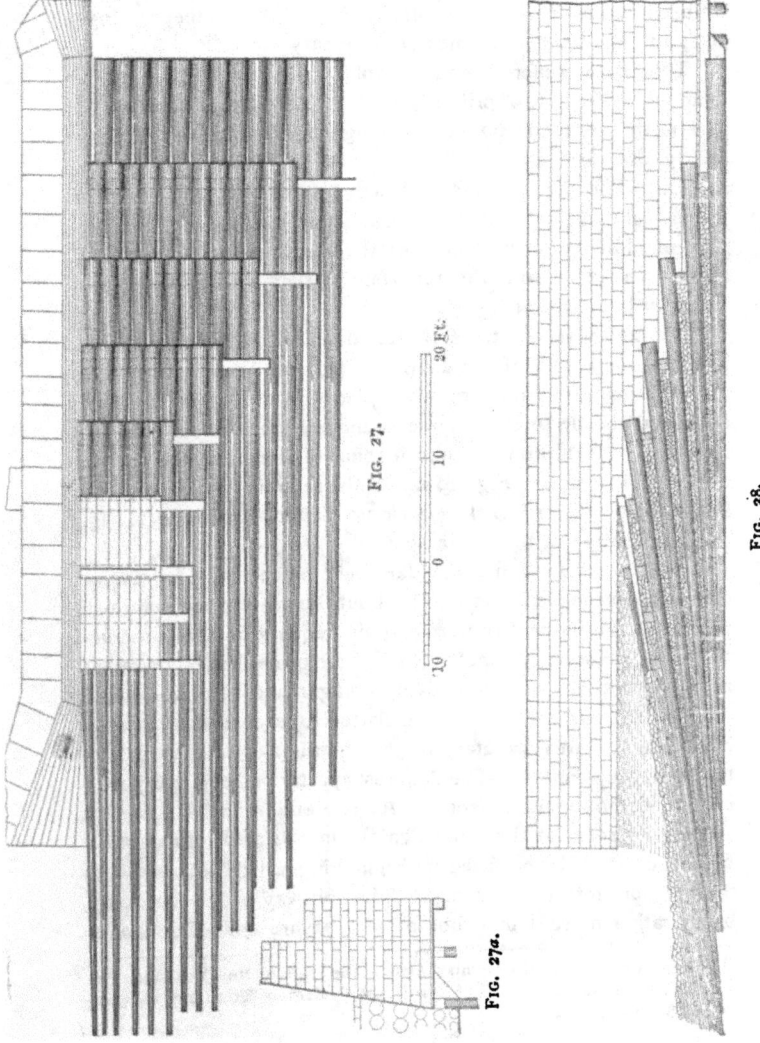

FIG. 27.

FIG. 27a.

FIG. 28.

20 Ft.

10

0

10

earth cannot be applied even with the awkward expedient of a timber running across the planking. With a steeper slope such an expedient would not be necessary.

This dam suffered considerable injury in the floods of October, 1869, and April, 1870. In 1871 * it was rebuilt and its length extended 160 feet, making the overflow or spillway nearly 600 feet long. Figs. 29 and 30 show the dam as rebuilt. The height was reduced to three layers of trees instead of five. The down-stream slope was entirely covered with planking, and the length of the apron extended. So far as can be learnt from the reports, the dam, in this form, has stood without further injury.

Fig. 27 shows the trees entirely divested of branches, while in the arrangement of Figs. 29 and 30 some branches remain. As the first tier of trees must be placed in the water, it would appear advisable to leave some branches on, in order that the tree may be held down by the loading of stone, till the second tier is put in place. Fig. 29 shows the retaining-wall extended in the form of a cribwork " docking " filled with stone. Fig. 29a is a section of the retaining-wall.

Fig 31 is a section of the dam built by the State of Massachusetts, on the Deerfield River, about the year 1864, to furnish power, in the form of compressed air, required in the construction of the Hoosac Tunnel.† It is more properly designated as a log dam than as a tree dam, being made of round logs, notched at their intersection and flatted to receive the planking. The drainage-area tributary to this dam is 234 square miles of mountainous country. The dam rests partly on rock and partly on gravel, underlaid by rock. At one end it abuts against a precipitous rock, at the other against an abutment of masonry, resting on gravel and faced with planking which is fastened to timbers built into the masonry. The spillway is 250 feet long, being rather more than 1 foot for each square mile of drainage-

---

* Report of the Canal Commissioners of the State of New York for 1871.
† Tenth Census Report U. S., vol. XVI.   Region Tributary to Long Island Sound, p. 109.

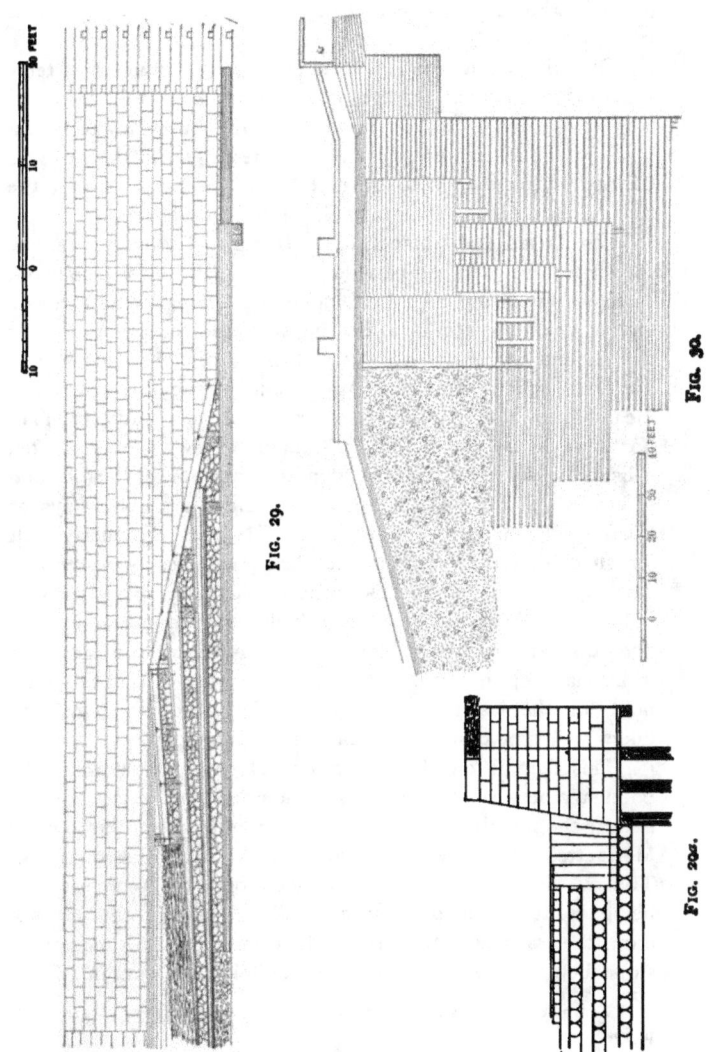

FIG. 29.

FIG. 30.

FIG. 29a.

ground, whereas the Olean dam just described has but 1 foot to 5 square miles, and the Schoharie Creek dam 1 to 2½.

On this stream the ice sometimes gorges and forms temporary dams. Such a gorge, obliterating the fall at this dam, is a serious danger to be provided for, the buoyancy of such a mass of timber tending to lift and derange the dam. This danger is understood to be met by a loading of stone, the spaces represented as vacant being filled with stone.

Log Dams and Lumberman's Dams.—The main overflow shown at Fig. 32 is a typical log dam, consisting of a cobwork of logs, supporting a deck of inclined planking or timber, one end of which projects over the logwork and the other rests on the ground and is covered by a bank of gravel. Such dams are often built remote from any source of supplies, with such means as can be transported over very rough roads, in small boats, or on the shoulders of workmen. Hewn timber is often substituted for planking, wooden treenails for drift-bolts and spikes. The lumberman's dams, so called, are not always or usually built for purposes of power. They are commonly built for sluicing, viz., to accumulate large volumes of water and discharge it at a very rapid rate, to facilitate the floating of logs. In other cases they are built to form ponds for the detention and sorting of logs, especially where the latter are to be made up into rafts. Fig. 32 * is a cross-section of a dam built for the last-named purpose. This, however, may be taken as a type of a class of dams for water-power.

This dam is on the Menominee River, which forms the boundary between the States of Michigan and Wisconsin. It is near the mouth of the river, about a mile above the town of Marinette. The site of the dam is a smooth limestone rock. The overflow or spillway is 557 feet long, the drainage-area being a little over 4000 square miles. Unlike the Hoosac dam, the transverse timbers are laid horizontal, the necessary slope being given to the deck by building the dam in offsets.

---

* Tenth U. S. Census Report, vol. XVII. Northwestern Watershed, p. 66.

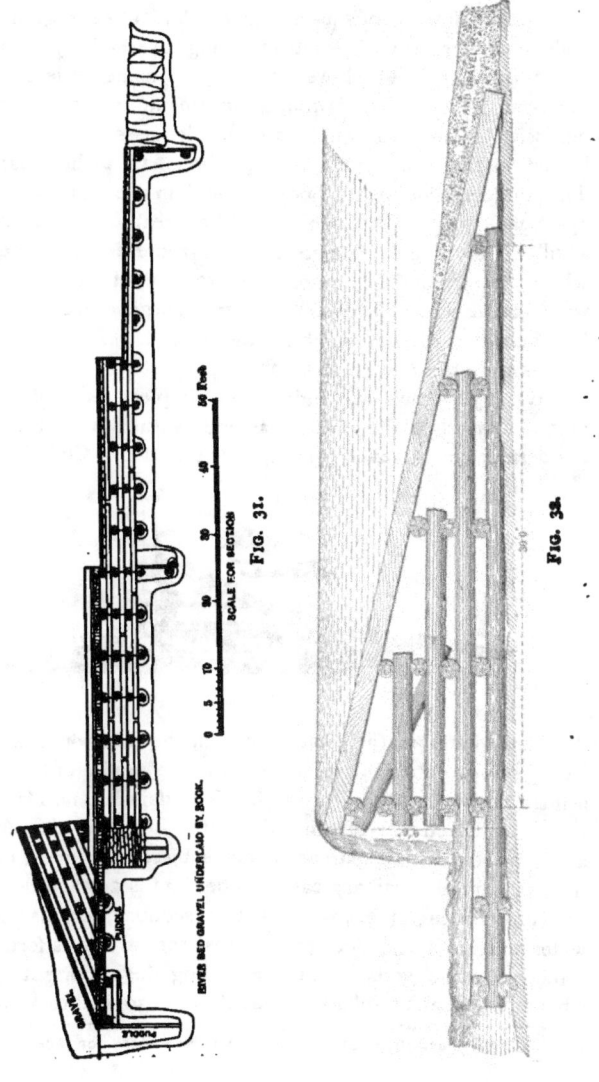

FIG. 31.

FIG. 32.

The deck is formed of square timbers laid close together, the ends which rest on the bottom being overlaid by a bed of gravel or earth. The lower log projects some 10 feet beyond the vertical face of the dam and sustains an apron, though the necessity for an apron on a limestone ledge is not clear. It is bolted to the rock by fox-wedge bolts 1¼ inches diameter. The timbers forming the body of the dam are laid about 8 feet apart each way. This dam would be greatly strengthened by a filling of stone. Constructed as represented it depends wholly for its stability upon the pressure of the water, and would be very readily destroyed by a gorge of ice or a jam of logs below it, such as to obliterate the fall. The total cost of this dam is stated to be $10 000.

Fig. 33 * shows the rudest and simplest form of lumber-man's dam that the writer has met with. It was built by lumbermen, as a sluicing-dam, at the outlet of Club Lake, on

FIG. 33.

the headwaters of the Black River in the State of New York. Fig. 34 shows the general situation and a profile of the stream a little below the dam, Fig. 34,*a*. The dam has no abutments and takes its chance of being flanked. Being at the outlet of a series of lakes, the stream is not subject to so great fluctuations as in the ordinary case. The dam is 267 feet long and contains sluices 20 feet wide, in two sections, for discharging water and logs. A foot-path across the stream is formed by spiking to the ends of the projecting logs upright plank, which sustain a foot-plank elevated some 30 inches above the

* Report of the State Engineer of New York for 1888.

crest of the dam. A constructive difficulty in this dam would
appear to be the limited bearing which the transverse timber
has on the longitudinal timber at the inner end. This can
be managed by using larger logs next the planking and notching deeply into them for the bearing of the transverse log, as
at *a*, Fig. 33. This gives the bearing for the log without
interfering with the bearing for the planking. No coffer-dam
is required for this work. The logwork and sluices can be

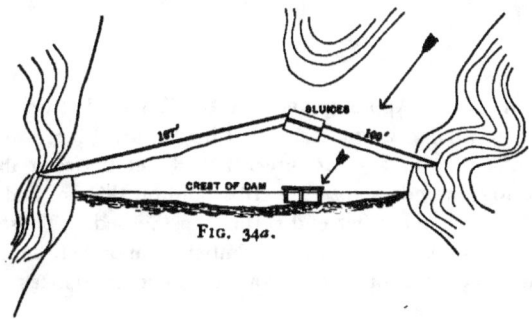

FIG. 34*a*.

FIG. 34.

built without any sensible obstruction of the flow of the water.
The dam would be the better to be entirely filled with stone,
but in any case a sufficient quantity of stone must be thrown
in to prevent rapid wear of the bottom during construction.
The bottom of the sluice is supposed to be at the level of
extreme low water. Upon throwing in the stone, the water
rises somewhat and will be partly flowing through the sluice.
The plankwork is now commenced, starting at the sluice,
which should occupy the deepest part, and working both ways.
As fast as the planking is applied the earthwork is put in,
which closes the passage under the plank. As the work proceeds, the water rises more and more, and when the shallower
parts of the stream are reached, it is flowing wholly through the
sluice. After completion of the planking and earth filling, the

closing of the sluice causes the water to rise and flow over the crest.

Fig. 35 * is another form of log dam, or "spar dam" as it is more commonly called in the lumber regions, of which there are, or were recently, two examples on the lower Fox River in

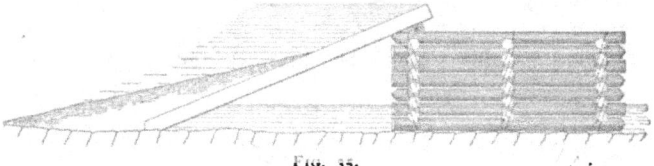

FIG. 35.

Wisconsin, one at Appleton and one at Neenah, both on beds of limestone rock. Log cribs, 8 feet by 16, are placed in line on the bottom, with a clear interval of 8 feet between them. A heavy log extends across the cribs, and on this are laid one end of the spars, the other end resting on the bed. These are hewed or sawed timbers. Round timbers cannot well be used, as, being smaller at one end than the other, interstices must

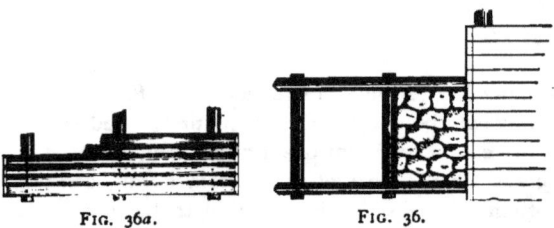

FIG. 36a.   FIG. 36.

exist too wide for the security of the earth filling. This construction could not judiciously be applied to a soft bottom, requiring expensive arrangements to secure the cribs from being washed out. Figs. 35 and 36 represent the cribs as filled with stone, and contemplate the use of mortar in the upper layers to secure the stone from being washed out by the overflow.

---

* Tenth U. S. Census Report, vol. XVII. Northwestern Watershed, p. 45.

A better method is to cover the cribs with a decking of logs confined by drift-bolts as at Fig. 36a.

Fig. 37 shows a mode of finally closing this dam, supposing it to be built without sluices or passages for the water. In that

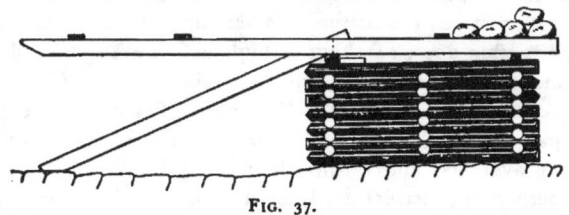

FIG. 37.

case, as the sparwork approaches completion, the water is concentrated into a narrow gap through which it pours with great violence. The insertion of the spars, one by one, becomes more and more difficult and dangerous, and finally impossible. The case calls for a method of suddenly closing a gap of some width. Such a method is indicated in the figure. We suppose the gap to extend from the middle of one crib to the middle of the next, viz., 16 feet in width. The cross-log is placed in position and confined as shown so as to be capable of a slight movement of rotation. A light log is placed across the down-stream end of the cribs. The spars for closing are some 16 feet longer than the others, and are held in a horizontal position by weights, one end resting on the down-stream cross-log, the other in the air. They are fastened by drift-bolts to the up-stream cross-log, and to one another by planks and flatted timbers, as indicated, forming one great gate or door. While resting in this position the free ends are cut to the right length and trimmed to fit the rock. When the time comes for closing the gap, stout ropes are attached to the crib and take a turn around the flatted timbers on the top of the gate, the free ends in the hands of workmen. The weights are shifted so as to lower the gate, and it is eased down into its position.

Lumberman's Dams in Water of Some Depth.—The log dams thus far described are supposed to be built in shallow water 2½ to 3 feet deep; this depth occurring in but a small part of the length. Even in this depth it may be sometimes advisable to put the lower timbers together, and build up till the work rests on the bottom. When the depth exceeds 3 feet, which it often does, reaching as high as 6 or even 10 in places, this mode of construction is indispensable. As an illustration we take the lower dam of the Menominee River Manufacturing Company* at Marinette. This dam is 700 feet long, resting on a gravel bed and raising the water 7 feet. In describing the method of construction, I use the words of J. L. Greenleaf, C.E., who collected information as to water-power, etc., in this region, for the Tenth Census. "The method employed was to build the cribs over their proper positions and sink them. Starting from the Wisconsin bank, four logs were floated down into position and held there, parallel with each other and 8 feet apart, with their length in the direction of the current. These covered an area 45 feet long and 32 feet wide. Across these was placed a course of sawed timber. Another section, entirely similar, was floated down adjoining the first toward the centre of the river, but leaving a space of 8 feet between them, and across these the cribwork was built locking one with the other. This process of floating in the sections and building up the cribwork was continued, until the position of the runway was reached, which was then built. In sinking the cribs, enough of the sections in the cribwork would be floored, so that when filled with stone the weight would suffice to carry the structure down to the bed, and the other sections were left open. When the cribs were sunk in position, the open sections were filled with stone, which filled up the irregularities in the bed under the crib, and gave a uniform bearing to the lower course of timber. After completing the runway, a crib 32 feet long was built next

---

* Tenth U. S. Census Report, vol. XVII. Northwestern Watershed, p. 67.

to it on the Michigan side, and then a space of 32 feet was left for the passage of the water. The process of building the sections and cribwork above them was then continued to the Michigan shore. The heavy body of water which now passed through the space of 32 feet left open, scoured the bed out to a considerable depth before it was closed. The method of closing it was to make an immense trap-door, with a very strong frame, hung on a horizontal axis, and at the proper time this was let down. The force of the current immediately forced it shut against the cribwork on each side, and loads of gravel, hay, etc., held in readiness were thrown in. And in two hours from the time of starting, the opening was closed. Back of this trap-door the cribwork was built up and the dam completed.''

A more judicious proceeding would have been to cover the bed at the gap with a strong mattress, overlapped a little by the neighboring cribs before concentrating the flow in it. The runway referred to was evidently for the purpose of discharging logs. The dam contained no sluice in the sense of a passage for large volumes of water to create a temporary rise in the stream for floating logs. Such a sluice would obviate the necessity for a special passage for the water. Both the dams just described and the dam of Fig. 32 were standing intact in May, 1898.

Dams of Sawed or Hewed Timber in Combination with Earth or Loose Stone.—Pile and log dams into which squared timbers enter have already been considered. Dams of squared timber without bearing-piles may be considered under two general heads:

1. Framed work deriving its stability mainly from the pressure of the water.

2. Cribwork rendered stable by a filling of loose stone.

Fig. 38 * represents a dam built on the Kankakee River near Wilmington, Illinois, in 1871, by Mr. E. S. Waters, the

* Tenth U. S. Census Report, vol. XVII. The Mississippi River and some of its Tributaries, p. 113.

engineer of the Kankakee Company, a corporation organized
to construct and maintain works for water-power and navigation on that stream.　The Kankakee at this point is a large
stream, having a drainage-area of over 5000 square miles.　It
is subject to some influences tending to moderate its floods, and
to others tending to aggravate them.　It issues from an immense
morass in northern Indiana called the Kankakee Swamp, which
has considerable influence in steadying the flow.　On the other
hand, its general course is westerly, which tends to the
augmentation of floods by the simultaneous melting of snow
and breaking up of ice.　For the last 25 miles above Wilmington its course is northerly, tending to retard the breaking up
and create ice-jams.　Rises of 3 or 4 feet are common, and
floods of 9 feet have been known.　The site of the dam was a
limestone ledge covered in places with gravel.　The dam was
1000 feet in length, creating a head of 15 feet.　The dam consisted of a series of frames, as indicated in the figure, 6 feet
apart, covered with planking, and backed on the up-stream side
with gravel.　The sills and the up-stream rafters, purlins,
and struts were 12 × 12 inches.　The down-stream rafters
were 10 × 12, and purlins 8 × 10.　The up-stream planking
was 3 inches thick, down-stream 2.　Each sill was confined
to the rock by six 1½-inch bolts.　The up-stream planking,
it is stated, extended to the rock, and the interstices were
closed with cement.　Fig. 38 is taken from the report.　Figures *a* and *b* are added by the author.

　　It is manifest that, so far as the sills are concerned, the
structure must be very imperfectly represented by Fig. 38.
No rock bottom was ever so smooth and level that such a frame
could be placed on it and bolted without some work of levelling
up.　38*a*, no doubt, represents the actual situation more
correctly.　Such frames could never set level without fitting
pieces of timber to the rock to form a bed for the sills, or
building a foundation of masonry.　It will be perceived that
the down-stream planking does not extend to the sill.　It is
understood that it originally did so, but that, dry rot appear-

ing, a portion of the planking was cut away, to admit the air under the dam. 38*b* is suggested as a better and stronger arrangement of the framing for the crest of the dam. It is understood that the purlins were bolted to the rafters, and it is presumed that all the members were so connected to the sills as to prevent the lifting of the structure in case of an ice-jam below, such as to obliterate the fall. The lifting effort in that

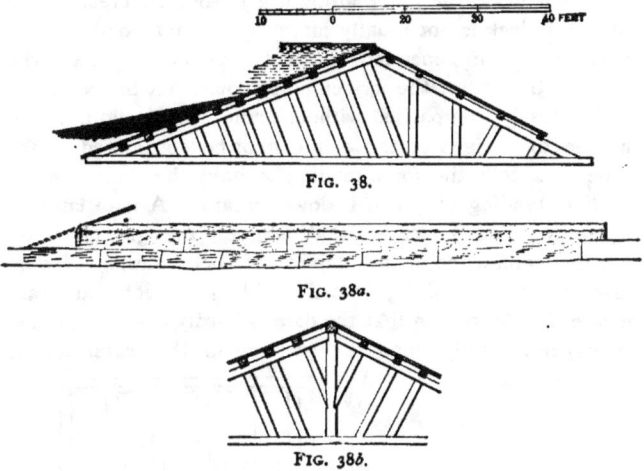

FIG. 38.

FIG. 38*a*.

FIG. 38*b*.

case might be much greater than the buoyancy of the timber. The jam and rise of water is liable to take place suddenly. The air confined under the decks having no opportunity to escape, the whole structure would tend to rise like an empty vessel submerged.

In 1883, the ice, being some 2 feet thick, broke up and aggregated in jams in the upper part of the stream, remaining intact in the reach immediately above the dam. In this condition several days of severely cold weather occurred, and broken ice froze together in great masses. This was followed by a warm rain lasting two days. The ice moved down the stream, breaking up the standing ice and accumulating in

masses which filled the entire valley.   A 20-foot railroad embankment is said to have been moved bodily out of line 3
feet by the ice.   When this action reached the dam, the ice
grounded upon the up-stream slope, accumulated to a height
of 20 feet above the crest, fell over upon the dam, and crushed
it like an eggshell.

Framed dams on a rock bottom are more commonly made
with but one deck, the water falling freely from the crest.   The
slope of the deck is not usually more than 2 to 1 nor less than
3 to 1.   If a dam depended wholly for its stability upon the
pressure of the water, the former slope could not be exceeded
or even closely approached without some risk of sliding.   In
such case the weight of the dam cannot be considered.   For
a slope of 2 to 1 the force tending to hold the dam down is
twice that tending to shove it down-stream.   A man familiar
with hauling heavy weights that drag on the ground could
hardly feel confidence in the stability of such a dam.   Nevertheless accidents resulting from the sliding of such dams are
very rare, for the reason that the dam is hardly ever so situated
as to depend wholly upon the pressure of the water for its

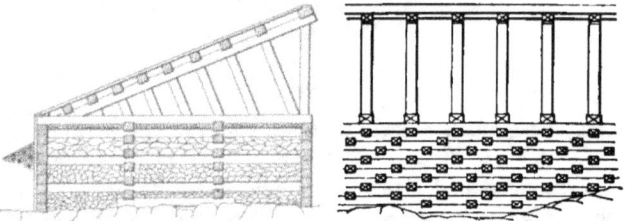

FIG. 39.   FIG. 39a.

stability, being either bolted to the rock, backed with gravel,
or loaded with stone,—sometimes provided with all these
elements of solidity.   The dam of Fig. 33 stood for many
years with a slope of 2 to 1, with very little else to confine it
than the pressure of the water.

Fig. 39 represents a frequent case of a dam upon a rock
bottom of very irregular form, much lower at the middle of the

stream than the sides. In this case a cribwork is built up and filled with stone to the highest point of the rock occupied by the dam. This cribwork is surmounted by a framed dam, and its width will depend upon the height and slope of the latter. This was the construction adopted for the dam of the Jackson Company on the Nashua River at Nashua, N. H., near where that stream empties into the Merrimac. The Nashua at this point has a drainage-area of 524 square miles. The dam, in connection with a canal of about 1000 feet length, creates a fall of 21.5 feet, though the framed part of the dam is but little over 10 feet high. The sills are 12 × 15 inches, the rafters and struts 12 × 14, the purlins about 11 × 12, the frames 4 feet apart. These dimensions are given from recollection and may not be quite exact. The spillway is 134 feet wide, and the water has been known to go over it to a depth of 8 feet 4 inches. This dam was built in 1878 and has never (1898) received any injury or required any repairs, so it is stated by the custodians of the dam.

Cribwork.—Loose stone, of the kind usually available in river construction, has but slight stability against the action of running water. It is true that a mass of loose stone is well adapted to sustain the pressure of water, and a dam composed of loose stone might stand for some time provided its crest were at a uniform height. The water flowing over it in a thin sheet of unvarying thickness might not have power to move any of the stones. The trouble with such a dam is that the displacement of a single stone, by the impact of a floating body or any other accident, is liable to destroy the work. The removal of a single stone strengthens the current at that point, and makes the removal of the second stone easier, the third still more easy. Every advantage gained by the water increases its power for mischief, and it soon concentrates into a torrent of irresistible force.

The ease with which heavy blocks of stone are moved by a current sometimes appears incredible. A stream of water impinging upon a flat surface perpendicular to its direction exerts a pressure represented by twice the head due the velocity, acting upon an area equal to the cross-section of the stream. The direct impact of the water is not the only force acting against the stability of the stone. By reason of its immersion or partial immersion in water, the latter loses one-third to one-half the weight of the part immersed. Moreover, on account of the equal transmission of pressure in fluids, the pressure of the impact acts often upon the entire under side of the stone and exerts a strong tendency to lift it off its bed, thus facilitating its movement.* Stone used to protect the bed and banks of a stream from the rush of water occasioned by a dam, is liable to the same objection. It can be taken away piecemeal, stone by stone. It is apparent that the application of loose stone, so placed that one stone derives no support from others, is very limited in dam construction, being restricted to the protection of earth from the wash of a moderate current. This object is greatly facilitated by the orderly arrangement of the stone, so as to present no projecting surface to the action of the water.

---

* The work of Ganguillet and Kutter gives the following formula for the velocity required to move stones or other heavy bodies :

$$v = 5.67 \sqrt{ag},$$

in which v is the velocity in feet per second, a the mean diameter of the body in feet, and g its specific gravity. This formula is ascribed to Chailly. It assumes that the body is immersed in water moving with the velocity v. Some writers, by forgetting this condition, have made very extravagant statements. They have found that stones, in order to remain permanently at the foot of a dam, must have a diameter more than half the height of the dam. For a diameter of 10 feet and specific gravity of 2.5 the above formula gives $v = 28.35$, being the velocity due to a head of 12.5 feet. In other words, a stone to withstand the wash of a dam 12.5 feet high must have a diameter of 10 feet. This statement would be correct if we could assume a stream of water of a depth equal to the height of the stone, moving with the velocity due the height of the dam, which, in the case supposed, is an impossible assumption. A depth of 4 feet on the crest of a dam would imply a velocity at that point of about 11 feet per second. When the sheet of water has reached the bottom and attained a velocity of 28.35 feet per second, it cannot have a depth of more than 18 or 19 inches, and it is only on a portion of the stone of this height that the water exerts a pressure represented by twice the head due the velocity.

The judicious use of random stone in river work usually requires them to be confined so that they cannot be separately washed away, and can offer their united weight against the pressure of the water. One of the most common devices to this end is cribwork. A crib is a great box made of square or round timbers, notched and bolted together, strengthened with cross-ties, and filled with stone. Next to mortar these are the most efficient means of uniting great quantities of stone into a coherent mass. It is much cheaper for that purpose than mortar, though that advantage is rapidly disappearing through increasing cost of timber and diminishing cost of cement. It can often be executed by sinking cribs in water where masonry would require expensive coffer-dams. For these reasons this kind of work is still extensively used in dams.

Figs. 40 and 41 * relate to a dam built across the Merrimac River at Sewalls Falls, near Concord, N. H. The river at this point has a drainage-area of about 2350 square miles. The formation at the site of the dam is a tough clay gravel, called by geologists "glacial gravel," underlaid, at a depth of 10 or 12 feet, by sand. It is in a somewhat rigorous climate and heavy ice is formed, although the formidable effects observed in rivers running northward are not to be apprehended. The overflow or spillway has a length of 497 feet. The dam, in connection with a canal about 1300 feet long, creates a fall of 16 feet, which, it is said, can by the aid of flashboards be increased to 19 or a little more. It is stated that, preliminary to the construction of the spillway, a line of log cribs was placed across the stream, 32 feet apart, to support a roadway for transport of materials, and in part to sustain the low coffer-dam required to turn the water through the sluices. After constructing the abutments, which were of rough masonry, the spillway was put in in sections, commencing at the abutments and working toward the centre. The first two sections, one on each side of the river, were each about 140 feet wide and contained the sluices for discharging the water during construc-

---

* *Engineering News*, vol. xxxi. p. 326; April 19, 1894.

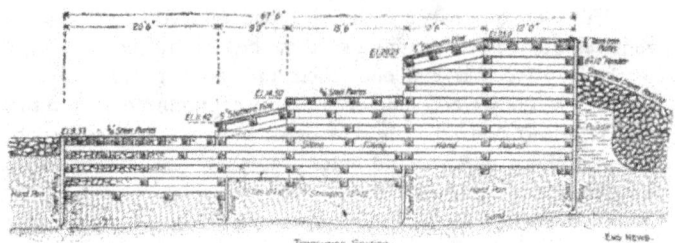

Transverse Section.

FIG. 40.

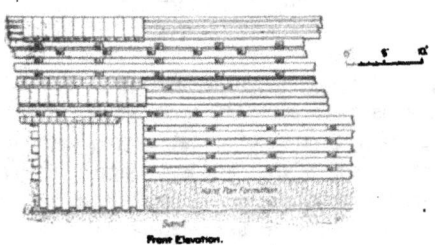

Front Elevation.

FIG. 40a.

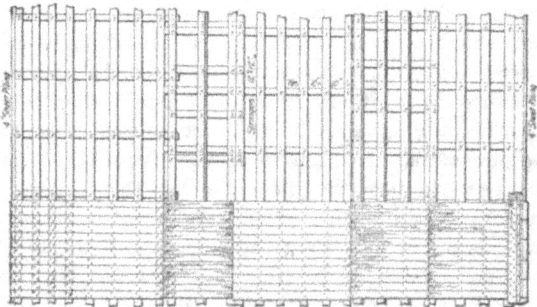

FIG. 41.

tion. A general idea of the nature of this work can be got from the figure. It is cribwork formed of 10- and 12-inch square timbers laid up without notching or locking, and confined with ⅞- and 1-inch square drift-bolts 20 and 30 inches long. It forms a series of pockets, 8 or 10 feet long and 7 or 8 feet wide, filled with stone which is understood to have been packed by hand with care. The crown and slopes were covered with 5-inch Southern-pine planking. The level platforms, which receive the falling water, were covered with steel plates $\frac{8}{16}$ inch thick. This novel feature gives a peculiar interest to this dam, as the greatly diminished cost of steel and iron has of late made such plates available for use in this class of work. The up-stream crest of the dam is also defended by a plate of the same thickness. The planks were not laid in contact, but with half-inch intervals in order to keep the interior woodwork constantly saturated with water. This dam was commenced in July, 1892, and completed in August, 1893, at a cost, as is stated, of $175 000. The general situation of this dam appears from the photograph, Fig. 42.

This dam depends for tightness wholly upon the row of 4-inch spiling along the up-stream face. The 4-inch line at the toe of the dam is inserted as a guaranty against undermining. The two intermediate 3-inch lines can hardly be regarded as essential. It is understood that the planks of this spiling were not tongued and grooved, but merely jointed on the edges in a planing-machine.

During the breaking up of the ice in the spring of 1895, one of the aprons or steps of this dam was crushed in and pounded to pieces for a length of 100 feet and over. It is stated by the superintendent of the works that the water went over the dam during that flood to a depth of 14 feet. This experience led to the strengthening of this apron by the addition of a 5-inch layer of planking. The injury was extended and increased in 1896. The abutments, which rested on wooden platforms on a sandy foundation, were very much damaged. The easterly abutment was undermined and badly tilted, and has since been rebuilt.   The westerly abutment was also injured by scour.

The introduction of open spaces in the deck-plank of this dam for the purpose of keeping the timber wet is not a feature to be commended.   It results in a constant stream of water, of considerable size, entering the structure at the uper end, and escaping at all crevices, near the lower end.   The wearing power of such a constant flow of water is very great, and in the course of time it is attended with highly injurious results to the timber.   The maintenance of the timber in a wet condition can be secured by filling the cribs with sand or gravel in addition to the stone, so as to form an unbroken mass from which the water does not readily escape.   Then, however tight the planking may be, we may count on the admission of sufficient water to keep the filling in a state of saturation.

The experience with this dam clearly shows the inherent weakness of a timber platform to sustain the impact of falling masses of ice, especially where so situated that there can be no depth of water over the platform.   The case is very different from that of Fig. 21, in which, during the high water accompanying the breaking up of ice, the latter would have to move at least 12 feet through the water before striking the bottom, and then in an inclined direction.   It has appeared to the author that there is no inherent difficulty in giving a curved outline to the descending face of such a dam, and the attainment of that object is well worthy of study.

Fig. 43 is given with that view.   The arrangement of the timbers presents no difficulty.   The only feature which has in any sense the character of an experiment is the bending of the deck-plank, an operation which, as is well known, is practised daily in other lines of construction.   The planking would go on in two lengths.   The upper length of 16 feet would require to be bent to a versed sine of less than 18 inches.   The lower length of 25 feet would require a versed sine of 26 inches. There does not appear to be any difficulty in securing these deflections in 3-inch plank.   They are small compared with

Fig. 42

83

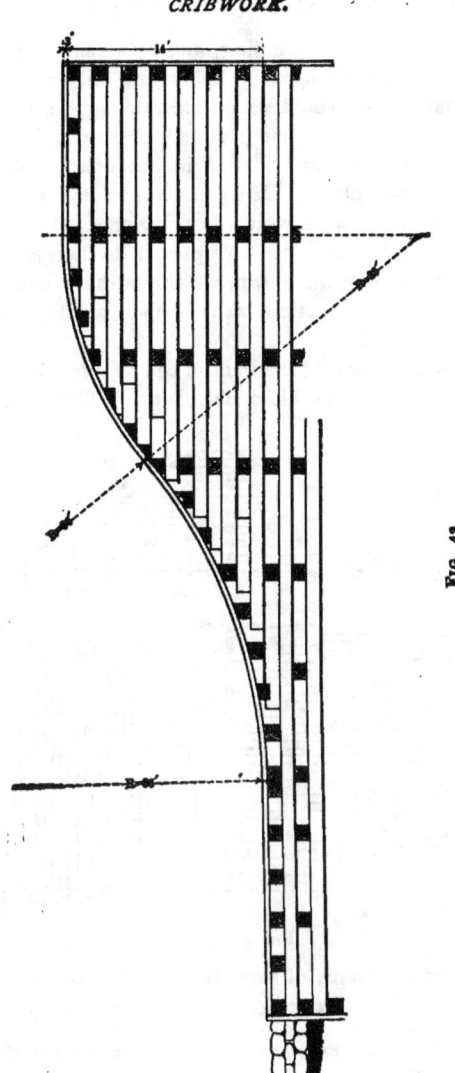

FIG. 43.

what is common in ship-building, and by steaming the plank all difficulty would disappear. Below the inflection-point of the curved face there would be no harm in applying a protection of iron, the wear on that portion being severe. Strips of flat iron 3 inches wide and 6 or 8 inches apart would be as good as continuous plates. It will be noticed in Fig. 43 that we have adopted a curve which leaves some 16 feet of the apron straight and level. Had we placed the tangent-point at the lower extremity of the apron, we should have had a curvature such that the deflection of a 16-foot plank would have been less than 6 inches.

Fig. 44 is a section of the dam of the Wauregan Mills * on the Quinnebaug River at Plainfield, Conn. The river at this

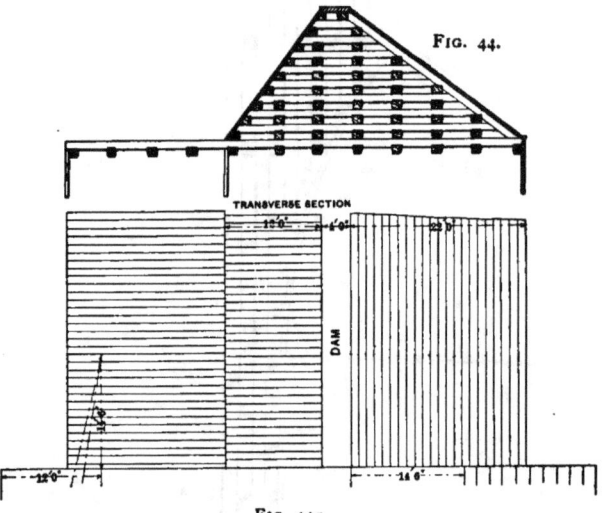

FIG. 44.

TRANSVERSE SECTION

DAM

FIG. 44*a*.

point has a drainage-area of something over 500 square miles. The spillway is 350 feet long. The abutments are of rubble

* Tenth U. S. Census Report, vol. XVI. Region Tributary to Long Island Sound, p. 34.

masonry with dressed face, rising 10 feet above the cap of the dam, which is a cribwork of square and flatted timbers filled with stone; the timbers being laid about 5 feet apart each way and drift-bolted at their intersections. The cap of the dam is about 17 feet above the apron. Two features of this dam are to be noticed: (1) The up-stream planking, which runs length-wise of the dam, contrary to the usual practice. This disposi-tion is secured by means of rafters running on the slant, notched and drift-bolted to the longitudinal timbers. (2) The very steep slope of the face, causing ice and floating bodies to strike with great force upon the apron, which consists of square timbers laid in contact. The angle of incidence being too great to permit falling bodies to glance, ice and logs must strike with the full force due the head, the blow probably being more destructive than a direct one.

Fig. 45 is a study of an arrangement for obviating the

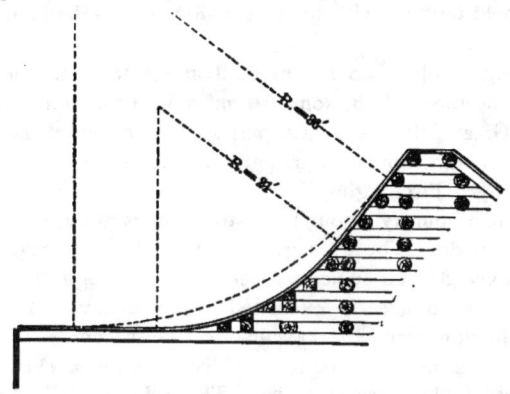

FIG. 45.

destructive action of falling bodies upon the apron, showing how readily it could be done with the use of iron plates applied to curved planking. The inclined face joins the horizontal apron by a curve of about 21 feet radius. The transverse timbers of the dam are extended and sustain longitudinal

timbers cut to receive the curved planks. There appears to be
no constructive difficulty in such an arrangement, except that
the radius is rather short for plank. The dotted line shows
the curvature that would very easily admit of applying planking to the face. This is a curve of 36 feet radius, on which a
16-foot plank would require to be bent something less than 11
inches out of line. With this arrangement the apron should
be somewhat extended. For the shorter radius it would probably be better to confine the iron plates direct to the timbers
without the intervention of plank. Since the above was
written, the author has been informed by Mr. J. A. Atwood,
agent of the Wauregan Mills, that a portion of the apron
actually was destroyed in 1893 or '94. Mr. Atwood says:
"The bed of the river, where the apron gave way, was mainly
quicksand. The part of the apron which was on a hard
bottom has remained in good condition. It is certain that no
dam should be built with so steep a slope where the foundation
is poor."

Fig. 46 * relates to a dam built about 1890 on the Bear
River in northern Utah, some 40 miles from the mouth of that
river at Great Salt Lake. Its purpose was to divert the water
of the river into canals for irrigation. It has an overflow 370
feet in length, discharging the drainage of 6000 square miles of
mountainous country, though the stream does not appear liable
to great floods. The average monthly flow from 1889 to '92
did not exceed 5500 cubic feet per second, though this figure
may have been greatly exceeded for some days. In June,
1894,† the flow reached 8000 cubic feet per second. The dam
is a cribwork made of 10 × 12 squared timbers, except the
bottom sills, which are 12 × 12. These sills are bolted to the
rock bottom by 1¼-inch anchor-bolts 3 feet long, four to each
sill, confined by fox-wedges. The three lower courses of
transverse timbers are horizontal. Above these the timbers

* *Engineering News*, New York, vol. XXXV. p. 83.
† Eighteenth Annual Report of the U. S. Geological Survey, Part IV,
p. 321.

are inclined at an angle of 30°, giving the down-stream face an angle of 60° with the horizontal and forming a right angle at the summit. The timbers are fastened to each other by drift-bolts, and the internal spaces filled with stone. The sills

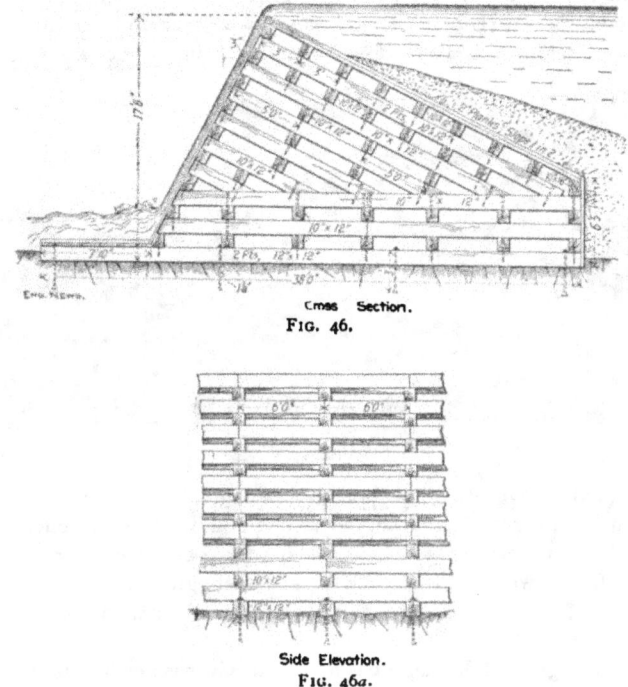

Cross Section.
FIG. 46.

Side Elevation.
FIG. 46a.

extend 7 feet 10 inches beyond the foot of the face-slope, and are planked to form an apron for the protection of the bottom. The rear slope of the dam is covered with two layers of 2-inch planking; the front slope with one layer of 3-inch. The up-stream vertical face has a single layer of 3-inch plank. A heavy bank of earth closes all interstices on the up-stream side. The dam joins steep escarpments of rock on either end, and no artificial abutments were required.   The photographic view Fig. 47 shows the general situation of the dam.

The section exhibits the dam resting upon firm rock.   This was the fact for the greater part of its length; but some 100 feet of the westerly part rested on gravel and boulders, through which the water found its way, and in 1891, soon after raising the water, the river was found running mainly under the dam,

FIG. 47.

instead of over it.   The breach was temporarily stopped with brush and gravel, and on recurrence of low water a concrete wall was inserted reaching down to rock, and the dam reinforced with boulders and backed with gravel, at a cost of some $5000.   It is understood that no further trouble has occurred.

The general drainage-basin of Bear River is north of latitude 42 and some 6000 feet above the sea-level.   It is presumable that heavy ice must occur.   The apron if constructed as indicated appears to be a very inadequate safeguard for such an emergency, especially for the portion resting on gravel.

The largest dam of cribwork in this country, perhaps the largest in any country, is the dam of the Holyoke Water-power Company, on the Connecticut River, at Holyoke, Mass., where that stream has a drainage-area of 8000 square miles.   The history * of this dam is obtainable from authentic sources, and it is believed that it will be instructive to give it in some detail.

A timber dam was built across the Connecticut at Holyoke in 1848, of a much less substantial construction than the one here called the crib dam. It was, as is stated, only expected to maintain a head of water for a few years, and then to serve as a coffer-dam for a more substantial structure. No plan or description is extant from which the details of this dam can be learnt. It is said to have had a base of 60 feet, with a height of 30, and to have been nearly vertical on the face. It was completed in November, 1848, and the gates were closed to raise the water on the sixteenth of that month, there being then a considerable rise in the river. When the water had risen to within 2 or 3 feet of the top, the greater part of the dam rolled over and was carried away.

It would be interesting to know the exact construction of this dam, and consequently the precise reason for its failure. A structure which fails is often more instructive from an engineering point of view than the one which stands; but it is not at all likely that any plan of the work was ever made. It is quite conceivable that, through defects in construction or design, the water came to full pressure within the dam, i.e., that the down-stream face held the water instead of the up-stream. This would relieve the slant face of the pressure tending to hold the dam down, and leave it free to turn over under the pressure acting on the down-stream face.

The crib dam was commenced and finished in the summer of 1849. It is a cribwork of square timber, mainly 12 × 17, put together as shown in section in Fig. 48. The transverse timbers are 6 feet apart and slope at an angle of $21\frac{3}{4}°$ with the horizontal, about $2\frac{1}{2}$ base to 1 perpendicular. The lower ends of these timbers were bolted to the rock by fox-wedge bolts, of which it is stated that 3000 were used. The back slope was covered with 6-inch hemlock planking and, toward the crest, by two and even three layers of the same. The crest was pro-

* See Transactions of the Am. Soc. of C. E., vol. xv. Paper No. 339.

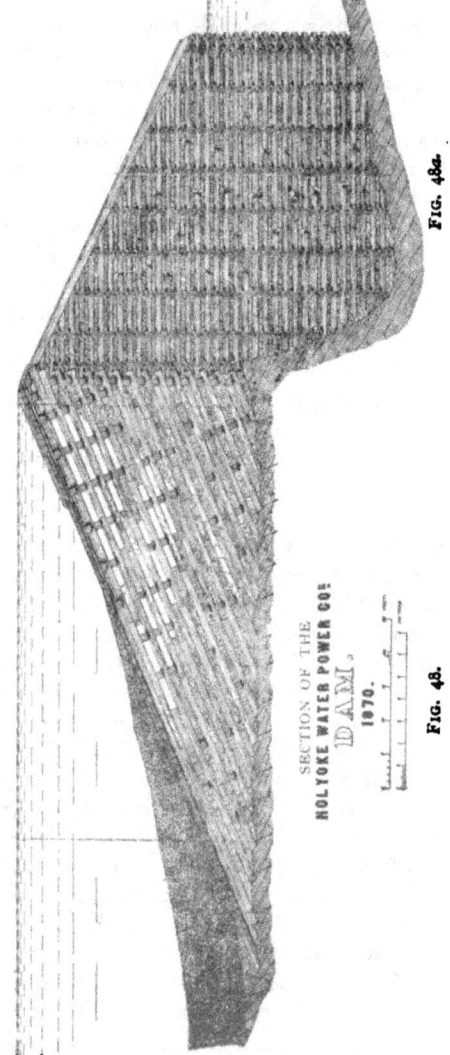

FIG. 48a.

SECTION OF THE
HOLYOKE WATER POWER CO:
DAM.
1870.

FIG. 48.

tected by sheets of boiler-plate. The abutments were of masonry and were very massive. A massive bulkhead or gate-house adjoins the canal and controls the flow of water into the system of canals; but this is not reckoned as a part of the dam. The rollway of this dam was 1017 feet long.

In the construction of this dam a low coffer-dam was built to exclude the water from half the site of the dam. This portion was commenced, raised above the ordinary water-level, and provided with sluices through which the water was turned when work was commenced on the other half. The commencement of the work was the bolting of three 15-inch square timbers to the rock at the foot of the down-stream face (see Fig. 48). These timbers were level in the longitudinal direction; their upper faces were in line with each other and inclined transversely at the angle of the slope, viz., $21\frac{3}{4}°$. This formed the starting-point of the work. The short transverse timbers were laid on, 6 feet apart, bolted to the rock, and drift-bolted to the 15-inch timbers; then a set of longitudinal timbers, then a set of transverse timbers, etc. The whole was filled to a depth of 10 feet with stone. Where the planking joined the rock, at the foot of the up-stream slope, a small bank of concrete was deposited. A bank of gravel was deposited at the foot of the up-stream slope, extending 70 feet up-stream and 30 feet or more up the slope. This bank was extended later. The gates were closed October 22, 1849, and the water rose and flowed over the dam. These gates were merely portions of the planking joined together and provided with hinges. They were 46 in number, measuring, each, 18 feet lengthwise of the dam and 18 feet on the slant side, occupying, with the necessary intervals between them, the entire length of the dam. The lower edge of the gate-opening was on a level with the top of the stone filling. During the construction of the dam the gates were held open by props. On completion, a man was stationed at each gate and, on a signal from the engineer, the props were knocked away and the gates fell simultaneously.

November 12th, following, a very striking phenomenon was

developed which has ever since been a subject of scientific
curiosity. The water going 6 feet deep over the dam, the falling sheet vibrated, and communicated vibrations to the air, at
the rate of 128 per minute. So strong were these vibrations
that windows at Springfield, 8 miles distant, rattled in harmony
with the vibrations of the water. The maximum depth on the
dam occurred in April, 1862, when the water stood 12 feet
6 inches deep on the crest. This was 2 feet 6 inches more than
had been assumed in the design of the work, and necessitated
the raising of the abutments and gate-house. This height of
12 feet 6 inches would indicate more than 140 000 cubic feet
per second going over the dam, being at the rate of 17 or 18
cubic feet per second for each square mile of drainage-ground.
Depths as great as 9 feet on the dam have occurred several
times.

This dam stood, in the form of Fig. 48, some twenty years,
during which time forces were continually at work threatening
its ultimate destruction. It stood upon a ledge of red slate
interspersed with seams of cleavage which dip down-stream at
an angle of 30° with the horizon. The texture of the stone
not being of the hardest, and the direction of the falling stream
such as to cause immense pressure in any open seams, a rapid
abrasion of the bed ensued. By the year 1866 this had become
threatening. A pit more than 20 feet deep had been excavated below the dam and was extending up-stream. Another
agency of destruction had also become formidable. Logs and
heavy blocks of ice, falling over the dam, were caught in the
rolling eddies which are always set up when a stream of water
falls into a pool, and battered against the face of the dam.
Logs were especially harmful, for they often struck the dam,
head on, with such force as to wedge themselves between two
horizontal timbers, where, the outer ends being acted on by
the water, they exerted a leverage which no strength could
resist.

In 1868 it had become apparent that immediate measures
were necessary to save the dam, and the company commenced

the construction of the crib apron shown in Fig. 48*a*. This was made of round logs 6 feet apart each way, notched together and fastened by drift-bolts and filled with stone. The part under water was built in separate cribs, each about 150 feet long and 50 feet wide, built afloat, floated into position and sunk. During such work, the sheet of water coming over the dam was confined to one-half the latter by flashboards, allowing work to go on upon the other half. The work being done in a low stage of the stream, this was easily managed. The superstructure of the apron, or part of it above water, was built continuously, without reference to the length of the separate cribs, so that the work consisted of a series of cells or pockets 6 feet square, less the thickness of the logs. These cells were filled solid with stone to the top and planked over with 6-inch hard-wood plank. The apron was commenced in 1868 and finished in 1870.

With this reinforcement, the dam stood satisfactorily for nearly ten years, but about 1879 new signs of weakness began to appear.

It was confidently predicted by the builders of the dam that the work was practically indestructible. They say in a pamphlet calling attention to their enterprise, in 1853: "The firm basis upon which the dam stands, . . . the strength and solidity of the structure, . . . the thick bed of gravel on the upper end of the dam, . . . the substantial manner in which the work was executed, . . . the effect of the water upon the material of which it is constructed rendering it practically indestructible . . . These things leave no room for doubt as to its permanence and security." Again, a contemporary publication says: "The dam leaks a little, but . . . this is considered no fault, as a sufficient leak is necessary to keep all the timbers bathed in water to prevent rot."

In adopting this opinion as to the preservation of the timber-work, the promoters of the enterprise committed an error. Timber immersed in water does not decay, for the reason that the water excludes the air. It is a mistake to suppose that keeping timber sprinkled or moistened with water, without excluding the air from it, will prevent decay. It is conceivable that a stick of timber in the dam might be so thoroughly bathed with water by leakage as to arrest decay, but this is a condition which cannot be expected generally. Many parts of the dam must be in a state not wet, but simply moist, that is, in a condition eminently favorable to decay, air and moisture being the twin agents of decomposition. Another error was committed in using hemlock for the exterior planking, that timber being very perishable under the influence of air and moisture.

In 1879 a whirling vortex appeared in the surface of the water over the up-stream slope of the dam, indicating the passage of water. Examination disclosed a break in the planking at this point. A very simple and effective expedient was adopted to repair this and subsequent breaks. A cofferdam of cribwork was made in the form of a box open at top and bottom. The top was square to the axis, and the bottom inclined at an angle equal to the slope of the dam, so that when resting on the dam the top was level. This was floated into position and placed on the dam so as to surround the break. As soon as a water-tight connection was formed between the crib and the dam, the water disappeared from the former, passing through the break in the planking, and men were able to enter the crib and remove and renew the defective planking. This method was by no means so easy of application as it appears. The back of the dam was originally covered to the height of 30 feet from the foot of the slope with gravel in which huge boulders were imbedded. The remainder of the slope had in the course of years become covered many feet in depth with sediment and river gravel containing trunks, branches, and roots of trees, which from long immersion become, so to speak, "water-logged" and do not float, but roll along the bottom till arrested by some obstacle. Different methods of removing these deposits were tried.* A jet of

---

* It is probable that a 15- or 18-inch centrifugal pump driven by a

water playing through a nozzle directed by a diver was effective, but slow. A road-scraper operated by lines from an anchored boat, and attended by a diver, was more satisfactory, but was attended with no little danger to the diver, viz., there was no telling when he might encounter a spot in the planking just ready to break through, in which event he would almost inevitably be drawn into the opening. One diver was, in fact, lost by being drawn into an unsuspected opening in the dam.

These breaks became alarmingly frequent in the succeeding years, and in 1884, under direction of Mr. Clemens Herschel, then engineer of the company, comprehensive and systematic repairs were undertaken. The water was excluded by the methods already described and by analogous methods. Much of the planking was found entirely decayed except a very thin crust of the upper surface, a result which should have been foreseen, since the under surface of the planking could by no possibility be maintained in a condition to exclude the air. The timber was in a better state of preservation, and only a small portion near the crest required renewal. The planking was wholly renewed above the level at which water stood within the dam. A diaphragm of 3-inch planking was inserted within the dam, extending from one abutment to the other, and from the crest to a level 25 to 28 feet below the same. The dam never having been entirely filled with stone, and much of the original filling having been washed out, the filling was completed, not with stone, but with gravel. After the completion of these repairs, in 1885, the dam appeared to be as good as new, and showed no further indication of weakness. In 1895 the company commenced the construction of a massive stone dam, to replace the one whose history we have been giving.*

---

powerful steam-engine, pump and engine mounted on a flat boat, throwing a stream of 30 cubic feet per second at a high velocity, would have been an efficient means of removing these deposits.

* The facts relating to the Holyoke dam are taken mainly from Clemens Herschel's paper, No. 339, in Trans. Am. Soc. C. E., 1886.

Dam on a Soft Bottom.—Fig. 49 is given, with its several detail figures, as a design for a dam of economical construction and reasonable chance of perma-nency on an alluvial formation, or a formation of sandy gravel, clay gravel, or soft and friable rock. A desirable condition is that the ma-terial excavated shall serve for the embankments. This mode of con-struction does not present to the mind the sense of security that would be desirable, though, if faith-fully executed, it is difficult to see where the cause of failure exists. Such a dam, moreover, might suffer failure two or three times and still be cheaper than a dam whose per-manency was free from question. The distinctive feature of the dam is a broad and deep basin on the down-stream side to receive the dis-charge and free the same from its commotions without any dangerous action upon the dam or its foun-dation. Fig. 49 is a section through the dam and basin. 49a is a general plan of the situation, indicating the mode of constructing the work. The remaining figures show various details to be referred to in due order.

Nearly one half the dam is under-laid by sluices through which the flow of the stream may be discharged while the other half is under con-struction. The body of the dam,

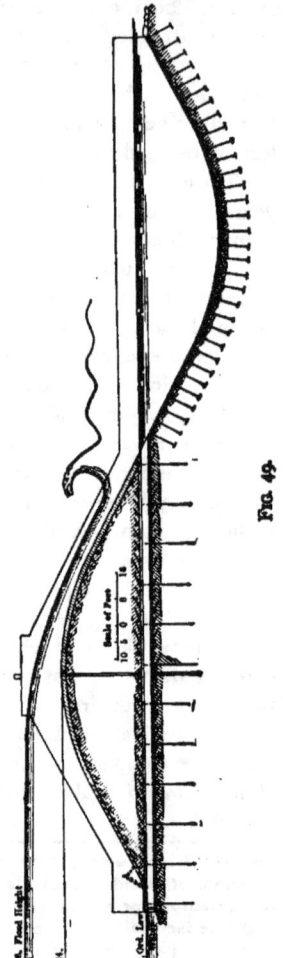

Fig. 49.

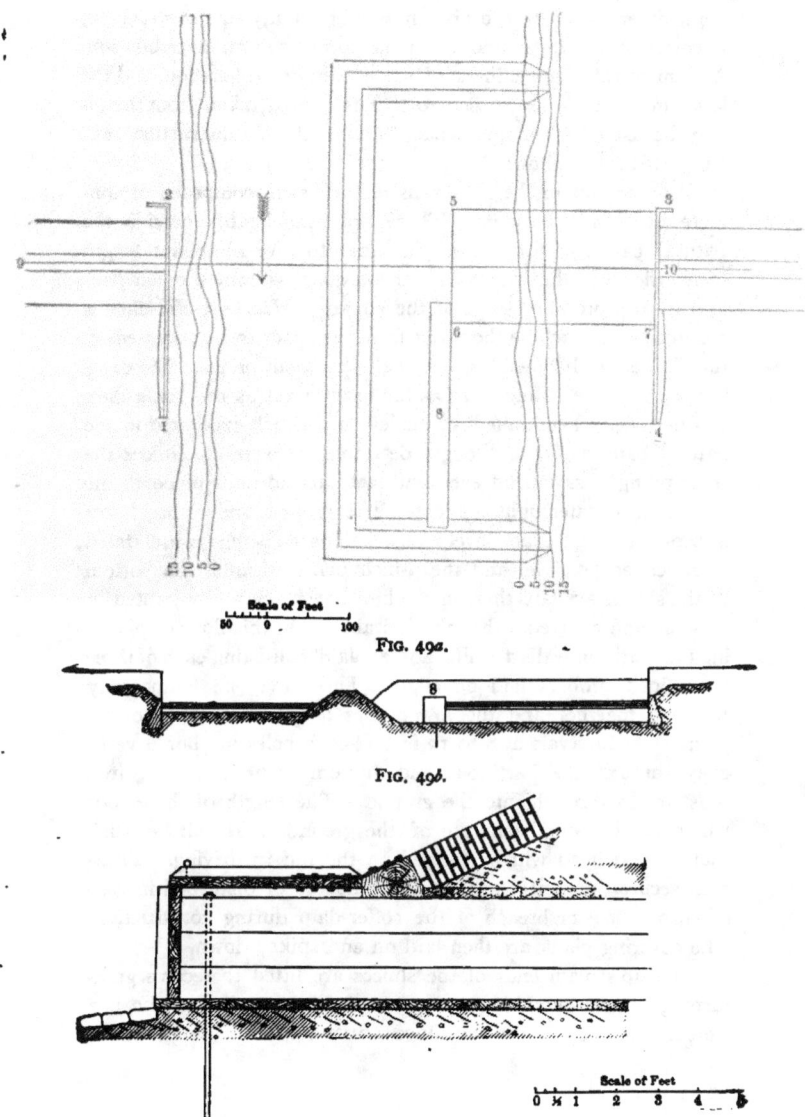

Scale of Feet
50   0   100

FIG. 49*a*.

FIG. 49*b*.

Scale of Feet
0 ½ 1   2   3   4   5

FIG. 49*c*.

being intended merely to give mass and solidity to the structure, is composed of any sound earth susceptible of firm consolidation. A dam of this kind admits of some latitude in location, and the latter may usually be chosen so that the earth taken from the pit may be used for embankments. The mode of construction contemplated is as follows:

We first put in the side walls of the basin, composed of concrete or rubble masonry. These are considerably outside the natural banks of the stream, in order to give abundant length of overflow, so that the work can be done without a coffer-dam and in any ordinary stage of the stream. Next a coffer-dam is constructed to exclude the water from the space to be occupied by the sluices, and something over half the basin or pit. We drive the row of sheet-piling 9, 10 as far to the right as the coffer-dam will permit, and continue it to the left in a trench excavated in the natural bank as far as thought desirable. Under the sluices the sheet-piling is sawed off even with the ground and joined to the sluices in a water-tight manner. The ground under the sluices is brought to an exact level, the filled parts being consolidated as much as possible, and the 3-inch plank forming the bottom of the sluices are laid thereon. They may be carefully jointed or tongued and grooved if thought desirable. On this floor of planking the partition-walls 1 1, Fig. 49*f*, are laid, consisting each of three 6×10-inch timbers laid on edge. These may be temporarily fastened together and the lower ones *toe-nailed* to the bottom plank. At intervals of 8 to 12 feet 1½-inch holes are bored vertically through the partitions and bottom plank and long iron rods driven through into the ground. The length of these rods will depend upon the nature of the ground. It will be such that a good head will be formed on the rod in driving. These rods secure the sluices against being floated or disturbed in case of an overflow or breach of the coffer-dam during construction. The covering plank are then laid on and spiked down.

The up-stream ends of the sluices are fitted to receive gates turning on horizontal hinges as shown at Fig. 49*c*. 1, Fig. 49*c*, is a timber extending across the sluices, drift-bolted to the partition-

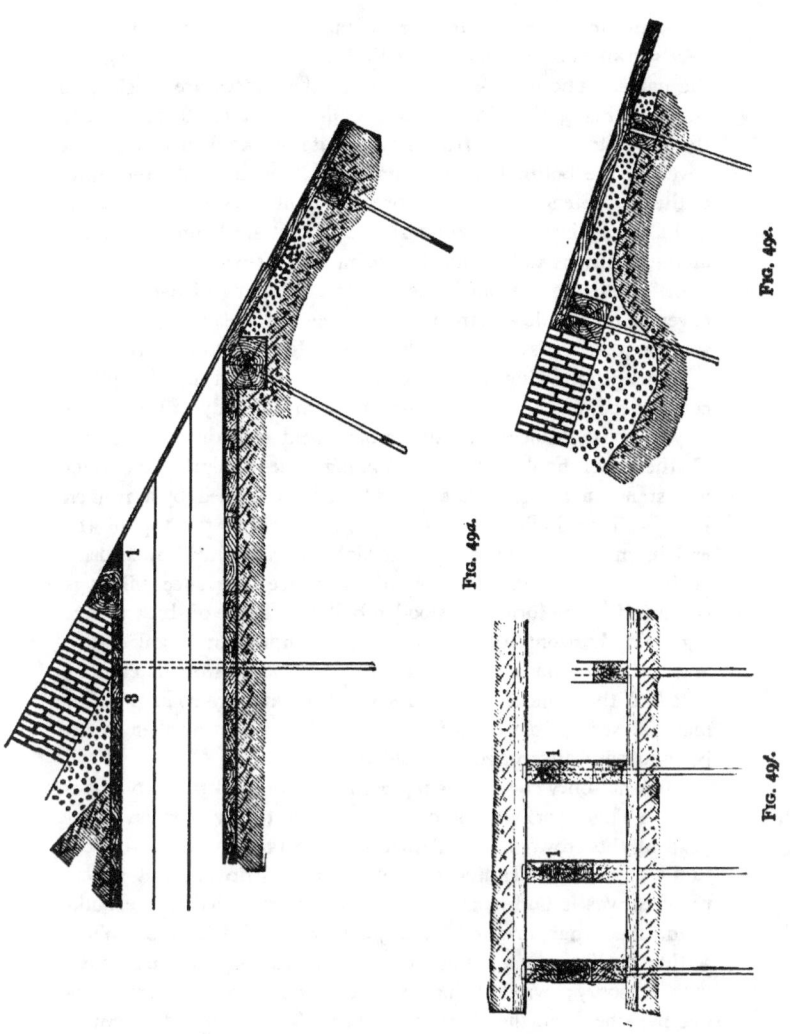

Fig. 49a.

Fig. 49b.

Fig. 49c.

walls and forming a footing for the masonry covering of the dam, here shown as brickwork.   The timber 1 also carries the hinges of the gates.   The down-stream ends of the sluices are finished as shown at Fig. 49*d*.   The partition-walls are slanted to conform to the down-stream slope.   Iron or steel plates are applied at this point covering the bottom of the sluices and lapping on to the lining of the pit.   Steel tongues also cover the slant faces of the partitions and abut against a triangular strip 1 of cast steel running across all the partitions and confined to them by drift-bolts.   The timber 2, drift-bolted to the partitions, forms the footing of the masonry covering on the down-stream end.   The triangular strip is necessary because the situation calls for an edge too thin for timber.

After completing the sluices, the excavation of the pit is commenced, the earth being used to form the body of the wasteway above the sluices, and the embankment extending to the left. To the left of the sluices the sheet-piling comes as high as the water will stand in the ground, above which it is capped by a 12-inch brick wall as indicated at Fig. 49*l*, reaching to the top of the embankment and one or two feet higher as an additional security against overflowing.   On top of the sluices the sheet-piling is continued in the form of a wooden bulkhead made of thick planking laid horizontally and confined by uprights; plank being added, one by one, as the work is raised.   The work is consolidated to the utmost by the use of water and by rolling.   For material susceptible of compression a heavy steam-roller might be used with advantage on such work.

The masonry covering is represented as of brickwork 16 inches thick.   This work is not quite closed on to the wooden bulkhead at the crown of the dam till the work has had an opportunity to settle.   Pending the settlement a strip of stout waterproof canvas is laid over the earth, one edge nailed to the bulkhead, the other inserted in a joint in the brickwork.   When settlement has ceased this opening is cleaned out and filled with masonry; water being excluded by flashboards, first from one half the dam, then from the other.   The down-stream covering rests upon a layer of broken stone, so that water which passes

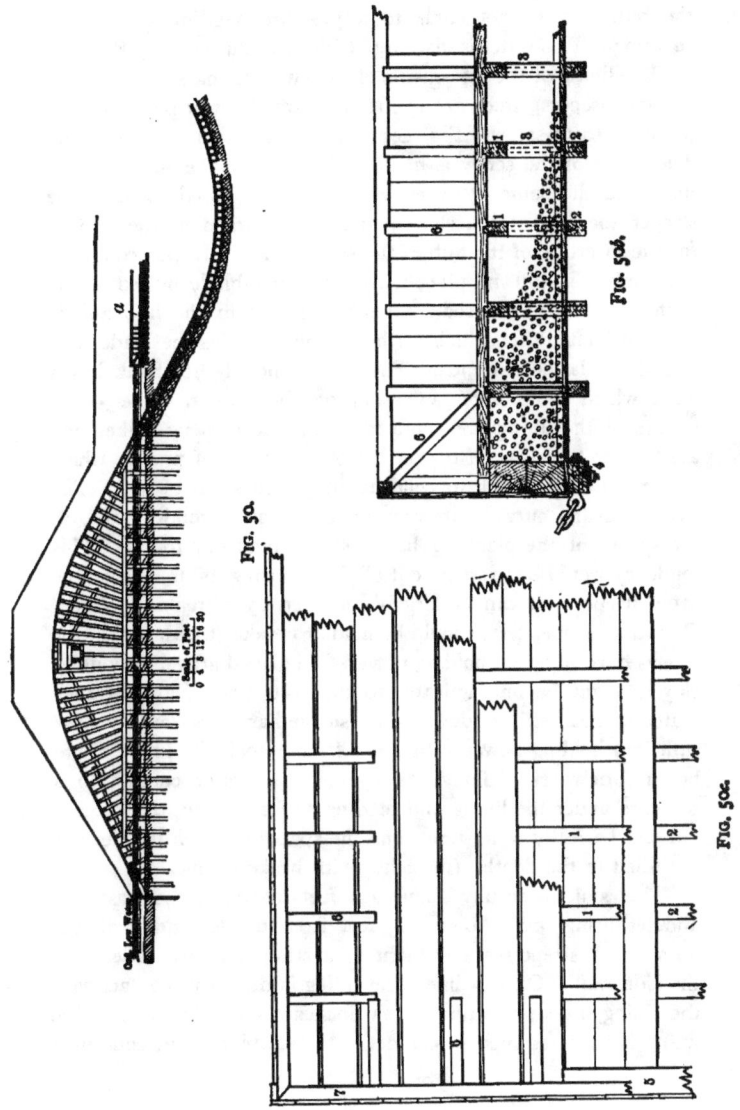

Fig. 50a.

Fig. 50b.

Fig. 50c.

the bulkhead cannot come to a pressure tending to lift this covering. This water is discharged into the sluices at 3, Fig. 49*d*.

The lining of the pit consists of 3-inch planking spiked to 9- or 10-inch sleepers, which are confined by small screw-piles, or more properly earth-screws, Fig. 49*g*, shown as about 8 feet long. The blade of the screw is made separate from the shank and is joined to the same by a screw. Holes are bored through the sleeper the size of the shank, and are enlarged at the bottom for the insertion of the hub of the screw-blade. In this condition the timber is put in place with the screw-blade buried. The shank is then introduced, enters the socket in the hub and is screwed home, after which further turning carries the blade with it and the latter descends. The upper end of the shank has a head which bears upon a washer on the timber. The 3-inch planks of the lining are confined to the sleepers by spikes and are bent to the curve of the pit bottom, a radius of 75 feet, which requires a 16-ft. plank to bend about 5 inches out of line. The sleeper on the curved part requires to be quite solidly fixed, as the spring of the planking has a strong tendency to lift. This tendency would be irresistible if all the abutting joints were on the same sleeper. It can be largely annulled by " breaking joints." Of course proper tests should be made to make it certain that the screws have sufficient holding power. In glacial gravel or boulder clay it might be impracticable to turn the screws, and in such material rods driven down with sledge-hammers might have sufficient holding power. In loose friable rock the lining might be of brickwork. To guard against the water coming to a pressure under the lining and tending to force it up, the latter is not laid perfectly water-tight, and the spaces between the sleepers are filled to the depth of 6 inches with broken stone.

The sluices occupy about 250 feet in width, the wasteway and pit being 600. About 265 feet from the left side wall the lining and sleepers are brought to a straight line parallel with the side wall. On this line sheet-piling is driven in contact with the lining, and projecting a few inches above the same. The lining is then resumed in contact with this piling and continued as far as the condition will permit. Over the line of piling a coffer-dam 8, Fig. 49*b*, is built, up to the level of the side wall. The water is then turned through the sluices and excluded from the right half of the dam and pit. After the pit is finished the

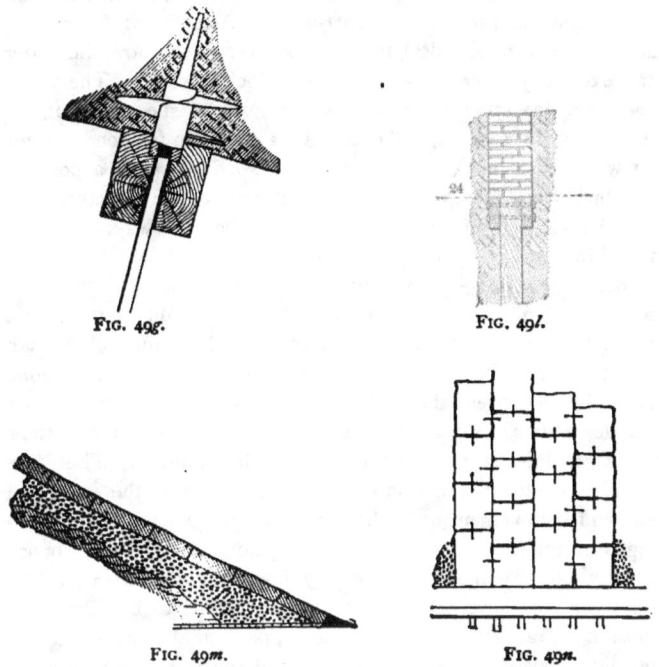

FIG. 49*g*.   FIG. 49*l*.

FIG. 49*m*.   FIG. 49*n*.

coffer-dam 8, is removed, being constructed with a view to ready removal. We need not dwell on the details of this work. Large stones should not be left in the pit or permitted to pass through the sluices, as the constant motion due to the action of the water would wear the lining.

Fig. 49*e* represents the down-stream footing of the masonry covering, outside the sluices. The up-stream footing descends two or three feet into the natural ground. Outside the side walls of the pit the ground slopes up to the natural level. These slopes together with the adjacent slopes of the embankment are covered with heavy stone pavement resting on a layer of broken stone or pebbles. It must be conceded that a work of this character would be liable to great damage should an unusual flood occur while in process of construction. After being finished in accordance with the design, it is not easy to see how the water can exert any dangerous action upon the structure. The water goes over the spillway in a smooth unbroken sheet, encounters the water of the basin, and expends its energy in friction of water on water; neither water, ice, nor driftwood being in a position to administer a dangerous blow to any part of the structure.

The use of brick for the pavement of the wasteway, especially the down-stream slope, might not be advisable in a rigorous climate, not so much on account of the disintegration of the brick-work as from the possible heaving of the earth under the action of frost. Some kinds of earth are more liable to this action than others. Where it is to be apprehended the construction Figs. 49*m*, 49*n* might be preferable. A layer of broken stone is applied on the down-stream slope. This is covered with large granite slabs of as regular shape as can be got out by splitting. These are fastened together with iron cramps deeply sunk in the stone and embedded in cement or lead, forming a continuous flexible covering over the down-stream slope, susceptible of slight disturbance at any point without in any way impairing its integrity. All joints except those near the foot of the slope to be filled with mortar. Cracks would, no doubt, appear in this pointing, but no dangerous crevices could occur, and the work would always be accessible in low water for repairs.

Fig. 50 with its several detail figures represents a dam of the same general form as the preceding. but constructed without laying the pit dry, or dividing it into two parts by a high coffer-dam. It also avoids the expensive sluices of Fig. 49, but calls for a greatly increased amount of timber-work in the body of the dam. This dam is supposed to raise a head of 24 feet on a

stream of 3000 square miles drainage area, requiring a wasteway
some 620 feet long.

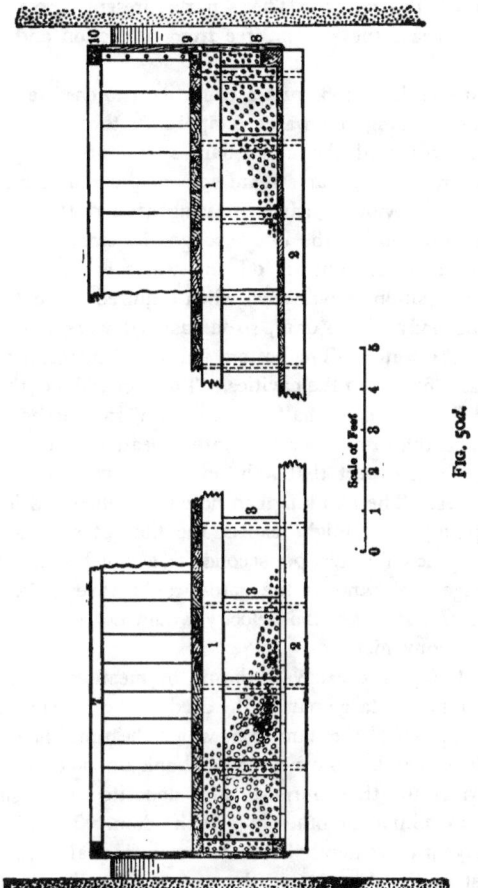

Scale of Feet

FIG. 504.

As in the former case the first step is to build the abutments,
which, being located outside the natural banks and requiring no
coffer-dams, can be executed without reference to the stage of

the stream. Opposite the body of the dam, the abutments go
to the natural soil or a little lower; opposite the pit they go below
the bottom of the latter. These parts present a reasonably
smooth face toward the stream, free from projection and irregularities.

The next step is to construct a coffer-dam to enclose one half
the dam site, turning the water temporarily through the other
half. Under cover of the coffer-dam a row of sheet-piling is
driven on a line with the up-stream ends of the abutments, there
joining a line previously driven, passing around the back side
of the latter and joining the line through the embankment (Fig.
49*l*). The gates for shutting off the water are also executed
and left in a position to be closed when required; also the pile-
work for the body of the dam, so far as that work requires the
exclusion of the water. The bottom is levelled by scraping down
the ridges and filling up the cavities. This being done, the water
is excluded from the other half of the stream, and turned through
the gates, and the same operations are repeated for the opposite
half of the river, except that it is not necessary to provide this
half with gates. The planking can here be applied and joined to
the sheet-piling. We might expect the flow of such a stream
to be under 3000 cubic feet per second for as much as five months
of an average year, and as the gates would have at least 1000
square feet of waterway, the velocity would not be sufficient to
cause any inconvenience.

The next step is to excavate the pit by means of a dredge or
similar machine. A large part of the dredgings can be transmitted
directly to the half of the dam from which the water is excluded.
The remainder can be conveyed to the bank by some of the well-
known devices for that purpose, and deposited in a situation
convenient for filling the other half of the dam. The pit is supposed to have a clear depth below low water equal to the height
of the crest of the dam above the same, and, allowing for the
space occupied by the lining, the excavation will come somewhat
short of the quantity required to fill the dam. We now spread
a light layer of loose stone along the up-stream edge of the pit,

opposite the open gates, to guard against temporary abrasion, and proceed to apply a protective lining to the pit. This consists of an immense mattress or apron, built afloat over the position it is to occupy, and sunk. This being the distinctive feature of this type of dam, it must be described in some detail. *a*, Fig. 50, shows it afloat, in the position where it is built; Fig. 50*c* shows it in plan; Fig. 50*b* is a vertical section at the up-stream end, Fig. 50*d* a section athwart. It is put together in the following

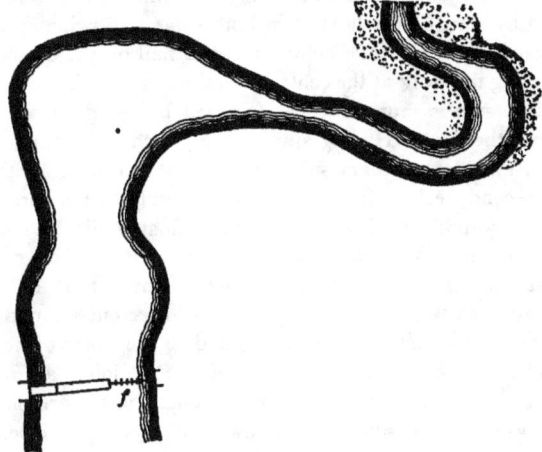

FIG. 51.

manner: In putting in the pile-work of the dam, a seat is prepared for the up-stream end of the apron, so that it will rest in a position to receive the planking of the dam. The up-stream keelson 4 of the float is confined by chains to the pile work, and consists of timbers spliced together and reaching, when complete, from abutment to abutment, with sufficient clearance to avoid fouling. The successive keelsons are brought into position and temporarily attached, afloat, by strips to 4. The lower sheathing or flooring resting on the keelsons 2 2, consisting of clear stuff 1¼ inches thick, planed, tongued and grooved, and slightly, but not fully, seasoned, is then put on and nailed down. The keelsons under one of these planks will have sufficient buoyancy to hold up a workman. After the planking has extended somewhat it will bear any number required to work. The outside plank of the floor is a runner, 3 inches square, for the attachment of the upright planking or curbing, which consists of the same kind of lumber as the floor. If the entire floor were completed from abutment to abutment, there would be a difficulty in nailing on this curbing on the ends of the float, as there would be no room to swing a hammer. This trouble is avoided by starting the float from both sides and swinging the uncompleted parts clear of the abutment to nail on the curbing, before joining the work at the centre. In the drawing the curbing on the right side is represented as fastened by wood-screws inserted from the float. Having applied the 3-inch runners on the right, left, and down-stream sides of the float, we tack another over the second keelson from the up-stream side to temporarily exclude the water. This gives the whole float a displacement of some 4 inches and secures it against sinking under the weight of men and materials. We may now put on the "front log" 5, which consists of two $12'' \times 8''$ timbers, the lower one secured to the keelson by lag-bolts, the upper secured to the lower by drift-bolts. The deck-beams 1 1, consisting of $4'' \times 6''$ joists, are now put in, resting on the stanchions 3 3. These latter are short pieces of 3- or 4-inch square joist, sawed off square and bored from end to end. These are set erect over the keelsons, the deck-beams properly bored resting thereon. An auger of proper size is passed through beam and stanchion and bores through floor and keelson. Long lag-bolts, with heads and washers, are then passed through and screwed solidly into the keelson, binding the latter firmly to the deck-beam. The curbing is then completed, the upper ends of the planks being attached to runners, 7, sustained by braces, 6, resting first upon the deck-beams and finally upon the deck-plank The whole vessel is now filled with broken stone, or with the gravel dredged from the bottom, which brings the top of the deck-beams some 24 inches below the water-line. The deck, consisting of 3-inch plank, is now fastened down, the braces 6 6 being shifted as the work proceeds. These plank are not necessarily in close contact.

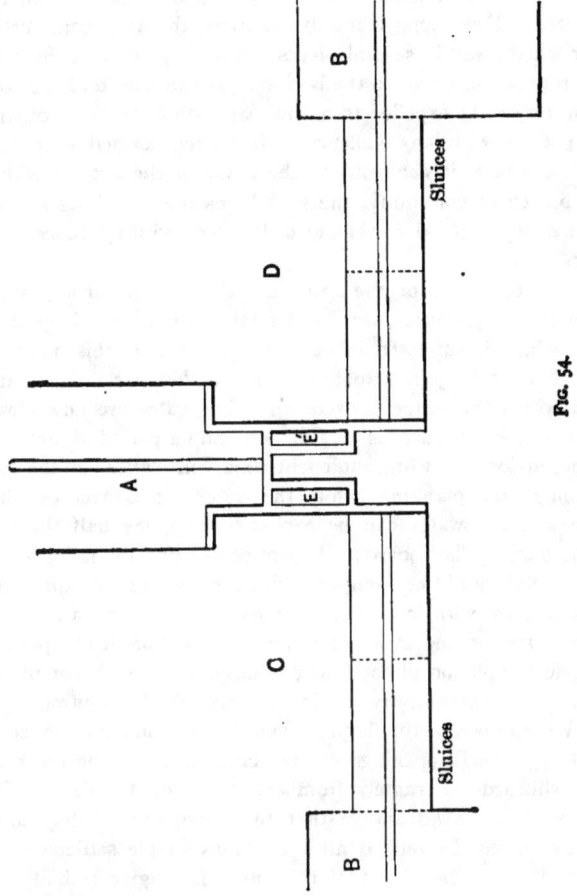

They may, if thought necessary, carry strips or plates of iron to resist the wear of ice or drift coming over the dam. The float

is now hauled into the proper position and securely fastened
to the piling of the dam. The curbing on the up-stream and
down-stream sides is checked so that it will break off even with
the deck. Everything being in readiness, the up-stream curbing
is knocked away in several places simultaneously, and the entire
structure settles down to the bottom. It can sink no faster than
the water can be expelled from the space below, and cannot strike
the bottom with any violence. Not being formed to resist a
bending strain, it conforms to the curve of the bottom without
any breach of continuity, the stanchions merely tilting so as to
allow a slight lateral movement of the deck with reference to the
floor.

The deck-plank of the apron lap on to the front log by half
the width of the latter, leaving a footing for the planking of the
dam, which is supposed to be now completed to this point. It
may at once be joined to the apron on that part of the dam
from which the water is excluded. The gates are now closed.
If the dam creates a considerable pond and a period of low water
is chosen for this work, sufficient time will elapse for the completion of the planking before the water comes over the dam.
Otherwise the water can be excluded from one half the crest
of the dam by flashboards. Before completing the filling of the
dam gravel should be flung in under the head of the apron and
worked down with long poles or rods, so as to form a solid bed
for it to rest on and avoid subsequent dislocation at this point.

The completion of the filling is supposed to be accomplished
by means of a tramway coming in through the abutment just
below the crown of the dam. There is no insuperable objection
to filling through openings in the planking, excluding the water
by flashboards alternately from each half of the dam. This
method has the advantage that the space immediately under
the crown can be refilled after the unavoidable settlement has
taken place. The object of the filling is to give stability and
mass to the dam, to increase its water-tightness, and to preserve
the planking and purlins by insuring their constant immersion.
The planking of the dam is represented as curved to conform

to the outline of the latter, viz., to a radius of about 76 feet, on which a 16-foot plank would deflect 5 inches from a straight line, a degree of curvature perfectly practicable.

The inclined struts supporting the purlins are secured to the pile-caps, as the spring of the planking, when first put down, would tend to lift the purlins. The struts are secured to the latter by dovetail mortise-and-tenons with wedges. Otherwise the framework need not involve the highest grade of timber or of carpentry. Although it requires the strength to sustain the pressure of the water, it is of the character of scaffolding rather than permanent framework, and is superseded when the filling is completed.

The filling in contact with the planking of the down-stream face should be gravel or broken stone, to guard against the washing out of the earth by streams which might gather near the crown and flow down under the planking.

Dam in Tidal Waters.—The construction of a dam to close a tidal inlet or estuary is an engineering problem of peculiar difficulty, and the difficulty increases in proportion to the magnitude of the basin created by the dam. In dams upon ordinary streams the interruption of the current and the raising of the water to flow over the dam is an operation beset with difficulty, but in the case we are considering the difficulties are greatly magnified by the volume seeking ingress and egress at every tide, as well as the alternately changing direction of the flow. The natural flow of the stream in this case bears usually but a trifling proportion to the volume of the tide. The difficulties may not disclose themselves till the channel is half closed, and may not become serious till two thirds or three fourths of the structure is raised. From the point where the velocity begins to act injuriously upon the bed the difficulties rapidly accumulate, and when the opening is narrowed so as to create a head of several feet the rush of water becomes uncontrollable.

The difficulty of this operation is of precisely the same nature as is encountered in closing a sea-dike, except that the latter is not usually founded below low tide. The closing of a breach in a sea-dike, however, is a more nearly parallel case. Such a breach once formed rapidly enlarges. The sea rolls through the opening on the flood-tide, and the basin or polder protected by the dike discharges into the sea on the ebb. The enlargement of the breach proceeds till the current is so much reduced that it ceases to act upon the dike or the bottom. When that time arrives the ground occupied by the dike is scoured away to a depth of several feet, which remains covered at low tide. The only difference between this case and the one we are considering is that we have a stream entering the basin which is not necessarily a factor in the case of dike-break. It will at least be worth our while to consider the methods adopted in the closing of a breach in a dike. Sea-dikes are little known in this country, and there cannot be said to be any established practice in regard to such structures. Such works have existed from time immemorial in the low countries of Europe, and definite methods of procedure have established themselves. The following method of repairing a dike-breach is given by Storm Buysing, an eminent engineer of the Netherlands.* He assumes an ordinary tide oscillation of 13 feet, a depth of 15 feet at low tide after the scouring has ceased, and a width of 460 feet at the level of low tide. No work of repair is undertaken till the condition has become permanent, i.e., till abrasion has ceased. It is sometimes necessary to moderate the current through the breach by artificially widening the same.

The first step is to bring the bottom to a uniform level by filling up the deepest parts and removing the higher parts. It is then covered, except a strip of 10 or 12 feet wide along the centre line of the dike, with mattresses or mats. These are rafts of brush about 3 feet deep enclosed between two sets of stringers or poles, one above and one below, bolted or bound together through the brush. Buysing's method does not contemplate a length or width of more than 60 feet for these mattresses, but

---

* Buowkundige Leercursus, 1854, I, p. 659 et seq. Here quoted from Hagen. Seeufer und Hafenbau, 1863, I, 310.

in the operations on the Mississippi River, mattresses were used which would cover the entire area here assumed. The bottom covering extends well up on to the slopes of the dike, and reaches 30 feet seaward and landward beyond the succeeding layers, forming an apron to receive the overfalling water; the extreme width of the covering being some 230 feet. Except on the slopes which the dike naturally takes at each end of the breach, the mattress work is kept level, so as not to concentrate the flow of water. Each mattress is narrower than the preceding, forming, on the outside of the dike, steps which outline a slope of $1\frac{1}{2}$ or 2 to 1. At the central space the interval between the mattresses increases slightly as the work rises. This space is filled with good binding earth. The latter, after reaching the level of low water or nearly that, is kept covered with light mats of brush, except when adding earth, to protect it from wash.

The mattresses are built upon inclined ways, at convenient points on the shore, launched at high tide, floated to their destined locations, sunk by being loaded with gravel, and heavily loaded after they are down, being at the same time rammed and beaten in order to work the loading into the brush, so that it cannot be removed when the water runs over the work. After the work has risen above low water it is accessible at every tide.

When the work is raised to within 4 feet of the level of spring-tide it cannot be raised higher by this method, as the mattresses cannot be floated to their positions with safety. The work is suspended till neap-tides, which rise not more than a foot or two above the top, if at all, the water on the inside of the dike standing even with the top of the work. The sea-level will now be below the top for eight or nine hours at every tide. During this interval, on a quiet day, the earth bank can be raised so as to exclude the tide entirely, and before the ensuing high tides the whole work can be made secure.

Sometimes the interior basin is so great that the current through the breach is too violent to allow the work to be carried on. The basin is considerably below the tide when the latter reaches its culmination, and there is a powerful flood-current through the breach during the stand of the flood.   After the ebb commences the sea comes into equality with the basin, but only for a moment; the powerful flood-current is abruptly changed into a powerful ebb-current, and no opportunity occurs for sinking the mats. In this case there is but one thing to be done, viz., to construct an independent dike, dividing the basin into two parts to diminish the volume of water flowing through the breach.   Cases have arisen in Holland where no less than three such interior dikes have been required before the work on the breach could proceed.

For depths at low tide equal to or approaching that assumed above, it is not probable that American engineering can offer any substitute for European methods.   For lesser depths it would be more in accordance with our ideas to proceed with the work of construction as far as it can be carried without dangerously increasing the scour, and then suddenly stop the flow allowing the interior basin to fill, and discharge over a wasteway. The following method may be adopted where the low-water depth does not exceed 6 feet.   We assume an average tide of 10 feet, a spring-tide of 12, a neap-tide of 7, and a storm-tide of 15 feet.   The estuary of a river is usually much wider than the inland channel, so that the dam has a great length in proportion to the volume of water to be discharged over it.   No difficulty is encountered in commencing the work, or in carrying it on, by any method deemed suitable, whether by the aid of coffer-dams or by deposit of material in water, till a large part of the length is completed.   Even should abrasion manifest itself, it can be readily checked by a light deposit of broken stone, quarry chips, or any available material.   The abutments, the wasteway for carrying off the inland waters, the sluices, races and wheel-pits may all be completed before disturbing the regimen of the stream. The completed portion is supposed to terminate in a vertical face, and to have a line of sheet-piling through the centre, the portion to be closed lying between the vertical face and the completed abutment.   The bottom is supposed to have been brought to a uniform depth by dredging off the high parts and filling up the low.

A line of sheet-piling is driven along the centre line of the dam, properly joined to the completed parts, and cut off, at a uniform height, a few inches above the bottom. These piles are represented as 6 inches thick, but at intervals of 13 feet, centre to centre, a 12-inch square pile is driven, forming a part of the line. On each side of the last-mentioned pile three or four piles of the same size are driven in contact, after the manner of sheet-piling, forming a short line at right angles to the main line. At each end of this line a brace is applied, footed on a cap-sill, which rests on a bent of round piles, as indicated in Figs. 52, 53, forming a trestle, very firmly bound together by iron straps as indicated. The square pile next the line of sheet-piling on the land side rises to the height of storm-tide; the next 2 feet lower; the remainder 4 feet lower still. The two inner square piles are joined to each other by a 4-inch plank, through which go long bolts uniting the groups of piles. The remainder of the space between the two adjacent piles forms the recesses or grooves for the gates that are to close the opening. The gate is put together resting on the wedges *d*, which are sustained by blocks bolted to the piles. It is formed of 12-inch square timbers. How these are fastened together need not now concern us. A round log, *a*, rests in bearings formed on the top of the high piles. This is banded at the ends and has gudgeons similar to those in the shaft of the old-fashioned water-wheel. The gate is attached to this log by chains, each chain taking several turns round the log, and the parts of the latter to which the chains are attached being of uniform diameter. At the middle of the log a drum, *c*, is formed, to which a rope is attached, taking a turn around the spar *b*, and made fast on a cleat. Planks are laid from trestle to trestle for workmen to stand on.

When these preparations are all completed and the time of low tide approaches, the wedges *d* are knocked out, leaving the gates suspended by the chains. At dead low tide there is a period of several minutes with no current except that due to the inland stream, and when the tide is fairly turned this also ceases. Reliable men familiar with boat-craft are chosen to ease down the gates, and at a given signal the ropes are unwound from the cleats and the gates dropped. The gates are sufficiently high so that the buoyancy of the immersed part is overcome by the part above water. Weights can also be added if thought desirable. When the gates are down there is an interval of five or six hours in which to put in the remaining timbers for raising the bulkhead to the proper height. Now the current through the pass entirely ceases. The basin rises till the affluent waters are discharged over the wasteway, and the work can go on to completion, or, if the convenience of the work requires it, sluices, controlled by tide-gates, may be used to prevent the inland waters from rising inconveniently high while the work is in progress. Fig. 51, p. 109, shows the general situation contemplated in the preceding. This is not a case of the utilization of tidal power strictly so called. That subject is considered under the head of Development of Natural Water Powers (p. 402).

CHAPTER IV.

DAMS OF MASONRY.

A DAM is ordinarily intended or desired to be a permanent structure. The same reasons which lead to its construction demand its continued existence. Dams of timber and earth or timber and loose stone can only be regarded as temporary. They either decay, wear out, or are destroyed by floods in a limited number of years. Such dams have played an important part in the development of industry hitherto, and will continue to do so in future; but there is a constantly increasing demand for permanence in this class of structures, and we may look with confidence, in the future, for a constantly increasing proportion of masonry dams. This expectation is founded upon these considerations:

1. The increasing cost of timber.
2. The diminishing cost of masonry.
3. The greatly diminished rate of interest.

The increase in cost of timber, though very decided, is not so great as might have been expected. Accessible forest lands have been stripped of timber, or have greatly increased in price, and resort is had to distant lands; but these tendencies are largely counteracted by the great improvements that have taken place in means of manufacture and transportation. Moreover, the stripping of timber from forest lands does not destroy its power of reproduction. Timber, although a crop of slow growth, is none the less a usufruct of land, which comes to maturity in due time, and the increased attention now being directed to the care and preservation of forests will, it is to be hoped, insure reasonable supplies in future.

The diminution in the cost of masonry is more marked than the increase in the cost of timber. The steam-drill, steam-hoisting, channelling and dressing machinery, improved explosives, lower rates of transportation, have not failed of their due effect upon this branch of industry. Improvements in the manufacture of cement, also, have diminished the cost of this element. These tendencies have been partly met by higher rates of wages, but the net result is a material diminution in the cost of masonry.

The rate of interest in the United States may be fairly stated to have diminished by one-half since 1850. At that time, the United States government, rich corporations, and prosperous municipalities in the older parts of the country paid 6 per cent for money, while in the newer parts fabulous rates prevailed, 12 per cent being no uncommon rate on municipal bonds of undoubted validity. At present the United States government and the State and municipal governments have no difficulty in borrowing at 3 per cent.

It may be worth while to introduce here certain simple formulas in regard to computation of interest.

Let P represent a principal or sum placed at interest or payable at a future time, r the rate of interest, n the term of years considered. Then the

Amount of P at the end of 1 year is $P(1 + r)$;

$$2 \text{ yrs. is } P(1+r)(1+r) = P(1+r)^2 ;$$
$$3 \text{ `` `` } P(1 + r)^3 ; \text{ etc.}$$

Amount at the end of n years is $P(1 + r)^n$.

Call P_1 the present value of P dollars due in n years, r being the annual rate of interest. Then

$$P_1(1 + r)^n = P, \quad \text{whence} \quad P_1 = \frac{P}{(1 + r)^n}. \quad \cdot \quad (14)$$

Let P_2 represent the sum of money which, at the rate r, will produce P dollars at the end of every period of n years, in perpetuity.

We must have $P_2(1 + r)^n = P + P_2$,   or

$$P_2\{(1 + r)^n - 1)\} = P,$$

whence

$$P_2 = \frac{P}{(1 + r)^n - 1}. \quad . \quad . \quad . \quad . \quad (15)$$

An annual payment of P dollars running for n years at the rate of interest r:

The first payment amounts at the end of n years to $P(1 + r)^n$;

second " " " " " " $P(1 + r)^{n-1}$;

third " " " " " " $P(1+r)^{n-2}$, etc.

The last payment, " " " " " $P(1 + r)$.

The total $S = P\{(1 + r)^n + (1 + r)^{n-1} + \text{etc.} + (1 + r)\}$.

Multiply this series by $\dfrac{(1 + r) - 1}{(1 + r) - 1}$. This does not alter its value, but reduces its form to

$$S = P\frac{(1 + r)^{n+1} - (1 + r)}{r}. \quad . \quad . \quad (16)$$

This is for the case in which the first payment is made at the beginning of the first year, and the nth or last payment at the beginning of the last year. If the first payment were made at the end of the first year and the last at the end of the last year, the result would be

$$S = P\frac{(1 + r)^n - 1}{r}. \quad . \quad . \quad . \quad . \quad (17)$$

These results may be made equally applicable when the interest period is other than a year, as a quarter, a month, etc., by adopting the proper values for r and n. Thus in the case of a monthly payment P, continuing for 30 years at 6 per cent per annum, or $\frac{1}{2}$ of 1 per cent per month, compounded monthly, the formula would be

$$S = P\frac{(1.005)^{360} - 1}{0.005}. \quad . \quad . \quad . \quad (18)$$

To those who prefer graphic methods the diagam Fig. 55 will be interesting. Let CA represent a sum of money. Draw $A1$ perpendicular to CA and make its length such that

C_1 = the amount at the end of 1 year at the assumed rate of interest, i.e., $C_1 = CA(1 + r)$. Then draw 1 2, 2 3, 3 4, etc., respectively perpendicular to C_1, CA. C_2 will be the

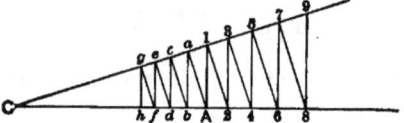

FIG. 55.

amount at the end of two years, C_3 at the end of three, etc.; C_n at the end of n years. In like manner draw Aa, ab, bc, cd, etc. Ca will be the present value of the sum due one year hence, Cb the present value due in two years, Cc due in three, Cd due in four, etc., at same rate of interest. Annuities can be summed up with a pair of dividers, as the reader will readily perceive.

Let us now see how these considerations would affect the choice between a masonry and a timber dam, fifty years ago as compared with the present time. We will take as an example the dam built at Holyoke in 1849, and we will proceed upon the following assumptions:

1. That a timber dam with a suitable apron, to avoid the great expense of 1868, might have been built in 1849 for $300 000, and we will assign $2000 per annum as the cost of maintaining the dam.

2. That the duration of such a dam might have been assumed at fifty years, and that the average rate of interest during that period might have been taken at 6 per cent.

3. That a masonry dam to last indefinitely, without repairs, would have cost $800 000.

On this basis how would the question have stood in 1849 ?
$800 000 at 6% would amount in 50 yrs. to $14 736 000
$300 000   "   "   "   "   " $5 526 100
$2000 per an.   "   "   "   " $615 520
Rebuilding dam at end of 50 years $350 000—$6 491 620

Balance in favor of timber dam. $8 244 380

So that if the company in 1849 had been possessed of $800 000 applicable to the building of a dam, and could have been assured of 6 per cent per annum during the ensuing period of fifty years, by adopting a wooden dam they might, in addition to having a new dam at the end of that period, have been better off by 8¼ millions.

How would the question stand at present? We may assume that a substantial dam of masonry could be built now for $550 000, and that not more than 3 per cent can be obtained, as an average, on large sums of money during the next fifty years. We will take the present cost of a timber dam at $350 000, and the annual repairs at $2000, as before.

$550 000 at 3% amounts in 50 years to... $2 411 140
$350 000   "   "   "   "   " $1 534 350
$2000 per annum "   "   "   " $232 360
Rebuilding........................—$350 000 $2 116 710

   Balance in favor of timber dam............ $294 430

The computation still shows in favor of a timber dam, but the advantage is slight compared with that of 1849.

In the ordinary case, and especially on a soft bottom, a timber dam cannot be expected to last fifty years. Were this the fact, there would be hardly any conditions that could justify, from a financial point of view, the construction of a masonry dam. The present value of $1000 due fifty years hence is

    at 6%.................... $54.29
    at 4%.................... $140.71
    at 3%.................. . $228.13

Leaving annual repairs and changes of value out of account, these figures represent the excess of value, per $1000, of a thing that will last forever over a thing that will last fifty years. That is to say, with the above reservation, counting interest at 4 per cent, it would be as judicious to spend $85 929 for a timber dam to last fifty years as $100 000 for a stone dam to last forever.

Let us assume a masonry dam costing $100 000 to last

forever, without any repairs. How would it compare in point of economy with a timber dam to last twenty-five years at a first cost of $60 000, and $500 per annum for repairs?

1. In the newly settled parts of the country, where interest can be taken at 7 per cent: To put the second dam on the same footing as the first, we must add to the first cost a fund to furnish the annual maintenance, and another fund to renew the dam every twenty-five years.

Masonry dam.......................... $100 000
Timber dam, first cost.................. $60 000
Fund to yield $500 per annum........... $7 143
  "   " renew the dam every 25 years..... $13 552—$80 695

    Balance in favor of timber dam..............$19 305

2. In the older parts of the country where interest can be taken at 4 per cent:

Masonry dam.......................... $100 000
Timber dam, first cost.................. $60 000
Fund to produce $500 per annum........ $12 500
  "   " renew the dam every 25 years.... $36 014—$108 514

    Balance in favor of masonry dam..............$8 514

Compare a masonry dam to cost $100 000 and last forever with a brush dam to last ten years, at a first cost of $30 000 and an annual outlay of $1000 for maintenance.

1. Interest at 7 per cent:

Masonry dam.......................... $100 000
Brush dam, first cost.................... $30 000
Fund to produce $1000 per annum........ $14 286
  "   "   " $30 000 every 10 years... $31 024—$75 310

    Balance in favor of brush dam...............$24 690

2. Interest at 4 per cent:

Masonry dam.......................... $100 000
Brush dam, first cost.................... $30 000
Fund to produce $1000 per annum........ $25 000
  "   "   " $30 000 every 10 years ... $60 250—115 250

    Balance in favor of masonry dam..............15 250

These computations are misleading only in so far as they fail to take account of progressive changes in the rate of interest and cost of work during the periods considered. They serve to show the principles on which such calculations should proceed. There are other considerations, however, in favor of permanent dams. The failure of a dam usually carries with it the stoppage, for a longer or shorter period, of industries dependent on it, and entails losses far in excess of the cost of renewal. Moreover, ethical and æsthetic considerations often enter into the question. The builders of an important dam may desire to transmit it as a memorial of themselves to the remote future, and are sometimes willing to spend more upon its construction than a close consideration of pecuniary interests would require.

Description of Some Stone Dams.—Fig. 56 is a section of the dam across the Merrimac River at Lowell, Mass. The river here has a drainage-area of 4085 square miles. The dam crosses the stream in an oblique and irregular line, giving the overflow a total development of 1093 feet. In 1852 the water reached the depth of 13 feet 7 inches on this dam. The site of the dam is a very firm refractory rock, overlaid by glacial gravel of a hardness equal to some formations that are classed as rock. The general surface of the rock is not more than 16 feet below the level of the crest of the dam; at the shores it rises considerably above that level. At one point a fissure or cavity occurred filled with gravel, where the excavation for the dam reached to the depth of 38 feet without finding rock. A dam of some sort has existed at this point for a century or more. It was rebuilt in its present form, a part in 1846–7, and the remainder in 1875–6. The part built in 1846–7 was not laid in mortar except the rear slope and the cap-stones, so far as necessary to exclude the water. Serious objections to this mode of construction were developed before the remainder of the work was commenced. The freezing and consequent expansion of the water in the cavities of the masonry gradually worked out some of the stones of the down–

stream face. Stones were noticed projecting 6 or 8 inches
beyond the face of the dam.   Warned by this experience, the
portion subsequently built was laid in full mortar except the
ashlar, which was laid without mortar and grouted with pure
Portland cement.   This was done, it is understood, that the
stones might rest in contact with each other, and not be liable
to derangement in case of the subsequent disintegration of the

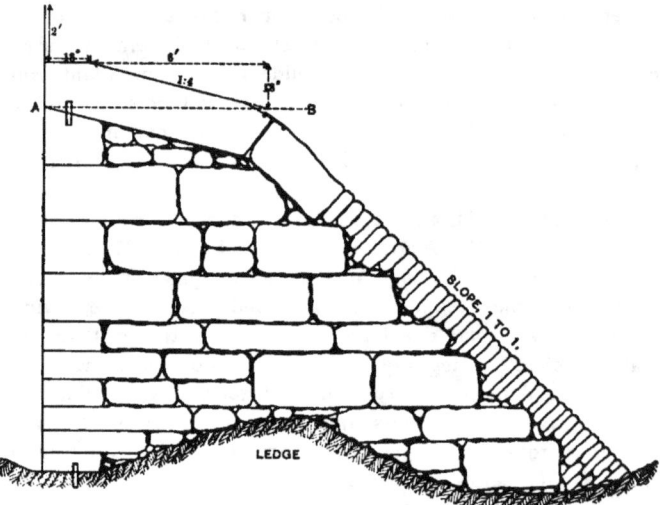

FIG. 56.

mortar. The rubble-work was laid in mortar of Newark
cement mixed 1 to 1.   The ashlar work, including the cap-
stones, was laid to quarter-inch joints.   The joints of the cap-
stones, above the line *AB*, were filled with melted sulphur.
The dowels were also imbedded in sulphur.   The cap-stones
were 2 to 3 feet wide and were laid on an incline, with the
view of avoiding any shock from floating bodies.   A face 18
inches wide was cut and hammered on the cap-stones at the
down-stream end for the application of the flashboards.

The writer examined this dam in 1895 after the new part

had stood about twenty years. It was perfectly intact and had sustained no injury of any kind. The sulphur in the cap-stone joints had withstood the action of frost. It was noticed that both bed and build joints of the face were largely devoid of mortar. A thin stick could be thrust into these joints to a depth of 18 inches, and on withdrawing it, water followed, bringing disintegrated mortar. Mortar used in rubble-work was perfectly sound and of flinty hardness. The only comment suggested by this examination was this: It appeared to me that the method of laying the ashlar dry and grouting the joints was not one to be copied. The work, to my mind, would have been better if the joints had been wider and filled with stiff mortar.

The total cost of this dam is given as $114 000. This dam checks the current of the river for a distance of about 18 miles, forming a pond of about 1120 acres.

Fig. 57 is a section of the Essex Company's dam across the Merrimac River at Lawrence, where the stream has a drainage-area of 4600 square miles. In describing this work we cannot do better than adopt the exact language of the Tenth Census Report:*

"The development of the water-power at this place dates from the year 1845, when the owners of the land and water-power were incorporated as the Essex Company with a capital of $1 000 000. Steps were at once taken toward the utilization of the power, and on July 5, 1845, the contract was concluded for the building of the substantial dam which holds back the waters of the river and diverts them into the channels of industry. The structure is built in a curve, the chord being 900 feet, and the centre ordinate 14.97 feet. Its maximum height is 40½ feet, and its average height about 32 feet, the breadth at the base being about 35 feet. The front face has a batter of 1 in 12; the top or capping is level for 3 feet from the face, then slopes back or up-stream 1 foot in 3 feet for 12 feet,

---

* Tenth U. S. Census Report, vol. XVI. Water-power of Eastern New England, p. 25.

beyond which the back is stepped off at a slope of 45°. Its
section is shown in Fig. 57. It is composed entirely of solid
stone masonry resting on a rock foundation, which is stepped
off to receive it, the front of the dam being secured by blasting
out a trench in the rock along the entire length, in which the
first course of granite masonry is laid, all the stones being

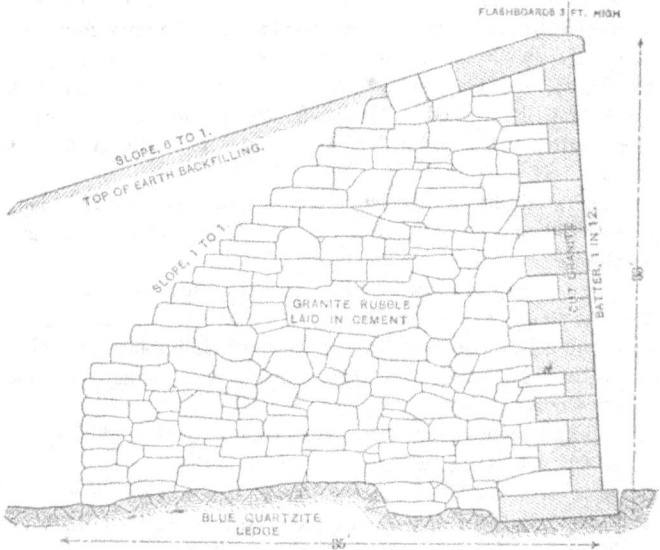

FIG. 57.

headers and the next course above being all stretchers dowelled
to the foundation-course. The face of the dam is all of dressed
stone, the headers and stretchers of each course being dovetailed together, and the capping-stones being dowelled to each
other and to the next facing-course below. The remainder of
the dam is of rough stone laid in cement, with a back-filling
of earth sloping 6 to 1. The level of the water may be raised
3 feet above the crest of the dam by means of flashboards, in
lengths of 16 feet, resting against 1¼-inch iron bolts inserted

at intervals of 20 inches in holes drilled in the capping-stones. The top of the dam is 34.12 feet on the company's scale. At the south end of the dam is a substantial wooden fishway. The excavation for the dam was begun August 1, 1845, the first stone laid September 19 of the same year, and the structure completed in 1848, its construction having occupied almost exactly three years. The rock excavation in preparing the foundation amounted to 1700 cubic yards; the quantity of masonry laid in cement was 29 000 cubic yards. The surface of hammered granite was 148 000 square feet. The cost of the dam, including coffer-dam and all incidentals, was $250 000. The agent and engineer of the Essex Company was Charles S. Storrow, who designed the dam and canals and the general development of the water-power, and he was efficiently aided in carrying out the work after April, 1846, till its completion, by Captain Charles H. Bigelow, as engineer in charge of construction.

"The dam crosses the stream in an oblique direction, and on each side there are extensive wing walls, that on the south side being 324 feet long, and that on the north side 405 feet, making with the overfall a total length of 1629 feet. There is a canal on each side of the river. That on the north side, which is the principal one, is 5330 feet long, 100 feet wide at the upper end, and 60 at the lower end," etc., etc. We shall have occasion to refer to these canals later.

This dam has now stood full fifty years, meeting in all respects the expectations of its builders. It is not understood to have suffered any material damage from any cause during that time. The profile of the dam makes it abundantly massive. Suitable calculations would show that it derives no additional stability from its curved outline.

It is not easy to see what rational purpose is served by the bank of earth on the up-stream side. The whole work being laid in cement mortar, it cannot be presumed that any necessity existed for this deposit for the prevention of leakage, while there can be no doubt that it increases the static pressure tending to overturn the dam. In any stage of the water such that injury from floating bodies is to be apprehended, this arrangement tends to increase the risk of such injury by increasing the velocity with which such bodies approach the dam, and the consequent shock with which they strike it.

As may be gathered from the foregoing, the spillway of this dam is 900 feet long, while that of the Lowell dam is longer by nearly 200 feet. In the great flood of 1852 the water is recorded as having stood 10 feet deep on the former and 13 feet 7 inches on the latter. This seeming anomaly is probably explainable in this manner: The Lowell dam is nowhere more than 16 feet above the bed of the stream, and at that time was comparatively lower than at present, the rock being covered with its deposit of gravel, protected by a pitching of heavy stone. The dam during the flood was "drowned," the water being but a few feet higher above than below. Moreover, a considerable stretch of the Lowell dam runs parallel with the current and, in such a stage of the river, counts for little as regards the discharge. The Lawrence dam, on the contrary, is substantially straight, nearly square with the current, and too high to be drowned.

Fig. 58 is a section of the dam constructed in 1890–93 across the Colorado River at Austin, Texas, designed and partly executed under direction of the author. The river at this point has a drainage-area of 32 000 square miles according to the most reliable maps. It is a typical Southern stream, the ordinary low-water volume running from 600 to 1000 cubic feet per second, and this appears to be sustained wholly by springs issuing from the rock. A simultaneous rise of all the tributaries is very rare, but when this occurs it results in an enormous flood. A flood is said to have occurred in 1869 which raised the water 45 feet at Austin, and brought down such a volume of drift as to form a permanent raft farther down the stream, which remained in place many years. The flow at that time must have been 200 000, possibly 250 000, cubic feet per second. Floods approach with such rapidity as to jeopardize the lives of people who are fording the stream. After the crisis of the flood is past it rapidly subsides, and in a few days the river reverts to its normal condition, except that the water remains turbid for a long. time. The water falls a little lower in summer than in winter, but there is no distinctly high-water season. The aggregate flow of the stream prob-

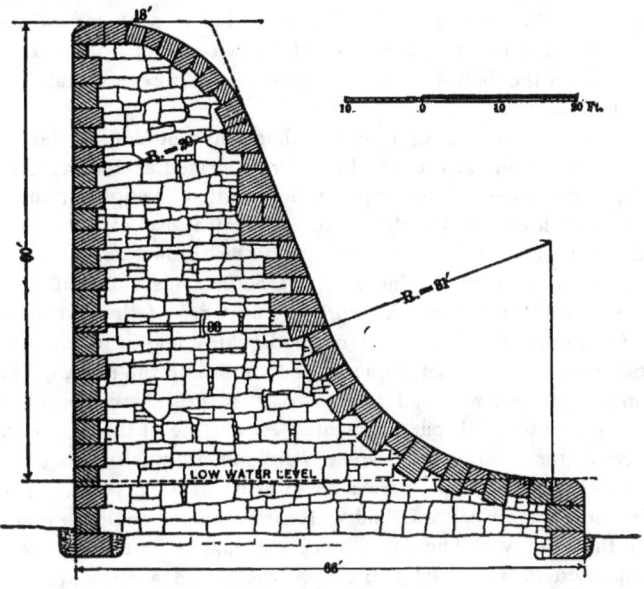

FIG. 58.

ably does not amount to a tenth part of what is to be expected from an equal drainage-area in Northern latitudes.

The river here flows in a formation of shelly lime rock lying in nearly horizontal strata. It has worn a canyon 45 or 50 feet deep and some 1200 feet wide, about half this width being filled with argillaceous deposit brought down by the stream. The dam raises the water 60 feet above the low-water stage and has an overflow about 1140 feet long. It forms a lake extending some 20 miles. Although the question of the abrasion of the bottom was here of subordinate importance by reason of the small flow, the base of the dam was spread out into a broad apron or toe to shield the rock from the direct impact of the stream. It was contemplated by the engineer to extend this apron by a bed of concrete if found necessary. A flood such as occurred in 1869 would probably raise the water 15 feet on the crest of the dam, but in such a flood the water would be 45 feet deep on the down-stream side, and the action on the bottom would be probably less severe than in a lower stage.

The original design of the dam did not contemplate a rounded outline at the top, but a crest formed as indicated by the dotted lines. The expectation was that it might in future become desirable, for the more economical use of the water, to use flashboards 6 or 8 feet high to be thrown down on the approach of floods. The rounded outline was substituted by the board controlling the work. The faces of the dam were laid in granite which occurs of good quality and in inexhaustible quantity at Burnet, some 75 or 80 miles up the river. The up-stream face was laid in courses 36 inches thick, the joints being pointed with pure cement. At the toe of the dam very heavy stones were used, some weighing as much as 6 tons. Dowells and cramps were also inserted at this point. The interior of the work was laid with such stone as could be found in the vicinity. The entire work was laid in Portland cement supposed to be mixed 3 to 1. A cable with a clear span of 1350 feet was used in the construction of this work. The entire dam contains some 90 000 cubic yards of masonry and cost a little over $600 000. This work is described in the technical journals, especially the *Engineering News.**

Figs. 59, 60, 60*a*, and 60*b* represent the dam now in course of construction and nearly complete (January, 1899), at Holyoke, Mass., to replace the wooden dam already described.

* *Engineering News*, N. Y., vol. xxix, p. 87; also numerous articles in succeeding volumes up to the period of its destruction in April, 1900.

Fig. 60*a* shows the position of the new dam with reference to the existing wooden dam. It will be noticed that the toe is so formed as to give an upward direction to the escaping water. The facing of the dam, both up- and down-stream, is to be of granite. The specifications require the level crest of the dam,

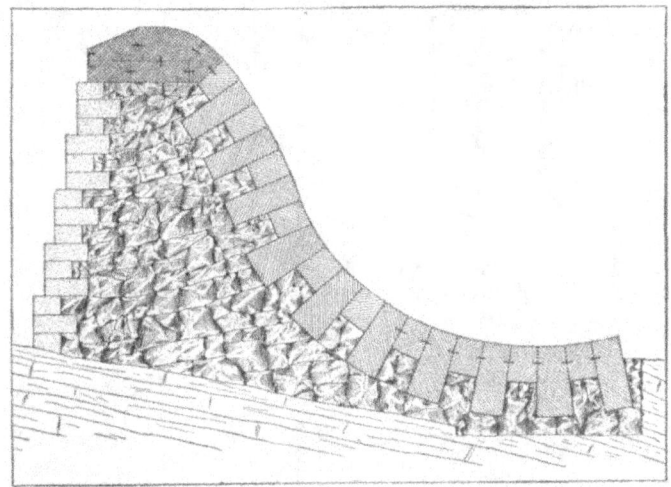

FIG. 59.

and the face for 3 feet below the same, to be fine-hammered. In the down-stream face masonry all contact surfaces are to be cut to lay a quarter-inch joint. The up-stream face is to be of split granite cut to half-inch joints. The interior is to be of rubble composed of stone found in the vicinity of the dam.

The writer sees no reason to doubt that a safe and permanent dam can be built in accordance with this design. He will, however, offer a few comments upon some of its features.

1. The masonry of the toe appears to be sunk in the natural rock, and the stones which stand on edge or on end appear to derive their stability from their backing of natural rock. This arrangement appears to assume what cannot be realized. The toe-stones necessarily lie all at the same level.

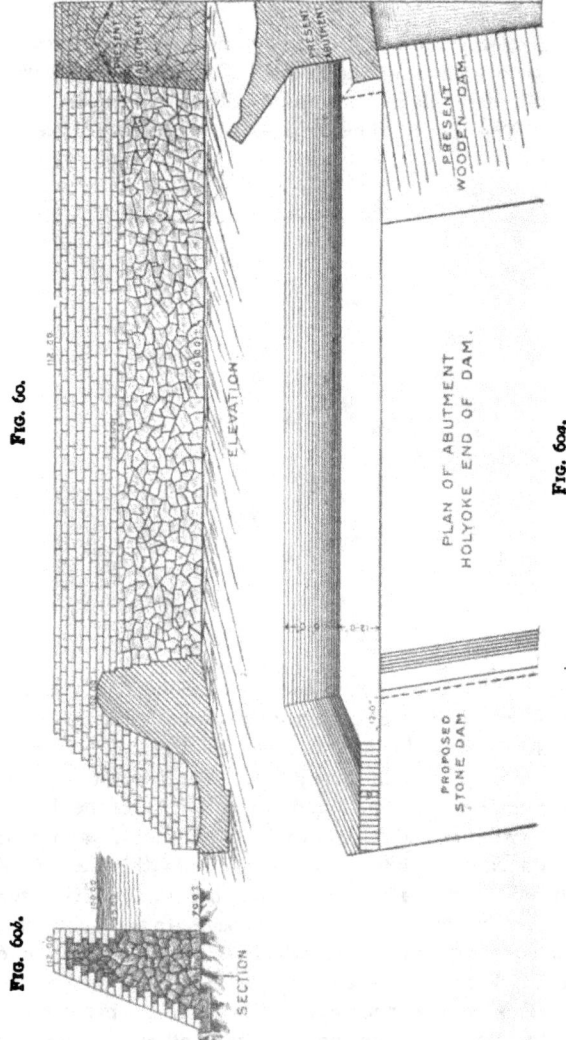

FIG. 60.

PRESENT ABUTMENT
ELEVATION
PRESENT ABUTMENT
PRESENT WOODEN DAM.
PLAN OF ABUTMENT
HOLYOKE END OF DAM.
PROPOSED STONE DAM
SECTION

FIG. 60a.

FIG. 60b.

The bed-rock cannot be assumed to lie at a uniform level. Its contour cannot be known till coffer-dams are built and the work is commenced. It will at some points lie at or below the bottom of the outer toe-stone. In such a situation it would apparently be necessary to introduce a supplementary mass of masonry. The disposition adopted at the toe of the Austin dam appears preferable as being more conformable to the actual shape of the bottom.

2. I am not able to perceive any advantage in the upward turn at the extremity of the apron. It does not diminish the velocity with which the water leaves the dam, the commotion which it undergoes in coming to rest, or its consequent wear and abrasion of the bottom. The water after leaving the dam comes to rest by a series of whirling movements. The violence and amplitude of these movements depend upon the initial velocity of the water, not at all upon its initial direction. On the other hand, the disadvantages of this feature are obvious. Heavy bodies coming over the dam are acted on by centrifugal force, which tends to separate them from the convex face, and to hold them more firmly against the concave face. On reaching the lowest point of their journey, if the remainder of the dam runs level, they simply press upon it with their weight, if not immersed, and if immersed, exert no pressure on it. The upward turn of the masonry brings them into strong contact with it, and gives them a frictional hold, tending to separate the last stone from the rest. Should the backing fail or wear out, the outer stone would be very likely to separate, and a slight separation would greatly increase the violence of the shocks. Against this action the dowels represented in the drawing have little effect.

3. Nothing can be more fanciful than any attempt to conform the face of the dam to any law of movement of the water, as, for instance, to give it the form of the trajectory of a projectile. This necessarily involves an assumption as to the velocity of the water. If the curve is correct for that assumed velocity, it is incorrect for all others. The only object of the

curve is to shield the bottom from the direct impact of the
water.　It is just as well to secure this result by arcs of circles as
to waste time in conforming the work to a more complex curve.

4. It is a mistake to suppose that a curved face will secure
this work against the shock of floating bodies.　In high stages
of water, where there is a depth of 8 feet or more on the toe
of the apron, a reverse current will be set up which will cause
bodies to strike the curved face with considerable force.

5. No part of the dam is so free from liability to injury as
the crown, and no part stands less in need of extraordinary
reinforcement.　Water and floating bodies approach this part
with very low velocity.　The velocity becomes rapid only in
high water when the crown stands in no danger of being
touched by floating bodies.　At this point water and floating
bodies have no power for mischief.　This power is acquired in
going over the dam.　It is in the lower part of the work that
strength and solidity are required.　In the light of these principles it is not easy to see the necessity for the great precautions
against injury to the crown of the dam.　The steps or offsets
on the up-stream face of the dam were designed for convenience in controlling the water during construction.　At least
they are said to have served that purpose.　The existing dam
serves as a coffer-dam to exclude the water from the work in
progress.　In low stages the flow of the stream can be turned
through the water-power canals, and either drawn by the mills
or allowed to go to waste.　The old dam, however, admits a
large amount of leakage which has to pass the dam under
construction.　This leakage goes through a gap in the stone
dam, till it becomes necessary to raise the masonry.　Then
the gap is closed by a coffer-dam, and the water is raised and
turned through a second gap.　The steps afford a convenient
footing for these coffer-dams.

Fig. 61* is a section of the dam recently constructed by the
Hudson River Power-transmission Company across the Hudson
at Mechanicsville, N. Y.　At this point the river has a drain-

* *Engineering News*, N. Y., vol. XL. p. 130.

age-area of about 4500 square miles. An island separates the river into two channels, the easterly of which is occupied by the spillway with a length of 800 feet. The westerly channel

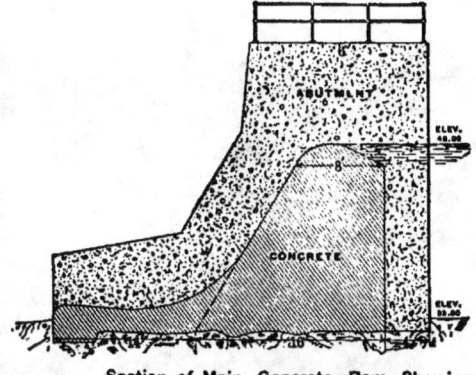

Section of Main Concrete Dam Showing Overfall and Section of Abutment.

FIG. 61.

is closed to the free passage of water, by the power-house in connection with a dam rising above the level of high water, and provided with waste-gates for emergencies. Waste-gates are also inserted in one of the abutments of the main dam, i.e., the spillway. There are sixteen of these waste-gates in all, each 4 feet wide and 6 or 7 feet high. The figure shows the spillway in section, and the abutments in elevation. The rock at this point is of a somewhat friable character, and, as will be noticed, the spillway has a curved outline, the toe or apron being so formed as to give a slight upward direction to the escaping water. It extends to some distance beyond the tangent point of the curve in a straight line. This is preferable to the form of the Holyoke dam, as the centrifugal force ceases to act at the tangent point of the curve. This work is specially interesting for the reason that the dams, abutments, and power-house foundations are composed wholly of concrete. It will be interesting to observe how this material stands the wear of

water and ice on the curved part.   Fig. 62 is a photographic
view of the spillway in the easterly channel with water coming

Fig. 62.

Fig. 63.

over, and Fig. 63 is a view of the power-house occupying the
westerly channel, from the up-stream side.

A Dam and Bridge Combined.—There is nothing incongruous in the idea of combining a bridge with a dam, the
roadway resting on the abutments and, if necessary, on piers
springing from the dam and apron. Such structures are not
so common as might be expected considering the obvious
advantages of the arrangement, especially where flashboards
are used, the bridge offering a safe and convenient mode of

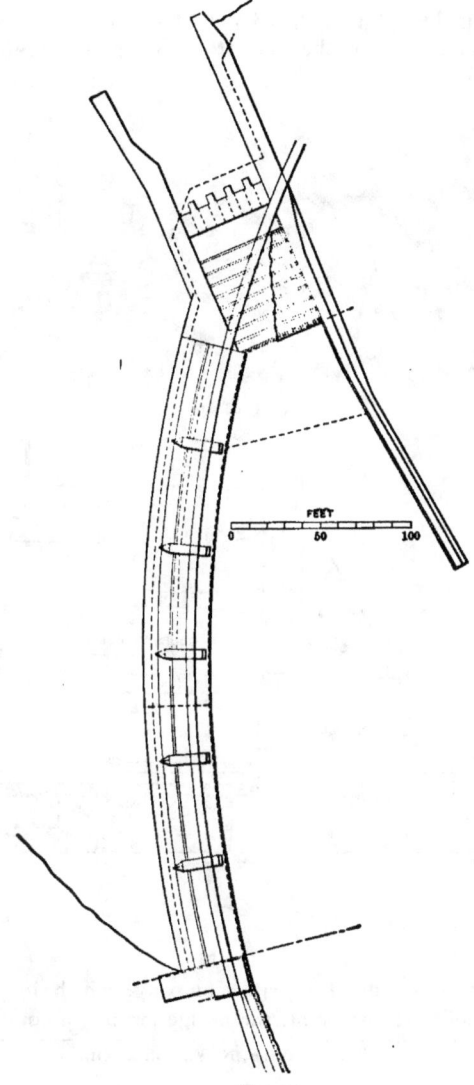

FIG. 64.

handling the latter.   Figs. 64 and 65 * show a construction of
this kind erected on the Blackstone River at Lonsdale, R. I.,

FIG. 64a.

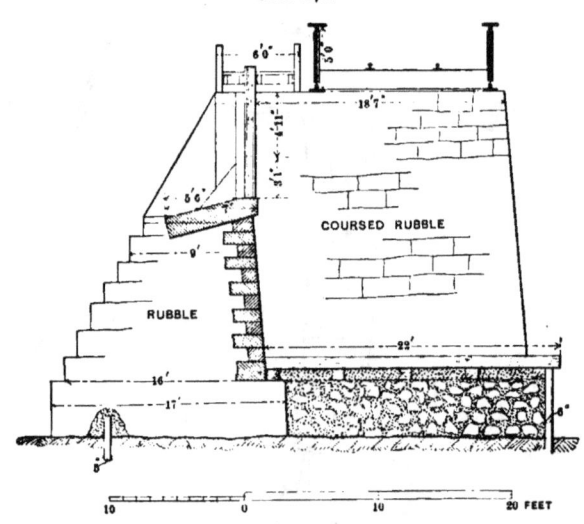

FIG. 65.

in 1893–4, by the Lonsdale Company, which has extensive
cotton-mills at this point, the bridge forming a communication

---

* *Engineering News*, N. Y., vol. xxxiii. p. 166.

between mills on opposite sides of the river. Fig. 64 shows
the work in plan, and Fig. 65 in section. Fig. 64*a* is a
down-stream view of east half of dam with gate-chamber.
The dam raises a head of 12 or 14 feet. It consists of a spillway of rubble masonry, faced with granite ashlar, furnished
with a massive apron. Five piers each 5 feet in thickness rest
partly on the apron and partly on the dam, and sustain a
single-track railroad suited to the heaviest locomotives,
together with a footway. The bridge floor is about 8 feet
above the crest of the dam. Uprights reach from the crest of
the dam to the floor of the footway to sustain the flashboards,
and the footway carries hand-gearing whereby the latter are
removed and put in place. Topographical considerations
require a curve in this part of the track, which is the reason for
the curved outline of the bridge. The Blackstone, at this
point, has a drainage-area of some 435 square miles.

CHAPTER V.

APPENDAGES OF DAMS.

Flashboards.—The height of a water-power dam is usually limited by agreement with millowners above, by decree of court, by legislative grant, or by some other condition beyond the control of the owners. It is usually subject to the condition of doing no injury to the next dam above, even in a stage of high water; a condition which necessitates the wasting of some fall between the consecutive dams in seasons of low water. During high stages the water will rise to a greater or less height on the spillway, depending on the length of the same, and the extent of drainage-ground above. Without considering such floods as occur at intervals of a generation or more, there is a stage which may be called "ordinary flood height," i.e., such floods as may be expected once or twice every year. Below this level the banks and shores of the stream are of no value. No crops can be produced, no fences or buildings can be placed. No material injury can accrue to any interest by the permanent maintenance of the water at this level, provided the barriers be removed and free passage given to the water in time of flood. These conditions give rise to the use of flashboards.

The right to maintain flashboards can usually be obtained readily from riparian proprietors, and is of great value to the owner of the dam, not only increasing his available head in low stages of the stream, but enabling him to hold to a larger extent the flow of nights and Sundays for use in working hours. This privilege, however, is always granted under the expressed or implied agreement that the passage of water over the weir shall be left free during high stages of the stream, otherwise the privilege of applying flashboards would simply amount to the privilege of raising the dam so many feet higher. The fulfilment of this obligation requires a form of flashboard barriers susceptible of ready removal without reference to the stage of the stream, a condition but imperfectly fulfilled by the method in common use, which consists generally of boards resting against iron pins let into the dam as shown in Fig. 66. This figure represents a system of flashboards 4 feet high, which is the greatest height ever attempted by this method so far as the author is aware, heights of 2 feet being most common.

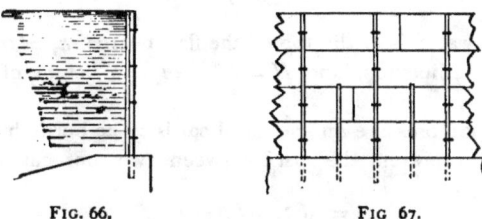

FIG. 66.   FIG 67.

This arrangement holds the water well enough; but if caught in a flood, there is no relief till the pins are broken down or bent over by the pressure of the water or by ice or drift, in which event the boards go down-stream. When the boards are up, with no water behind them, as often happens, they are liable to fall off or be blown off by the wind if not fastened to the pins. Fig. 67 shows one mode of fastening them, viz., by small staples driven into the boards and embracing the pins. Rough unplaned boards are used either butted together or overlapping at the ends. The joints are tightened by throwing in horse-dung, sawdust mixed with sand, or similar material.

At most water-powers, notice of the approach of a flood may be given in time to allow of the removal of the flashboards and pins, especially if a trained crew of men are kept for that purpose. Otherwise the only guaranty of a clear waterway for the flood is the bending down of the pins. For this reason, the diameter and number of the pins should be such as to hold the water with safety up to the top of the flashboards, and to yield when the water rises much above that height.* To determine the proper size and spacing of the pins, we must consider the mechanical principles of the case.

Let h_2 represent the height of water above the crest of the dam; h_1 represent the height of water above the top of flashboards. The height of the centre of pressure of the flashboards above the crest of the dam is

$$H = h_2 - \frac{2}{3}\frac{h_2{}^3 - h_1{}^3}{h_2{}^2 - h_1{}^2}.$$

When the water is at the top of the flashboards, $h_1 = 0$, $h_2 =$ height of flashboards, and $H = h_2 - \tfrac{2}{3}h_2 = \tfrac{1}{3}$ height of flashboards.

The total pressure on the flashboards to be borne by a pin will be, if l represent the space between two consecutive pins,

$$P = \tfrac{1}{2}62.5l(h_2{}^2 - h_1{}^2).$$

The bending moment on a pin therefore will be $HP = \tfrac{1}{2} \times 62.5l(h_2{}^2 - h_1{}^2)H$. This is the external bending moment. It is met by the internal bending moment of resistance, $m = \dfrac{\pi}{32}fd^3$, in which $f =$ the maximum strain on the fibres in pounds per square inch, and $d =$ the diameter of the pin in inches.

Suppose flashboards 2 feet high held by 1-inch pins 2 feet apart. The pressure against a pin with water at top of flash-

---

* The total pressure on the flashboards appertinent to one pin is $62.5l(h_2 - h_1)\cdot\dfrac{h_2 + h_1}{2} = \tfrac{1}{2} \times 62.5l(h_2{}^2 - h_1{}^2) = P$. The pressure on an element dh of the flashboards is $62.5lhdh$. Its moment with reference to the surface is $62.5lh^2dh$. Total moment $= 62.5l\displaystyle\int_{h_1}^{2h_2} h^2dh = 62.5l(h_2{}^3 - h_1{}^3)\tfrac{1}{3}$. Let X represent the depth of the centre of pressure, then $PX =$ total moment, or

$$X = \frac{62.5l(h_2{}^3 - h_1{}^3)}{62.5l(h_2{}^2 - h_1{}^2)}\frac{2}{3} = \frac{2}{3}\frac{h_2{}^3 - h_1{}^3}{h_2{}^2 - h_1{}^2}, \quad \text{whence} \quad H = h_2 - \frac{2}{3}\frac{h_2{}^3 - h_1{}^3}{h_2{}^2 - h_1{}^2}.$$

boards is $\frac{1}{2} \times 62.5 \times 2 \times 4 = 250$ pounds. The external moment is $12 \times \frac{1}{2}h_2 \times 250 = 2000$ inch-pounds. The internal moment is $\frac{\pi}{32}f$; say $\frac{1}{10}f$. $\therefore f = 10 \times 2000 = 20\,000$ pounds —a strain which would be admissible under the circumstances.

Suppose the water to rise 2 feet above top of flashboards. $h_2 = 4$, $h_1 = 2$. Pressure on a pin $= \frac{1}{2} \times 62.5 \times 2 \times 12 = 750$ pounds. $H = 4 - \frac{2}{3}\frac{64-8}{16-4} = 0.8889$ feet $= 10\frac{2}{3}$ inches.

External moment $= 750 \times 10\frac{2}{3} = 8000$. $\therefore \frac{\pi}{32}f = 8000$, and $f = 80\,000$ nearly. As ordinary iron does not possess this strength, the pins would break down before the water had risen 2 feet above the top of the flashboards.

Again, consider 4-foot flashboards held by $1\frac{1}{2}$-inch pins 1 foot apart. $h_2 = 4$, $h_1 = 0$. Pressure $= \frac{1}{2} \times 62.5 \times 4 \times 4 = 500$. $H = 16$ inches; moment $= 8000$. $\therefore \frac{\pi}{32} \times f \times 3.375 = 8000$, and $f = $ say $10 \times 8000 \div 3.375 = 23\,700$. For the purpose in view there would be little risk in subjecting iron to this strain. When the water rises 2 feet above the flashboards $h_2 = 6$, $h_1 = 2$, $P = \frac{1}{2} \times 62.5 \times 32 = 1000$. $H = 6 - \frac{2}{3}\frac{216-8}{36-4} = 1.6667$ feet $= 20$ inches. External moment $= 20\,000$ inch-pounds. Internal moment $= \frac{\pi}{32}f \times 3\frac{3}{8}$.

$\therefore f = \frac{32}{\pi} \times \frac{20\,000}{3.375} = $ say $10 \times \frac{20\,000}{3.375} = $ nearly $60\,000$ lbs.

This result shows that while pins of ordinary iron would stand safely with water at top of flashboards, they would fail before the water had risen much over 2 feet above the top. Where a height of 4 feet is attempted the pins are often of different lengths, alternately 2 and 4 feet as at Fig. 67.

For regulating the height of water at wasteways in canals which are not subject to river fluctuations the form of flashboards indicated at Figs. 68 and 69 is often used. These consist of stout boards, 4 to 6 feet in length, placed horizontally and provided with handles. Piers are erected on the weir, built of plank 6 or 8 inches thick, set edgewise one above the other,

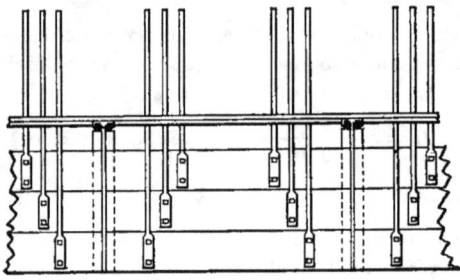

FIG. 68.

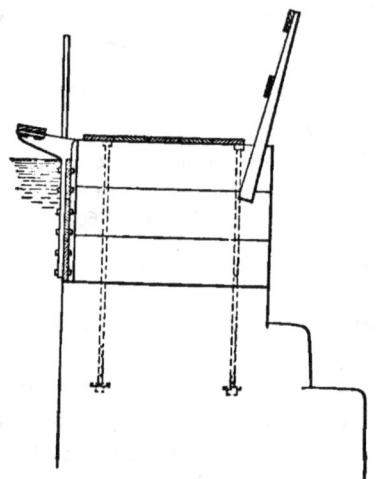

FIG. 69.

and confined by long bolts anchored in the dam. The piers sustain planks, forming a foot-bridge for the use of attendants. The interval between the handles of the boards is narrowest for the upper ones and wider for the lower. In front of the boards is a rest for the workman's foot, otherwise he could not exert his full strength in lifting the boards against the friction incident to the pressure of the water. When he lifts a board he passes it under his feet, stepping between the handles, and leans it against the railing in rear, so that the boards are always in the order required to go upon the dam.

This plan of wasteway would not ordinarily serve for a natural stream, as the flood conditions would require the foot-bridge to be placed too high for convenience of handling the boards. A dam provided with flashboards of substantially this form has, however, existed at Hyde Park on the Neponset River for many years and has worked satisfactorily. The Neponset flows from extensive meadows and is not liable to great freshets. The owners of the dam are obliged to limit the height of water to avoid injury to meadow-owners above. The combination of a highway or railroad bridge with a dam, as adopted at Lonsdale, R. I., Figs. 64 and 65, admits the use of mechanical tackle for handling flashboards and is susceptible of more extended application than it has yet received.

The removal of a system of flashboards sustained by pins, on the approach of a flood, is not always possible. The ordinary operation of such a dam, in low water, is this: The water is usually found at the top of the flashboards in the morning, giving a reserve of water in the pond. During the day the water is drawn from the pond faster than it enters, and gradually falls. Late in the day it becomes safe for men to go upon the dam to remove the boards. If at that time it is known that the water will be high on the following day, the boards may be removed. If heavy rains occur during the night while the flashboards are up, the water may not fall during the ensuing day sufficiently to permit their removal, and, the river continuing to rise, they must go down-stream.

Many water-power streams are subject to ordinary floods of 6 feet and more, and no material injury to any riparian interest would result from the maintenance of flashboards of that height provided free way could be given in time of flood.

The maintenance of this height by the method of pins would probably be impossible, and, if possible, the loss of the flash-boards at every freshet would be a serious matter.   Fig. 70 is a design for a system of flashboards to maintain this height, and, at the same time, to admit of being lowered in any stage of the river without danger to the workmen.   This is given as a suggestion worthy of study, not as a device sanctioned by experience and usage.   Fig. 70 shows the method as applied

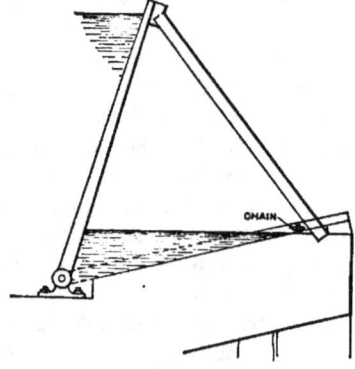

FIG. 70.

to an existing dam.   Its application to a dam specially designed with that view would be simpler.   The barrier consists of a line of shutters hinged to the up-stream crest of the dam and supported when up by props resting near the down-stream crest.   When down, the shutters lie flat on the dam and are very little exposed to injury.   When the shutters are up, a chain is stretched along the top of the dam between the props and shutters, and runs from the abutment to a capstan located down-stream on the bank, Fig. 70*a*.   By winding in on the capstan, any desired strain can be brought upon the outside prop.   When this yields the strain comes on the next prop and the shutters can then be thrown down in succession.   Fig. 70*a* shows the situation in plan.   It shows the arrangement that would be proper in case of an angle in the dam, as sometimes occurs. A trapezoidal pier would be built up, forming an abutment for the shutters to close against. The prop at the angle would be under a considerable strain from the tension on the chain, but not so great as the exterior prop, and the latter would yield first. After the shutters have been thrown down by reason of high water, they are not required to be

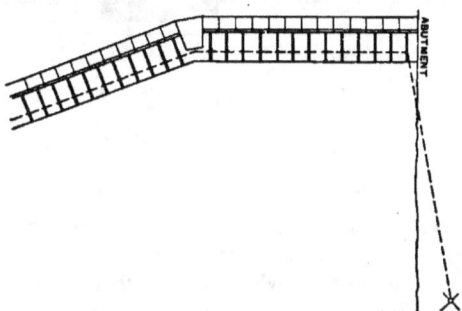

FIG. 70a.

raised again till the stream has fallen to a stage allowing the entire flow, at certain hours, to be drawn by the mills. At such a time the water is below the cap of the dam, and the raising of the shutters and adjustment of the chain is perfectly simple. When the water-pressure is off the shutters they might be liable to the same difficulty as is met with in the case of flashboards supported against pins, viz., being lifted off the props by the pressure of the wind and allowing the latter to fall. This would be obviated by employing the form of hinge shown at Figs. 70b and 70c, which limits the movement of the shutter, and allows it to be strained up with considerable force before inserting the prop. The fall of the shutter is cushioned by the body of water beneath it, which can never fail to be present when the necessity arises for throwing down the barrier. When lying flat on the masonry the shutter can receive no injury from floating bodies. The only injury to be apprehended from the fall of the shutter is from extraneous bodies lying on the dam. The props need not be sacrificed as are the boards in the ordinary arrangement, as they can be attached to the main chain and hauled in with it.

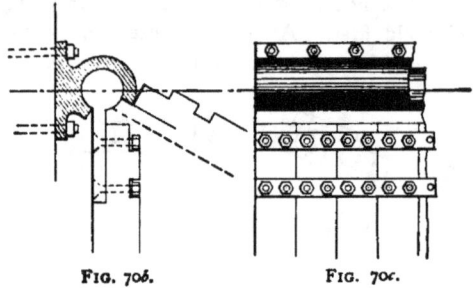

FIG. 70*b*.   FIG. 70*c*.

Fishways.—Certain species of fish, dwelling habitually in salt water, are capable of existence in fresh water, and are endowed with an instinct which impels them to ascend rivers at certain seasons of the year, there to deposit their spawn, after which they return to salt water. The young also return to salt water, and after reaching the age of reproduction, two, three, or four years, return to the grounds where their own existence commenced, and revisit the same grounds annually thereafter. This fact has been unmistakably verified by the United States Fish Commission.

The chief varieties of fish possessed of this characteristic are salmon, shad, and alewives. So imperious is the migratory instinct that no common obstacle can arrest the upward progress of these fish. They make their way through foaming rapids, and easily surmount a perpendicular fall of considerable height. At Currytunk,* on the Kennebec River, there is a fall of 16½ feet, as near perpendicular as the movement of the water will permit, the water falling into a pool of great depth. Salmon have been repeatedly seen to jump this fall. They apparently start from a point deep in the pool, and, on reaching the surface, have a velocity which carries them 10 or 12 feet

* Report of U. S. Commission of Fish and Fisheries, 1872-3.

out of water. They thus strike the stream where it has not acquired a great velocity, and are able to swim to the upper pool in the descending stream, the movement of the fish being more rapid than that of the water. Ordinarily 5 or 6 feet is as high a fall as a salmon can be expected to jump, and probably somewhat less for the other varieties.

Migratory fish ascend the streams in the spring and early summer. The salmon comes earliest and generally remains the longest, not being ready to spawn till late in the season. Salmon and alewives go farther up the stream than the shad. Alewives push up the smallest streams, evidently in quest of quiet water. Fishermen take advantage of these annual migrations to capture large numbers of the fish. The erection of a high dam across a stream operates as an effectual barrier to the ascent of fish, and is looked upon as a great grievance by those who have depended upon the fisheries.

The exclusion of migratory fish from the upper reaches of streams has long attracted public attention, and this was one of the facts which, thirty or forty years ago, led to the appointment of State fish commissions charged, among other things, with the duty of enforcing the construction and maintenance of fishways at dams. Thirty-two of the States now have executive bodies of this kind, in addition to the United States Commission of Fish and Fisheries acting on behalf of the general government. These agencies have succeeded in enforcing the erection of fishways at most of the important dams, though they have not always been successful in securing their maintenance.

The erection and maintenance of fishways is less imperative now than formerly, as, by the methods adopted by the United States Fish Commission, the eggs of fish in immense numbers are deposited and hatched in the upper reaches of streams, compensating in some measure, though not entirely, for the loss incident to the stoppage of the annual migrations.

A fishway is a series of passages leading from the upper to the lower pool, large enough to be easily traversed by the fish,

and so arranged as not to give the water a velocity sufficient
to make the passage difficult. From a 20-foot dam, for
instance, the water falling freely or running down a straight
incline reaches the lower pool with a velocity of some 36 feet
per second, a velocity which no fish can move against. Leading the water down, however, by a series of falls of 1 foot each,

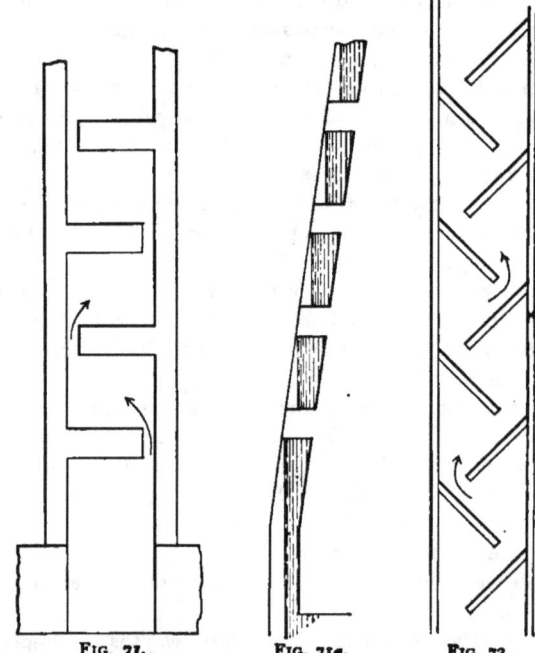

FIG. 71.   FIG. 71a.   FIG. 72.

it nowhere attains a velocity greater than 8 feet per second, a
current which any fish can ascend with ease. This principle
gives the key to the construction of the fishway.

The earliest form of fishway is indicated at Fig. 71, and in
this form was applied to certain English rivers some sixty years
ago. It consisted of a channel leading down from the crest of

the dam at an inclination of about 1 in 7. Cross-walls or partitions divided the channel into a large number of pools, communicating with one another by passages through the walls. In this form it was called a salmon-ladder, being designed with special reference to that fish, and worked very successfully.

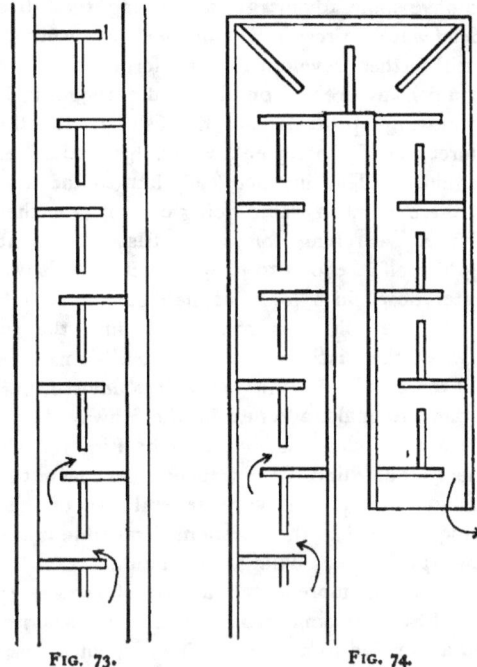

FIG. 73.　　　　　　　　FIG. 74.

The figure shows the fishway as forming a part of the dam, and of massive construction as was frequently the case. The application of the fishway to English rivers in this form is attributed to James Smith of Deanston, England, a civil engineer of considerable note in the early part of the 19th century. On American streams the fishway is more commonly built of timber, either as a cribwork loaded with stone or bolted to the masonry or rock.

The chief modification which this model has undergone in its application to American streams is in the direction and form of the partitions.   In the form shown at Fig. 72 it is called the Swazy fishway.   The inclined direction of the partition is thought to have some advantage, as offering the fish a larger space of dead water to rest in when he feels so disposed.   This feature is still further developed in the form of Fig. 73, which represents a fishway erected on the Androscoggin in Maine in 1870.   In moving up the stream the fish has no other guide than the direction of the current, which his instinct leads him to swim against.   This instinct leads him to the foot of the dam when there is much water going over it, and there he is apt to remain, searching for some passage around it or exhausting himself in efforts to jump over it.   A fishway leading the water down in a long incline discharges so far from . the dam that the fish does not readily find the entrance, especially when the discharge of the latter is small compared with that of the dam.   For this reason it is advised, in the case of a high dam, to make a return in the fishway, bringing the discharge nearer the dam, as indicated by Fig. 74, where the fish will more readily find the entrance.   Of course the reader will understand that in this very general view of the subject we are giving no heed to the question of possible injury to the structure by water or ice coming over the dam.

Fig. 75 shows in more detail a form of fishway approved by the United States Commission of Fish and Fisheries.   It is represented as applied to a dam of cribwork, but is susceptible, with suitable modifications, of application to a dam of any construction, and requires no further explanation than is contained in the following directions of the Commission:

Slope of fishway should not be steeper than on a ratio of 1 vertical to 4 horizontal.   Intake, or up-stream end of fishway, should be amply large and placed not less than 1 foot lower than the crest of the dam.

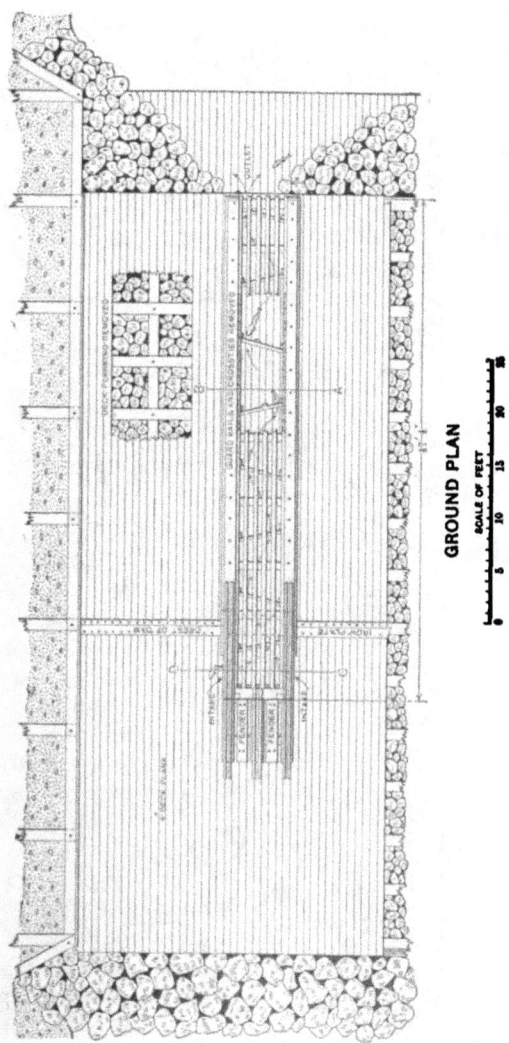

GROUND PLAN

SCALE OF FEET

FIG. 75.

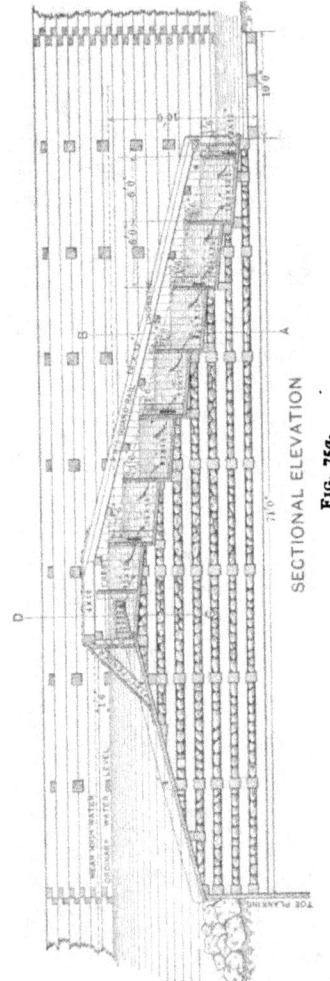

SECTIONAL ELEVATION

FIG. 75a.

Outlet should be below low-water level, and so located or constructed that fish are naturally led to it when ascending the stream.

There should be relatively deep water, with an unobstructed flow below the outlet of the fishway. An ample discharge of water should attract the fish to the outlet.

There should be plenty of light admitted in the fishway, and its construction should be such as to be readily inspected and cleaned of any débris lodging therein.

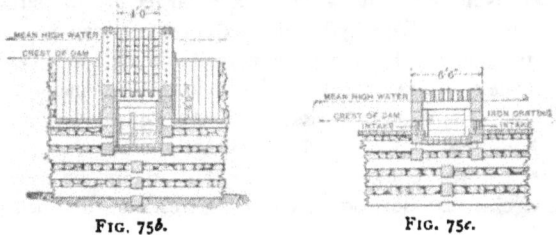

FIG. 75*b*.   FIG. 75*c*.

The floor of the compartments should be laid slightly inclined, and the bulkheads somewhat obliquely across the fishway, so that the current of water passing through the compartments can more readily clear the same of sand, mud, and rubbish.

There should be no regulating-gates or other devices at the intake which necessitate the services of an attendant.

The apertures in the bulkheads should increase progressively from the lower to the upper ones, to insure overflow from compartment to compartment.

The flow of water should be abundant, forming small waterfalls over the bulkheads, so that the fish may either jump from one compartment to the next above or may dart through the apertures in the bulkheads.

While the flow of water through the apertures may reach a velocity of 10 feet per second, there will be relatively quiet water in the compartments, thus furnishing a resting-place for the ascending fish.

To maintain the operation of the fishway at an average high water the same as at the ordinary stage of the stream or river, the uppermost compartment is made somewhat longer, and a central bulkhead is inserted having its crest at the high-water level.

The fishway may be constructed of wood or masonry and iron; it may follow a straight line or be built in angles and curves, as the local conditions may require.

The size of the fishway depends principally on the volume of water available, and can be made larger or smaller than shown on plan. The hydraulic head between two successive compartments must be so chosen as to obtain a current velocity through the apertures of not to exceed 10 feet per second. At low stage of the stream or river, with the fishway flowing full, there should be a liberal discharge over the crest of the dam.

The fishway should be built very strong and be well protected against the destructive effects of freshets, drift, ice, etc.

CHAPTER VI.

MOVABLE DAMS.

DAMS which can be set up and thrown down at will have been developed and brought to a high state of perfection, during the last fifty years, in connection with river navigation. Their application in projects of water-power is limited to those cases where dams are required to be lowered, in whole or in part, in time of flood. As already observed under the head of Flashboards, the devices in use for this purpose are crude and imperfect. There are cases in existence, and doubtless many more to arise hereafter, in which some applications of the principles of movable dams might judiciously be made to this purpose. For this reason it is believed that a glance at the leading types of movable dams will not be a waste of time.

Fig. 76 shows the principle of the bear-trap dam, which

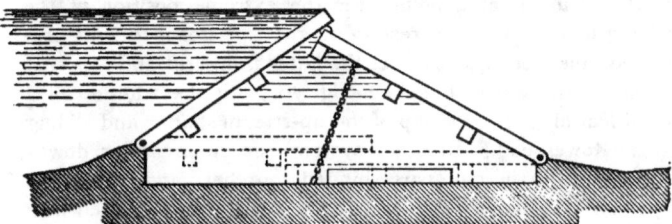

FIG. 76.

found many applications for purposes of navigation on the rivers of Pennsylvania in the early part of this century. Two continuous shutters or leaves extend entirely across the channel to be closed, and abut against smooth perpendicular walls at the ends, the walls forming the sides of the channel. When the dam is not raised these leaves lie in a recess in a mass of masonry on the bed of the stream, the up-stream leaf folded over the down-stream. The two shutters, in connection with the bed and end walls, form, at all times, a closed chamber from which the water of the stream is excluded. To erect the dam, water from a higher level is admitted to the chamber through an opening in the abutment shown in the drawing. Of course there is a good deal of leakage, but with a sufficient volume of water the dam will rise till the down-stream leaf reaches the limit of its chain. To lower the dam, the supply of water is cut off and the chamber is put in communication with the down-stream channel. In the simple form shown in the figure this dam has a marked advantage over every other form of movable dam, in its freedom from liability to derangement from the intrusion of drift and sediment—a consideration of primary importance in structures of this kind. This device has been extensively applied, and almost every application takes a new form. Often the two shutters are hinged together at the summit and the foot of the up-stream shutter slides on a smooth platform, the down-stream shutter turning on a fixed hinge. Another modification has the same elements, with the addition of a third leaf or shutter. This is hinged to the bed of the channel, at a point near the extreme position of the sliding foot, its free end resting on the up-stream leaf, its presumed object being to protect the latter from drift and deposits. Another arrangement is like Fig. 76, with the addition of a third leaf hinged to the top of the up-stream shutter and sliding at the lower end. In other systems the up-stream or down-stream leaf is in two parts hinged together, and these fold up when the dam is down. These again are combined with sliding leaves, etc.

The Thénard Shutters.—This combination consists of two sets of shutters extending entirely across the stream and hinged to the bottom. The up-stream set fall up-stream, and their movement is limited by chains. The down-stream set fall down-stream, and when up are sustained by props. The distinctive feature in the working of this system is that the erection of the up-stream set arrests the entire flow of the stream, causing the water to drop on the down-stream side, so that the set of shutters designed to hold the water can be raised and secured before the water rises so as to overflow the up-stream set and interfere with that operation.

Fig. 77 shows the principle of this system as applied to a permanent dam of masonry, that being the first application

FIG. 77*b*.   FIG. 77*a*.

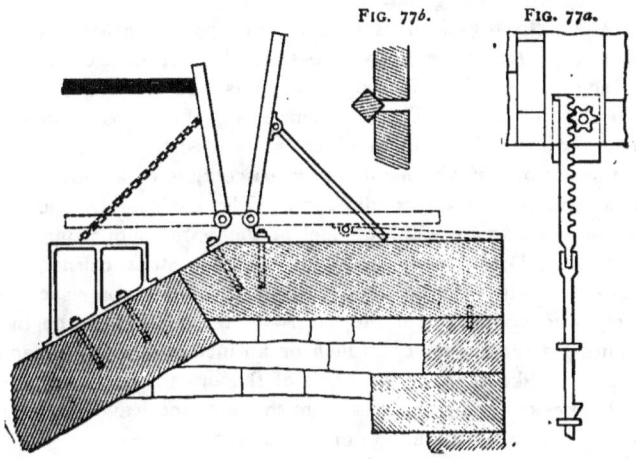

FIG. 77.

which it received and, so far as the writer is aware, the only one. When the water is not raised, both sets of shutters lie flat on the dam. The up-stream set is latched down securely and the water flows freely over both sets. To each set of shutters pertains a tripping-bar, not shown in the cut, reaching the whole length of the dam and terminating at the abutment, where it is coupled with devices enabling it to be pulled endwise. A pull on the up-stream tripping-bar releases all the up-stream shutters, which on being released are instantly seized by the current and set erect so far as the chains will permit.

The entire flow of the stream is now stopped and the dam becomes accessible till the water has risen so as to overflow the shutters. In this interval attendants erect the down-stream shutters, placing the feet of the props in recesses prepared to receive them. When the water overtops the up-stream shutters and fills the space between the two sets, the up-stream set fall and are latched automatically. The entire flow of the stream goes over the down-stream shutters, till the latter are thrown down by the tripping-bar.

The Tripping-bar is shown at Fig. *77a*, consisting of a stout rod extending entirely across the channel to be closed by the shutters, confined in proper bearings and carrying a series of projecting studs. The abutment end of the rod is a toothed rack gearing with a pinion which is operated by a hand-wheel on the top of the abutment and is susceptible of a movement of several feet in either direction. The studs of the down-stream bar engage with the feet of the props supporting the shutters, and draw them successively off their steps, causing the shutters to fall. These studs are spaced so as to come successively into contact with the props—not all at once, the bar requiring a movement of an inch or an inch and a half to each prop. In like manner the studs of the up-stream bar release the latches confining the up-stream shutters, and leave the latter free to rise under the action of the current.

The Thénard shutters have worked very well in the situations to which they have been applied, viz., as a supplement to the height of a permanent dam. Their application to the general purpose of a movable dam, where it would rest upon a mass of masonry level with the natural bed of the stream, is more doubtful. In the former situation, as soon as the up-stream shutters are erected the down-stream set is accessible for raising. The permanent dam creates something of a pond and the water rises slowly, giving ample time for the erection of the down-stream shutters. In the latter case the conditions would be different. The shutters when down would lie in a depth of 3, 4, or 5 feet of water, and to raise a head of 12 feet would have to be from 15 to 18 feet in length. The erection of the up-stream shutters would be attended with no difficulty. That being done, the water would commence to rise on the up-stream side and fall below. Some time must elapse before the down-stream shutters become accessible to commence the work of raising. Should any accident occur or any unforeseen difficulty arise to delay the raising of the down-stream set till the water commences to run over the up-stream, then the entire movement would be blocked. The down-stream set could not be raised nor the up-stream set lowered. To obviate this difficulty it is necessary to provide a weir to discharge the entire flow of the stream in moderate stages without allowing the water to overtop the shutters. These difficulties have prevented the application of this system to the general purposes of the movable dam.

The reader will of course understand that dams of the construction described cannot be operated without a considerable interval or clearance between the consecutive shutters, and that a great volume of leakage must result. These gaps are closed by square pieces of scantling, as indicated at Fig. 77*b*, the pressure of the water holding the scantling firmly in its place.

We come now to the two forms of movable dam which have received the most extended application; the needle dam and the wicket dam. These have one feature in common, viz., each requires a temporary foot-bridge for its erection and removal.

The Poirée Needle Dam, so called from its inventor, M. Poirée, an officer of the French engineer department, was first applied in France about the year 1834. A general idea of the device will appear from Fig. 78, which is an elevation of the dam in process of erection, Fig. 78*a* a plan, and Fig. 78*b* a cross-section. It is supposed to extend across a channel of any required width with vertical side walls or abutments. One of the abutments has a recess into which the adjoining trestles fall. The construction and manipulation of

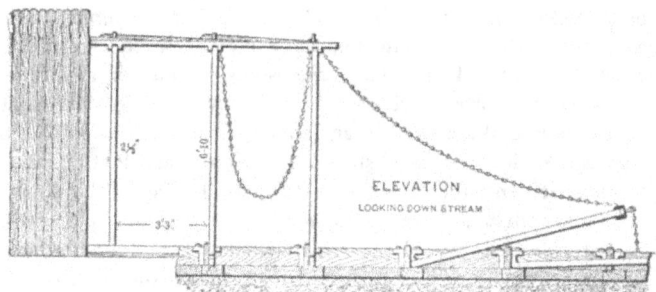

FIG. 78.

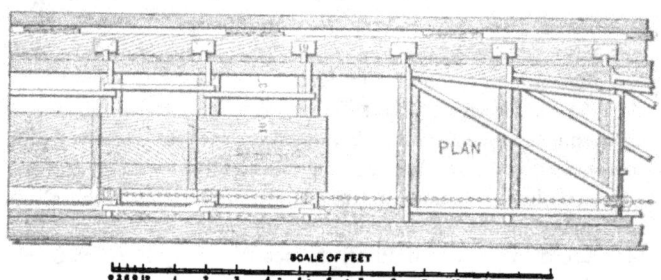

SCALE OF FEET

FIG. 78*a*.

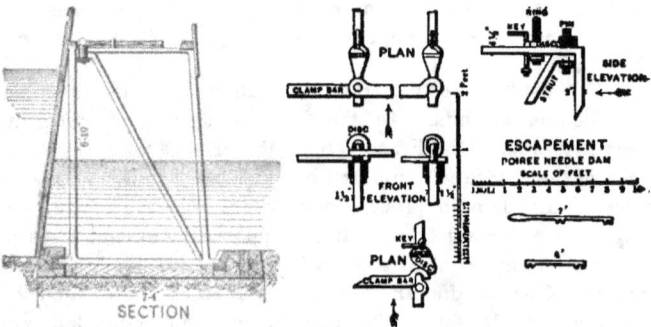

FIG. 78*b*.

the dam cannot be better described than in the language of the inventor:[*]

"It consists of a row of trestles placed parallel with the current, turning around their bases fixed to the floor, and connected with one another in the upper part, when they are upright, by clamps or bars having claws at the end. Wooden needles, resting against the up-stream side on a sill at the bottom and on the bars at the top, form the wall which arrests and sustains the water. When all the trestles are bedded they present no obstacle to navigation above the sill of the floor. Each trestle is shaped like a trapezium. The two bases are horizontal; the lower base ends in journals which fit into two boxes of cast iron; the upper base carries the planks of a service bridge. The up-stream side is vertical, the down-stream one sloping. The inside is furnished with a brace or with other bars according to the strain to be supported. At the head of the trestle is a bolt which carries on its upper side a washer against which rests the up-stream bar, and having on the lower end a cap against which is fitted, on one side the curved claw of the hook which unites each trestle to the preceding one, and on the other side the end of the hook which joins it to the one following. Each hook is provided at its extremity with a chain which serves for working it, and the end of the chain is fastened to the cap of the preceding trestle. In order to allow the trestles to be easily worked by two men, they are placed 3.28 feet apart and are only 6.23 feet high, 2.56 feet wide at the top, and 4.92 feet wide at the base. The thickness of the iron is 0.12 foot, and the weight of each trestle is 242 pounds, without the bars and hooks. (This refers to the Decize dam, built in 1836.) When it is necessary to raise the dam, two men take the chain which hangs along the abutment, raise the first trestle, place its hook in the ring fixed in the masonry, lay the two planks on the foot-bridge, and fasten the trestle to the

---

[*] See Thomas on Movable Dams, Trans. Am. Soc. C.E., vol. XXXIX. p. 439. The description is quoted from De La Grené, Cours de Navigation Intérieure, p. 175.

coping by the front and back bars. They work in the same
way with the rest of the trestles. The skeleton of the dam
being up thus, the two men proceed to fill it in by placing the
needles, one by one, first one space apart to break the current,
then close together to make the wall as tight as possible. If
it is desirable to lower the water, the two men take the needles
away one at a time and lay them on the back part of the foot-
bridge. If it is desirable to remove the trestles, the needles
are taken to the storehouse, the bars and planks of the last bay
are removed, then the hook is raised which joins the last
trestle to the last but one, and it is allowed to fall, the shock
being lessened by means of the chain fastened to the hook.
The same method is pursued with each bay. When each
trestle is laid down and the chain stretched, a ring of particular
construction placed at a conveniently determined distance
should be on the right of the screw-ring on the trestle still
standing; if such is not the case, it shows that the trestle is not
on the bottom.''

The above description conveys a fair comprehension of the
subject, though it is not quite clear, by reason of certain details
referred to not being fully expressed in the drawing. The
drawing represents the earliest application of the system, which
did not contemplate a head of more than 3 feet. It has been
applied on a larger and larger scale till in its later applications
it raises a head of 12 or 13 feet. Many modifications have
been introduced. In the higher dams an additional bar has
been introduced above the sill to give the needles an intermediate bearing-point. In some cases the foot-bridge has been
made continuous, consisting of plates of sheet iron jointed to
the tops of the trestles, which in that case must be raised all
at once, requiring a powerful crab for that purpose. In some
a track is laid along the foot-bridge on which runs a truck to
fetch and carry the needles and other materials.

One of the disadvantages of this system is that it requires a
weir separate from the movable dam, to discharge the flow of
the stream and prevent the overflow of the foot-bridge.

Another is the large amount of extraneous material to be handled at every manœuvre of the dam, and kept in storehouse when the dam is down. Another is the liability to derangement from drift. Should drift run while the dam is being lowered, or be lodged against the dam on commencing to lower it, it is very liable to become entangled in the trestles and prevent them from being lowered, in which case, should the river continue to rise, and drift to increase, the dam is likely to be destroyed or greatly injured.

The Chanoine Wickets.—This form of movable dam, shown in Fig. 79, consists of a series of wickets suspended at the centre, each resting on a "horse." The horse is a quadrangular frame, consisting of two legs jointed to the bottom and rigidly united by a horizontal crosspiece. A prop is articulated to the crosspiece, and when the horse is up, the foot of the prop rests on a step. When it is thrown off the step, the horse and prop lie extended on the bottom. The horizontal crosspiece terminates in journals which unite the frame to the wicket by means of journal-boxes on the back of the latter. The wicket is thus susceptible of a considerable movement while resting on the horse. A sill, suitably formed and secured to the bed of masonry on which the dam rests, runs entirely across the channel. When the wickets rest on this sill, the channel is closed and the water rises. When the up-stream end of the wicket is raised so as to bring the latter into a horizontal position, the movement of the water is but slightly obstructed and the head or lift disappears. When the prop is tripped or thrown off its step, the horse with its wicket and attachments falls and all lie flat on the bottom, allowing boats to float over them.

There are two methods of raising and lowering the dam. In the earlier applications of the system the wickets were raised by a hoisting device carried on a boat, and were thrown down by a tripping-bar. The arrangement for unstepping the prop is indicated by Figs. 80, 80*a*, and 81. Fig. 80 is a plan and Fig. 80*a* a section of the hurter or step and slide pertaining

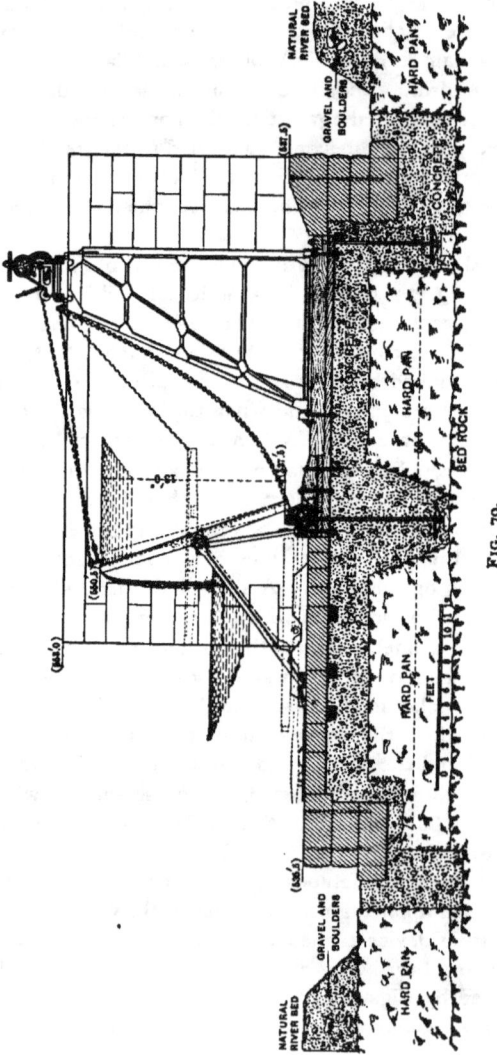

FIG. 79.

to the prop when operated by the tripping-bar. The latter is shown in section with its stud, ready to take hold of the foot of the prop and draw it off its seat, at Fig. 80a. The shape of the slide is such that, while the horse is falling, the foot of the prop is shunted into the path which it is to follow when the

FIG. 80.

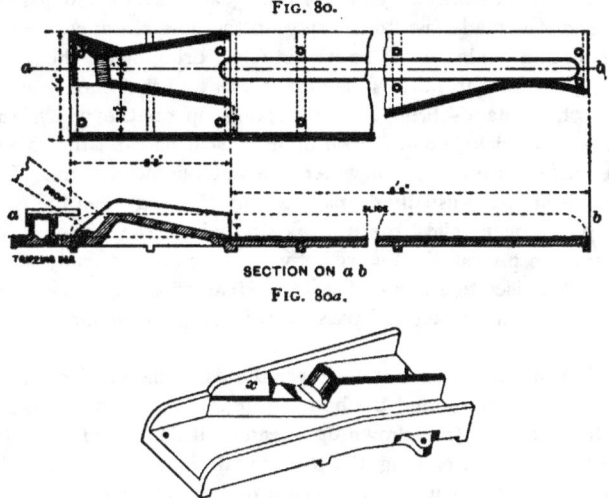

SECTION ON a b
FIG. 80a.

FIG. 81.

horse rises. When the horse is erect, the foot of the prop falls into its seat on the hurter.

As represented in Fig. 79 the system involves a service-bridge similar to that described for the Poirée needle dam, consisting of a series of trestles lying ordinarily on the bottom and susceptible of erection at will. With the service-bridge, the tripping-bar is dispensed with and a different form of hurter is used. Fig. 81 shows the up-stream end of this hurter, the remainder being identical with the form of Figs. 80 and 80a. In this form the horse is tripped by being drawn a little further up-stream. This movement causes the foot of the prop to fall into its descending path, in which position it slides freely when the horse is lowered, and, as in the former case, is shunted so that when the horse is down, the foot of the prop lies in the path which it is to follow in rising.

When it becomes necessary to raise the dam, the service-bridge is set up; on this bridge a light railroad-track is laid on which runs a truck carrying hoisting-gear. The hoisting-rope or chain is made fast to the ring in the up-stream end of the wicket and hauled in. The wicket rises, bringing up the horse and prop till the latter falls into its seat on the hurter, when the whole stands firmly. By slacking off on the winch, the end of the wicket comes down upon its seat on the sill. When the wickets are all up, the water rises and overflows them. The service-bridge is usually dismantled and lowered when the dam is up. It is manifest that the axis on which the wicket turns may be so placed that the wicket will swing automatically when the water rises to a certain height. Heavy floating bodies may also swing the wicket and pass over, leaving the wicket to right itself.

When occasion arises to lower the dam, the service-bridge is set up as before, and the hoisting-gear is attached to the top of the wicket. It is drawn up-stream a little as indicated by the dotted lines, causing the prop to fall into its return path. Then the wicket may be eased down or let go by the run, which is perfectly safe as the resistance of the water prevents any shock. After lowering the wickets the service-bridge is dismantled and lowered, leaving the channel free for the passage of boats.

In order that the wickets may work with certainty, a clearance of an inch or two is required, occasioning considerable leakage. In works of navigation this is not objectionable so long as there is plenty of water. When the leakage threatens to diminish the head it is stopped by the insertion of square scantling as at Fig. *77b*.

CHAPTER VII.

RESERVOIR-DAMS. STORAGE-RESERVOIRS.

THE requirements of modern industry are best subserved by a uniform supply of power, a necessity in no wise conformable to the natural flow of streams, as is apparent in Table 1. The flow of all streams not controlled by natural or artificial reservoirs is liable to at least twentyfold variations. In fact no such stream except the Mississippi in its lower reaches shows variations so small as twentyfold. For smaller streams the variations increase as the drainage-area diminishes till we come to streams of two or three square miles, in which the variations are absolutely infinite, such streams running, at times, entirely dry. This statement only applies to the northeastern section of the United States. In the southwesterly parts streams commanding hundreds of square miles are often dry for months.

The distribution of the flow of streams throughout the year is stated on page 4.

These figures would be modified somewhat by the latitude. In northern Maine we should put the maximum flow about a month later and in Pennsylvania about a month earlier than given on page 4. This statement shows in a strong light the chief defect of water-power, viz., its extreme variability. A stream furnishing 1000 horse-power during the five wettest months could not be counted on for more than 200 during the driest, viz., in July. In reality not so much as this, since the figure for July, 2 per cent, being the average of a series of years, there will be Julys in which the flow will fall below 2 per cent, and days in the same month on which it will fall lower still.

An industrial plant operated wholly by water-power, and subject to the necessity of running at a uniform rate, could not make rational use of more than 20 per cent of the total flow of a natural stream.

There are two modes of meeting this difficulty. The more common one, which will be considered later, is by the installation of a reserve of steam-power for reliance during the dry months. There are situations, however, where the construction of storage-reservoirs, to hold the superfluous waters of wet months for use during the dry, is more economical than steam-power, and still other situations in which such a proceeding is advisable though it may not wholly obviate the use of steam-power.

Storage-reservoirs, or reservoirs designed to hold the water of seasons of abundance for use during seasons of scarcity, are built for several different purposes, viz.: (1) for municipal water-supply; (2) for irrigation; (3) for navigation properly so called, that is, for the supply of canals or to maintain the depth of water in natural channels; (4) for log-driving; (5) for water-power. The first-named purpose is closely connected with the health, comfort, and prosperity of great communities, and warrants expenditures which would generally be unjustifiable in connection with water-power. Nevertheless these reservoirs are valuable in the present connection as exhibiting the principles of construction perhaps more clearly than any other class. Reservoirs for the four last-mentioned purposes may all be classed together, except that the fourth class is usually provided with outlet-sluices larger than are necessary for any other purpose.

A reservoir-dam has these parts:

1. The body or mass, which gives solidity to the dam and enables it to resist the pressure of the water.

2. The skin or diaphragm, which arrests the movement of the water through the dam. This element is sometimes wanting when the body of the dam is of uniform and homogeneous material.

3. The outlet-sluice, which enables the water to be discharged according to necessity.

4. The wasteway, for the discharge of superfluous water.

Location of Reservoirs.—The earth's surface presents to the general view a series of ridges, a conformation referred by geologists to shrinkage of the earth's crust in cooling from a molten state. This agency has defined the great salient features of topography, continental elevations, mountain ranges, and the general course of streams. The surface thus rough-hewn has been carved, eroded, and brought to its present form by the agencies of air, water, and frost. The air has operated chiefly by oxidizing certain elements of exposed rocks and reducing them to soils and gravels. Water has diversified the contours by carrying away soils, gravels, and rocks, and, acting through the immeasurable periods of geologic time, has trenched the earth with channels and canyons. Gaining access to highly heated rocks, it has reduced them to sands. Set in motion by the winds, it has worn down rocky shores, tossed the fragments to and fro upon beaches, and ground them to clay which it has spread over wide areas to be subsequently hardened into rock. It has nourished the myriads of insects whose microscopic shells falling to the bottom have aggregated into great geologic strata. Frost also has played its part in the disintegration of rocks by congealing the water in their crevices and causing it to expand with irresistible force. Likewise in the formation of glaciers which have crept slowly over the land armed with rocky fragments firmly grasped to aid their erosive action, and finally, under changed climatic conditions, have dissolved in enormous floods and deposited the accumulations of thousands of years in glacial moraines.

A storage-reservoir is almost invariably constructed by throwing a dam across the valley of a stream, with the addition sometimes of dikes or embankments to close lateral outlets. In forming a judgment of the availability of any reservoir-site, some knowledge of the principles of geology is of the greatest importance. The engineer should make himself acquainted with the geological history and character of the region.   These are usually to be learned from the geological surveys which have been conducted under authority of the several States, and the more comprehensive survey now going on under authority of the general government.   Such knowledge, if it serve on other purpose, will at least enable the engineer to conduct his examinations understandingly.   To the trained geologist the exterior configuration of the ground is a pretty good indication of what lies below, and a knowledge of these principles shows the engineer where to look for the faults, dislocations, and anomalies of structure which sometimes impair the value of reservoirs.

The most favorable locations for reservoir-dams are the outlets of existing lakes, or of marshes, which are ancient lakes that have been filled up by river sediment and decayed vegetation.   Reservoirs for municipal purposes are usually forced by considerations of proximity and availability into other situations, and also avoid marshes for sanitary reasons.   Irrigation-reservoirs do not occur in regions where lakes and marshes abound.   Dams for these purposes, therefore, often occupy narrow valleys with steep declivities and abrupt sides, and are necessarily very high to impound any considerable quantity of water.   To show the enormous difference in cost between different sites, it may be mentioned that the system of reservoirs constructed by the United States Government on the headwaters of the Mississippi, by damming the outlets of several lakes, cost at the rate of nine dollars per million cubic feet of storage capacity, while the reservoir now contemplated for the metropolitan water-supply of Massachusetts, to be formed by a dam across the valley of the Nashua River, will cost not less than one thousand dollars per million cubic feet.

The entire northern region of the United States abounds in lakes and marshes or wet meadows.   When these occur near the headwaters of the streams they command but little drainage-area and are of little value for the purpose in view.   Their occurrence at points where the stream has attained a considerable volume is rarer, but still not uncommon. In forest regions during the early stages of the country's settlement such sites are usually seized upon by lumbermen for sluicing-dams. Later, after the exhaustion of the timber, they are applied to purposes of water-power.

A natural lake is often formed in the course of a running stream by the slow rising of the rock through volcanic agencies, in a line crossing the stream. The wearing of the rock by the action of the stream keeps pace, in some degree, with the rising of the ground, so that we now find the outlet underlain by firm rock flanked on either side by solid walls of rock. At Marble Falls, on the Colorado River in Texas, this process is clearly traceable. The river flows for a mile or more between two walls rising almost vertically to a height of some 200 feet, the dip of the exposed strata showing the history of its formation unmistakably. Above this canyon the valley spreads out into a wide expanse, and at the foot of it are immense banks of loose rock. The surface rock on the top of the cliffs is worn by water. Localities of this character form the best sites for reservoir-dams. On the other hand lakes frequently owe their origin to the obstruction of river valleys by glacial deposits, and in this case the bed-rock is usually found covered to a greater or less depth by gravel or sand.

On a rock bottom or where the rock is accessible at no great depth it is usually advisable to build a dam of masonry, though where economical considerations control and the height is moderate a structure of timber and stone is not inadmissible. For a dam on a sand or gravel formation an earth embankment is more commonly adopted.

Embankments.—Fig. 82 represents a loose homogeneous mass of earth resting on a formation of similar material, and exposed to the pressure of water. The water is in constant motion through the embankment, but it moves with immeasurable slowness as compared with its flow in an open channel of equal cross-section. An open channel 100 feet wide, 6 feet deep, and connecting two bodies of water a mile apart and differing by 1 foot in elevation, might carry 1200 millions of gallons per day. The same channel filled with coarse sandy gravel might perhaps pass 600 gallons per day; that is to say, 2 million times as much in the first case as in the second.

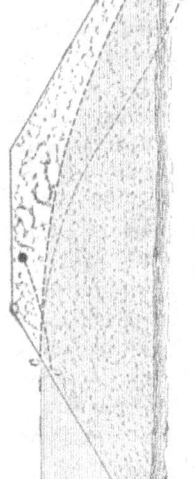

FIG. 82.

From this slow movement of water through the dam it results that the latter is partly in a state of saturation. Below the line *cd* the embankment is saturated with water in a state of slow movement. If this line intersects the down-stream slope, the dam is in a critical condition. The water will ooze forth, gather into a rill and carry away the earth. The longer this action continues the more energetic it becomes, and unless something intervenes to moderate the flow the destruction of the bank is only a question of time. The natural agency that may intervene to check the percolation is the action of sediment borne by the water, which being deposited on the up-stream face gradually diminishes the permeability of the mass, and alters the line of saturation, bringing it more into the position *cb*. It would appear from these considerations that an embankment might be safe on a permeable formation which would be entirely unsafe on a rock bottom; since, in the latter case, all the water that enters the dam must escape near the down-stream toe, and cannot fail to act injuriously. On the other hand, it would appear that, laying aside all question of the wearing away of the bank by movement of water on the up-stream side, of the overflowing of the same, of outlet passages, or of burrowing animals, the necessary and sufficient condition of stability is: The line of saturation must not meet or closely approach the down-stream slope. It also appears that the more impervious material should be applied at the up-stream face, the more loose and permeable at or near the down-stream. In preliminary examinations with reference to a proposed embankment, no more instructive experiment can be made than the determination of the line of saturation.

An attempt * was made to determine this element in reference to embankments proposed for the Wachusett Reservoir for the metropolitan water-supply of Massachusetts. A flume was made of close-fitting plank, some 80 feet long, 8 feet wide, and 8 feet high. This, being in a horizontal position, was closed by a dam of the material to be used. Water was admitted on one side, and the percolation observed, together with the position of the line of saturation. This was an interesting though, for the purpose in view, a not very instructive experiment. The important fact is the line of saturation when the dam rests upon the ground that it is destined to occupy, since the permeability of that ground is a most important factor. It is obvious that a homogeneous embankment, not absolutely impermeable, resting on an impermeable diaphragm of plank, would be in a condition to insure its destruction.

Homogeneous embankments of porous material resting on ground of like character may be made safe against rupture though not against leakage by adopting sufficiently large dimensions. But in wide and high embankments the effect of rain falling directly thereon must not be lost sight of. Ruts of wheels or foot-paths may cause the surface water to gather into streams of considerable size and produce dangerous gullies. In great storms the volume of water falling on the embankment is liable to be increased by waves splashing over the same.

Embankments have sometimes been destroyed by burrowing animals. The muskrat enters the bank at e, Fig. 82, works her gallery in an ascending direction to a point f, above the line

---

* *Engineering News*, vol. XLVII. p. 366.

of saturation, where she constructs a safe and comfortable habitation. She sometimes reaches a point too near the exterior slope, and a rise of water breaks through and starts a stream running through the bank.

Fig. 83 represents a form of low embankment very common in New England fifty to one hundred years ago. It is confined by two side walls of split stone laid without mortar, which appear to have no other object than to prevent the slopes from spreading out as wide as they otherwise would. Above the walls the earth was carried up in slopes generally of $1\frac{1}{2}$ to 1 to a width of 9 to 12 feet on top. There is no reason to suppose that the earth was consolidated in any other manner than by the passage of teams and carts. This construction was never applied to a head of more than 10 or 12 feet. A failure of this form of embankment has never come to the writer's knowledge.

Fig. 84 shows the more common form of earth embank-

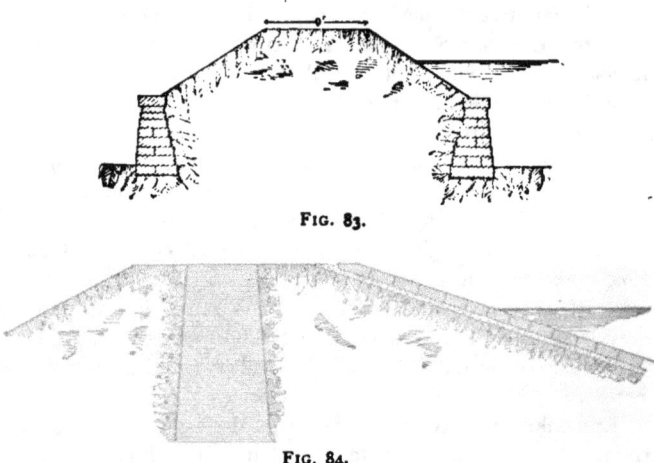

FIG. 83.

FIG. 84.

ment, especially a high embankment, by which I would imply a height of not less than 25 feet. It shows a wet slope of 3 to 1, a dry slope of 2 to 1, a width of 15 feet on top, exclusive of pavement, the summit 4 feet above high water, and is provided with a puddle-wall in the middle 3 to 6 feet thick at the top and widening about 1 foot in 6 downward. In considering the material of embankments, we divide all natural earths into two general classes, viz., binding and non-binding, i.e., those which are susceptible of consolidation by pressure and those which are not. It is well known that sand, and gravel composed of sand and pebbles, cannot be consolidated by pressure, neither does this material retain the shape impressed on it by pressure, although when removed from its natural bed and redeposited it undergoes considerable settlement in the course of time. On the other hand, material containing a sufficient percentage of clay can be moulded into any desired shape and is very susceptible of compression and consolidation, qualities which fit it for use in water-tight embankments. Pure clay, though one of the most impervious of earths, has disadvantages in application to situations liable to be alternately wet and dry. In drying it shrinks greatly and presents crevices through which the water, on the occasion of a sudden rise, is liable to find its way, and, before the material becomes saturated, to enlarge the passages beyond the possibility of closing. The same reason which makes pure clay inapplicable to bricks, tiles, and pottery, forbids its use in embankments. It is generally not difficult to find in the immediate vicinity of a proposed dam material suitable for embankments.

The traditional mode of forming embankments, omitting for the present the consideration of the puddle-wall and the sluices, is as follows: First remove all vegetable soil, tree-roots, and loose materials from the site. Spread the embankment earth in layers 6 to 9 inches deep, keep it moistened and roll it with a heavy roller. Commence with the lowest part of the work and maintain the layers approximately level, except that it is thought advisable to give them a slight inclination toward the centre of the embankment. It is not certain,

however, that any advantage results from this disposition.   As
to the amount of moisture to be supplied, the earth should by
no means be reduced to a pasty condition, but should be moist
enough to be readily squeezed into any desired form in the
hand, and should not readily lose its form.

As to the slopes, they should not in any important work
be greater than 2 to 1.   It is more common to give the water-
slope a declivity of 3 to 1 and even 4 to 1, but for really suitable material there seems no good reason for a flatter slope
than 2 to 1.   The water-slope is usually paved with heavy
blocks of stone.   The necessity for this feature depends upon
the exposure of this face to the action of waves.   Where the
water-slope faces a broad and deep expanse of water the pavement cannot safely be omitted.   Where the body of the reservoir is at some distance from the dam, approached by a narrow
channel, the pavement is less important.   In the former case
the height of the embankment should not be less than 5 feet
above the crest of the wasteway.   Pavements usually consist
of rectangular blocks of split stone 12 to 18 inches thick, resting on a layer of broken stone about 6 inches thick.   This
arrangement is usually adopted in municipal reservoirs, and is
in that case partly designed to prevent the water from becoming muddy under the action of winds, as well as to prevent the
wearing of the embankment by waves acting through the
crevices of the pavement.   This feature may be omitted, for the
purpose in view, when the material of the embankment contains
a large proportion of pebbles.

Puddle.—It is the ordinary practice to provide an embankment with a specially impervious layer, stratum, or diaphragm
composed of puddle.   This is usually prepared by adding a
proper proportion of pure clay to the material of which the dam
is composed.   In this operation a pug-mill similar to that
used in brick-making may be employed with advantage.   The
puddle is brought to the consistency of brick-clay when ready
for moulding, and generally no attempt is made to consolidate

it after depositing. A puddle-wall is the name applied to a core of puddle carried up through the embankment and keeping pace with the latter while under construction. A trench, in this case, is sunk along the centre-line of the embankment to a sufficiently firm and impervious stratum. Often great expense is incurred in this feature of the work, and the trench reaches down 50, 75, or even 100 feet to solid rock, in which case the wall or core is of uniform thickness below the surface of the ground.

The impervious diaphragm sometimes consists of a layer of puddle spread on the water-face of the embankment and joining a puddle-trench near the inner toe. This is preferable on some accounts to the puddle-wall, especially during construction, as the arrangements for depositing the puddle-wall are something of an embarrassment and hindrance to the rest of the work. The face-puddle is not applied till the embankment is finished, the work is exposed to view, and it is easier to detect defects. In contract work this is a point of importance. The objection that this mode is more expensive is not well founded. It is true that the area of the slope is two or three times that of the midsection, but in this situation the depth of puddle need not be great; 18 inches is as effective as any greater depth. Face-puddle with pavement should not be applied on a slope steeper than 2 to 1, to avoid risk of slipping. Where pavement is applied to face-puddle the layer of broken stone should not be omitted. Where the ground is favorable to the driving of sheet-piles, that is, free from large boulders and buried timber, a row of sheet-piling may advantageously take the place of the puddle-trench. Face-puddle has a great advantage over a puddle-wall as regards strength of the embankment. This is especially true when the body of the embankment is composed of porous gravel, and reliance is placed mainly on the puddle for tightness. In the case of the puddle-wall, the full pressure of the water acts against the latter, and only half the mass of the embankment is available to resist this pressure. In fact the down-stream half of the embankment has to resist the pressure of the up-stream half in addition to the full pressure of the water.   This pressure moreover acts horizontally, tending to shove the lower half of the embankment down-stream. In the case of face-puddle the pressure is resisted by the entire mass of the dam, and the downward component of this pressure tends to hold the dam in place.

The Outlet.—This is the most difficult feature of the earth embankment.   The failure of the Dale Dike Embankment near Sheffield, England, in March, 1864, by which some two hundred persons lost their lives, was traced to the breaking of the discharge-pipes through the settlement of the embankment. More failures of reservoirs have occurred from defective outlets than from any other cause. It is to be expected that some settlement will take place in the embankment, as well as in the ground on which it rests. If the outlet-pipes are laid upon unyielding supports such as piles, the latter are liable to cause ruptures or cavities in the embankment. When cast-iron pipes are used they should be put together with a view to some yielding in the joints.   They should always be laid on natural earth, not on filled ground, and especially not on ground partly

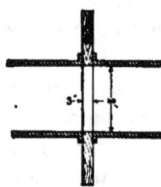

FIG. 85.

filled and partly natural. Broad flanges are applied at intervals to prevent the water from following the outside of the pipe, which it is very liable to do, especially along the under side, where it is difficult to consolidate the earth. The arrangement of Fig. 85 has been suggested as combining a broad flange to prevent the creeping of the water along the pipe with a considerable yielding of the joint to conform to the settlement of the earth.   The pipe is put together with flanges, and at each joint is inserted a disk of 3-inch pine plank, with an opening in the centre, the exact size of the pipe, and an external radius some 18 inches greater. This is fitted to the joint after the pipe is bedded, and is bored with holes for the

bolts that unite the consecutive lengths.   The wood is always
wet and so preserved from decay, and it admits of considerable
compression without losing its integrity.   In a 30-inch pipe put
together in 12-foot lengths, a compression of $\frac{1}{4}$ inch in the wood
implies a deflection of more than 2 inches in the pipe, which is
greater than ever need be expected.   The gate or valve for
controlling the discharge is placed sometimes on the land side,
sometimes on the water side.   The former has great advantage
in point of simplicity and accessibility, but in this case the pipe,
above the valve, is under full pressure, and any crack is liable
to cause great injury to the embankment.   The placing of the
valve on the water side usually involves a mass of masonry
built up to the high-water level, and accessible by a bridge
from the embankment.   In municipal reservoirs of later construction this masonry usually takes the form of a gate-chamber,
near enough to the crown of the embankment to be accessible
without a bridge.   Such chambers * are often of very complex
construction and need not be considered here.   In some of
the older reservoirs the outlets were very simply and efficiently
managed by extending the pipe to the foot of the water-slope

---

* It is a singular fact that almost all important municipal reservoirs,
storage or service, of recent construction, have been provided with complex and expensive arrangements for drawing water from different depths,
by opening orifices at surface, mid-depth, full depth, etc., according to
requirement. It may be a fact that the water of mid-depth is at times
preferable to that of the bottom or surface, and *vice versa*, but it appears
to me that the engineer who expects the discharge of an orifice at mid-
depth of a reservoir to be confined to water lying at or near that depth
is presuming somewhat upon the obliging disposition of this element.
When an orifice is opened in the vertical side of a reservoir, any particle
of water nearer the orifice than the nearest boundary of the reservoir
moves directly toward the orifice, whether its path be upward, downward,
or horizontal. There is no conceivable reason why it should move more
readily in one direction than another. So that an orifice at mid-depth,
after discharging a hemisphere of water whose radius is the semidepth, is
discharging an exact mixture of all the horizontal strata, top, bottom, and
middle. An orifice at the surface or bottom would arrive at this condition after discharging a quarter-sphere whose radius is the full depth.

and there providing it with a flap-valve operated by a chain
running up the slope to a windlass on the crown of the embankment. The placing of the gates on the water side of the
embankment is coupled with the great advantage that the
interior of the sluice can be rendered accessible for inspection
if of sufficient size to admit an observer, which it would be in
any work of importance.

In embankments under 30 feet in height siphon-pipes have
been successfully employed, and have shown in some respects
decided advantages over every other mode of discharge. The
pipe has its influx near the foot of the inner slope, protected
by a screen to exclude floating bodies, passes up the inner and
down the outer slope, discharging into a basin sufficiently
extensive to moderate the commotion of the water. It is
advisable to cover the pipe with puddled earth both on the
inner and the outer slope, not only for protection against frost,

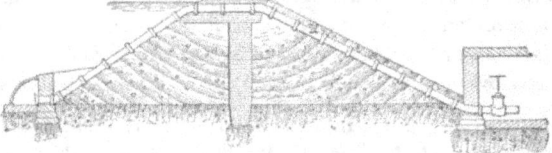

FIG. 86.

but to prevent the influx of air through the joints. When the
crown of the reservoir is above the highest level of the water
no gate is necessary, the flow being established by the exhaustion of air and arrested by its admission. A gate or valve,
however, is generally inserted for the purpose of regulating the
quantity of water discharged. It is manifest that the height of
the crown of the siphon above low water cannot exceed the
height due to atmospheric pressure, viz., about 34 feet. Practically it is considerably less, and 25 feet is about as great a
lift as can be adopted. Fig. 86 * shows an outlet of this form

---

* Description of the Bann Reservoir by Mallet in Weale's Quarterly Papers
on Engineering, vol. VI. part I. The figure is taken from Hagen's Wasserbau-
kunst, Pl. LXXVI.

as applied to one of the Bann reservoirs near Dublin, Ireland. The success of this method requires efficient means constantly in readiness for exhausting the air from the summit of the siphon. Air enters through crevices which would hardly show a trace of leakage of water. A small orifice under atmospheric pressure will pass twenty-eight times as great a volume of air as of water. In addition to this, absolute tightness of the pipe does not obviate the necessity for the withdrawal of air. Water absorbs air under pressure, and eliminates it when the pressure is diminished. Water passes the crown of the siphon under a reduced pressure and is constantly giving out air, which must be constantly withdrawn in order to maintain the efficiency of the siphon. The best proceeding is to adapt to the efflux of the siphon a small simple water-wheel, driving an air-pump, fitted by means of a small pipe, to draw the air from the summit of the siphon. Even this arrangement does not obviate the necessity of manual labor to set the water in motion after it has been stopped and the siphon partly filled with air.

The Wasteway.—This feature is the safety-valve of the embankment, its function being to discharge the surplus water and prevent the water from rising so as to overtop the embankment, which would result in the speedy destruction of the latter.

The width of the wasteway must be sufficient to accommodate the discharge in the greatest flood that is to be expected, without rising above the top of the embankment. These dimensions may be determined by the rule of Chapter II relative to wasteways except in so far as modified by the accumulation of water in the reservoir while rising on the wasteway, as is pointed out later in this chapter.

In considering the accumulation of water and its effect upon the height to which the water may be expected to rise upon the wasteway we need not confine ourselves to the reservoir immediately appertinent to the dam, but may take account of swamps and low grounds further up the stream which operate to hold back the water and diminish the depth of overflow.

The extent to which the width of the wasteway is affected by the area of the reservoir will appear from the following: Suppose two reservoirs of equal capacity; the one of shallow depth, covering 1000 acres, the other of great depth, covering 200 acres.   The areas are taken at the level of the wasteway, and extension of area consequent upon rise is disregarded. The top of the embankment is supposed to be 5 feet above the sill of the wasteway.   Suppose we find reason to expect, as a maximum, 10 000 cubic feet of water per second continuing for three hours, then subsiding to 1000 cubic feet per second at the end of the sixth hour.   Assuming the water to be at the level of the wasteway at the commencement of the flood, it is manifest that the first-named reservoir would take the entire volume of the flood, making no account of the discharge, without raising the water more than 4 feet, and that a wasteway capable of discharging 1000 cubic feet per second at the latter height would be abundant.   On the other hand, the 200-acre reservoir would not be safe without a wasteway capable of discharging at the danger-level, which we will call 4.5 feet above the wasteway, a quantity equal to the total inflow.   This may be shown by a little computation.   A wasteway discharging 10 000 cubic feet per second at a depth of 4.5 feet would discharge at any other depth, h, $10\,000\dfrac{h\sqrt{h}}{4.5\sqrt{4.5}}$.   On this principle we compute the discharge of such a wasteway for different depths as follows:

0.5 feet........ 370.4 cu. ft. per sec.	3.0 feet....... 5443.3 cu. ft. per sec.
1.0 " ....1047.6 " "	3.5 " .... 6859.4 " "
1.5 " ....1924.5 " "	4.0 " .... 8380.5 " "
2.0 " ....2963.0 " "	4.5 " ....10000.0 " "
2.5 " ....4140.9 " "	

We assume the water to stand even with the wasteway at the commencement of the flood.   The average discharge while rising the first half-foot may be taken as $\frac{1}{2}$ of 370 = 185.   The time occupied in rising the first half-foot will be

$$\frac{200 \times 43\,560 \times \frac{1}{2}}{10\,000 - 185} = 444 \text{ sec.}$$

$$\text{From 0.5 to 1} = \frac{4\,356\,000}{10\,000 - \frac{1}{2}(370.4 + 1047.6)} = 469$$

1 to 1.5	512
1.5 to 2	577
2 to 2.5	676
2.5 to 3	836
3 to 3.5	1132
3.5 to 4	1830
4 to 4.25	3585

Total time . . . . 10 061 sec.

$$= 2^h\ 47^m\ 41^s.$$

We thus have the water 4 feet 3 inches above the crest of the wasteway before the termination of the flood, and must admit that the wasteway is none too large.

It is a very favorable condition when the wasteway can be placed on natural ground at a distance from the embankment. Such situations are not uncommon. Reservoir-dams are always placed where the ridges enclosing the valley approach closely to one another. Not infrequently it occurs that a depression in one of the ridges may be made available for a wasteway. Cases even occur in which the wasteway is in rock, while the rock at the dam-site lies far below the surface. Where the conformation necessitates placing the wasteway on the embankment it is constructed by spreading a bed of masonry on the exterior slope with two side walls to limit its width. Whether it should be narrow and massive or broader and less massive will often be a debatable question. In the former case, the embankment must rise higher than in the latter, which is coupled with the disadvantage that in the event of a disaster from insufficiency of the wasteway there would be a greater volume of water set free for mischief than in the latter case. The wider the wasteway the less occasion for massiveness in its bed.   If we suppose a wasteway of such width that the water never need rise more than 2 feet above its crest, there is no reason to doubt that a thickness of 12 inches of the best concrete would be perfectly safe.   The water in this case would take a depth of about 16 inches, after passing the crest, and the descending sheet would have a thickness of not more than 3 inches after a descent of 30 feet.   It must be remembered in this case that water running down a smooth slope has no power to injure its bed.   Its power for mischief is developed by meeting with obstacles, and fully developed when its motion is arrested at the end of its descent.   It is also to be remembered that water in this case never carries ice or floating bodies, and further that the wasteway, carrying water only at rare intervals, is at all other times accessible for examination and repairs.   The possibility of water getting under the bed and coming to a pressure sufficient to crack or derange it is to be carefully guarded against.   Wasteways are sometimes made with a narrowing channel designed to maintain the depth uniform while the velocity increases.   This method concentrates the entire energy of the stream at its outfall upon a small area, and calls for massive work to resist the abrasion.

Fig. 87 is a section through the wasteway built on the embankment.   At the foot of the slope, the ground is excavated to the level of 2 or 3 feet below low water, the excavation extending the entire width of the wasteway, and the bottom protected against the wash of the same.   This protection may consist of heavy stone or a bed of concrete, but the form here contemplated is a lining of timber; 8- or 9-inch square timbers are laid down 2 or 3 feet apart, and confined by long iron rods driven through them into the ground.   A flooring of 3-inch plank is laid upon these timbers, and along the up-stream end of this platform the foot-wall is built.   The sluices discharge through this foot-wall, which rises above low water.   The top of this wall forms the footing of the wasteway, which consists of a concrete bed bounded by two side walls running up the slope. The toe of this wasteway might advantageously take a form to give the discharge an upward turn, as such a thin sheet leaping into the air would break into spray and fall harmlessly. Floating bodies are not here to be apprehended. Outside the wasteway the earth rises to the top of the side walls. The side walls continue to the inner slope of the embankment. The floor connects with the puddle-wall or the face-puddle, and if there is no puddle, it should terminate in a wall of masonry sunk 4 or 5 feet into the crown of the embankment.

In constructing a reservoir-dam commanding a large drainage-area, the occurrence of a flood during the process of construction is a contingency not to be overlooked. The discharge of the outlet even under full head is but a small fraction of what might be expected in a flood. A dam of masonry resting on rock can be overflowed at any stage of its construction without danger, but the passage of water over an unprotected embankment would inevitably cause its destruction.

Fig. 88 indicates a mode of constructing an embankment so as to be prepared for the occurrence of a flood at any stage of its progress. The toe of the embankment is constructed as shown at Fig. 87. As fast as the embankment rises, timbers

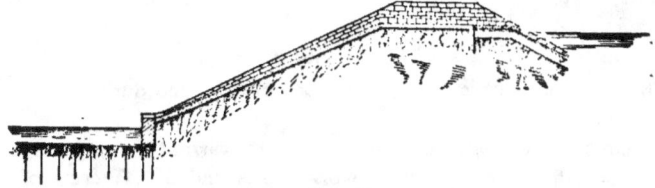

FIG. 87.

or sleepers about 9 inches square are imbedded in the outer slope, and confined by long iron rods driven into the embankment. The timbers extend the width of the work, shown by Fig. 88, and the wider this is, the better for the purpose.

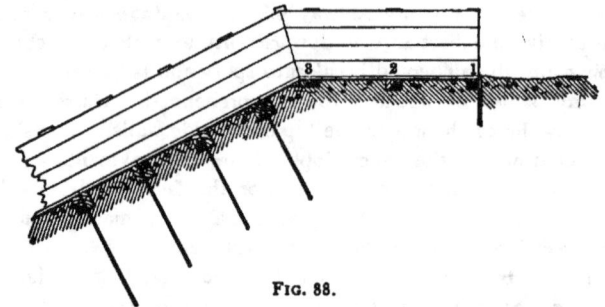

FIG. 88.

The sleepers are covered with a flooring of 2- or 3-inch plank, in lengths only sufficient to reach from one to another. This flooring is terminated by side walls of plank, sustained by uprights mortised into the sleepers and braced therefrom, the sleepers being continued beyond the side walls for that purpose.

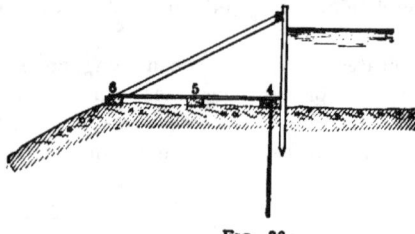

FIG. 88a.

The embankment is kept level from end to end during construction.

On the occurrence of rains such as to warrant the expectation of a flood, construction work is suspended. A crew of men proceed to lay down timbers 1, 2, and 3 on the top of the embankment, and cover them with plank joined to the completed planking as shown. A second party drives a row of tongued and grooved spiling, 2 or 3 feet deep, against the up stream face of timber No. 1, sawing them off and spiking them to the timber. A third lays down the timbers 4, 5, and 6 from the wasteway to end of embankment. A fourth drives the row of 3- or 4-inch spiling on that part of the work, Fig. 88*a*. A fifth puts in the bracing. The ordinary construction gang furnishes laborers sufficient for this emergency. Carpenters are summoned as soon as it arises, at prices that will command their services. All the timber required is kept in readiness from the beginning, together with all necessary tools and a supply of rubber coats and boots, so that men may be induced to work without regard to weather or working hours.

This temporary work may serve the purpose of a wasteway for several years after completion of the reservoir, but must be replaced by masonry as soon as decay appears.

CHAPTER VIII.

ROCK-FILL DAMS.

IN the mining operations which commenced in California some fifty years ago, and, later, in the necessity for irrigation which arose in this and neighboring regions, many projects for collecting, storing, and conducting water on a large scale have been executed. These were at first controlled by men of very limited scientific knowledge, though of great energy and natural sagacity. The traditions of the engineering profession have been largely disregarded, and though many failures have been incurred and great losses sustained, the result is that works have been executed and methods of construction adopted, and have stood the test of time, which no engineer with a reputation to maintain would have dared to recommend, and which in the older sections of the country would have appeared too bold and hazardous for the investment of money even if so recommended. Among these works are dams often more than a hundred feet in height, built of loose rock and sustaining an impervious skin to prevent the passage of water. These dams occupy deep ravines worn in the rock formation, which are usually dry for several months in the year, so that there is usually time to start the work, to lay the sluices, and to raise the bank high enough to create a head capable of passing a good flow of water before the stream commences to flow. Of course the sluices cannot be made capable of carrying off the greatest flow of the stream, and such a work while under construction is, to a greater or less extent, at the mercy of a flood, a risk which has to be accepted or provided for, though it is to be observed that this work is susceptible of much more rapid construction than the ordinary earth embankment.

The wasteway of such a dam is usually cut in the escarpment of the ravine, passing around the end of the dam, a few feet below the level of high water. Fig. 89,* *a, b, c, d*, shows a rock-fill dam recently constructed at Escondido, San Diego County, California, on the Von Segern branch of San Eligo Creek. The drainage-area of this creek was very small, not over 8 square miles. The main supply of the reservoir was obtained from the San Luis Rey River by means of a tunnel through the intervening ridge. The work was thus free from serious risk of damage from high water during its construction. The dam is 76 feet high; 380 feet long on top, 100 feet on bottom; 140 feet thick at bottom, 10 at top. The slopes of the dam are $\frac{1}{2}$ to 1 on the water-face, 1 to 1 on the down-stream face for the upper half, $1\frac{1}{4}$ to 1 for the lower half. The front face to a thickness of 15 feet at bottom and 5 feet at top was laid by hand. The remainder consists of loose angular granite blocks up to 4 tons weight, dumped from cars and handled to some extent by derricks. Tracks for tram-cars were carried across the dam on trestles, which were lengthened upward as the dam rose, leaving the posts imbedded in the rockwork. Along the up-stream face a trench was excavated 3 to 12 feet deep, 5 feet wide, into the bed-rock, and filled with rubble masonry laid in Portland cement, forming a footing to connect the plank facing with the bottom and sides of the ravine. Timbers 6 × 6 inches were imbedded to a depth of 4 inches in the slanting face of the dam, running up and down, 5 feet apart. To these timbers plank are spiked to form the water-tight skin of the dam. The plank are 3 inches thick on the lower third, 2 for the middle, and $1\frac{1}{2}$ for the upper. This plank facing is carried up to a height of 3 feet above the top of the dam, 9 feet above the sole of the wasteway. The space between the planking and the wall is packed with concrete. The sluice is a 24-inch vitrified sewer-pipe imbedded in con-

* Eighteenth Annual Report U. S. Geol. Survey, Part IV, Pl. XLIX.

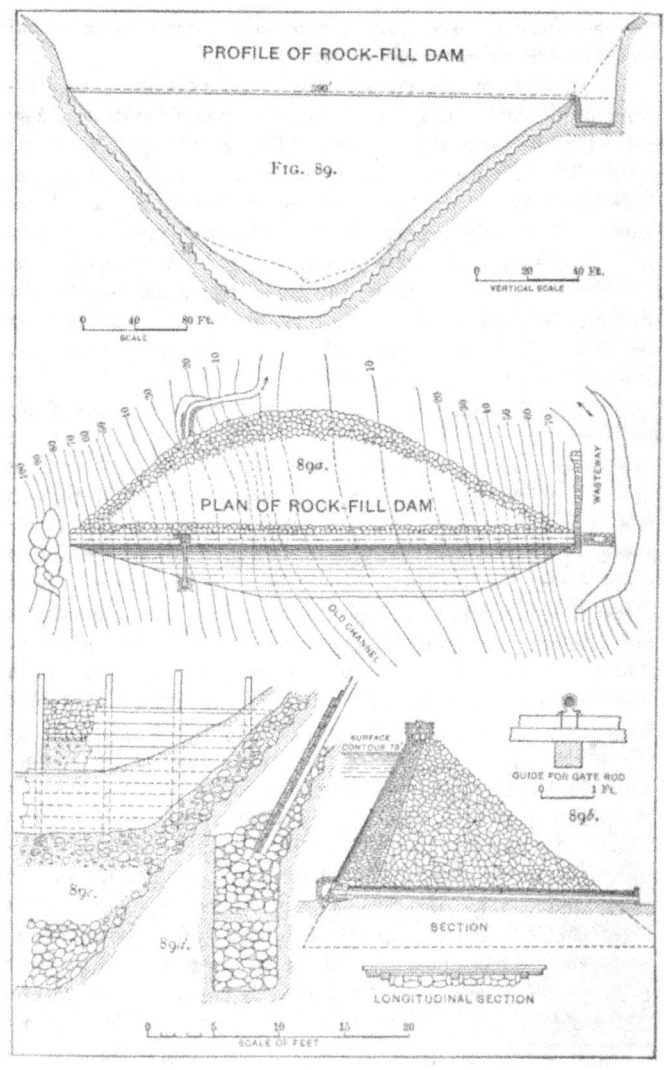

PROFILE OF ROCK-FILL DAM

FIG. 89.

PLAN OF ROCK-FILL DAM

crete. It is closed by a sliding gate which is operated by a rod running down the inclined face of the dam. Fig. 89 is a section of the ravine and wasteway, 89*a* plan of same, 89*b* section of dam through sluice. 89*c* shows plank facing, 89*d* rubble trench and footing. This dam has been in use since 1895.

Figs. 90, *a*, *b*, *c*, *d*, represent a dam of a still more startling character, the impervious skin consisting of a steel diaphragm extending entirely across the valley from top to bottom. This dam was under construction at last accounts and is presumed now to be completed. The details are obtained from the same source as in the former case,* and we cannot do better than use the language of the Report:

"One of the most interesting and remarkable dams now under construction is located 20 miles southeast of San Diego, California, about 10 miles back from the coast, on Otay Creek. The stream here cuts through the great porphyry dike which traverses San Diego County from north to south nearly parallel to the coast-line. The dike is several miles in width, and is crossed and cut into by all the streams of the county that reach the ocean, affording sites for the Sweetwater and the La Mesa dams, already built, and others farther north that are projected. The dam is being built to store water for irrigation and domestic supply on Coronado Beach and the regions south and east of the head of the Bay of San Diego. Its ultimate height above the stream-bed is to be 130 feet, and it will be completed early in 1897. It is a simple embankment of loose stone, dumped in, without any portion of it being laid by hand as a wall, and depending for water-tightness on a central core of steel plates riveted together after the fashion of a large tank, forming a web-plate across the canyon from wall to wall, filling the entire cross-section. It was originally intended to build a masonry dam at this place, and a foundation was laid for that purpose, 62 feet thick at the base. This masonry reaches down to a depth of 31.4 feet below zero contour, into an irregular 'pot-

* Eighteenth Annual Report U. S. Geological Survey, Part IV, p. 637.

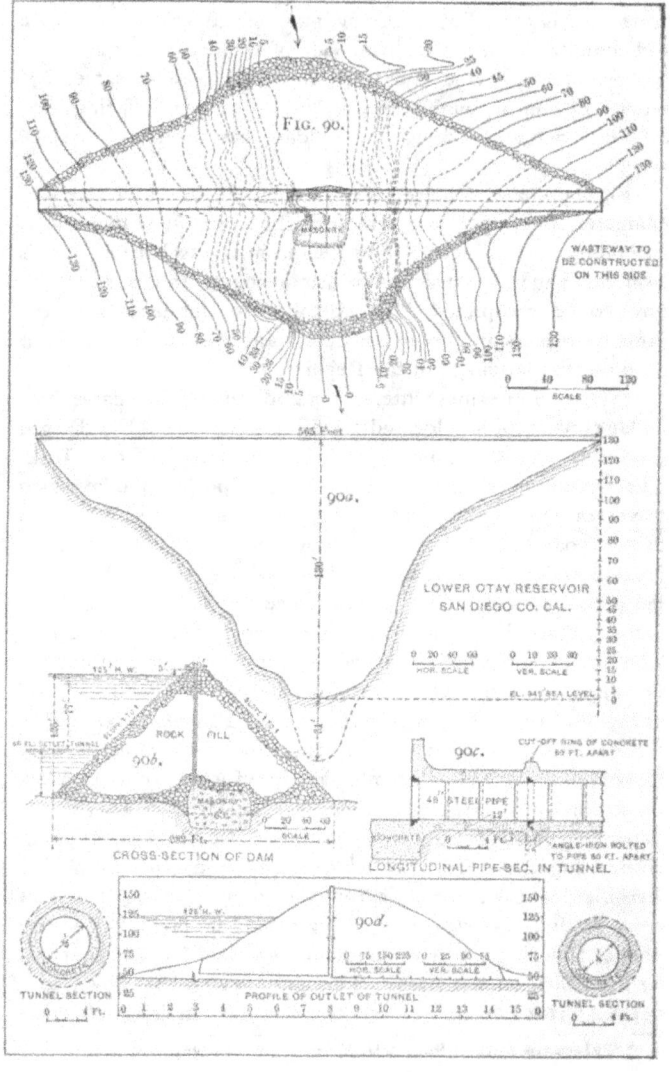

hole ' excavated in the softer portion of the bed-rock under
the stream-bed, and it was carried up to 8.6 feet above zero
before the plan was changed, its length being about 100 feet
at this level. The up-stream side of this masonry was built
as an obtuse angle of about 164°, and on this foundation and
6 feet back from the face, the line of steel plates was begun,
following the same central angle as the masonry foundation.
The plates were 5 feet wide and 17.5 feet long, and the three
bottom courses were 0.33 inch thick. From 28 to 50 feet
height they are one-fourth inch thick, and above 50 feet they
are 8 feet wide and 20 feet long. The plates were riveted in
position, chipped and calked, and afterwards smeared with hot
asphaltum and covered both sides with burlap saturated in the
same material.

"Starting at the foundation, a rubble-masonry wall was
built up each side of the plates, 6 feet thick at bottom, battering up on both sides in a height of 8 feet to 1 foot which thickness was continued to the top. This masonry was somewhat in
the nature of concrete, as it was rammed in between a mould of
p'anks and boards placed on each side of the plate, with a large
rock, often the full thickness of the wall, imbedded in the
mortar. Its function was evidently to stiffen and steady the
web-plate and protect it from injury from the loose rock piled
against it. At the ends the plates were carried into a trench
excavated into the solid rock, and fastened to anchor-bolts set
into the rock, the masonry being expanded to a greater width
for a few feet at the sides. The depth and width of the trench
were quite irregular, depending upon the direction and position
of seams in the rock.

"The expansion of the plates, after they were riveted
together, gave them a very irregular alignment, and before they
had reached the 50-foot level the angle in plan had almost
entirely disappeared and the sheet was in a practically straight,
but wavy, line from side to side. In order to straighten up in
this way the sheet must be inclined from the vertical at some
points, and this inclination may be as great as 7 feet in the

total height. The possible effect of the settlement of the great
mass of stone in the rupturing of the plate when the reservoir
is filled and one-half the wall is enveloped in water must be
regarded with some measure of concern. The uncertainty as
to the strains set up in the central core and the impossibility of
calculating them in advance must be urged against this innova-
tion in dam-construction, although the experiment will be
watched with genuine interest by the engineering profession.

"The stone was quarried immediately below the dam on
the right bank, and was transported by means of a Lidgerwood
cableway, the cable having a diameter of $2\frac{1}{4}$ inches and a span
of 948 feet between towers, crossing the canyon at an angle
of about 60° with the axis of the dam. The head tower was
100 feet high, the tail tower down-stream was 40 feet in height,
and a direct line between them crossed the site of the dam 260
feet above the bed of the stream. The cableway has a maximum capacity for carrying 10 tons weight, under which load
the deflection is 88 feet. It has therefore not been necessary
to move the cable during construction, from the time of its
erection in 1894 till the completion of the dam. The distribution of the rock either side of the line of the cable has been
made by powerful derricks, and latterly by a small auxiliary
cable stretched parallel with the line of the dam and anchored
to cars movable on parallel tracks on either side of the valley.

". . . The total volume of the dam to the 130-foot line
with top width of 20 feet and side slopes of 1 to 1 on each side
is approximately 140 000 cubic yards. . . . The only outlet
to the reservoir has been made by means of a tunnel, 1150 feet
long, through a gap in the ridge 1000 feet west of the dam.
The bottom of the tunnel is at the 50-foot level. The material
encountered in the tunnel was hard-pan and cemented gravel,
bone-dry. For 500 feet from the inner heading it was lined
with concrete to a clear circular diameter of 5 feet, the concrete
being from 12 to 18 inches thick and plastered with cement
mortar. At the end of this section a shaft 104 feet in depth
reaches to the surface, to admit of the operation of a gate across

the conduit. Outside of the shaft a 48-inch steel pipe is laid
to the outside; this is surrounded with 1 foot or more of Portland-cement concrete, filling the space between the pipe and
the irregular surface of the tunnel, with collars of concrete
every 25 feet, and 1 to 2 feet deep all around the pipe. There
are no pipes or openings of any sort through the dam.

"The wasteway is to be located on the left bank, some
hundreds of feet away from the dam, and discharging at a safe
distance below the foot of the embankment. . . .

"The watershed of Otay Creek above the reservoir is about
100 square miles in area; but as its average altitude is not over
1500 feet, the precipitation is light and the run-off insufficient
to fill the reservoir except in occasional years. In dry seasons
there is no flow whatever. To make up for this shortage and
to fill the reservoir regularly the company is planning to divert
the water from Cottonwood Creek, a larger and more reliable
stream on the south and lying next to Mexican territory, by a
conduit 12 miles in length from the diverting-weir, at what is
known as the Barrett dam, to Dulzura Pass, where the water
will drop over into the Otay drainage."

A general idea of the dam is shown by Fig. 90, which is
a plan of dam showing contours of ravine; 90*a*, a condensed
cross-section of ravine; 90*b*, section of dam; 90*c*, tunnel pipe;
and 90*d*, section of ridge showing tunnel.

Fig. 91 is a cross-section of a rock-fill dam built across the
Rio Pecos in 1889–90, near the town of Eddy, New Mexico.

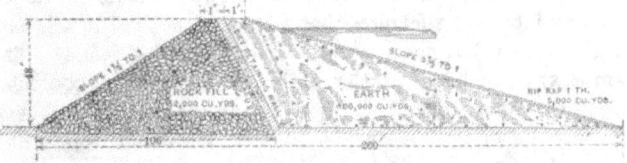

Fig. 91.

The peculiarity of this construction is that it depends for water-
tightness upon a bank of earth. The rock-fill is 48 feet high

in the middle, with a down-stream slope of $1\frac{1}{2}$ to 1 and an up-stream slope of $\frac{1}{2}$ to 1, being faced with a dry retaining-wall. It has a top width of 10 feet. Against the rock-fill on the up-stream side rests a bank of earth with a slope of $3\frac{1}{2}$ to 1, and a top width of 10 feet, making the total top width of the dam 20 feet. The Pecos River at this point has a flood-volume of over 40 000 cubic feet per second and has been observed as low as 200. In 1893 a flood occurred in excess of the capacity of the wasteway. The water ran over the crest of the dam and speedily destroyed it. It was immediately rebuilt, raised 5 feet higher, and provided with additional wasteway capacity.

Embankments made by Sluicing.—Another method of constructing reservoir-embankments has been developed exclusively in California in the course of operations incident to placer-mining, especially to the sluicing down and washing away of hills and banks of auriferous gravel. The method consists in sluicing the material of the dam into position by the aid of running water. La Mesa dam, near San Diego, Cal., holding a head of 60 feet of water, was constructed in this manner. As a clear exposition of this method, I quote from the above report a description of the construction of a dam at Tyler,[*] Texas:

"The experience which led up to the construction of the La Mesa dam and the planning of the other structures described was obtained by Mr. Howells by the building of a dam in Tyler, Texas, by the same method in 1894, which developed so many interesting features that a brief account of it will here be given, as it has never been described in print before. The dam is 575 feet long, 32 feet high, and contains 24 000 cubic yards, the inner slopes being 3 to 1, and the outer 2 to 1, with a 4-foot berm on the inside 10 feet below the top. This impounds 1770 acre-feet,[†] covering 177 acres. The maximum depth is 26 feet. All the material used in the dam was sluiced

---

[*] Eighteenth Annual Report, etc., p. 654.
[†] The acre-foot in an acre one foot deep.

in from a neighboring hill at a cost of 4¾ cents per cubic yard, including the plant and all appurtenances of the reservoir. The water was pumped through a 6-inch pipe from the old city pumping-station on the opposite side of the valley from the hill which supplied the materials. This hill is 150 feet high, and the pipe terminated about half-way up from its base, where a common fire-hydrant was placed, to which was attached an ordinary 2½-inch hose, with a nozzle of 1½ inches diameter. From this nozzle a stream was directed against the face of the hill under a pressure limited to 100 pounds per square inch. The washing was carried rapidly into the hill on a 3 per cent grade, which soon gave a working-face of 10 feet or more, increasing gradually to 36 feet in vertical height. By maintaining the jet at the foot of the cliff it was undermined as rapidly as it could be carried away by the water.

"The material found in the hill consisted of a soft, friable sandstone, infiltrated with ochre of varying shades—yellow, brown, and red—alternating with clay and sand, the whole overlain by a sandy loam soil from 2 to 6 feet deep. Experiment and observation led to the conclusion that 65 per cent of the entire mass washed into the dam was sand and 35 per cent clay.

"In beginning the work a trench 4 feet wide was excavated through the surface soil, down into the clay subsoil, a depth of several feet, and this was first filled with selected puddle-clay sluiced in by the stream. Then the form of the dam was outlined by throwing up low sand-ridges at the slope lines, which were maintained, as the dam rose in height, by men with hoes. A pond of water was thus maintained over the top of the dam, the water being drawn off from time to time, either into the reservoir or outside, as preferred. The material was transported from the bank in a 13-inch sheet-iron pipe, with loose joints, stovepipe fashion, extending from near the face of the bluff, where the jet was operating, across the centre line of the dam. These were so arranged as to be easily uncoupled at any point, so as to direct the deposit where required to build

up the embankment uniformly.   It was found that the quantity
of solids brought down by the water varied from 18 per cent
in solid clay to 30 per cent in sand.   Sharp sand does not flow
as readily as rounded sand or gravel, and is improved in
delivery by an admixture of clay and stones.   The entire
amount of water pumped, computed by the percentages of
solids given, must have been less than 20 000 000 gallons
The limitation of the nozzle-pressure to 100 pounds per square
inch is thought to have restricted the duty of the water used to
considerably less than might have been accomplished with
higher pressure.   The entire cost of the dam with all its accessories is given at $1140, which must be regarded as a marvel
of cheapness, and gives an average cost per acre-foot of
storage-capacity of reservoir formed by it of 65 cents.   The
dam is reported to have no apparent defects and gives satisfactory service.   Mr. L. W. Wells was engineer and foreman in
charge of the work, from whose memoranda, furnished by
Howells, consulting engineer, the foregoing description has
been compiled.''

It is asserted that embankments built in this manner are
firmer, less liable to settlement, and more impervious than those
built in the ordinary manner.   It is well known that the action
of water in excavating, moving, and depositing earth has
recently received very extended application in dredging.   It
has also been applied with great success in railroad work,
many embankments having been recently put in on the
Northern Pacific and Canada Pacific roads, by sluicing at a cost
as low as 5 cents a cubic yard.

CHAPTER IX.

RESERVOIR-DAMS OF MASONRY. .

MASONRY dams for impounding water differ from those designed to create a head for water-power in two ways: 1. As to the sluices for discharging the water, which are not an essential feature of the latter class.   2. In adaptation to the discharge of flood-waters, which generally forms no important modification of the former, this class usually commanding a comparatively limited drainage-area.   This difference becomes more marked as the height increases, and, at the limit of height, the latter class usually has dimensions greatly in excess of what the pressure would call for, and its form is controlled mainly by the necessity of protecting itself from the action of water in motion, while the former is designed mainly with reference to the pressure.   In other words, the latter conforms in shape and dimensions to the dynamic action of water, the former to its static action.   The limit of height in a water-power dam may with limited exceptions be taken at 40 feet, while storage-dams rise to a height of 150 and 200.   The Quaker Bridge Dam now under construction, pertaining to the New York water-works, is intended to reach a height of 265 feet above the rock.

No instance of the failure of a high masonry dam from the crushing of the stone has come to the writer's knowledge. Such dams have failed, but the failure was due to other causes. Engineers therefore have no experimental light to guide them in this class of structures, and so far as the true principles of construction are concerned are groping in the dark.   The danger of failure, like all other dangers encountered in the dark, is avoided by giving it a wide berth.   Engineers know that the application of certain rules will insure a safe structure.   To what extent the dimensions so ascertained might be diminished without passing the limits of safety they do not know, though they generally feel that such dimensions are excessive.   Even on the Pacific slope, except in some of the curved dams we do not find the same boldness in dams of masonry as appears in other types.   So disastrous are the consequences of failure that suggestions of diminished massiveness backed by the ablest opinions would not be likely to find favor with those who have the responsibility for such structures.

One of the most important recent works of the kind under discussion is the Quaker Bridge Dam * already alluded to. Engineers of eminence were employed on the preliminary investigations.   They discussed the subject with great thoroughness, and their reports must be held to embody the best modern practice in this branch of construction.   Among the rules and dicta adopted in these discussions are the following:

1. Although the work is intended to be laid throughout in Portland-cement mortar, tensile strength is ignored in all calculations relative to the masonry.

2. Compressive strains are limited to 15 tons per square foot.

3. The resultant of the forces upon any horizontal joint must fall within the middle third of the joint.

Let us endeavor to discover the rational basis of these rules.

1. *As to Tensile Strength.*—It is customary on all important works to test the tensile strength of cement.   This operation is regarded as a necessary detail of the work.   It is conducted by skilful assistants, the results carefully recorded and tabulated.   The introduction of any large quantity of worthless or inferior cement is not an admissible supposition, except as the result of wilful fraud.   In testing established brands of cement it is no uncommon thing to find in 500 con-

---

* Report of the Aqueduct Commission, Quaker Bridge Dam, 1889. New York.

secutive tests not one sample differing by 25 per cent from the average. It may be assumed that an average strength of 400 pounds per square inch in Portland-cement mortar mixed 2 to 1 is attainable after two years' exposure. Strength of masonry, however, depends upon the adhesion of mortar to stone as much as upon its cohesion. On this point we have some 100 tests recently made at the St. Mary's Lock, on the strength of concrete beams. These were made not wholly as tests of strength, but to test many other points of interest. All the results were consistent and free from anomaly, and a fair deduction from them is that a strength of 400 pounds per square inch is attainable in concrete and rubble masonry with mortar of Portland cement mixed 2 to 1 after two years' exposure.

As to the assumption that no reliance can be placed upon the tensile strength of masonry in a dam. Suppose a high dam so constructed as to call for tensile strength to the extent of 100 pounds per square inch on the up-stream side. Make the extreme supposition that one barrel of the cement in every ten is worthless, the good cement having an average strength of 400. It is manifest that entire lack of strength in a few square yards of horizontal joint would have no effect. To endanger the dam we must assume at least 100 square yards of some horizontal joint devoid of tensile strength. As such work is usually laid up in layers some 3 feet thick, and should contain at least a half-barrel of cement to a cubic yard of masonry, this supposition would imply that fifty barrels of the worthless cement have come together in some part of the dam. The slightest consideration will show that such an event, except as the result of fraud, is too remote to be considered. To speak mathematically, its probability is represented by a fraction of which the numerator is unity and its denominator is the fiftieth power of 10—for all human purposes an impossibility.

No distinction is more important in engineering than that between a theoretical possibility and a degree of probability entitled to practical consideration. It is within the bounds of possibility for any important building to be struck by a meteoric body from the planetary spaces. That event is, however, not sufficiently probable to warrant any special modification in the design of the building.

In like manner, the supposition that the strength of any continuous portion of the work calling for 50 barrels of cement could be as low as 200 pounds per square inch would imply that 25 barrels of worthless cement had found its way into that portion. Excluding the idea of wilful fraud, such an event is too improbable to be considered. Nevertheless, among the possibilities which must enter into such calculations, wilful fraud, especially in contract work, cannot be wholly excluded. A sudden frost, also, followed by a thaw may destroy the tenacity of the freshly laid mortar on a considerable area of the work. It would be impossible, however, to adopt dimensions liberal enough to cover all possible results of fraud and negligence.

2. *As to the Compressive Strength of Stone.* —The crushing of a solid body by a force uniformly applied on all its sides is not possible or conceivable. We might, for instance, conceive a block of stone enclosed in the cylinder of a hydraulic press and surrounded by water under a pressure of 100 000 pounds per square inch without suffering the slightest injury, while the same block exposed to a pressure of 10 000 pounds per square inch on two of its opposite faces would be crushed to fragments. In this case the compressive force resolves itself into components which overcome the tensile and shearing strength of the material. The crushing of a stone imbedded in mortar and located in the interior of a wall is not to be apprehended, neither is that of the mortar itself in the joints of the stones. It is only the exterior stones which are liable to fail from this action. Nevertheless the crushing of the exterior stone leaves the adjacent stones exposed to the same danger in greater measure, and the structure is destroyed in detail.

In dealing with bridges and similar structures, where the failure of a single piece would imperil the whole work, wide margins of safety are usually allowed. The destruction of an important bridge might result from the failure of a single bar; and inasmuch as, in a great number of pieces, a defective one will sometimes occur, common prudence requires that no piece should be strained beyond a fourth or fifth part of its presumable strength. In dealing with large masses of masonry, however, where failure would imply the simultaneous yielding of at least a hundred contiguous stones, it is manifest that the necessity for a wide margin of security does not rest upon the same rational basis. In this case, as in the case of tensile strain, we must inquire what probability there is of such a simultaneous yielding. Suppose that tests sufficiently numerous and complete show that we may assume nine stones out of every ten to have a strength of 5000 pounds per square inch, and make the extravagant supposition that every tenth stone is worthless. It is obvious that this supposition would make the stone entirely inapplicable to a structure whose safety depended upon a single stone. Yet it would affect the value of the stone to but a comparatively slight extent in a massive dam. The probability of a hundred defective stones coming together in such a work is too slight to enter into any rational computation.

The most reliable tests of compressive strength made in recent times are those of General Q. A. Gillmore, published in the Report of the Chief of Engineers, U. S. A., for 1875. This embraced the principal varieties of stone used in construction in the United States. Also those of Mr. Thos. Hudson Beare, reported in the minutes of proceedings of the Institution of Civil Engineers of London, vol. CXVII. p. 341 *et seq.*, embracing the series of building-stones in use in Britain. Mr. Beare's tests were made upon $2\frac{1}{4}$-inch cubes, General Gillmore's upon 2-inch cubes, and he found good reason to believe that the strength per square inch increases with the size of the piece, though no testing-machine has yet been made massive enough to crush a stone of the size used in practical construction. The results are stated in tons (2240 pounds) per square foot.

General Gillmore found:

99 specimens of granite,   highest 1541 tons, lowest 497
43   "   " limestone,   "   1600 "   "   221
12   "   " marble,   "   1284 "   "   488
62   "   " sandstone,   "   1136 "   "   251

Mr. Beare found:

88 specimens of sandstone, highest 1090 tons, lowest 171.5
16   "   " dolomite,   "   642 "   "   281
86   "   " limestone,   "   1075 "   "   58
49   "   " granite,   "   1440 "   "   766

It is to be observed with reference to both these series of tests that the difference between the highest and lowest results represents the difference between the strongest and weakest quarries in the country.   No such difference is to be looked for in the stones available for any particular work which come all from the same quarry or the same formation.

3. *As to Limiting the Resultant to the Middle Third of the Base.*—Let *ABEF*, Fig. 92, be a vertical section of a dam, *AB* indicating any horizontal joint. We may suppose the section taken at the highest part of the dam, which is supposed to rest on firm rock.   For convenience we regard the section as representing a length of the dam equal to unity, and restrict our view to he forces acting on that length. If the conditions of stability are satisfied for this portion of the dam, they will be satisfied for the entire dam.

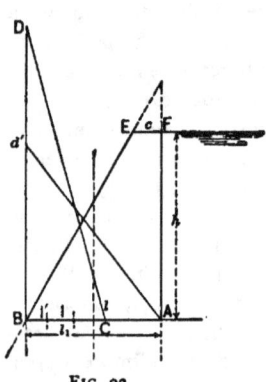

FIG. 92.

Let h = the depth of water above *AB*;

$l_1 = AB$, the length of the joint transversely to the dam;

l = the distance from *A* to the centre of pressure on the joint *AB*;

$s =$ the specific gravity of the masonry;

$c =$ the width of dam at top EF;

$W =$ the weight of masonry above AB.

We shall find it convenient to use one cubic foot of water as the unit of pressure. Thus if the volume of masonry above AB be represented by $\frac{1}{2}h(AB + EF)$, the total weight will be represented by $\frac{1}{2}h(AB + EF)s = W$. The moment of the pressure of the water with reference to A is h times $\frac{1}{2}h$ times $\frac{1}{3}h = \frac{h^3}{6}$. We will suppose the dimensions to be such that the joint AB is relieved of pressure at A. The total weight of the dam still rests on AB, but is not uniformly distributed. The pressure is o at A, and increases uniformly toward B, where it is greatest, the average pressure per square foot being $\frac{W}{l_1}$. If we make $Bd' = 2\frac{W}{l_1}$, the area of the triangle ABd' represents the total pressure, and the pressure at any point in AB is represented by the vertical through that point terminated by Ad'. The centre of pressure is the centre of gravity of ABd', $\therefore l = \frac{2}{3}l_1$. The maximum strain is $2\frac{W}{l_1}$ and no tension exists.

The equation of moments with reference to A is

$$*\frac{l_1 + c}{2}hs \cdot \frac{2}{3}l_1 = \frac{h^3}{6} + \frac{l_1 + c}{2}hs \cdot \frac{1}{3}l_1,$$

or $\quad l_1^2 + cl_1 = \frac{h^2}{s}$, whence $\quad l_1 = -\frac{1}{2}c \pm \sqrt{\frac{h^2}{s} + \frac{c^2}{4}}$. (19)

Passing to figures and putting $h = 100$, $c = 10$, $s = 2.5$, we get $l_1 = 58.44$. In the above and following equations of moments we assume for simplicity the profile as a triangle so far as to regard the vertical through its centre of gravity as

* The first member of this equation is the pressure of the dam on its base, or the pressure of the base on the dam, which is supposed to act at the extremity of the middle third.

acting at $\frac{1}{8}l$ from A. This leads to but slight error, and that on the side of safety.

Total weight on AB is $\dfrac{l_1 + c}{2} hs = 34.22 \times 250 = 8550.$

Maximum compressive strain $= d = 2\dfrac{8550}{58.44} = 293$, being a little more than 9 tons per square foot (short tons).

Now suppose the horizontal pressure relative to the length of joint to be greater than we have assumed, carrying the resultant pressure beyond the middle third. Two cases are to be considered. 1. Assume that the material of the dam is not susceptible of tensile strain. The base of the triangle of pressure will be $3(l_1 - l)$. The average pressure will be $\dfrac{W}{3(l_1 - l)}.$ The maximum $d = \dfrac{2W}{3(l_1 - l)}.$ There will be no strain on the joint from A to C, where $BC = 3(l_1 - l)$. The strain at any point in BC will be represented by a vertical through the point terminated by CD. The condition becomes unsafe when $BD = d$ exceeds the safe compressive strain on the masonry.

Still regarding the masonry as devoid of tensile strength, let us find the value of l_1 corresponding to a compressive strain d on the down-stream end of the joint. We have

$$\frac{3(l_1 - l)}{2}d = \frac{l_1 + c}{2}hs, \quad \text{or} \quad l_1 - l = (l_1 + c)\frac{hs}{3d},$$

whence
$$l = l_1 - (l_1 + c)\frac{hs}{3d}. \quad \ldots \ldots \quad (20)$$

And, since the moments tending to turn the system in one direction about A must be equal to those tending to turn it in the opposite direction,

$$\frac{l_1 + c}{2}hsl = \frac{h^3}{6} + \tfrac{1}{8}l_1\frac{l_1 + c}{2}hs.$$

Substituting the value of l from (17),

$$\frac{l_1 + c}{2}hs\left(l_1 - (l_1 + c)\frac{hs}{3d}\right) = \frac{h^3}{6} + \tfrac{1}{8}l_1\frac{l_1 + c}{2}hs,$$

which reduces to the form

$$\left(2 - \frac{hs}{d}\right)l_1^2 + 2c\left(1 - \frac{hs}{d}\right)l_1 = \frac{h^2}{s} + c^2\frac{hs}{d}. \quad . \quad (21)$$

Passing to figures, $h = 100$, $c = 10$, $s = 2.5$, $d = 512 = 16$ short tons per square foot,

$$l_1^2 + 6.77 l_1 = 2678, \quad l_1 = 48.48.$$

2. Regard the masonry as susceptible of tensile strain, and call the maximum strain of compression d, of tension f. The centre of pressure arising from the weight is at the extremity of the middle third; the excess is met by elastic strains. The equation of moments with reference to the centre O Fig. 86 is

$$\tfrac{1}{2}dl_1\frac{l_1}{6} + f\frac{l_1}{2}\cdot\frac{l_1}{6} + \frac{l_1 + c}{2}hs\frac{l_1}{6} = \frac{h^3}{6},$$

whence $\quad (d + f + hs)l_1^2 + chsl_1 = 2h^3. \quad . \quad . \quad (22)$

Substituting $h = 100$, $s = 2.5$, $c = 10$, $d = 282$, $f = 230 =$ about 100 pounds per square inch, we get $l_1 = 49.62$. Showing that we gain nothing by taking account of the tensile strength.

The maximum compressive strain is $d + f = 512 = 16$ short tons per square foot. It will appear from the above that the engineering canon which requires the centre of pressure at the base of a dam to lie within the middle third cannot be regarded as an inflexible rule, and for dams not more than 100 feet high leads to excessive dimensions. The reason alleged for this rule is that a pressure exceeding this limit causes tensile strain. This is true if the masonry is susceptible of tensile strain. It is not true in the sense that a lack of tensile strength would, in that condition, endanger the dam. We see that the pressure can be pushed far beyond this limit in a dam devoid of tensile strength without passing the limits of safety.

In Fig. 93, *ABEF* represents the outline of the dam, *BD* the total compressive strain at *B*, and *Af* the tensile strain at *A*. Of the compressive strain *Bd'* arises from the weight, and $Be = Af = d'D$ from the tensile strain. I, where $AI = \frac{2}{3}AB$, is the centre of pressure disregarding tensile strain. i, where $Oi = \frac{2}{3}OA$, is the centre of tensile strain. i', where $Oi' = \frac{2}{3}OB$, is the centre of compressive strain disregarding the weight.

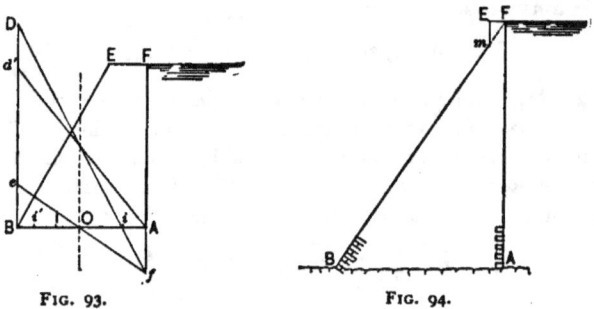

FIG. 93.   FIG. 94.

We may here advert to the fact that in a dam of triangular section with one face vertical, the centre of pressure goes through the extremity of the middle third with no water on the up-stream side.

The computations are much simplified if we suppose the dam to terminate in a sharp crest, Fig. 94. Constructive reasons usually forbid such an arrangement, but after determining the dimensions on that supposition it is easy to add the small triangle mEF to the profile, which has but slight effect and acts to increase the stability.   The equation of moments is

$$\frac{2}{3}l_1\frac{shl_1}{2} = \frac{h^2}{6} + \frac{2}{3}l_1\frac{shl_1}{2}, \quad \text{whence} \quad shl_1^2 = h^3,$$

and $$l_1 = \frac{h}{\sqrt{s}}. \quad . \quad . \quad . \quad . \quad . \quad (23)$$

This is deduced on the supposition that the centre of pressure is at the extremity of the middle third in all horizontal sections, and eq. (23) being the equation of a straight line, l_1 and h being the coordinates of the down-stream face, we conclude that this face is, subject to the above condition, a straight line. The maximum compressive strain at any depth h is $\frac{shl_1}{l_1} = sh = d$. The compressive strain, therefore, is proportional to h. The crushing pressure on the down-stream end of any horizontal joint when the reservoir is full is the same as that on the up-stream end of the same joint when the reservoir is empty. It is also to be noted that the centre of pressure on every horizontal joint is at the down-stream extremity of the middle third when the reservoir is full, at the up-stream extremity when the reservoir is empty. If we adopt 32 000 pounds per square foot as the maximum crushing strain, the greatest height of a dam with a straight down-stream face would be, for $s = 2.5$, $h = \frac{512}{2.5} = 204.8$. Below this depth, in order not to exceed the crushing strength d, the down-stream face would require to be curved. Above this depth, also, some saving of material might be effected by drawing the down-stream face within the line determined by eq. (23). Considerations of this character lead to the curved faces often adopted in high dams.

The following approximate method may be used to compute the outline of a dam which will bring any assumed constant strain upon the masonry of the down-stream face. It is obvious that a strict adherence to this condition would bring the masonry to a sharp edge at the top, which is inadmissible, but after computing the outline upon this supposition any desired small addendum can be made to the profile to obviate the difficulty. Making $c = 0$ in eq. (20), it becomes

$$l = l_1\left(1 - \frac{hs}{3d}\right). \quad \cdot \quad \cdot \quad \cdot \quad \cdot \quad \cdot \quad (24)$$

(21 becomes on the same assumption

$$l_1 = \frac{h}{\sqrt{\left(2 - \frac{hs}{d}\right)s}}. \quad \cdot \quad \cdot \quad \cdot \quad \cdot \quad \cdot \quad (25)$$

Equation (25), regarded as representing a line of which h and l are the coordinates, would give an approximate outline of a dam subject to the assumed condition, and would not pass the limit of safety at any point.   It would materially overestimate the weight and would not be sufficiently exact.

It would be fanciful to commence the curvature at the top of the dam.   The face may be assumed to run straight to any desired depth, which we will here assume at 50 feet.   At this point, taking $d = 512$, $s = 2.50$, we find, by eq. (25), $l_1 = 23.864$.   We suppose the work to consist of courses 2 feet in depth, designated as course 1, course 2, etc.   The lengths of the joints are designated l_1, l_2, l_3, etc.

The weight of masonry above l_1 is $23.864 \times 50 \times \frac{1}{2} \times 2.5 = 1491.5$.

The moments tending to turn the system to the left about A, Fig. 92, are:

$$1. \quad \frac{h^3}{6} \qquad = 20\,833$$

$$2. \quad 1491.5\frac{l_1}{3} = 11\,864$$

$$\text{Total..} \quad 32\,697$$

Moments tending to turn to the right $3l\dfrac{l_1-l}{2}d$.   Here we put for l its value as given in (21), making the moment 32 697.

Adding a 2-foot course, the value of l_2 is approximately

$$23.864 + 2\frac{l_1}{h} = 24.818.$$

The mean length of course 1 is $l_1 + \dfrac{l_1}{h} = 24.341$.

Weight $= 2 \times 27.371 \times 2.5. = 121.7$

We now find the moment as increased by the addition, viz. :

$$\text{Water-pressure} = \frac{52 \times 52 \times 52}{6} \qquad = 23\,435$$

Masonry above l_1................... 11 864

Course No. 1 $= 121.7 \times \tfrac{1}{2} \times 24.34 =$ 1 481

Total........................ 36 780

Total wt. of masonry above $l_2 = 1491.5 + 121.7 = 1613.2$.

$$\text{Lever-arm of moment} = l = \frac{36\,780}{1613.2} = 22.80.$$

$$l_2 - l = \frac{1}{3}\frac{1613.2}{d} \times 2 = \tfrac{1}{3} \times 6.30 = 2.10.$$

$\therefore$ True value of $l_2 = 22.80 + 2.10 = 24.90.$

l_3 approximate $= 24.90 + l_2 - l_1 = 25.94.$

Mean length of course 2 $= 25.42.$

Weight of course 2 $= 25.42 \times 2 \times 2.5 = 127.10.$

Weight above $l_3 = 1613.2 + 127.10 = 1740.30.$

Moments above l_3:

1. Water-pressure $\quad = \dfrac{54 \times 54 \times 54}{6} = 26\,244$

2. Above $l_2 \qquad\qquad = 11\,864 + 1481 = 13\,345$

3. Moment of course 2 $= 127.10 \times \dfrac{25.42}{2} = $ 1 615.4

Total turning to left............... 41 204.4

Lever-arm of moment turning to right,

$$l = \frac{41\,204.4}{1740.3} = 23.68.$$

$$l_3 - l = \frac{2}{3}\frac{1740.30}{512} = 2.27. \quad \therefore\ l_3 = 23.68 + 2.27 = 25.95.$$

And so following, giving the ordinates l corresponding to the assumed abscissas, h for the curved outline of the down-stream face.

The ingenious reader will, no doubt, work out for himself a similar method founded on the assumptions of eq. (22).

General formulas and computations upon this subject necessarily rest upon a very imperfect basis. They assume a uniform height of the dam, and take account of a portion only one foot thick measured in the direction of the dam's length. Such dams are usually built to close ravines and are of full height only for a short distance in the middle, diminishing to nothing at the ends. The formulas treat this portion of the dam as though it derived no support from its connection with the remainder, and as though the latter derived no support from its connection with the sloping sides of the ravine. The methods and computations can be fully developed only in connection with some particular case.

For facility of computation we have taken s in the preceding calculations $= 2.5$. This would only be justifiable for the heaviest stone and the best work; ordinarily s would be less than here assumed.

All the above computations imply a triangular section, vertical on the up-stream face. The approximate method just presented gives the outline of a dam in which there is no tensile strain; no compression exceeding d, which is constant on the curved part of the down-stream face. Compression may exceed d on the up-stream face when reservoir is empty. Resultant confined to middle third of base. Eq. (23) implies straight faces, no tension, compression not defined; resultant confined to middle third of base. Eq. (25) assumes straight faces, no tension, compression not exceeding d when full; resultant confined to middle third.

CHAPTER X.

EXAMPLES OF HIGH DAMS.

FIG. 88 and several following figures are examples of high dams which have apparently been built without much regard to statical principles, but with the sole purpose of getting in material enough.

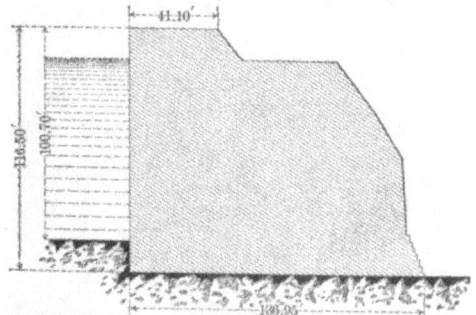

VAL DE INFIERNO DAM.
FIG. 95.

Fig. 95 is a section and Fig. 96 a plan of the Val de Infierno Dam, in the province of Sorca in Spain, built for purposes of irrigation. Figs. 97 and 98 refer to the Gileppe Dam built at Verviers in Belgium, for purposes of water-power and water-supply. It was built of rubble masonry laid in mortar. Fig. 98 shows that the dam had a curved outline in plan as a supplement to the strength of its enormous cross-section. Comparing this section with that of the Otay rock-fill dam,

217

Fig. 90, and observing that, in the latter case, it is only the down-stream half of the dam which resists the pressure, we see that the head of water is sustained, in the latter case, by a mass

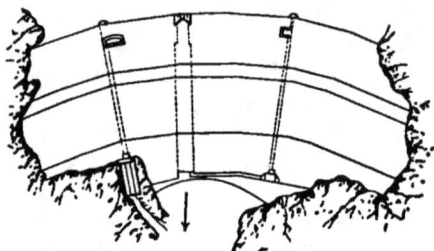

VAL DE INFIERNO DAM
FIG. 96.

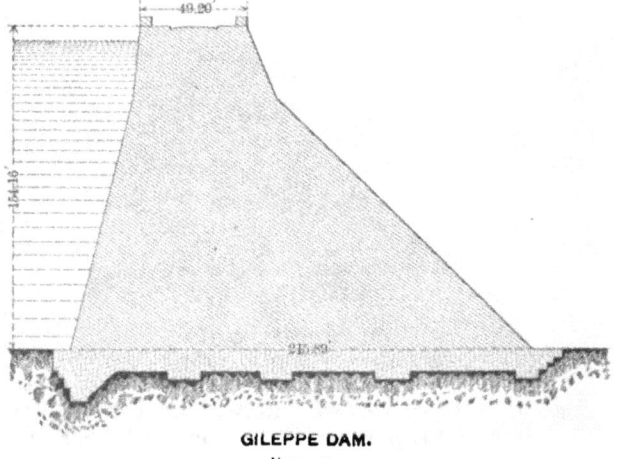

GILEPPE DAM.
FIG. 97.

of loose stone no greater than that of the masonry in the former. The dimensions of this dam were well understood to be unnecessarily great by its builders, but were demanded by the interests which would have been jeopardized by its failure.

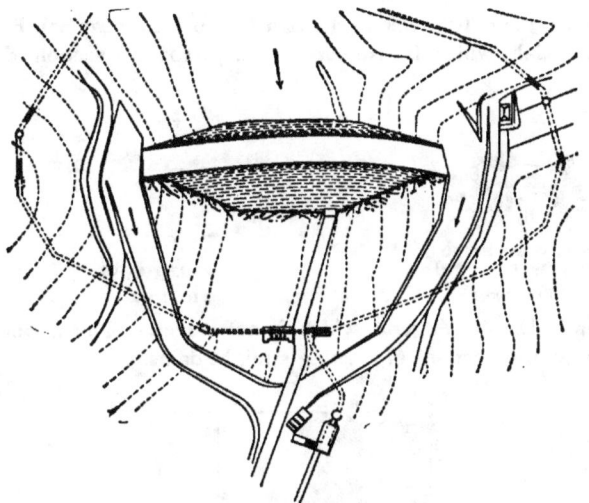

GILEPPE DAM
FIG. 98.

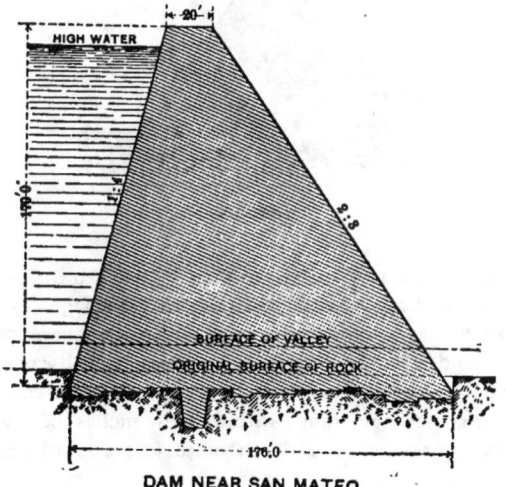

DAM NEAR SAN MATEO
(SAN FRANCISCO WATER WORKS)
FIG. 99.

Figs. 99 and 100 refer to the San Mateo Dam near San Francisco, Cal., which is also curved.　Fig. 102 is a section of the

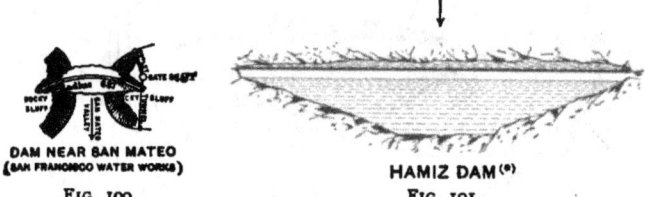

DAM NEAR SAN MATEO
(SAN FRANCISCO WATER WORKS)

FIG. 100.

HAMIZ DAM.[a]

FIG. 101.

Hamiz Dam in Algeria, which is not curved.　This is one of the lightest sections of existing straight dams.

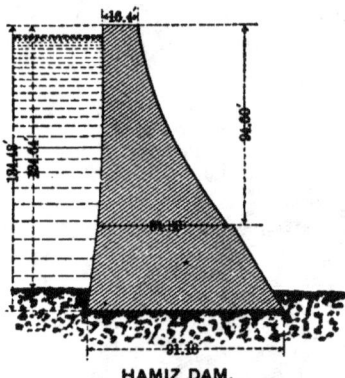

HAMIZ DAM.

FIG. 102.

Figs. 103 and 104 * relate to Bear Valley Dam, San Bernardino County, Cal., already referred to (see page 39).　This dam commands a drainage-area of about 56 square miles and is not intended to be overflowed, having a spillway separate from the dam, as appears at *a*, Fig. 104.　It is understood that the water sometimes stands within a few inches of the top, in which case the stone at a depth of 48 feet would be subjected

* Eighteenth Annual Report U. S. Geol. Survey, Part IV. p. 683.

to a pressure of $335 \times \dfrac{48}{16} \times 1000 \times \dfrac{1}{8.5} \times \dfrac{1}{2240} = 53$ tons per square foot of masonry, making abstraction of the support

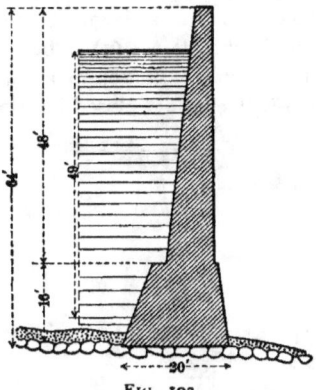

FIG. 103.

which the masonry derives from the thicker course below. The masonry is a rough granite ashlar with hearting of rubble, all laid in cement mortar.

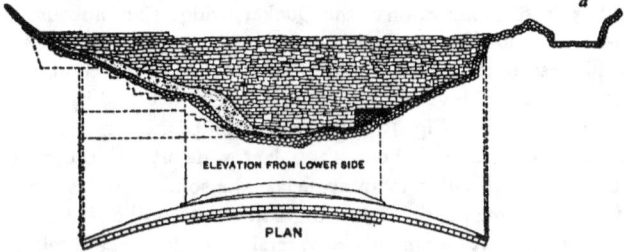

ELEVATION FROM LOWER SIDE

PLAN

FIG. 104.

Sweetwater Dam,* Fig. 105, on the river of that name, near San Diego, Cal., has a profile somewhat less than would be indicated by eq. (18), with $s = 2.25$, being 96 feet high, 46 feet thick at bottom, and about 10 feet effective thickness at

---

* Eighteenth Annual Report U. S. Geol. Survey, Part IV. p. 688.

the top.   The dam, however, has a curved outline at the top, while eq. (18) contemplates a straight outline.

The Folsom Dam * on the American River in California is noteworthy for two things: (1) as being the only straight high masonry dam in the Pacific States; (2) as being the only existing application of a movable shutter to a water-power dam. It has a wasteway 180 feet wide, closed by a movable shutter

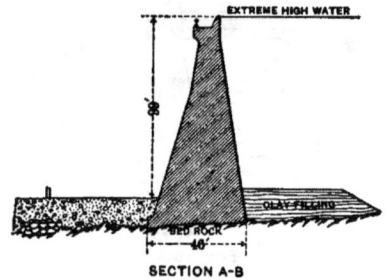

SECTION A-B

FIG. 105.

6 feet high, operated by hydraulic jacks.   This shutter is opened during floods, and remains closed at other times.†

Fig. 106‡ is a section of the Quaker Bridge Dam now under construction for the water-supply of New York, believed to be the highest dam in the world.   The section is taken at the lowest point of the valley, where the rock is overlaid by 80 to 87 feet of gravel.   Fig. 107 is an elevation of the same.   This dam closes a ravine about 1300 feet wide at the proposed water-level.   It will form an artificial lake some 16 miles long, 165 feet extreme depth, containing some 5000 millions of cubic feet of water.   It commands a watershed of 361 square miles, already reservoired to the extent of some 1800 millions of cubic feet.   The greatest recorded flood at the dam-site is 1070 million cubic feet in 24 hours, an average of about 600 cubic feet per second.   The section is the one adopted by the board

* Eighteenth Annual Report U. S. Geol. Survey, Part IV. p. 687.
† 1900.
‡ Report of the New York Aqueduct Commission, 1889.

of aqueduct commissioners in 1889, and was the result of extended studies by engineers of reputation.

Were the down-stream slope prolonged to the level of the base, this dam would have a base width of 225 feet. This width is reached in the view of limiting the compressive strain

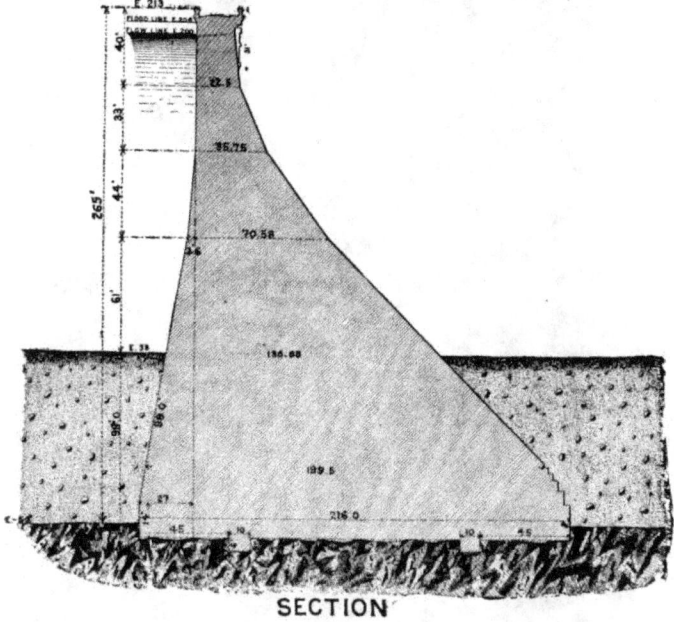

SECTION

FIG. 106.

to 30 000 pounds, 13.4 tons per square foot. The curve or batter of the up-stream face is arrived at on the same principle. The relation of eq. (20), putting $s = 2.5$, would make the width corresponding to $h = 265$ feet, $l_1 = 167.6$ feet, and the crushing strain at the base $d = 662$ feet $= 18.4$ tons per square foot, with no tensile strain on any part of the masonry and the centre of pressure within the middle third at every horizontal section.

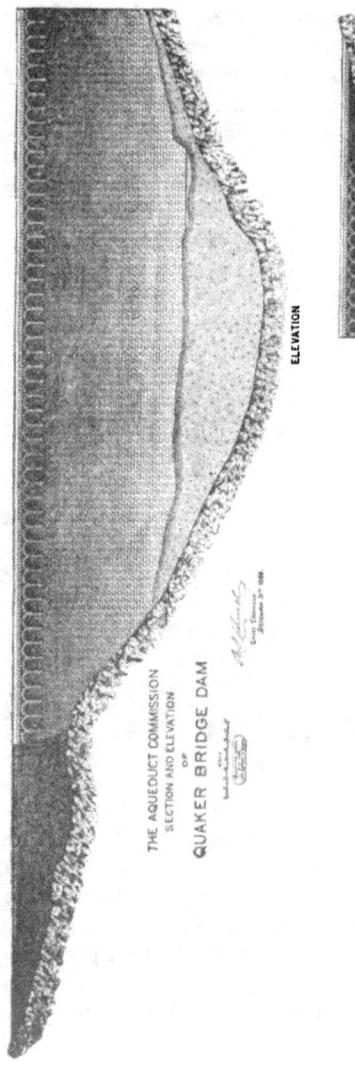

THE AQUEDUCT COMMISSION
SECTION AND ELEVATION
OF
QUAKER BRIDGE DAM

ELEVATION

FIG. 107.

It may be assumed that at every such work granite of 1000 tons average strength, limestone of 900, or sandstone of 600 are always obtainable. The strength assumed above is less than one-fiftieth of the first, about a fiftieth of the second, and less than a thirty-second of the third.

Conceding in the most unqualified manner that, in a structure of this magnitude and importance, stability should be assured beyond the possibility of a rational doubt, the question still remains: Can a doubt of the ability of masonry to sustain the fiftieth or even the thirty-second part of the ultimate crushing strength of the stone be regarded as a rational doubt ?

To what extent the weight of the masonry below water may be regarded as diminished by buoyancy is a question which has occasioned some perplexity. In a mass of uncemented stone, as the filling of cribwork, there is no question but that the effective weight of the stone is diminished by that of the water displaced. In the case of a dam resting on the bottom, united into a monolithic mass by mortar and joined to the bottom by the same means, the question takes a different aspect. Mr. James B. Francis, who reported upon this work some time about 1885, treated the submerged portion as of diminished weight. He made experiments to demonstrate his view. In a mass of concrete or mortar exposed to high pressure he inserted a pipe provided with a pressure-gauge, and found after a time that the gauge showed the full pressure. This result was to be foreseen and certainly shows that there will be an upward pressure at the base of the dam. If a series of vertical pipes were inserted in the dam in a line transverse to its length, each pipe reaching from top to bottom, then assuming the masonry to be perfectly homogeneous with itself and perfectly united with the bottom, the water would stand in the up-stream pipe very near the reservoir level, and in the down-stream pipe very near low-water line, diminishing with an approach to regularity from one to the other. This shows unquestionably that there is a pressure at the base of the dam tending to lift it; but we must reflect further that this pressure could not exist if the mortar did not have the tensile strengtn to resist it. The perplexity arises from an attempt to reconcile two incongruous assumptions: 1. That the mortar is to be regarded as devoid of tensile strength; 2. That it is to be supposed capable of holding water and sustaining the pressure thereof. It appears to me that if we debit the dam with the upward pressure of water, we must credit it with sufficient tenacity of mortar to resist the same.

If we suppose the dam simply to rest on the bottom, without being united with it at all points, a barrier to stop the underflow becomes necessary and the position of this barrier becomes important. If placed at the up-stream face, it excludes the reservoir pressure from the base of the dam; if at the down-stream face, it exposes the whole base to that pressure and vastly increases the overturning forces. There can be no rational purpose in placing this barrier below the up-stream face, and there can certainly be no rational purpose in inserting a second barrier as shown in the figure. If the latter does not serve its purpose of arresting water, it is useless. If it does serve that purpose, it is pernicious in exposing the dam to the upward pressure of the head in the reservoir.

Discharge of a Wasteway.—Where water flows through a rectangular notch in the vertical face of a dam, which, for simplicity, we suppose to have no sensible thickness, it approaches the dam from every direction that can exist in a hemisphere, toward the centre. Arriving near the plane of the orifice, it changes its direction, and issues at right angles to the latter. This change of direction gives rise to the phenomenon of contraction, making the issuing stream considerably smaller in dimensions than the orifice through which it has passed. This action makes the flow of water through an orifice very complex and not amenable to analysis. The flow over weirs of certain conventional form has been made the subject of experiment, and formulas which will be referred to later have been devised, representing the discharge with a sufficient degree of accuracy for practical purposes. A wasteway is generally different from the ideal weir, and the weir formula requires some modification in its application to this case. Where the entrance to the wasteway is rounded or expanded toward the reservoir, the effect of contraction may be disregarded, and in that case we can compute the flow with considerable accuracy. We will consider first the flow over an ordinary dam, the width of the top, which must be regarded as the length of the channel of discharge, being so small that the frictional head or resistance to movement may be disregarded.

Let h = the height of water in the reservoir above the sole of the wasteway;

v = the velocity with which the water issues from the wasteway;

x = the depth corresponding to the velocity v;

Q = the discharge;

w = the width of the channel of discharge.

Then we must have

$$v = \sqrt{2g(h - x)} \quad \text{and} \quad Q = wx \sqrt{2g(h - x)}. \quad (26)$$

$h - x$ in this expression determines the velocity of the stream. x determines the cross-section. On what principle will the stream resolve itself into these two elements? The division will be governed by a principle of universal application in nature, viz., the result of any given expenditure of energy is the greatest that is possible consistent with the existing conditions. In other words, Q must be a maximum with reference to x, a condition expressed by $\dfrac{dQ}{dx} = 0$. From (23),

$$\frac{dQ}{dx} = w \sqrt{2g}\left(\sqrt{h - x} - \frac{1}{2} \frac{x}{\sqrt{h - x}} \right) = 0,$$

whence $\qquad h - x = \dfrac{x}{2} \quad \text{and} \quad x = \tfrac{2}{3}h. \quad . \quad . \quad . \quad (27)$

Therefore $\quad v = \sqrt{2g\dfrac{x}{2}} = \sqrt{gx} \quad \text{and} \quad Q = wxv = wx \sqrt{gx};$

and since $v = \sqrt{2g\dfrac{h}{3}}$,

$$Q = \tfrac{2}{3} wh \sqrt{2g\dfrac{h}{3}} = 3.087\, W h^{\frac{3}{2}}. \quad . \quad . \quad . \quad (28)$$

Eq. (27) shows that one-third of the head is expended in velocity; the remaining two-thirds determines the cross-section. Eq. (28) is only true for the case in which the width of the dam or length of the channel of discharge is too small to have any influence upon the discharge.

Formula (28) may be tested by comparison with results of experiments made by James B. Francis in 1852 on the flow of water over a weir intended to represent a dam. The width, which we represent by w, was about 10 feet (9.995). The width of the level crest, which we represent by l, was only 3 feet, joined on the up-stream side by a slope of some 3 to 1. The flow was actually measured in a reservoir, viz., a disused lock-chamber.

		Cubic Feet per Second.
For h = 0.5872, Q by measurement = 13.385,	by eq. (24) =	13.883
0.7903, " " 20.892,	" "	21.672
0.9767, " " 28.914,	" "	29.782
1.3252, " " 46.183,	" "	47.069
1.6338, " " 64.346,	" "	64.434

Questions often arise as to the time required for the water in a reservoir to rise or fall with a given width of wasteway and given influx of water. A method of solving such questions by rough approximation has been already indicated (see page 159). We will now attempt to deduce a more exact formula for such cases.

Let A = area of reservoir surface in square feet, not considering the change in A incident to any slight rise or fall of the surface;

w = width of wasteway in feet supposed to admit of the application of formula (27);

Q = constant influx of water in cubic feet per second;

h = depth on wasteway in feet;

$t =$ time in seconds corresponding to h, reckoned from the moment the water commences to flow over the wasteway;

$h_1 =$ depth corresponding to the quantity Q.

We have the relation $(Q - 3.087wh^{\frac{3}{2}})dt = Adh.$

$$\therefore \frac{dt}{A} = \frac{dh}{Q - 3.087wh^{\frac{3}{2}}}.$$

Put $h = x^2$, $3.087w = a$, $\dfrac{Q}{a} = b^3$. Then

$$h^{\frac{3}{2}} = x^3, \quad dh = 2xdx,$$

and
$$\frac{a}{2A}dt = \frac{xdx}{b^3 - x^3} = \frac{xdx}{(b - x)(x^2 + bx + b^2)}. \qquad (29)$$

Assume $\dfrac{x}{b^3 - x^3} = \dfrac{l}{b - x} + \dfrac{mx + h}{x^2 + bx + b^2}$, whence

$$x = l(x^2 + bx + b^2) + mx(b - x) + h(b - x).$$

Putting coefficients of the like powers of x separately equal to 0, we find

$$l = \frac{1}{3b}, \quad m = l = \frac{1}{3b}, \quad n = -\frac{1}{3}, \quad \text{and (26) becomes}$$

$$\frac{a}{2A}dt = \frac{dx}{3b(b - x)} + \frac{(x - b)dx}{3b(x^2 + bx + b^2)},$$

or
$$\frac{3ab}{2A}dt = \frac{dx}{b - x} + \frac{(x - b)dx}{x^2 + bx + b^2},$$

or
$$\frac{3ab}{2A}dt = \frac{dx}{b - x} + \frac{1}{2}\frac{(2x - 2b)dx}{x^2 + bx + b^2}$$

$$= \frac{dx}{b - x} + \frac{1}{2}\frac{(2x + b)dx}{x^2 + bx + b^2} - \frac{1}{2}\frac{3bdx}{x^2 + bx + b^2}.$$

By integration,

$$\frac{3ab}{2A}t = -\log(b - x) + \tfrac{1}{2}\log(x^2 + bx + b^2) - \frac{3b}{2}\int\frac{dx}{x^2 + bx + b^2}$$

$$= \log \frac{\sqrt{x^2 + 2x + b^2}}{b - x} - \frac{3b}{2} \int \frac{dx}{\left(x + \frac{b}{2}\right)^2 + \frac{3b^2}{4}}$$

$$= \log \frac{\sqrt{x^2 + bx + b^2}}{b - x} - \frac{3}{2} b \frac{2}{b\sqrt{3}} \int \frac{\frac{2}{b\sqrt{3}} dx}{1 + \left(x + \frac{b}{2}\right)^2 \left(\frac{2}{b\sqrt{3}}\right)^2}$$

$$= \log \frac{\sqrt{x^2 + bx + b^2}}{b - x} - \sqrt{3} \tan^{-1} \frac{2x + b}{b\sqrt{3}} + C. \quad . \quad . \quad (30)$$

To determine C, the constant of integration, we must remember that $x = \sqrt{h}$, $b = \sqrt{h_1}$, and, since $h = 0$ when $t = 0$, eq. (30) becomes in that case

$$0 = \log \frac{\sqrt{h_1}}{\sqrt{h_1}} - \sqrt{3} \tan^{-1} \frac{1}{\sqrt{3}} + C.$$

$$\therefore C = \sqrt{3} \tan^{-1} \frac{1}{\sqrt{3}} = \sqrt{3} \text{ arc } 30°,$$

and we have

$$\frac{3abt}{2A} = \log \frac{\sqrt{x^2 + bx + b^2}}{b - x} + \sqrt{3} \tan^{-1} \frac{1}{\sqrt{3}}$$

$$- \sqrt{3} \tan^{-1} \frac{2x + b}{b\sqrt{3}} \quad . \quad (31)$$

Putting $h = h_1$, (31) becomes

$$\frac{3a\sqrt{h_1}}{2A} t = \log \frac{3\sqrt{h_1}}{0} + \sqrt{3} \tan^{-1} \frac{1}{\sqrt{3}} - \sqrt{3} \tan^{-1} \sqrt{3}.$$

The first term of the second member is infinite and the two second terms reduce to

$$\sqrt{3}\left(\tan^{-1} \frac{1}{\sqrt{3}} - \tan^{-1} \sqrt{3}\right) = \sqrt{3}(30° - 60°) = -0.88625.$$

This shows that, in mathematical strictness, the efflux can never become equal to the influx, though h can become equal to any assignable fraction of h_1. Let h be assumed $= ch_1$, where c is less than unity; then (31) becomes

$$\frac{3abt}{2A} = \log \frac{\sqrt{ch_1 + \sqrt{h_1}}\sqrt{ch_1} + h_1}{\sqrt{h_1} - \sqrt{ch_1}} - 0.88625 \quad \text{(approximate)}$$

$$= \log \frac{\sqrt{c + \sqrt{c} + 1}}{1 - \sqrt{c}} - 0.88625. \quad \ldots \quad (32)$$

Assume, as at page 158, $A = 200$ acres $= 8\,712\,000$ square feet, influx $= 10\,000$ cubic feet per second, $h_1 = 4.5$, $h = 4.375$, whence $c = 0.9722$, $\sqrt{c} = 0.9860$.

A discharge of 10 000 cubic feet per second under a head of 4.5 feet implies a wasteway $\dfrac{10\,000}{3.087 \times 4.5\sqrt{4.5}} = 339.35$ feet in width. $\therefore w = 339.35.$

$$\text{nat log } \frac{\sqrt{.9722 + .9860 + 1}}{1 - .9860} = \text{nat log } \frac{1.7199}{.014} = 4.81026$$

$$\sqrt{3}\left(\tan^{-1}\sqrt{3} - \tan^{-1}\frac{1}{\sqrt{3}}\right) = \underline{0.88625}$$

$$3.92401$$

$$t = 3.92401 \times \frac{2 \times 8\,712\,000}{3 \times 339.35 \times 3.087\sqrt{4.5}} = 10\,256 \text{ seconds.}$$

We found by the rough method employed (page 187) that 10 061 seconds would be required to raise the water to a height of 4.25 on the weir. The last result is, no doubt, more correct.

CHAPTER XI.

FAILURES OF HIGH DAMS.

THE successful execution of a high dam is not necessarily an instructive example to engineers, especially if success be attained by excessive and inordinate use of material and by expense far beyond any rational requirement. In that case the example is misleading, as inspiring the belief that such dimensions and outlays are unavoidable, although it must be freely conceded that a great excess in such a work is better than a small deficiency. The failure of a dam, however, when its causes are clearly understood, cannot be otherwise than instructive to the engineering profession and to the community at large, however disastrous to those immediately concerned. We will therefore, assuming the reader's approbation, devote this chapter to an account of some of the more conspicuous and instructive instances of the failures of high dams, commencing with earth embankments.

The Efflux has usually proved to be the weak point of earth embankments, and it is here that breaks have usually started. Imperfect consolidation of the bank and the absence of any effective special skin or diaphragm to stop the water sometimes allows the whole mass to become "waterlogged," every part of the work becoming saturated and under pressure. The water appears on the outside and gathers into rivulets. The whole becomes semifluid. Masses slough off the outside and the dam flows away. Defective wasteways allowing the water to rise till it overtops the embankment are also a common cause of failure.

Very instructive was the failure of the Berthaud Reservoir,* one of the most important of those pertaining to the central canal of France. This resulted neither from defective efflux, improper construction, nor insufficient wasteway. On the 14th of April, 1829, a flood occurred which taxed the wasteway to near its full capacity, and while the water was at its highest, and but little below the top of the embankment, a violent wind carried the waves over the latter. The reflex action of the waves destroyed the pavement on the interior face, and the. water which went over wore down the earthwork till a breach was made.

A small reservoir in Massachusetts failed, some twenty-five or thirty years ago, from a cause that might operate in any such work, and should not be forgotten. A tract of woodland was flowed and the trees were killed. In the course of time a large quantity of driftwood accumulated from the trees that fell into the water. During a period of high water and high wind this was carried into the wasteway in such masses as to choke the latter and raise the water over the embankment.

The Bradfield Reservoir,† near Sheffield, England, built for the water-supply of that city, and also for water-power, had an embankment about 90 feet in height, 13 feet wide, and 1250 feet long on top, with slopes of $2\frac{1}{2}$ to 1. A puddle-wall, about 4 feet thick at top, 18 feet at the lowest ground-level, extended from top to bottom, from end to end, and 60 feet into the ground. The remainder of the embankment was earth dumped from carts without any attempt at consolidation. The efflux was by means of cast-iron pipes at the base of the embankment, about 500 feet in length, surrounded by clay puddle. The reservoir was being filled for the first time, and on the 12th of March, 1864, the water had risen to within 5 or 6 feet of full height, the contents amounting, at that time, to something over 100 million cubic feet. A slight vortex suddenly appeared in the surface of the water, indicating a leak. Workmen were

---

* See Genie Civil, vol. XXVII.   † *Ibid.*

immediately summoned to arrest it, and a messenger was sent to warn the inhabitants. So rapid was the enlargement of the opening that the workmen barely escaped with their lives and the escaping waters outstripped the messenger. In 30 minutes the entire volume of the reservoir had escaped. On one great farm lying in the track of the flood, out of 103 persons only 3 escaped. The flood reached Sheffield at midnight, without warning, and inundated all the streets. Some 70 000 cubic yards of the embankment was carried away and deposits 7 or 8 feet deep were made in the streets of the city. The number of houses destroyed in whole or in part was 798, and 238 persons lost their lives.

Messrs. Rawlinson and Beardmore, two engineers who took part in the official inquiry into this disaster, attributed it to the disposition of the efflux-pipes, or perhaps more properly speaking to the fact that pipes through the embankment were used for the efflux, which they say should not have been done in a work of this magnitude. The writer is also of opinion that in so important a reservoir the efflux-pipes should be laid in a tunnel or deep cut in the natural ground. The difference in strength between an embankment with a puddle-wall and one with a facing of puddle should be attentively considered. In the former case the down-stream half of the embankment sustains the entire pressure of the water in addition to the pressure of the up-stream half. In the latter case the entire strength of the embankment is opposed to the pressure of the water. In addition to this the pressure in the former case acts horizontally; in the latter, at right angles to the slope. In an unconsolidated embankment like the above a well-jointed diaphragm of heavy timber would be preferable to a puddle-wall.

An earth-bank failure occasioning much perplexity to engineers occurred at Portland, Me.* This was the service-reservoir of the Portland Water Company, a comparatively unimportant work, partly in excavation and partly in embank-

---

* *Engineering News*, N. Y., vol. XXX, 1893.

ment, the extreme height of the latter being some 45 feet. The reservoir held some 40 feet depth of water, about 20 million gallons. The embankment was built with slopes of 1½ to 1. The inner slopes were faced with puddle of the extraordinary thickness of 6 feet surmounted by a thin layer of broken stone, and the latter by a substantial block pavement. The embankment rose 5 feet above the water-level, and there had a thickness of 10 feet including the puddle, which, above the water-level, occupied the whole thickness of the bank. The work rested upon a very refractory clay gravel, and the embankments had been formed of material obtained from the excavation, deposited in thin layers and very thoroughly rolled. At one angle of the reservoir was the drain-pipe, 8 inches diameter, at the level of the bottom, and imbedded some 18 feet in the natural ground. Also the overflow-pipe leading from the water-level on the inside to the drain-pipe level on the outside. Both these pipes lay in the diagonal line of the junction of the two embankments. The reservoir was built in 1889 and had been in use till the time of the accident. On the 6th of August, 1893, at 5.30 in the morning, some persons, happening to be early astir, noticed a considerable stream of water issuing from the ground at the above-mentioned angle. Instant alarm was given to the endangered houses, but before all the people could be made to understand their danger the whole volume of the water was in motion and several persons were drowned.

The only rational theory of this break that could be framed was that the pressure of the water on the two embankments tended to separate them on the exact diagonal line of the two pipes; that consequently the earth in this plane had less coherence and offered less resistance to the passage of water than elsewhere. This fact, combined with the natural tendency of the water to follow the pipes, probably occasioned the disaster. The superiority of natural methods of consolidating earth over artificial methods was clearly seen here in the fact that, though the current was able to cut a gap 30 feet wide through the embankment in a few minutes, it made but slight impression on the natural gravel.

A conspicuous instance of failure resulting clearly from improper construction was that of the Mill River Dam,* at Williamsburg, Hampshire County, Mass., on a branch of Mill River, a small stream entering the Connecticut at Northampton. It was built in 1865 by the Williamsburg Reservoir Company, a corporation chartered for the purpose, the stock in which was held by manufacturers whose mills are located on the stream below, the reservoir being built for the purpose of giving them a larger supply of water during the dry season.

The dam was 43 feet extreme height, 500 or 600 feet long on top, with a wasteway 33 feet wide in natural ground. It was an earth embankment with slopes of $1\frac{1}{2}$ on 1. It had a wall of rubble masonry through the centre longitudinally, 2 feet thick at top and widening downward, at the rate of 1 inch per foot. The discharge was through an iron pipe laid in the lowest part of the ravine. The site of the work was a glacial gravel of very firm texture overlaid by a few feet of porous gravel. No engineer or person so designated was employed on the work. A committee of the manufacturers entered into a contract for its execution, and an occasional visit of some one of the members was the only inspection maintained. Gross defects of workmanship were disclosed by the accident. The wall was not laid in mortar, but built up dry and grouted. The cavities were not filled and the mortar was not good. The loose gravel was not removed from the central part of the site, and the wall did not extend to the impermeable gravel. No proper means were used for consolidating the embankment.

On the 16th of May, 1874, between 7 and 8 A.M., the water being 4 feet below the top of the dam, masses of earth were observed to slide from the outer slope of the embankment. A milkman, on his morning round, stripped the harness from his horse, mounted, and started at full speed down the valley,

---

* Report of Committee in Trans. Am. Soc. C. E., 1874.

giving warning to the inhabitants and the mill people, which doubtless saved many lives. In 20 minutes three-fourths of the reservoir contents, then amounting to about 100 million cubic feet, were discharged, drowning 143 persons and destroying property to the amount of about a million dollars.

This was a very conspicuous instance of failure from improper construction, and the only wonder is that the dam stood so long. The immediate cause of the disaster was the attainment of that state of complete saturation which reduced the earth to a semifluid condition as shown by the sloughing off of masses of earth. Why this condition was deferred so long is not easy to say. The water had many times stood higher than on the day of the disaster. Ordinarily a bank destined to fail from this cause fails on the first trial. If it survives that trial, causes are constantly in operation which make it safer, viz., natural settlement and tightening by sediment borne in the water. The impounded stream, however, was of remarkable clearness, bearing no appreciable sediment.

The South Fork Dam.* —The crowning reservoir disaster of the nineteenth century was the failure of the South Fork Dam, in Cambria County, Pennsylvania, on the headwaters of the Conemaugh, a tributary of the Allegheny. This reservoir covered, at ordinary high-water level, about 407 acres; at 5 feet above, 457. It had a drainage-area of 48.6 square miles, extending to the summit of the Alleghenies, the surface of the full reservoir being about 1600 feet above the tide-level. On the same stream some 3 miles distant is the important city of Johnstown, containing at the date of the accident something over 20 000 people. The natural declivity of the stream from the outlet of the reservoir to Johnstown is something over 400 feet. This reservoir was built by the State of Pennsylvania as a feeder to its canal system, being completed in 1852 or early in 1853, and it was used for that purpose till 1857. It was formed by an embankment across a narrow ravine, with a

---

* See Paper 477, vol. XXIV, Trans. Am. Soc. C. E.

wasteway in the natural rock, around the end of the dam. The dam was about 70 feet in extreme height; the inner half was of earth with a slope of 2 to 1, the outer of loose heavy stone with a slope of 1½. The discharge was through a brick culvert founded upon the rock, reaching to within 65 feet of a gate-chamber at the toe of the embankment, and communicating with the same by five 24-inch cast-iron pipes. From the gate-chamber arose a tower communicating with the embankment by a bridge and containing the mechanism for operating the sluices. The inner slope and crown were covered with a pavement 15 or 18 inches thick resting on a layer of stone fragments. The earthwork was of suitable material very thoroughly compacted, and the whole work was faithfully executed.

In 1857 the reservoir, together with the canal to which it pertained, was transferred to the Pennsylvania Railroad Company, and continued to be used by that company. Shortly after the transfer, leaks occurred where the cast-iron pipes entered the culvert, and in July, 1862, there was a serious break at this point which washed out the upper end of the culvert and part of the embankment. The reservoir being only partly filled at this time, the contents were discharged through this breach with no serious damage. The abandonment of the canal being then under contemplation, the company did not repair the embankment, and it remained for a long time in this condition. The reservoir went out of the company's possession in 1875, and in 1880 the property, consisting of 500 acres of land covering the reservoir-site, came into possession of an organization called the South Fork Hunting and Fishing Club of Pittsburgh, Pa. This club discontinued the outlet-sluices, removed the pipes, gates, and tower, and maintained the reservoir at a uniform level, discharging the surplus water through the wasteway.

The original specifications of the dam provided for two wasteways of the aggregate width of 150 feet, or for a single wasteway of equal width. This provision was not carried out.

The wasteway existing at the date of the disaster was about 120 feet wide at the upper end, 176 feet long, and 69 feet wide at the lower end, being a channel cut in the rock with the bottom level. The specifications also required the crest of the dam to be 10 feet wide and 10 feet above the bottom of the wasteway. The width was adhered to, but in adapting the dam to a roadway the height was reduced to not over 8 feet above the wasteway. The latter was, moreover, obstructed by the supports of a bridge appertinent to the roadway and by screens to prevent the escape of fish. In other words, the wasteway was only half that originally contemplated and had been encroached on in three ways: by closing the sluices, by lowering the embankment, and by the said obstructions. The owners established a clubhouse at the dam with keeper and attendants, and used the property as a place of recreation. Such was the condition on the 30th day of May, 1889.

On the 30th and 31st of May and the 1st of June a rainfall of extraordinary severity occurred in this part of Pennsylvania, extending into Virginia, West Virginia, and Maryland, over an area of some 20 000 square miles. The Potomac rose to an unprecedented height, being 16 feet deep on the U. S. dam at the head of Great Falls on June 2d. The very able committee of the American Society of Civil Engineers * who investigated this disaster collected all available records of rainfall upon the entire storm-area, embracing about fifty stations. No station existed on the South Fork drainage-area, but the record of contiguous stations indicated 6 or 8 inches on that area. It appeared that the rain must have fallen at times during the night of May 30–31 at the rate of $\frac{3}{4}$ inch an hour. The declivities of the drainage-area were very steep and shed water into the streams with great rapidity. On the morning of the 31st the influent stream was a roaring torrent and the water was rapidly rising on the wasteway. Workmen were collected and efforts made to open a passage for the water

---

* Trans. Am. Soc. C. E., vol. XXIV. p. 421.

around the opposite end of the dam from the existing wasteway, and a channel was formed which the water rapidly widened and soon discharged a large volume. A slight breast, was also raised on the dam by plough and shovel which retarded the disaster. By 11.30 A.M. the water was flowing over the entire length of the dam, and at 3 P.M. a breach occurred which emptied the reservoir in the course of 45 minutes.

The city of Johnstown was already inundated to a large extent by the unprecedented flood, some of the streets being 10 feet under water when the break occurred, which made it impossible for the inhabitants to escape with readiness even had timely warning been given. When the reservoir flood arrived, bringing houses, bridges, trestles, and other wreckage with many people clinging to it, an enormous mass of débris lodged against a massive arched railroad bridge which spanned the stream and still further raised the water. This great raft, constantly augmented by the arrival of entire houses with their inmates, took fire and burned amid scenes of terror and desperation which those who witnessed did not readily forget.

That this disaster was due to an insufficient wasteway is entirely clear. The committee estimate the maximum influx, which did not occur till after the disaster, at 10 000 cubic feet per second. The wasteway could not have discharged a quantity exceeding what would be indicated by formula (28), with $w = 70$ and $h = 8$, viz., $3.087 \times 70 \times 8 \sqrt{8} = 4889$ cubic feet per second. The execution and maintenance of the work according to the original specification, viz., 150 feet width of wasteway 10 feet below the crest of the dam, would in concert with the sluices undoubtedly have saved the work.

A failure from an insufficient wasteway is more disastrous than from any other cause, the quantity of water liberated in that case being the utmost capacity of the reservoir.

Although this disaster chiefly emphasizes the necessity for an ample wasteway, the case is no exception to the general statement that the sluices are the critical point of an earth embankment, the dam having once before failed at that point.

The most reliable estimate places the number of lives lost in this disaster at 2142, and the property loss at between three and four millions of dollars.

High masonry dams are not exposed to so many causes of failure as earth embankments, and failures of the former are much more rare than those of the latter class. Such failures may always be ascribed to one of three causes: (1) inadequate dimensions; (2) defective foundations; (3) improper materials or workmanship.

The Puentes Dam in Spain.*—Fig. 108 is an example of a failure from defective foundations. This dam was built, toward

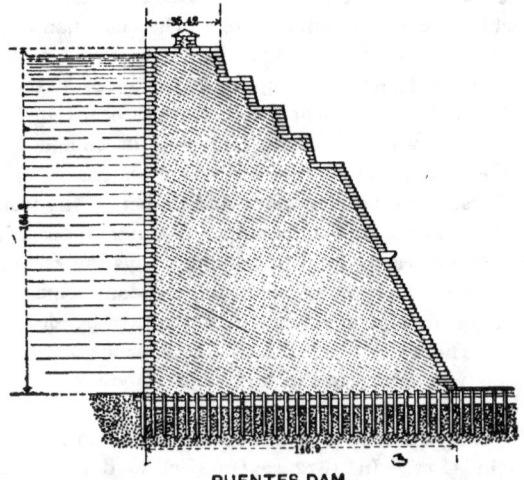

PUENTES DAM

FIG. 108.

the close of the 18th century, for purposes of irrigation. It closed a valley on a branch of the Segura River in the province of Murcia, which contained in the lower part of its cross-section a deposit of earth, while rock appeared at a higher level. Instead of excavating to the bed-rock throughout, it was

---

* Genie Civil, vol. XXVII.

decided to rest the central part of the dam on piles, while the flanks rested on rock. The piles penetrated about 22 feet, but in the centre did not reach the rock. The section of the dam, as will appear from the cut, was abundantly, in fact unnecessarily, heavy. In the centre of the dam was a well for operating the sluice-gates. It was intended to hold the water to a height of about 160 feet, but the dam had stood for eleven years without being filled to a height of more than 100 feet, and during this time sediment had accumulated to a depth of over 40 feet at the outlet. In April, 1802, heavy rains nearly filled the reservoir. About half-past two P.M. of April 30th volumes of water were observed to issue near the wooden platform which received the discharge. The custodian of the dam was sent for, and a messenger was despatched to warn the inhabitants of Lorca, a village of importance on the stream. About 3 o'clock a violent explosion was heard in the gate-well, and the water suddenly increased in volume. Soon a second explosion shook the earth, and an enormous mass of water burst forth, bearing piles and timber of the foundation. A third explosion and a mountain of water appeared, "of a color red like fire frightful to behold," says the record. The reservoir was empty in an hour, and the dam remained in the form of an arch of some 200 feet span, suspended over the chasm. The water reached Lorca in advance of the messenger, who was driven to the hills, and some six hundred persons were drowned.

Fig. 109 is a section of the Habra Dam * in Algiers, completed in 1871. In 1872 several feet in depth of the weir yielded to the pressure of the water, allowing a great quantity to escape, some 200 000 cubic feet per second, which did little damage. December 16, 1881, a flood went 8 feet deep over the weir, and the unexpected pressure caused a portion 360 feet long, 50 feet deep, to separate from the structure, liberating a volume of water which drowned some four hundred

* Genie Civil, vol. XXVII. p. 268.

people, although warned of their danger. It is stated that the
stone employed was not homogeneous, that the sand was too

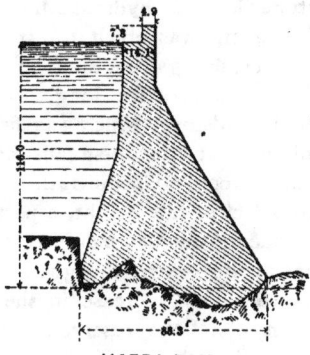

HABRA DAM

FIG. 109.

fine, and that the lime, made on the ground, lacked hydraulic
activity.

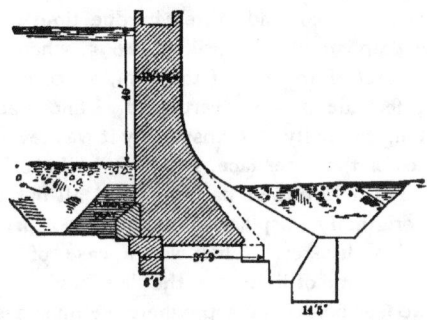

FIG. 110.

The Bouzey Dam,* Fig. 110, was situated near Epinal in
the valley of the Avière in France, being intended for the
supply of the Eastern Canal (Canal de l'est). It was com-

---

* Genie Civil, vol. XXVII. p. 281.

menced about 1878, commenced to fill in November, 1881, with a capacity of some 250 million cubic feet. It was built of dressed sandstone laid in hydraulic-lime mortar. The specific gravity of the stone was about 2. It was a sandstone conglomerate of the crushing strength of 295 to 550 tons per square foot. The dam was 1700 feet long, resting on a formation of the new red sandstone, which was fissured and permeable. A guard-wall, at the up-stream face of the dam, reached down to solid rock. On the 14th of March, 1884, * water being within 8 feet of full height, a portion of the dam 444 feet in length suddenly assumed a curved form, being some 15 inches out of line at the centre, though still remaining vertical, and there was a sudden increase in the flow of springs below the dam. In 1885 the reservoir was emptied and examinations were made. It was found that the dam had separated from the guard-wall. Repairs and additions were made at the base of the dam as indicated in the figure, in which the dotted lines represent the original work. The base was greatly widened and carried down to rock as shown. In 1889 the sluices were closed and water in due time rose to full height. The dam stood till April 27, 1895, when, the water being within 2 feet of the top of the dam, a portion 594 feet long and 34 feet deep was overturned. The fracture was nearly level longitudinally. Transversely it was level to within some 13 inches of the outer face, where it broke off when the dam tilted. The overturned part included all but 90 feet of the portion originally displaced. About 100 people were drowned by this disaster. This was a case of dimensions carried to the extreme of lightness, the dam having but 18 feet thickness at 40 feet below the top, where we must assume that the water was liable to stand. Formula (20), taking 2 for the specific gravity, would call for a thickness of 28 feet. It is probable that the dam would have stood had the bottom been properly secured at first.

<hr />

* Inst. Civil Engineers, vol. cxxv. p. 461.

Although a wedge 13 inches wide broke off the outer edge of the dam, it cannot be affirmed that the accident resulted from the crushing of the stone. The latter was merely an incident of the overturning. When the entire weight of the detached portion came upon the outer edge of the standing part such a result was unavoidable.

This accident suggests a way not ordinarily foreseen in which such a dam may fail. We will suppose that in the course of construction, when the work is at a certain uniform height, the contractor practises some familiar trick, and lays an entire course of the work with worthless cement, or that the work is stopped for a season, and, when resumed, the succeeding course is not bedded in mortar. To give the exposed part of the work a good appearance he carefully fills and points the down-stream joints with good cement. The water, on filling of the reservoir, enters the defective joint freely, but is stopped on the down-stream edge. The consequence is that the superincumbent part is exposed not only to the horizontal pressure on the up-stream side, but to an upward pressure on the bottom, more than doubling the destructive forces taken account of in designing the dam.

As to the possibility of practising such a fraud, this fact is within my knowledge: A contractor building the piers of an important railroad bridge was short of cement. A neighboring sand-pit furnished sand very similar in appearance to the cement he was using. He gathered his empty cement-barrels, took them in the night to the pit, filled, headed, and returned them to the work, and, mixing the contents with water and ordinary sand, put it into the work unquestioned by the inspectors.

Dam Across the Colorado River at Austin, Texas.—The failure of this dam having occurred in April, 1900, as this writing * was drawing to a close, some notice of this event appears proper, especially as the writer was for a time connected with this work as engineer. Austin is the capital of the State of Texas and had, at the time this dam was commenced,

---

* First Edition.

about 15 000 people. The river and dam are described page 130, Fig. 58. The people of the town had been persuaded that a water-power could be created here greatly conducive to their prosperity, and, assuming honest and competent management, such was no doubt the fact. The work was undertaken under charge of a board of works consisting of eleven members, with the mayor as chairman. As vacancies were not filled, it was speedily reduced to seven or eight. The writer examined the situation in March, 1890, and in May of that year was appointed engineer of the board. A contract was let in the fall of 1890 for the construction of the dam and canal in accordance with the plans of the engineer, a canal being thought necessary to reach ground suitable for the pump-station, power-house, and expected industrial establishments. The winter of 1890–91 was spent in the construction of a railroad and in excavating for the canal and dam-site, the latter requiring a broad and deep cut through the bed of argillaceous gravel which filled about half the rock canyon of the stream to a depth of 40 or 50 feet. These excavations disclosed some unexpected features. The escarpment of rock against which the dam abutted on the east side rose abruptly to a height of 50 feet or more above low water, and where the up-stream face of the dam met it, consisted of hard firm rock. Near the toe of the dam a seam cropped out some 2 or 3 feet thick, of extremely pulverulent material, such as might be handled with a shovel. This was perhaps 10 feet above low water. Above this the rock was sound to the top, with this exception: the canal excavation disclosed a number of pockets which may be described as vertical holes 6 or 8 feet diameter, filled with a very shelly material through which water readily disappeared. Elsewhere the rock was firm and impermeable, puddles of rainwater standing on it for days till they evaporated. The procedure necessitated by this feature was obvious to ordinary foresight, viz., to carefully search out, excavate, and plug up these shelly cavities, and to dig out the outcropping seam for

3 or 4 feet back from the face of the rock, and close it with masonry.

In the second place the rock disclosed by the excavation for the dam was less firm than could be desired. It lay in strata, hard and soft. There seemed to be no advantage in going deeply into it. It was decided to build the dam as planned and later, after observing the action of the water, to extend the apron with a bed of concrete.

After making some progress with the masonry other alarming symptoms, though not of a physical character, disclosed themselves. The contractor had gained such a footing with the chairman and members of the board that no efficient control by the engineer was possible. No hearing could be obtained for any proposition to enforce the stipulations of the contract by law, or to annul the same, though the latter measure was strongly urged. It became apparent that a first-class or even a tolerable piece of work could not be expected. In addition to this, the mayor and chairman of the board had developed a conspicuous talent as a hydraulic engineer. Taking a part of the office force under his own personal direction, he applied himself to getting up a different plan of development, the leading feature of which was a power-house located close to the toe of the dam. Experts, so called, were summoned to confirm the chairman's views, and performed that duty with singular unanimity and promptness, receiving, in each case, a large amount of profitable employment at the city's expense, after rendering their reports. After the adoption of this project, the engineer, fully satisfied that a successful issue of the enterprise under such control was impossible, resigned his position and disclaimed all further responsibility for the work. A procession of engineers in charge of the work followed one another in rapid succession, directed and controlled by the mayor who acted under the nominal advice of one of the aforementioned experts. In finally raising the water, it found its way through one of the above-mentioned pockets, issued below the dam, and washing out the soft stratum, undermined and destroyed the head-gates and power-house, so far as the latter was completed, occasioning a loss of about $100 000. This accident wrecked the enterprise financially, as it destroyed confidence in the permanence of the works. On completion of the power-house, a stream of water of some six million gallons a day was spouting through its foundations and continued till its final destruction.

In April, 1896, the author wrote to the then mayor of Austin pointing out the danger to the dam to be apprehended from the location of the power-house: first, that the discharge of the wheels, following the toe of the dam for a distance of some 600 feet before reaching the open river, tended directly to destroy the footing of the dam; second, that it tended indirectly to the same result by preventing the measures of protection originally contemplated. He suggested inquiry and examination. No heed was given to this warning. A well-known citizen of Austin states that he, while fishing near the dam in 1897, found abrasion so far advanced that he ran his fishing-rod 8 feet under the toe. T. U. Taylor, Professor of Engineering in the State University at Austin, states that, in 1899, he was unable to reach the bottom with means at his command, though reaching to the depth of 10 feet below ordinary low water. To any one comprehending the situation such reports should have struck the ear like the cry of "Fire!" It appears that no official notice was taken of the matter. In the spring of 1899 a flood went over the dam fully equal to that in which it failed.

On the 7th of April, 1900, with 10 or 11 feet of water going over the dam, a portion of the latter, some 600 feet in length, parted from the remainder and slid down the stream, a part of it remaining in an erect position. Under the sudden rush of water the power-house collapsed and the inmates, eight in number, were drowned.

That there was risk and hazard in doing such a work by contract was fully realized; but, on the other hand, its execution by day-labor appeared beset with great difficulties, since

every workman would have had his backer in the city government, and the votes of workmen in city elections would have
been made contingent upon employment at high wages and
light work.   Weighing these difficulties and fully counting on
the efficient support of the board of works in the event of
attempted fraud, the engineer gave his adhesion to the principle
of contract work.

There are honorable contractors; men who aim to do their
work faithfully and to acquire the confidence of the community,
and who do not bid lower prices than the work can be honestly
done for.   There is another class of contractors; men trained
from their youth in deception, trickery, subornation, perjury,
and every villainous instrumentality of fraud that can suggest
itself to the mind of man.   A work of this kind controlled by
men entirely inexperienced in such matters, is very apt to fall
into the hands of a man of the latter class: one who underbids
honest contractors, in the expectation of '' beating his way ''
through the work.   In such a work calling for at least 50 000
barrels of Portland cement, the withholding of 15 or 20
thousand barrels is of itself a good profit, and it is very difficult
for the engineer to convince people of this character that such
a proceeding brings the safety of the work into question.
Assuming such men to be honest and conscientious, they
usually have great reluctance to undergo the delay incident to
the enforcement of legal remedies, and in the hands, of a
shrewd, shameless, persistent, and determined knave, will
stand much imposition before nerving themselves to such a
step.   Where these persons are not honest, but are in the
receipt of money from the contractor, either directly or in the
guise of business profits or professional fees, the case is still
more hopeless.   When bribery is confined to inspectors, it can
usually be detected by an intelligent engineer.   Their natural
demeanor changes.   They become negligent in duty, reticent
and evasive in communication.   Their conduct betrays the
influence of bribes almost as readily as it would betray the influence of drink.   But when an engineer has reason to believe

that his employers are defrauding their constituents by corrupt collusion with contractors, his only alternative is to abandon the work as he would abandon fire and pestilence.

A Reservoir Flood. —An opportunity * occurred in India, a few years ago, to deliberately observe the bursting of a reservoir and the progress of the flood down the stream. Early in September, 1893, the river Bireh Ganga was completely blocked by a landslide near Gohna in Garhwal, in latitude 30° 22' north, longitude 79° 32' east. An immense mass of earth and rock slid from the neighboring hill and filled the valley of the stream for a length of 2 miles, and, rising to a height of 800 feet above the river-bed, completely stopped the flow of the stream, with a drainage-area of 90 square miles, so that nearly a year elapsed before the water raised to run over the top. This is a mountain stream with a slope of 250 feet to the mile. The matter came in charge of the corps of engineers for India, who, although no human power could avert the final catastrophe, did what they could to ameliorate it. They erected telegraph-lines to warn the people, and placed marks to show the probable height of the flood at points on the stream. They dismantled bridges and surveyed the future lake, which when ready to overflow had a capacity of 16¼ billions of cubic feet.

From the data available, necessarily very imperfect, the engineers assigned May 15, 1894, as the probable date of the breach. After completing the survey of the lake, and becoming better acquainted with the flow of the stream, the date was changed to August 15. Telegraph-stations were established at all important points, and warnings were disseminated as to the condition of the lake, as, that the breach might be expected in a fortnight; that it might be looked for in 48 hours; that it was liable to occur at any minute, etc. Early in August the water escaped freely by percolation. The breach occurred at 11.30 P.M. on the 25th. At 4 A.M. of the 26th, the water had fallen 390 feet and more than 10 billions of cubic feet had

---

* Journal of the Society of Arts, London, March, 1896, p. 431.

escaped. The water rose 260 feet above normal level immediately below the dam, 160 feet at a distance of 13 miles, and
it reached that height at the latter point in 25 minutes after the
rupture. At a distance of 20 miles it commenced to rise in 47
minutes; at a distance of 30 miles it commenced to rise in 80
minutes, and reached the extreme height of 130 feet in 2 hours
15 minutes; 51 miles down it commenced to rise in 2 hours 30
minutes, and reached the maximum height of 140 feet in
4 hours 15 minutes; 72 miles down it commenced to rise
in 3 hours 45 minutes, reached maximum height of 42 feet in
5 hours 10 minutes; 150 miles down, commenced to rise in
9 hours 15 minutes, reached the maximum height of 11 feet in
12 hours. So complete were the arrangements that not a
single life was lost during the flood.

CHAPTER XII.

CANALS, GATES, ETC.

WHERE water-power is to be applied directly to manufacturing establishments, a canal usually leads from the dam, or from the pond created by the dam, down-stream, diverging from the shore according to the contour of the ground till a sufficient width is secured to admit of placing the mills and their appurtenances so that water can be drawn from the canal and, after passing the wheels, discharged into the river. The arrangement is substantially the same where a general power-house is adopted in connection with electrical transmission, except that, less room being required, a shorter canal is admissible. Often it occurs that, in order to secure the entire fall, the canal must have a considerable length irrespective of the requirements of the mills for room.

The dimensions of the canal are to be fixed with reference to the quantity of water to be conveyed by it and the loss of head that is considered admissible. Many considerations go to the decision of these two questions. The application to be made of the power determines the quantity of water immediately required, but the tendency always is toward a progressively increased use of water; and it is usually safe to assume that the ultimate use will be limited only by the capacity of the stream. Except in the rare instances in which reservoir capacity exists capable of reducing the flow of the stream to substantial uniformity, the use of auxiliary steam-power must be assumed. As the requirement for power increases, more water is used during the time that it can be obtained, using steam at other times. Should the requirements continue to increase without limit, we should reach a condition in which the benefit derivable from an increased use of water would not pay the cost and maintenance of the appliances for the use of it. Probably the greatest use that is ever made of water for power is such a use as does not permit it to run to waste more than four months in the average year. The ultimate capacity of the canal never need exceed what this condition would call for. It is not always necessary, however, to construct the canal with its ultimate capacity. It may be built with a view to future enlargement, although the acquisition of ground for the ultimate dimensions should in no case be deferred. The work of enlargement, while the canal is in use, is generally more expensive than the original work, but the saving of interest on the deferred part of the cost generally justifies the increased expense. Where steam-power is employed the enlargement is not inordinately expensive. This proceeding is sometimes justifiable on other grounds, for a canal running for many years with excessive dimensions usually fills up and entails large expense for cleaning.

As to the loss of head, or slope of the water-surface, this depends upon the velocity of the water, which cannot exceed the limit of safety for the bed and banks. Where the latter are in rock or are protected by side walls or pavements, the question resolves itself into one of relative expense. A high velocity implies diminished first cost and increased loss of head. Generally the limits of velocity in a water-power canal are between 2 and 4 feet per second. When the latter limit is exceeded we generally find the loss incident to diminished head outweighing the gain from diminished cost. Of late, however, there is a tendency to adopt higher velocities and to diminish the loss of head by a smooth lining.

Wasteway.—In a long canal of uniform cross-section, if the efflux is suddenly closed, the motion of the water continues till the slope of the surface is obliterated and the water comes to a uniform level throughout the entire length. If the influx is arrested at the same time as the efflux, the water will fall at the upper end of the canal as much as it rises at the lower end, viz., by one-half the total slope. If the influx is not arrested, the water throughout the entire length of the canal will rise to the level of the pond. These are the statical conditions, taking no account of the momentum of the water. The momentum of the water, that is, its tendency to continue in motion, will cause it to rise considerably higher at the lower end, and in order to avoid the necessity of raising the banks to a great height it is customary to introduce a wasteway at the lower end. This is usually controlled by flashboards which are removed when the wheels stop (see Figs. 62 and 63). Its capacity is such as to pass the entire flow of the canal without raising the water to a dangerous height. In considering the necessity for a wasteway, we must always contemplate the case of the efflux suddenly closed while the influx remains open, as such a case is always liable to occur and might lead to great damage if not provided for. The necessity for a wasteway depends upon the length of the canal, and slope and consequent velocity of the water. If its cost falls short of the cost of giving the necessary height to the banks, it is introduced; otherwise not.

Ice.—In a long canal subject to considerable fluctuations of level, ice often breaks up after forming to considerable thickness, and accumulates in the lower part of the canal in such quantities as to occasion great inconvenience if not sluiced away. A waste-weir may usually be utilized for this purpose, but an iceway is often necessary where a wasteway is not required. This consists of a passage leading from the canal to the river, controlled by flashboards, the bottom and sides well secured from abrasion, and having a platform on which men can stand to handle the flashboards and by means of pike-poles prevent the ice from gorging. After ice forms on a canal it is advisable to subject the surface to as little variation as possible.

Anchor-ice or mush-ice is often very troublesome in canals, races, and penstocks. Ice forms in water as minute crystals which rapidly unite when permitted, and form a continuous sheet. When water is in rapid motion this union cannot readily take place and the ice floats as a mushy aggregation. It appears, in this form, to have a higher specific gravity than solid ice and is very readily immersed. It fills the water to considerable depths, passes through racks, and is very troublesome in wheels. The Mississippi at St. Paul in severe winters is sometimes packed to the bottom, in parts of the channel, with anchor-ice formed in the rapids at St. Anthony. The condition favorable to the formation of anchor-ice is open water in violent motion. Where this exists not far above the intake of the canal, trouble is usually experienced. The best guaranty against anchor-ice is a large mill-pond flowing out all the rapids, together with canals of gentle slope and moderate velocity, which close readily on the advent of cold weather.

Sluices.—Every water-power canal should be provided with sluices for drawing off the water for repairs and for removal of deposits. These should be of ample size and placed at different points in the length of the canal, as the time for such operations is usually limited and cannot be wasted in waiting for the water to disappear. The time required to empty a canal or basin bounded by vertical walls through an orifice in the bottom is twice that required to discharge an equal volume of water at the initial rate of flow.

GATES.

The most universally applicable, manageable, and reliable gate is one moving straight up and down under the action of force sufficient to overcome the friction. We will first consider this class of gates, and afterwards the modifications required by special conditions. Large gates for canals and sluices are usually made of heavy plank or timber. There are two modes of uniting the plank, each of which has its advantages. Fig. 104 is the simplest method wherein the plank, cut to the right length, are placed edge to edge and fastened together by long iron rods with heads and nuts. These latter.

being sharply screwed up, unite the plank into a single piece fitting closely on its seat and free from leakage.

When a gate is closed but seldom, like the head-gates of water-wheels, or guard-gates designed to be closed only in time of high water, the seat can be cut in the stonework, as at *b*, Fig. 111. For a gate liable to frequent use, the wood sliding on stone wears too rapidly, and it slides on a metallic seat as at *c*. The wood may also be protected from wear by a metallic strip confined by countersunk bolts. A gate put

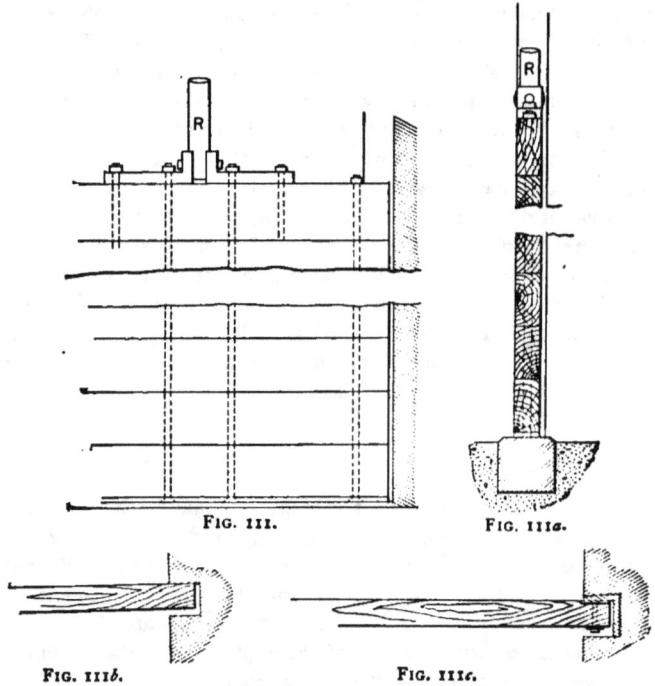

FIG. 111.   FIG. 111*a*.

FIG. 111*b*.   FIG. 111*c*.

together with long rods lends itself very readily to arrangements for lifting by hydraulic pressure, as indicated by Fig. 111, where *R* may be taken as the piston-rod of a hydraulic cylinder standing high above the gate. It is not always that sufficient room can be obtained for the hydraulic cylinder, and this arrangement is often inadmissible for other reasons. Neither is it advisable except in the case of the heaviest gates which require to be lifted against the full pressure of the water. Fig. 112 represents a common form of gearing for lifting wide gates.

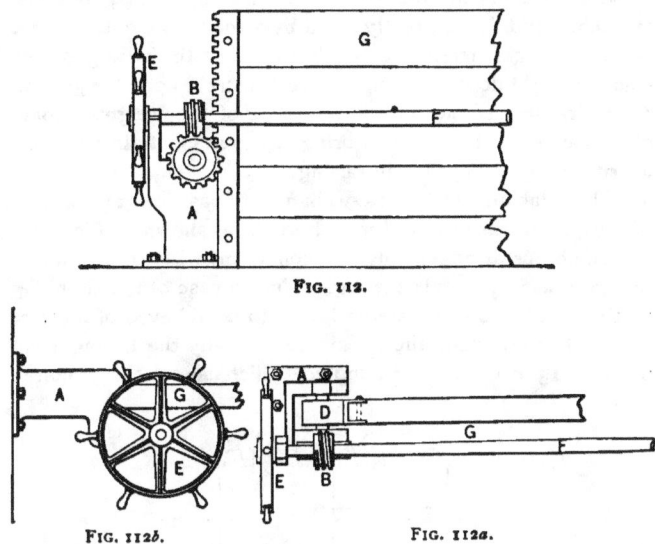

Fig. 112.

Fig. 112*b*.   Fig. 112*a*.

On each edge of the gate *G* is a toothed rack operated by a pinion *D*. The pinions are moved by worm-wheels *B* on the shaft *F*. At each end of the shaft there is a hand-wheel *E*, an arrangement which allows four men, if necessary, to work at opening the gate. The gearing is supported by two iron stands *A* bolted to the masonry. Gates of the width indicated at Fig. 112, operated in this manner, are not usually raised or lowered under full pressure. They usually close basins of small capacity like the forebays and penstocks of turbine-wheels. They are provided with wickets whereby the basin can be filled, relieving the gate wholly of pressure.   Then the only resistance to the raising of the gate is its weight.

Counterweight. — There is sometimes an advantage in attaching a counterweight to a gate, consisting of one or two heavy blocks of stone attached to the gate by chains which pass over pulleys above.   The head-gates of canals are obliged to be lifted under full pressure.   The starting of the gate is the "pinch," and this is greatly aided by a counterweight.   There is no frictional resistance to closing, and the raising of the counterweight does not bring so great a strain upon the gearing as the raising of the gate.   The size of the weights should of course be such as not to bring any greater strain upon the gearing in lowering than in raising.

The attachment of a heavy channel of cast iron to the edge of the gate reaching from top to bottom, as shown in Fig. 112, is a cumbrous arrangement, not consistent with the design of the gate, as it prevents the tightening in case of the shrinking of the plank.   For a gate designed to be relieved of pressure during its movement there is no reason why the arrangement of Fig. 113 could not be adopted, which shows the pinion of

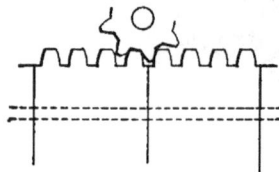

FIG. 113.

iron and rack-teeth of wood, the latter being cut in the ends of the gate-plank.   For a gate not less than 6 inches thick, made of white oak, maple, or other hard, close-grained wood, this mode of construction would not be inconsistent with a considerable pressure.

Often the hoisting-gears must be placed high above the passage or culvert to be closed by the gate.   For such a case Fig. 114 is a suitable arrangement.   Two wooden stems are united to the gate by brackets which receive the heads of the through-bolts, and to these stems the rack-segments are bolted. They are firmly united to each other by cross-pieces and braces not shown. In such a situation the gate must form a tight joint on the lintel or arch of the culvert as well as on its own seats. A gate of this construction was adopted by Mr. James B. Francis in his design for the gate-house of the Mine Run Canal at Nashua, N. H.

In the second mode of putting the gate together, the planks, instead of being confined by through-bolts, are bolted to one or two upright stems. Fig. 115 shows a small gate of

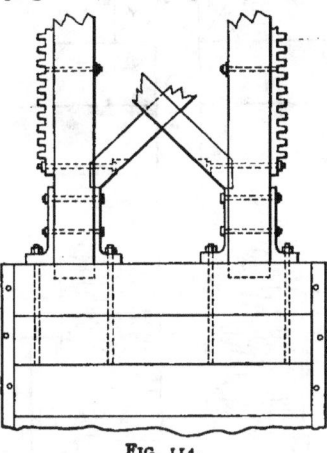

FIG. 114.

this construction, designed to control an opening through a bulkhead wall. Two beams fastened by anchor-bolts project over the wall, forming a "cat-head" which sustains the hoisting-gear. This consists of a shaft resting in bearings on the beams and sustaining between the latter a spur-gear which meshes with a toothed segment on the gate-stem. This passes between the beams, and its plain face bears against a roller. One end of the shaft sustains a capstan-wheel to be operated by a lever; the other a ratchet-wheel with a pawl. The gate slides on a framework of timber built into the masonry around

the opening.   The gate is shown with a projecting lip or sill
to limit its downward movement, and guides to hold it laterally,
only one of which appears.   These parts are sometimes dis-

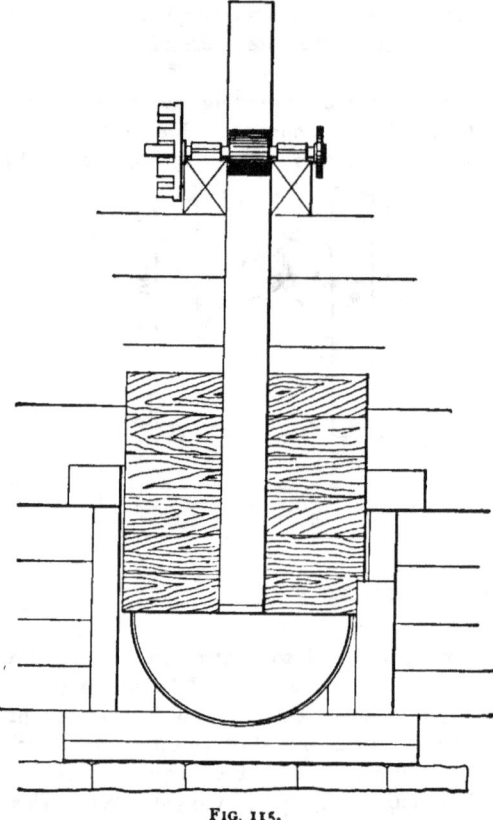

FIG. 115.

pensed with, but their omission is liable to lead to accidents and
derangement.   For instance, when the gate is down it is held
firmly by the pressure of the water, and the pawl is liable to

be out of place. In this condition, if the water is drawn off from the basin supplying the opening, the gate is liable to drop out of its connections. In this construction the planks are

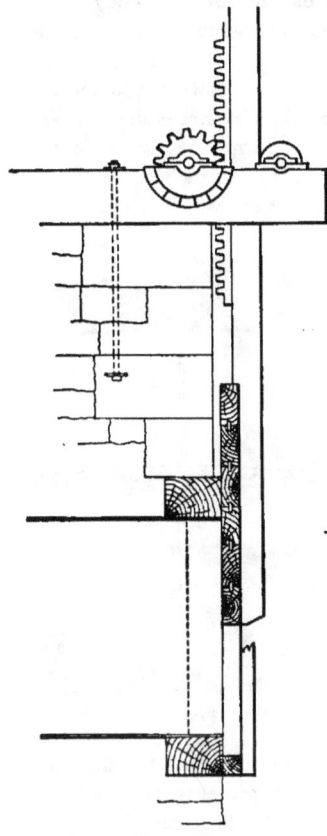

FIG. 116.

tongued and grooved, but this does not prevent leakage at the ends of the plank should there be any shrinkage. Head-gates sometimes stand open for weeks and months and are liable to

shrink.   When closed down in case of accident a slight leakage
is objectionable.   The toothed rack cannot, in this case, be
confined to the stem by screw-bolts unless the nuts are deeply
countersunk to avoid the roller.   They are shown as confined
by heavy split-head lag-screws which are almost as efficient as
screw-bolts.

Gates working up and down are sometimes objectionable for
extraneous reasons.   The necessary head-room is not always
obtainable, and they sometimes are objectionable as masking
windows and obstructing light when raised, as in the case of head-
gates they necessarily are while works are running.   In such

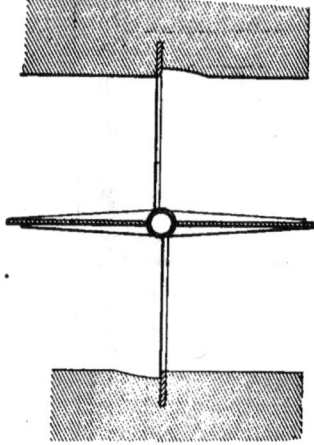

FIG. 117.

case, the best substitute is a gate turning on a vertical spindle.
Such a gate when open presents its edge to the current, and ob-
structs the opening by the diameter of the spindle.   It is operated
with less force than the sliding gate, but is generally liable to
greater leakage.   This form of gate, necessarily made of iron, is
shown in horizontal outline at Fig. 117, in vertical elevation at
Fig. 118.   It consists of a plate united to a strong hollow spindle

and stiffened by ribs. A plate extends across the culvert at the bottom and is confined by being built into the concrete floor. It contains the socket or seat for the foot of the spindle. A similar plate spans the culvert at the top, containing an eye for the passage of the spindle, and a broad flange or web built into the masonry of the arch-covering. These two plates carry the projecting lips or seats against which the web of the gate closes. They are also formed to unite with two vertical plates which extend up the sides of the culvert and are built into the masonry of the side walls. The entire frame could be made

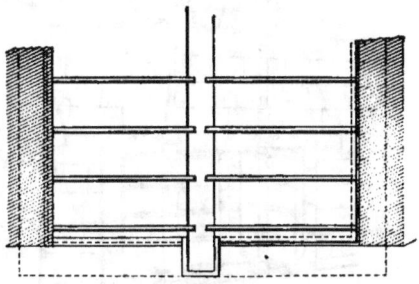

FIG. 118.

in one piece, but this is usually hazardous on account of the liability of such work to break in handling. For the highest degree of accuracy, all the surfaces which close together are faced with brass, which is trimmed and scraped to a close fit after the work is set up. The spindle reaches above the top of the masonry and is fitted with a worm-gear for operating the gate. For wider openings these gates are often arranged in pairs, each having its own spindle and worm-gear and both worked by the same hand-wheel. Fig. 119 shows a pair of such gates with separate hand-wheels. The only resistance to movement offered by these gates being the friction on the spindle, it is never worth while to adapt them to working by power. Such a gate is very liable to obstruction from gravel and small stones which get pinched between the closing sur-

faces. Such stones commonly gain access to the culvert by
rolling on the bottom, and may be excluded by forming a
pocket in the bottom of the culvert into which the stones fall
and from which they may be removed from time to time.

Gates working under great pressure and gates liable to
frequent closing are often arranged so as not to slide on their

Fig. 119.

closing surfaces.   The latter are inclined to the line of motion
of the gate, and are in contact only when the gate is closed.

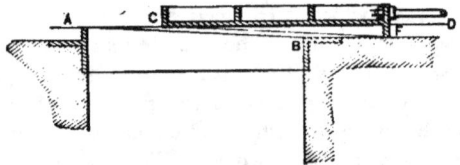

Fig. 120.

This arrangement is indicated by Fig. 120, in which the gate
slides upon the guide *AD*. The closing surfaces are *AB* and
CF, which do not touch each other except when the gate is
closed.   In this disposition there is no wear upon the contact

surfaces, and these can be adjusted with all possible nicety, without liability to derangement from use. Gates in pipes where the pressure is liable to be alternately in opposite directions have two closing surfaces and fit as a plug or wedge between two inclined seats.

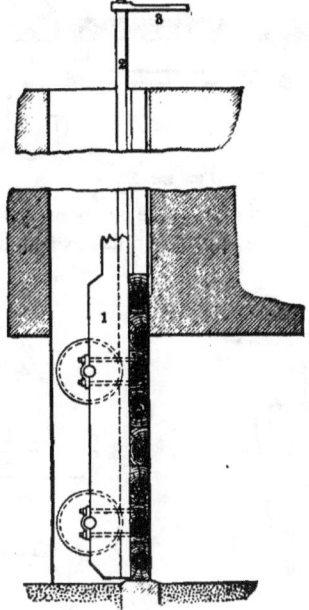

FIG. 121.

Figs. 121 to 123 show a form of gate running on wheels, with several devices for throwing the pressure on the wheel when in motion and on the seat when at rest. In these forms, instead of uniting the gate-planks by vertical rods it is more convenient to bolt them to vertical timbers, 1, Fig. 121. These timbers usually extend upward and are provided with toothed racks or other organs by which the gate receives motion.

In Figs. 121, 121*a*, the wheels are grooved, and the track on which they run is a rod of elliptical cross-section, the long diameter

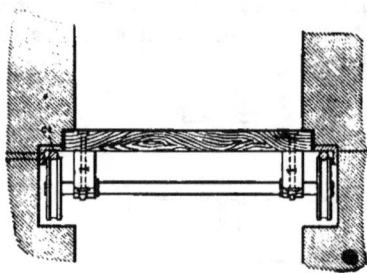

FIG. 121*a*.

of the ellipse exceeding the short one by about $\frac{1}{8}$ of an inch. When the long diameter of the rod is parallel to the face of the gate,

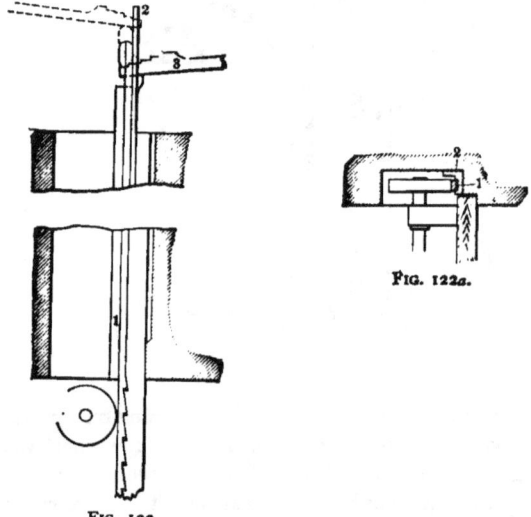

FIG. 122*a*.

FIG. 122.

the latter rests on its seat. When perpendicular thereto, the pressure of the water is borne by the wheel. Before raising the gate, a long wrench, 3, Fig. 121, is applied to the rod 2, and the latter is turned so as to throw the pressure upon the wheels, in which condition the gate can be raised with slight effort. A reverse movement of the wrench throws the gate back upon its seat, where the pressure of the water holds it. Usually it is not necessary to throw the pressure upon the seat while the gate is up. When the gate is down this is necessary in order to avoid leakage. It may not always occur that there is room to manipulate the long wrench required in the preceding disposition. Figs. 122, 122*a* show a different mode of shifting the pressure, involving a vertical instead of a rotary movement of the track-bar. The wheels, in

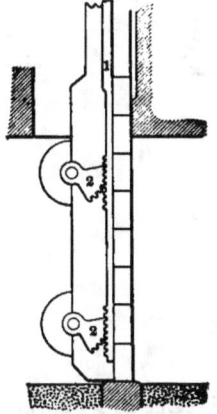

FIG. 123.

this case, are plain and they run upon the plain face of a rod, 1, of rectangular cross-section. The back of the rod is formed in a series of inclined planes and rests against a seat similarly shaped, so that a rectilinear movement of the rod 1 lifts the gate off its seat. The seat or bed-plate 2 against which the rod 1 rests is extended upward and has a long slot through which the lever 3 is inserted to lift the rod 1. Owing to the short "bite" of this lever a sufficient movement could not be communicated at a single traverse. This difficulty is obviated by giving a variable width to the lever, so that a number of effective traverses can be made with it. The same lever is used to depress the rod 1, and throw the pressure back on to gate-seat as indicated by the dotted lines, which will be readily understood. The rod 1 is called the wedge-bar.

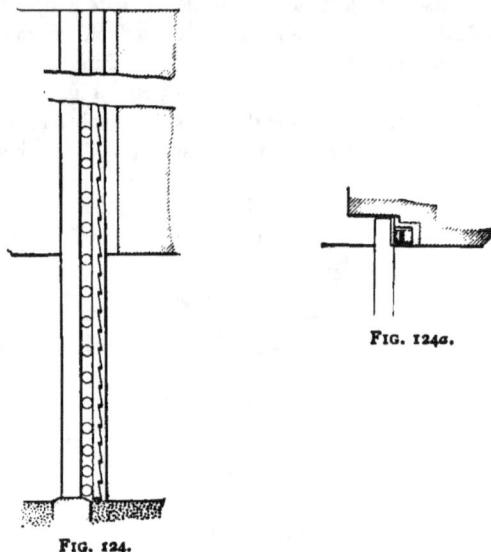

FIG. 124a.

FIG. 124.

In Fig. 123 toothed segments 2 2 are applied to the axles of the wheels, and gear with a toothed rack-rod, 1. The wheel is not fixed on the axle, and is slightly excentric with reference to the bearing of the latter. A vertical movement of the rod 1 rotates the axle and lifts the gate off its seat, in which condition it ascends readily. With the addition of a counterweight in either of the above systems, a pretty large gate may be readily moved by hand.

Figs. 124, 124a show a gate running on friction rollers, with a wedge-bar for shifting the pressure. The figures only show the principle of action, omitting details, which cannot be delineated on the small scale required for this work. The rollers are united into a train by two slender bars which are pierced to receive journals or trunnions formed on the ends of the rollers. The wedge-bar is rounded at the back and rests in a corresponding seat, so that the rollers can adjust themselves to the slight bending to which the gate is liable under heavy pressure. The wedge-bar is of steel, and the seat in which it rests is of bronze, not liable to rust.

A gate provided with this device cannot close against the crown or lintel of the culvert which it controls, but must extend above the water-level when closed.

It is obvious that none of these anti-friction devices are advisable in water carrying much floating trash.

Friction-rollers and wedge-bars were used in the great gates of the controlling works for the Chicago Drainage Channel.* These gates are 30 feet wide and sustain, when closed, a head of 16 feet or more, implying a pressure of 240 000 pounds.

Gate-houses.—The influx to an important water-power canal is usually controlled by a gate-house, consisting of a series of gates, with their hoisting-gear, enclosed in a building and covered by a roof. Fig 125 represents a gate-house designed by the author in a project for development of water-power at the Great Falls of the Potomac near Washington, D.C. The water-passages consist of twelve 6-foot sluices separated by 2-foot piers, and controlled by gates of the form shown in Fig. 111. The gates are supposed to be moved by hydraulic pressure, a cylinder, with piston and rod, standing over each gate. The gate is raised by admitting water, under pressure, below the piston and allowing it to escape above, and lowered by the reverse process. Where water, under sufficient pressure, can be obtained, as, for instance, from the water-mains of a town, this is unquestionably the simplest mode of operating heavy gates. It is even preferable in most cases where the power has to be generated by a special water-wheel at or near the gate-house, a pipe for conveying the water being so much simpler than the ordinary connection by shafting.

---

* See *Engineering News*, 1895, pp. 338, 387.

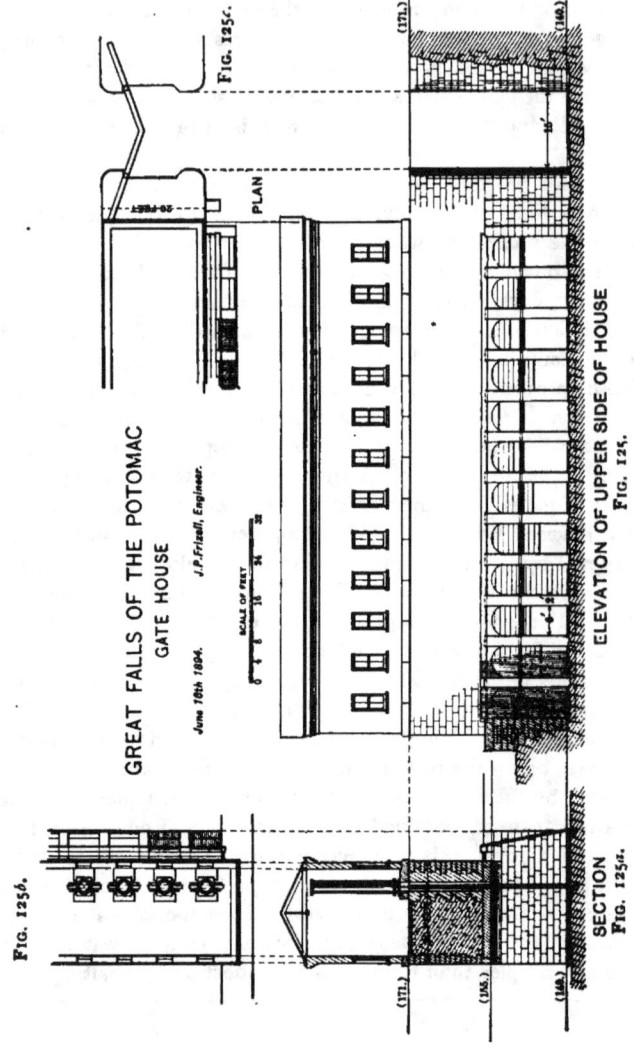

GREAT FALLS OF THE POTOMAC
GATE HOUSE

June 10th 1894.

J. P. Frizell, Engineer.

SCALE OF FEET

PLAN

FIG. 125c.

ELEVATION OF UPPER SIDE OF HOUSE
FIG. 125.

SECTION
FIG. 125a.

FIG. 125b.

Squaring-shaft.—A device called a squaring-shaft is sometimes applied to gates operated by hydraulic pressure. Its purpose is to prevent the gate from getting out of line in its movement and thus pinching or binding between its guides, which it is thought liable to do on account of the power being applied at a single point. This consists of a shaft turning in bearings attached to the masonry, and connected with the gate

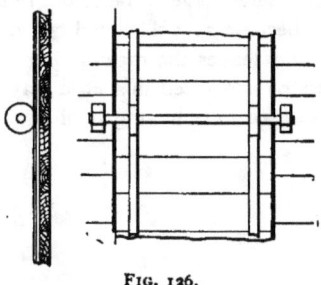

Fig. 126.

by two spur-pinions gearing with two toothed racks on the back of the gate, as indicated at Fig. 126.

Racks.—The piers supporting the gate-house, Fig. 125, are prolonged a few feet up-stream from the house, and sustain a platform and the racks for excluding floating bodies from the canal. This feature is not always introduced at gate-houses. It is in fact more commonly omitted, and floating bodies allowed freely to pass the sluices and lodge upon the racks appertinent to the different wheels. This course is pursued where the water is supplied to different lessees by a separate company or owner, who is under no legal obligation to exclude floating matters and prefers to throw that labor upon the lessees. Where the water-power is owned and used by the same party it is a question of relative economy whether the débris should be excluded at the influx or efflux of the canal. In certain seasons of the year, especially in time of falling leaves, great quantities of trash have to be raked out of the racks and sluiced back into the river. The head of the canal often offers better facilities for this latter operation than the wheel-racks. The racks consist of flat iron bars about 4 inches wide and $\frac{1}{4}$ inch thick, spaced about eight to the foot. These are usually obtainable from makers in sections containing 16 or 20 bars each. Otherwise they can be inserted in a slotted plate at the bottom and supported by a notched plate at the top and, if necessary, at intermediate points, being fastened by keys at the top. They have to be strongly supported, as the rack is liable to become so obstructed by trash that the canal becomes empty and leaves the rack acting as a dam. They must nevertheless be supported in such a way as will not interfere with the rake used in removing trash.

CHAPTER XIII.

HYDRAULIC MOTORS. WATER-WHEELS,

THE energy of falling water may be applied to the propulsion of machinery in different ways. An oscillatory motion may be secured by a lever turning on a joint at the middle and carrying a vessel at each end, the amplitude of the oscillation of the vessel corresponding to the fall of water. When the empty vessel reaches its full height it automatically opens a valve and fills with water; at the same time the full vessel reaches its lowest position and automatically opens a valve which empties it. Then the full vessel descends and the empty one rises, etc.

Water flowing through a long pipe is alternately brought to rest and set in rapid motion by the automatic action of valves. The momentum of the arrested mass causes it, while coming to rest, to exert a pressure greatly in excess of that due the head, and the expenditure of a large quantity of water under a small head is thus enabled to raise a small quantity to a great height.

By a recent invention, water from the upper pool of a mill privilege passes down through a vertical shaft sunk in the earth; turns at a right angle and pursues its course, for some distance horizontally; then ascends and discharges into the lower pool. It carries down air in the form of minute bubbles. In the horizontal part of its journey the air rises to the top of the passage, and accumulates under the full pressure due to the depth below the lower pool in a chamber, whence it can be drawn for use in driving machinery.

By far the most common mode of utilizing the energy of falling water is by causing it to act directly upon organs attached to a rotating shaft, which combination is called a water-wheel. Water acts to rotate a shaft in three ways: (1) by weight, (2) by impulse, (3) by reaction. Few wheels act wholly by either of these modes, but wheels take different forms according as one or other of these modes of action preponderates.

Water-wheels. —Although this term is properly applied to all hydraulic motors that rotate, there is a convenience in restricting it to such wheels as act mainly by the weight of the water, turn on horizontal shafts, and move with a low velocity, a velocity that has no necessary relation to the head under which the wheel works. We will here use it in this sense, reserving the term Turbine for wheels on which the water acts mainly by impulse or reaction and which move with a velocity having a definite relation to the head. These wheels are designated as overshot, undershot, or breast wheels according as they receive the water at or near the top, below the centre, or below the top and above the centre, running in the contrary direction to the overshot.

Figs. 127 and 128 represent the old-fashioned breast-wheel, very common in New England in the early part of the present century. It was not till near 1850 that it began to be replaced by the turbine. It consisted of a circular drum of a diameter nearly equal to the head, bearing on its periphery a series of vessels called buckets. The earlier forms were made wholly of timber, but later the shafts were of cast iron, together with the hubs or rosettes, as they were called, which sustain the radial arms. These rosettes were placed at intervals of about 6 feet on the shaft and this interval determines the length of the buckets. To the outer ends of the arms are attached a series of pieces corresponding to what are called "fellows" in ordinary carriage-wheels, which constitute the "shrouding" and form the ends of the buckets. The sheathing is bolted to the inside of the shrouding and forms the bottoms of the

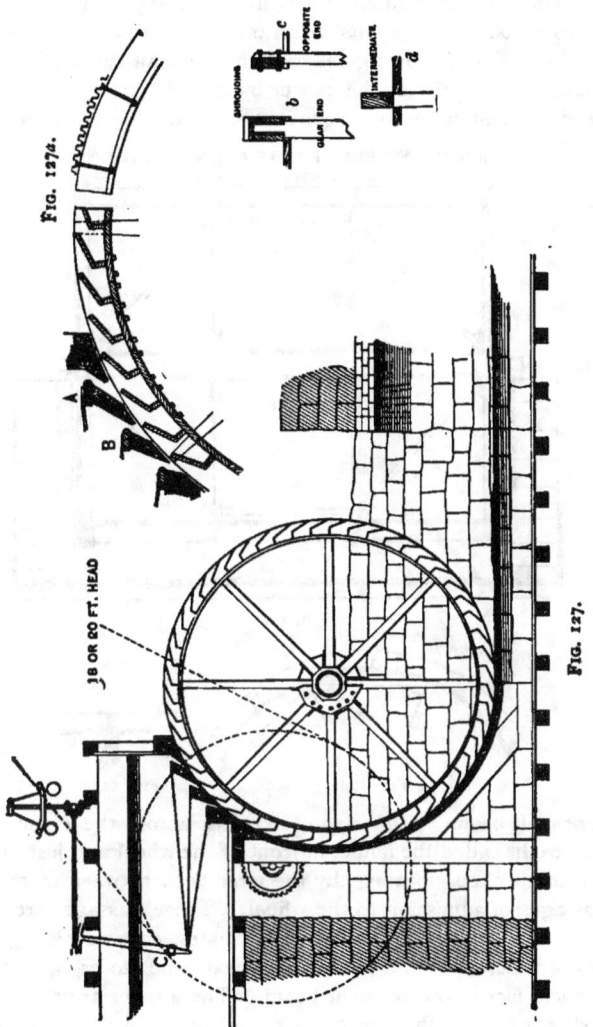

FIG. 127a.

SHROUDING

c
OPPOSITE
END

b
GEAR END

d
INTERMEDIATE

A
B

18 OR 20 FT. HEAD

C

FIG. 127.

buckets. To the end circle of the shrouding is attached the series of toothed segments which transmit the power to a small pinion. This, in turn, gives motion to a shaft carrying a large pulley which delivers the power by means of a belt, at a rate of speed suited to the requirements of the manufacture. A

BREAST WHEEL OF LATER CONSTRUCTION.
LONGITUDINAL SECTION.

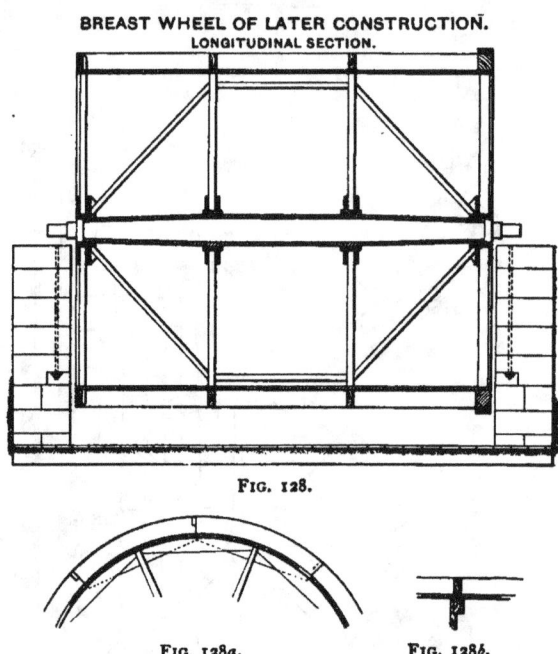

FIG. 128.

FIG. 128*a*.   FIG. 128*b*.

series of beams, Fig. 127*a*, extends across the channel of approach, called the flume, in front of the wheel and just clearing the buckets, forming, by the interstices between them, the passages of admission to the wheel. These passages are controlled by gates *A*, *B*, called wicket-gates, each consisting of a plank which slides on the top of a beam and closes against the vertical face of the adjacent beam. These gates are moved by rods connecting them with a series of arms on a rocking shaft

C which extends across the flume and receives a rocking movement from the regulator of the wheel. Below the gates is a "breast," viz., a concave cylindrical surface of planking concentric with the wheel, and as close to the same as is consistent with movement. This prevents the water from spilling out of the buckets till it has reached the lower level. Fig. 128 shows a longitudinal section of a similar wheel, and Fig. 129 indicates a different mode of admitting the water. The influx openings are controlled by rolling and unrolling a sheet of

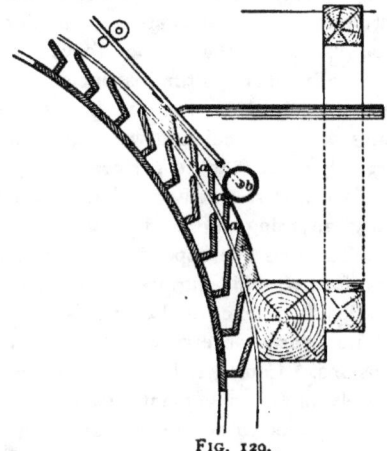

FIG. 129.

stout canvas or gutta percha, so as to cover and uncover them according to the requirements of the wheel.

We may notice some of the more obvious losses of effect incident to this form of wheel. It is often convenient in computations relative to the efficiency of water-wheels to regard the energy as represented by the head, and any loss of head as representing a proportional loss of energy. First, there is a loss at the entrance to the wheel. In a wheel acting on a head of, say, 18 feet the water enters under an average head of perhaps 4 feet and reaches the partly filled buckets with a velocity of some 16 feet per second. The periphery of the wheel moves with a velocity of 6 to 8 feet per second, say 7 feet. Here we undergo a loss of the head due to a velocity of $16 - 7 = 9$ feet per second, viz., 1.3 feet. The water escapes from the wheel with a velocity of 7 feet per second, and the energy due to this velocity is ineffective upon the wheel, involving a loss of 0.75 feet. Again, the wheel cannot be placed with its lowest point exactly at the level of low water. The water is above the point of low water for a large part of the year, and the wheel must be so placed as not to undergo too much obstruction from backwater. For this reason the wheel cannot be placed less than a foot above the lowest level of the water. Finally, there is the loss incident to the escape of water between the buckets and the breast, which, in the ordinary running of the wheel and working condition of the buckets, may readily amount to 10 per cent of the total power.

This form of wheel is so nearly obsolete that, considered as a motor, hardly anything more than a historical interest attaches to it. There is one aspect in which it often has a practical interest for hydraulic engineers. Water is often drawn under grants made fifty or one hundred years ago, entitling the user to so much water as is necessary to drive a breast-wheel of specified dimensions. Courts hold that the breast-wheel is referred to merely as a convenient means of defining the quantity of water intended to be granted, and that this quantity may be used on any motor whatever. It therefore becomes necessary to secure a judicial definition of the quantity of water intended to be granted, expressed in cubic feet per second. Such proceedings necessitate intelligent inquiries as to the construction, operation, and efficiency of the particular form of wheel referred to in the grant.

Fig. 130 and details represent an old form of undershot wheel, made entirely of wood,—such a wheel as used to be common in rural grain-mills of European countries. It is placed outside the mill, and a shaft extends through the wall, communicating motion to the machinery. The wheel is shown as working on a head of about 6 feet, and a wheel of this size would use not over 10 cubic feet per second. The water passes the wheel in a rectangular sluice concentric with the wheel, in which the floats revolve with as little clearance as is consistent with movement. The design is interesting as

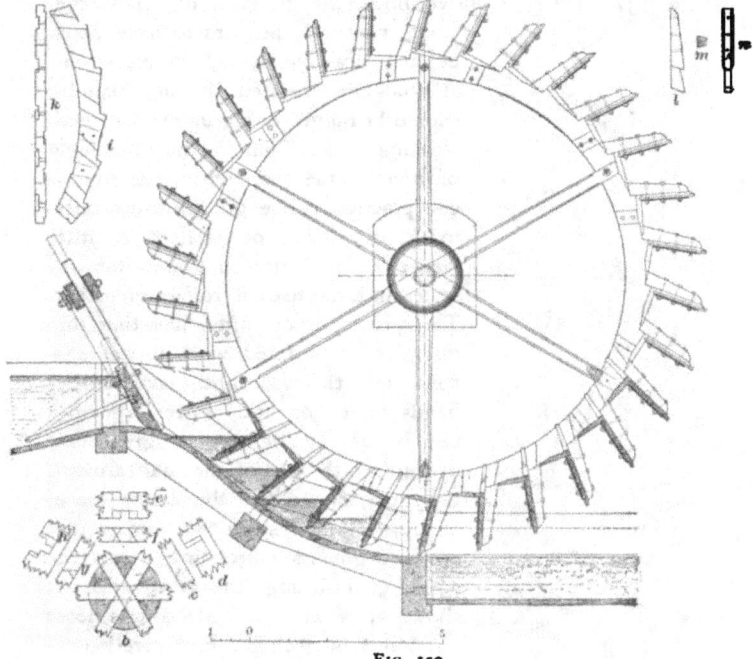

FIG. 130.

showing the methods of construction adopted in these old wheels. The shaft is a round log some 30 inches in diameter— not that such a diameter is necessary for strength, but that it is convenient for securing the arms of the wheel as well as those of the great wooden gear which drives the stones. The details *b*, *c*, *d*, *e*, *f*, *g*, *h* show the mode of inserting and confining the arms in the shaft.   Three elongated slots or mortises are cut through the shaft in which the arms are inserted, and locked together as indicated.   The first is only wide enough, length-wise of the shaft, to insert the arm, the second wider, the third wider still.   When the arms are inserted and locked together, the widest of the vacant spaces is filled with a wooden wedge. Fig. 131 shows, in dotted lines, the gudgeon on which the

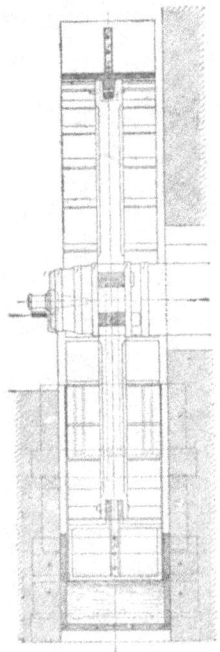

wheel runs.   It appears to have been pointed or wedge-shaped and driven in. A gudgeon inserted in this manner should be ragged or corrugated to avoid working loose.   This was not the mode of inserting the gudgeon in the Ameri-can practice.   The gudgeon was pris-matic in shape, or perhaps a little larger at the inner end, and was in-serted in a mortise cut from the outside. The vacant space, after inserting the gudgeon, was filled with a packing-piece, and this was tightly confined by bands driven on hot.   After applying the bands, small iron wedges were driven into the end of the shaft around the gudgeon to fix the latter more firmly.   *i* and *k* show the form of the crown segments (shrouding), and the mode of inserting the float-arms is shown at *m, n, l.*   The feed-gate does not move in grooves, but merely ex-tends from side to side of the channel, and is held against the pressure of the

FIG. 131.

water by two jointed rods, which allow it to be raised and low-ered with very little friction, though admitting of some leak-age.   Such a gate is tightened by cutting a groove in the end and inserting a piece of stiff rubber pipe which bears against the wall of the sluice.   The water should be admitted to such a wheel in an inclined direction, the downward component of its motion being equal to the velocity of the extremity of the float. This wheel is not subject to the same loss of head at the influx as the breast-wheel, but the leakage is somewhat greater in proportion, since it extends all around the perimeter of the float.

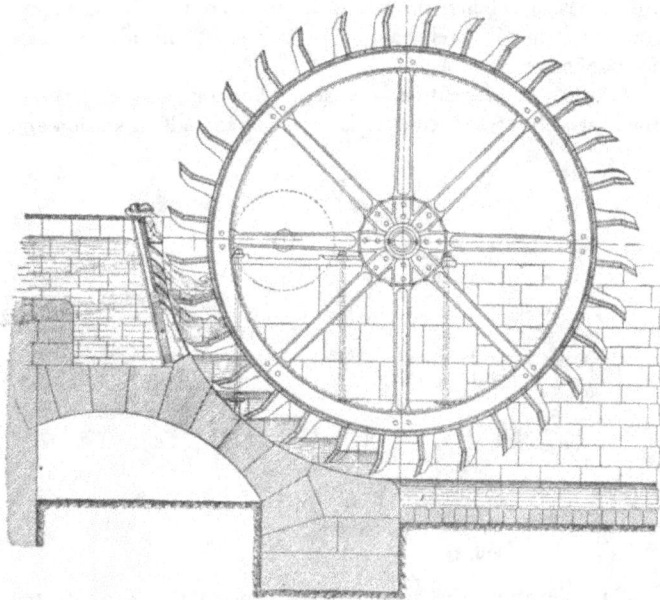

FIG. 132.

Fig. 132 shows a different form of feed-gate (often called the speed-gate) with inlet openings suited to a head of 10 feet or thereabouts. The sluice is of masonry, and the wheel and its attachments are of more permanent construction than in the last case. The shaft, rosettes, arms, and crown segments are of iron together with the float-arms; the sheathing and floats being of wood. It will be noticed that the water enters the wheel in an inclined direction, the vertical component of its

motion being about equal to or a little greater than the velocity
of the tip of the float.   Should the water enter the wheel horizontally, every float would strike the stream with great force,
accompanied by noise, violent strain on the floats, and loss of
useful effect.   The gate in this case slides in a groove, closes
the passages by rising and opens them by descending, the
lower passages being the last to be uncovered.   This is a more
modern form of wheel, having an iron shaft and iron rosettes
for confining the arms.

. Fig. 133 represents an overshot wheel of a very early type,
made wholly of timber.   It is represented half in section and

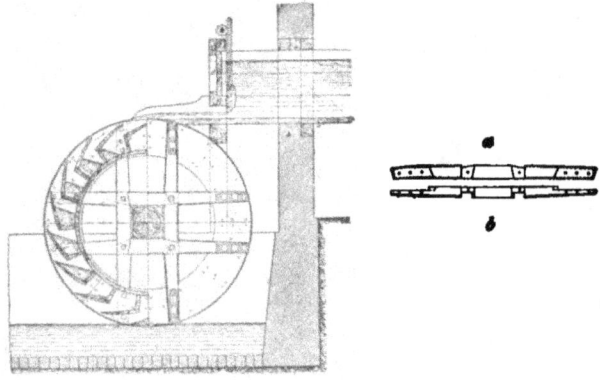

FIG. 133.

half in elevation, the spout which conducts the water to the
wheel being shown in section.   The wheel is mounted upon a
square wooden shaft which is provided with gudgeons and
banded in the ordinary manner.   The figure contemplates two
sets of arms, each set consisting of four pieces formed as at *a*
and *b*, locked together and confined to the shaft by wedges.
The shrouding, it will be noticed, is very deep, giving much
greater depth to the buckets than is required in the breast-wheel
or than is necessary to contain the water.   The wheel having
no breast to restrain the escape of water, this arrangement is

necessary in order to prevent the water from spilling out of the buckets till it has nearly reached the level of the lower pool.

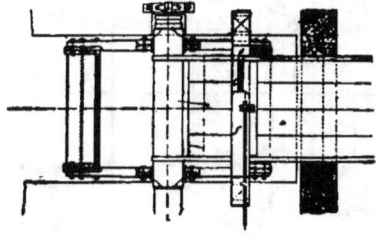

FIG. 134.

Fig. 135 shows a similar wheel of more modern construction. The shaft is of cast iron. The arms are like those of

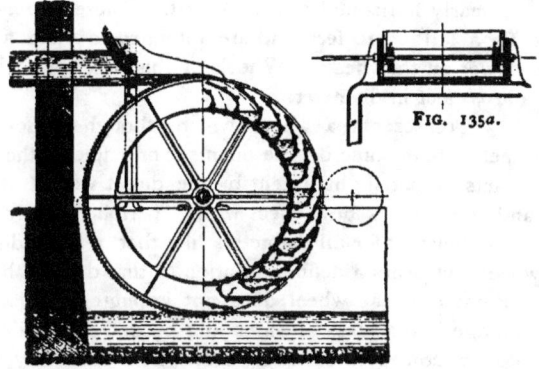

FIG. 135a.

FIG. 135.

large pulleys, each radiating from a central hub which is keyed on to the shaft. Each set of arms carries a broad web or ring which corresponds to the shrouding in a wooden wheel, and forms the ends of the buckets. One of these rings is toothed and drives a pinion. The buckets have great depth for the same reason as in the preceding case, and also for this further reason: the water enters the buckets in a nearly horizontal

direction, and, to avoid a shock at the entrance, the extremity
of the float must be nearly horizontal at the point where the

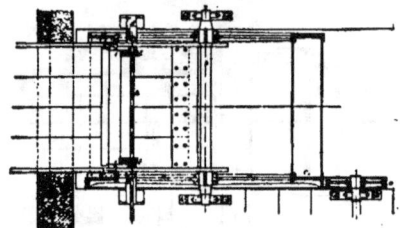

FIG. 136.

water enters.   The floats being of iron can conform to this condition better than those of Fig. 133.   They are prolonged by.
a curve till they are nearly tangent to the exterior circumference, i.e., nearly horizontal at the summit.   These wheels are
suitable for a fall of 10 feet and are intended to yield about
9 horse-power on that head.   Wheels of this description have
been made 50 feet in diameter.

Fig. 137 represents a wheel which, though in form a
water-wheel, acts in some degree upon the principle of the turbine.   It acts to but slight extent by the direct weight of the
water, and it gives its best effect with a peripheral velocity,
which is not constant for all diameters like that of the ordinary
water-wheel, but bears a definite relation to that due the head.
The water entering the wheel does not impinge upon a flat
vane or plunge into a mass of water in a bucket.   It glides up
a curved vane, comes to rest, and then glides smoothly back
and escapes.   It avoids the loss at the entrance which is inseparable from all the forms thus far considered, and it avoids
in part the loss at the discharge, since the water leaves the
wheel in a direction partly contrary to that of the wheel's
motion, and the quantity of energy carried away in the discharge is not great.   The wheel is shown as made of wood,
though such a wheel, if made now, would probably be made
of iron.   The shaft, arms, and shrouding are put together as

already described. The floats are inserted in grooves cut in
the shrouding and are firmly confined by through-rods which
draw the two sets of shrouding together and grip the floats.
Each float is composed of several plank, and is stayed at the
middle by an iron strap bolted to the back. The through-rods

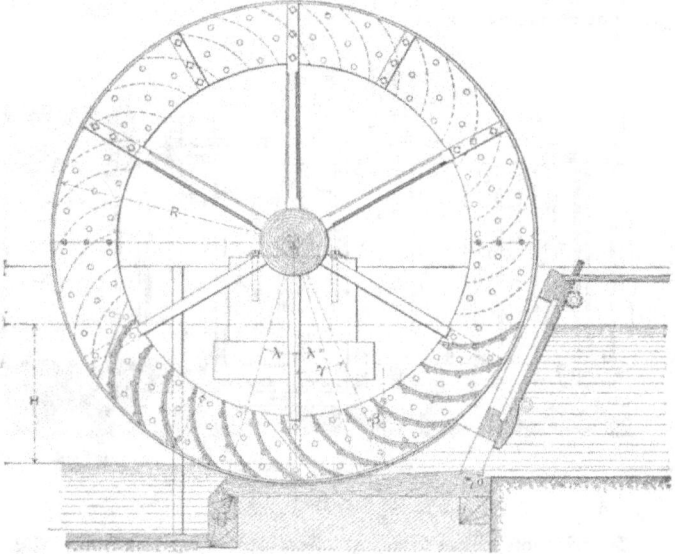

FIG. 137.

cause considerable obstruction to the water, but cannot be
avoided in a wooden wheel. In the iron wheel they can be
dispensed with. The proper position for these rods is close to
the back of the float. Situations sometimes offer in which this
wheel may, under existing conditions, be judiciously applied,
viz., a head not exceeding 6 feet, and an application not calling for a high velocity, as pumping, grinding, or the work
incident to a powder-mill. The best speed for the outer ends
of the floats is about 55 per cent of that with which the water
enters the wheel, though it can vary from 50 to 60 without

material disadvantage. The diameter of the wheel is three or four times the total head. The best efficiency found by the earlier experimenters was not over 60 per cent, but a wheel of this form erected at the powder-mill at Angoulême in France, about 1847, showed an efficiency of 68 to 75 per cent, and this efficiency was not materially diminished by a considerable amount of backwater.

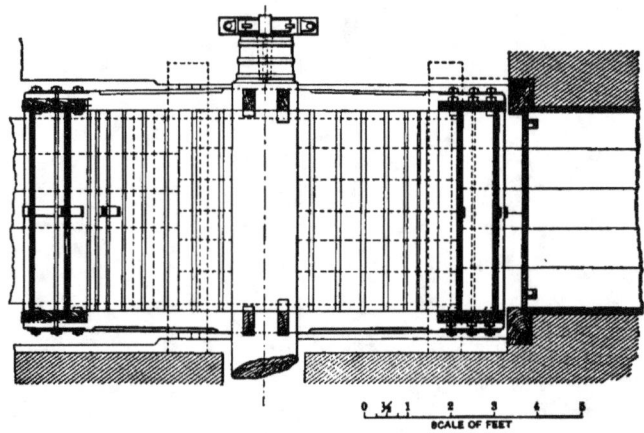

FIG. 138.

In addition to the losses of effect already pointed out, the disadvantages of water-wheels as compared with turbines are very great. They occupy a great deal more room than turbines of the same power, and must be enclosed in buildings to prevent obstruction by ice in winter. Backwater affects them injuriously, not only by diminishing the head, but by partly drowning them and causing them to expend their power uselessly in "wallowing," a source of loss from which turbines are wholly free. The necessarily low velocity with which they move necessitates cumbrous and expensive gearing to raise the speed to the requirements of industry, which is constantly calling for increased rates of speed. These numerous disadvantages have caused the water-wheel to be almost entirely superseded by the turbine.

TURBINES. GENERAL PRINCIPLES.

We come now to consider those wheels in which water acts mainly by impulse and reaction, and which run with a velocity having a definite relation to the head.

The action of water by impulse depends upon certain well-known mechanical principles. Force is required to impart velocity to water, and the velocity imparted is a true measure of the force exerted. Thus, if we find water issuing from an orifice with a velocity of 16.04 feet per second, we are certain that it issues under a head of 4 feet. When water is in motion, force is required to change the direction or velocity of its motion, and the change of motion is a true measure of the force. When the change of motion is due to the action of a vane we know that the pressure of the vane on the water is the same as the pressure of the water on the vane. Suppose a jet of water

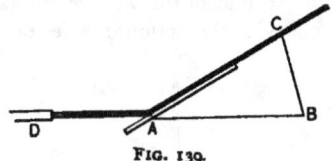

FIG. 139.

moving in the direction DB, Fig. 139, with the velocity AB, in feet per second. At A it encounters a smooth vane which so deflects it that it reaches C instead of B at the end of one second. It glides along the vane and glances off at the extremity with undiminished velocity, $AC = AB$. The effect of the vane is to impart to the water a velocity BC. To find the pressure of the water on the vane, which is equal and opposite to that of the vane upon the water, we reason thus: Gravity, acting freely for one second, would impart to the water a velocity of g feet per second. The pressure of the vane acting for one second imparts to it a velocity of *BC* per second.   Therefore the pressure on the vane is to the weight of water flowing in one second as *BC* to *g*.   If *a* represent the cross-section of the stream, and *w* the weight of a cubic foot of water, the weight of the water is *wav*, and the pressure is

$wav\dfrac{BC}{g}$ pounds.

The impulse of water upon a stationary vane is attended with no material loss of energy.   The water glides along the vane and glances off at the extremity in a direction tangent to the latter with substantially undiminished velocity.   There is in fact a certain friction between water and solids which has to be considered where large surfaces are concerned, but in these calculations such consideration would be superfluous.   When the vane moves, under the action of the water, a portion of the energy is imparted to the vane, and the energy of the stream is correspondingly diminished.   It is manifest that if the energy of the stream is wholly imparted to the vane, it must leave the latter with no absolute velocity.   That is, its velocity must be equal and opposite to that of the vane at the point of exit.

Let us now consider the action of a jet on a flat vane, per-

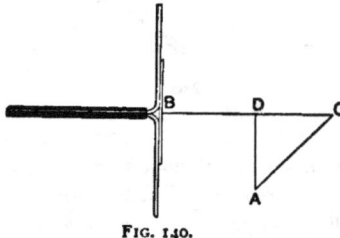

FIG. 140.

pendicular to its direction and moving in the same line.   In Fig. 140 let *BC* = *v* represent the original direction and velocity of the jet.   *BD* = *u*, the velocity of the vane.   Were the vane not present, a particle of water at *B* would have reached *C* at the end of one second.   By the action of the vane, the particle finds itself at the point A at the end of one second, DA being equal to $v - u = DC$. AC is the change of motion due to the action of the vane, and the pressure, regarded as parallel to AC, is $wav\dfrac{AC}{g}$; but the change of motion normal to the vane is DC, and the normal pressure is

$$P = wav\frac{DC}{g} = wav\frac{v - u}{g}. \quad . \quad . \quad . \quad (33)$$

The energy imparted to the vane is

$$Pu = wav\frac{u(v - u)}{g}. \quad . \quad . \quad . \quad . \quad . \quad (34)$$

This expression has its maximum value when $u = \frac{1}{2}v$, and becomes in that case

$$Pu = \tfrac{1}{2}wav\frac{v^2}{2g} = \tfrac{1}{2}wavh, \quad . \quad . \quad . \quad . \quad (35)$$

h being the head to which the velocity is due. In other words, the maximum energy that can be imparted to a flat vane normal to the stream is one-half that of the stream, or the maximum efficiency is 50 per cent, and the best velocity for such a vane is one-half that due the head.

There are two cases in which the energy becomes 0, viz.:

(1) When u is 0, i.e., when the vane does not move. In that case, by eq. (33),

$$P = 2wa\frac{v^2}{2g}. \quad . \quad . \quad . \quad . \quad (36)$$

That is, the pressure on the vane is twice that of the head to which the velocity is due.

(2) When $u = v$, i.e., when the velocity of the vane is equal to that of the stream.

A cup-shaped vane, Fig. 141, reverses the direction of the water's motion; so that, if such a vane be moving in the direction of the stream with the velocity u, the change of motion will be $2(v - u)$, and

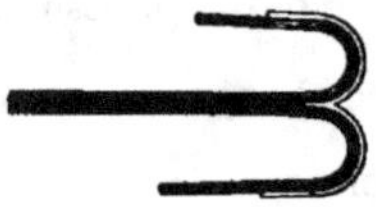

$$P = wav\frac{2(v - u)}{g}, \quad \cdot \quad \cdot \quad (37)$$

FIG. 141.

and the energy exerted on the vane is

$$Pu = wav\frac{2u(v - u)}{g}. \quad \cdot \quad \cdot \quad \cdot \quad (38)$$

As before, the expression has its maximum when $v = 2u$, in which case

$$Pu = wav\frac{v^2}{2g} = wavh. \quad \cdot \quad \cdot \quad \cdot \quad (39)$$

Or, the total energy of the stream is imparted to the vane, and the efficiency is 100 per cent. We have proceeded, however, upon assumptions which cannot be perfectly realized. In any practical application, the direction of the water's motion cannot be exactly reversed, the vanes and their attachments cannot move without friction, the water cannot approach and leave the vanes without velocity and consequent loss of head. The practical interpretation of this result is that floats of this form are consistent with the highest efficiency. Equation (37) shows that when the vane does not move, the pressure

$$P = 2wa\frac{v^2}{g} = 4wa\frac{v^2}{2g}. \quad \cdot \quad \cdot \quad \cdot \quad (40)$$

That is, the pressure on the vane is four times that of the head to which the velocity is due. As in the former case, the energy is 0 when $u = 0$ and when $u = v$.

To trace the application of these principles to different forms of vanes and to vanes which do not move in the same line as the water, would be out of place in a practical treatise. We are nevertheless even now in a position to notice two points of importance which are commonly lost sight of in the design of turbines:

1. The purpose of the vanes or floats, in an impulse-wheel, is, to effect the greatest possible change in the motion of the water. Their length need be no greater than is necessary to accomplish that change.

2. It is a condition of the highest efficiency that the water should leave the vane in a direction opposite to its motion, and with a velocity equal to that of the vane at the point of exit. Where the edge of the vane is radial to the wheel, different parts of it move with different velocities, and the fulfilment of this condition is impossible.

Reaction is the pressure exerted on the walls of a pipe or vessel from which water is discharged. Strictly speaking, the discharge of water from a pipe or vessel does not create pressure within the pipe or vessel from which it issues. It destroys the equilibrium of pressures previously existing. Suppose the pipe, Fig. 142, filled with water under pressure, and free to revolve about the centre C. When the orifice O is closed

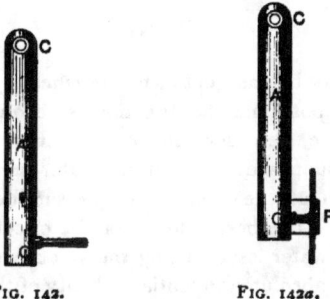

FIG. 142. FIG. 142a.

there is no tendency in the pipe to revolve. The water presses equally upon every part of the interior, and the force tending to turn it toward the right is exactly balanced by that tending to turn it toward the left. When the orifice O is opened the con-

ditions are changed. There is now a small area on one side
of the pipe relieved of pressure, while the pressure acts in full
force on the other side. The pipe will revolve around the
centre C in a direction opposite that of the stream.

Suppose the stream to impinge on a flat vane normal to its
direction, firmly attached to the pipe and in a plane passing
through the axis of rotation (Fig. 142a). The water will escape
radially with undiminished energy. A constant expenditure of
energy would be necessary to cause the pipe to rotate, because
movement of the orifice would increase the energy of the
escaping water. The stream therefore has no tendency to
rotate the pipe, and the reaction on the pipe must be equal to
the pressure on the vane, which we have found to be $2wah$.

Let Fig. 143 represent the rim of a wheel containing the
orifices $O\ O$, so disposed as to discharge water in a direction,

FIG. 143.

as nearly as may be, tangential to the wheel. We assume for
our present purpose that the direction is absolutely tangential.
Let a, as before, represent the cross-section of the stream.
The water is supposed to be at a greater pressure within the
wheel than without, the difference of pressure being represented
by the head h. The best velocity of the circumference is that
with which the water issues, being the velocity due the head h.
In this case the absolute tangential velocity of the water leaving
the wheel is 0. The energy imparted to the wheel is $2wavh$
= twice the energy of the water under the given head.

This does not imply that the wheel is capable of yielding
an efficiency of 200 per cent. In order that the water may
issue from the orifices while the wheel is in motion it must

receive a tangential velocity equal to that of the wheel, and to impart this velocity requires an expenditure of energy *wavh*. The wheel then, under the conditions supposed, is capable of exerting the energy $2wavh - wavh = wavh$, and from this must be deducted the several losses incident to motion, together with that due the deviation of the issuing stream from the direction of a tangent. We here ignore the centrifugal force developed in the water by its rotation.

Impulse- and Reaction-wheels.— Most wheels act partly by impulse and partly by reaction. In a wheel acting purely by impulse the water issues from orifices with the velocity due the head and impinges upon vanes. In a purely reaction-wheel the work of the water is finished when it issues from the orifices of discharge. A tangential or whirling velocity equal to that of the influx orifice is imparted to the water, not directly by the head, but indirectly through the action of the wheel. Neglecting friction and losses incident to the movement of the water, the energy imparted to the wheel by the water is twice that corresponding to the head and quantity discharged, but of this one-half or more is useless energy, being that expended in imparting the necessary tangential movement to the water.

The Pelton or Hurdy-gurdy wheel, to be described later, is an impulse-wheel pure and simple. The Barker mill, Fig. 142, and the forms of Figs. 143 and 144 are reaction-wheels

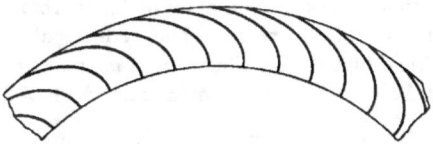

Fig. 144.

pure and simple, observing that in Figs. 142 and 143 the orifices of influx and discharge are the same. All wheels with guide-vanes may be regarded as acting partly by impulse and partly by reaction. The tangential velocity of influx is imparted directly by the head. The velocity of the efflux orifices is less than that due the head and greater than half the same. At full discharge they have mainly the character of reaction-wheels; at diminished discharge more the character of impulse-wheels. Impulse-wheels and reaction-wheels have this in common: their efficiency depends upon the change which they effect in the direction of the water's motion, perfect efficiency implying exact reversal. Perfect efficiency also requires the water to leave the wheel with no tangential velocity, a condition inconsistent with radial orifices of discharge.

Centrifugal Force playing an important part in the action of turbines, we may profitably devote a few words to that subject.

A mass whose weight is w revolving around a fixed point at a distance r therefrom with an angular velocity ω, tends to depart from the centre of rotation with a force $\frac{w}{g}\omega^2 r$. When the body moves in a path which is not a circle, the value of r at any particular point of its path is the radius of curvature at that point. When the size of the body is great compared with r, the centrifugal force acts with different intensities at different parts of it, and r must be measured to the centre of gravity of the body. When the rotating body is a mass of liquid, the pressure within the same increases with the distance from the centre of rotation. Consider a mass of liquid rotating around a centre lying within the mass. The centrifugal force acting on a cylindrical film whose thickness is dr and area unity is, when w is taken as the weight of a cubic foot of the liquid, $\frac{wdr}{g}\omega^2 r$. This force is transmitted to all the films lying outside, so that the total pressure at a distance r_1 from the centre is

$$\frac{w}{g}\omega^2\int_0^{r_1} r\,dr = \tfrac{1}{2}r_1^2\frac{w\omega^2}{g} = w\frac{r_1^2\omega^2}{2g}. \quad . \quad . \quad (41)$$

$r_1\omega$ is the velocity at the distance r_1 from the centre of rotation,

and $\dfrac{r_1^2\omega^2}{2g}$ is the head due that velocity.

Therefore the pressure at any point in the whirling mass is represented by the head due the velocity at that point. This is only true when the mass extends from the centre outwards. The pressure exerted by a whirling ring of water whose internal radius is r_1 and external r_2, Fig. 145, is represented by

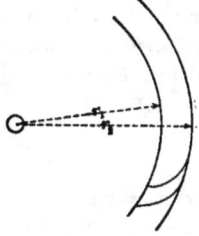

FIG. 145.

$$\frac{w}{g}\omega^2(\tfrac{1}{2}r_2^2 - \tfrac{1}{2}r_1^2) = w\frac{\omega^2}{2g}(r_2^2 - r_1^2), \quad . \quad . \quad (42)$$

i.e., by the difference between the head due the velocity at r_2 and that at r_1.

When the ring represents a turbine-wheel with water flowing through the buckets, it is manifest that equation (42) does not hold, because while the wheel is moving toward the right the water advances toward the left and ω does not represent the angular velocity of the water.* Errors have been committed by writers on the turbine in failing to take notice of this fact.

In the illustrations of the principle of reaction no account is taken of the effect of centrifugal force. The introduction of this element, though it does not affect the principle, makes the problem more complex. The full effect of centrifugal force appears in the arrangement of Fig. 142, in which a pipe filled with water under pressure revolves around a centre C, discharging from an orifice O, the entire mass of water being in rotation with uniform angular velocity. The pressure acting on the orifice, under this condition will be $2wah$, and the velocity of discharge will increase to $v = \sqrt{2g} \times 2h = 1.414\sqrt{2gh}$. If we increase the velocity u of the orifice so as to make $u = 1.414\sqrt{2gh}$, we develop a still greater centrifugal force,

---

* See Journal of the Franklin Institute, vol. CXVI. p. 92.

and so on; so that the condition of maximum efficiency $v = -u$ is impossible. If, for example, we attempt to determine the value of u on the asumption that it is equal to the velocity with which the water issues from the orifice O, we should have the equation

$$u = \sqrt{2g\left(h + \frac{u^2}{2g}\right)}, \quad \cdot \quad \cdot \quad \cdot \quad \cdot \quad (43)$$

or $u^2 = 2gh + u^2$, which is only possible when $h = 0$. If, however, we load the machine so as to maintain its velocity at that due the head, viz., $u = \sqrt{2gh}$, we should have for the velocity of the issuing stream

$$v = \sqrt{2gh + 2gh} = \sqrt{2gh}\sqrt{2}, \therefore v - u = \sqrt{2gh}(\sqrt{2} - 1),$$

and the loss of head, which represents the loss of power,

$$= \frac{(v - u)^2}{2g} = h(\sqrt{2} - 1)^2 = 0.1716h.$$ The power utilized, incidental losses excepted, is $0.8284h$, or the efficiency is 0.8284.

If we put the velocity $u = 2\sqrt{2gh}$, we have $v = \sqrt{2gh}\sqrt{5}$ and $v - u = \sqrt{2gh}(\sqrt{5} - 2)$. The loss of head is $(\sqrt{5} - 2)^2 h = 0.0556h$ and the efficiency is $1 - 0.0556 = 0.9444$. The practical interpretation of these results is that there is a loss of 17 per cent in the first case, and 5.6 in the second, inherent in the principle of the wheel. With higher velocities the computation would show greater economy, and with an infinite velocity we should have perfect efficiency; but it is probable that with any velocity above $u = \sqrt{2gh}$ the losses incident to the rapid motion of the water would outweigh the gain. Wheels of this kind, however, on account of their simplicity have found application in spite of their inherent lack of economy. Among them are Whitelaw's turbine and Barker's mill.

A wheel of the form indicated at Fig. 144, without guides, would operate wholly by reaction. The water, at its entrance to the wheel, would have a whirling motion, the tangential component of which is equal to the velocity of the inner circumference. We know this because otherwise the water could not enter the wheel. This whirling motion is not the direct effect of the head, but is imparted to the water mechanically by the action of the wheel. The pressure of the water does not diminish till the latter has entered the buckets. The energy of reaction exerted at the orifices of discharge exceeds the total energy due the head, but this, as we have already seen, is partly absorbed in imparting the whirling motion to the water before it enters the wheel.

CHAPTER XIV.

TURBINES.

THE Fourneyron turbine was first developed in France in the early part of the nineteenth century. M. Fourneyron, in 1834, received the prize of 6000 francs offered by the Society for the Encouragement of the Arts, for the construction of the best horizontal * wheel on a large scale. This was the first wheel erected by M. Fourneyron, which was at Pont on the river Ognon. This circumstance strongly directed the attention of manufacturers to wheels of this type, and before 1840 many such wheels of imperfect construction and small size had been introduced in this country. These, though less efficient than the ponderous water-wheels then in use, won their way by the decided advantages they offered in respect of compactness, rapidity of motion, and freedom from obstruction by backwater.

About 1840, Uriah A. Boyden of Massachusetts, an engineer of rare mechanical ability, applied himself to the study of this class of wheels. The time being ripe for discarding the old-fashioned breast-wheel, he had a very successful career in designing and erecting turbines of this type, in which he made so many improvements that they have been known in this country as Boyden wheels. Wheels of this type designed by Mr. Boyden and erected in the mills of the Appleton Company at Lowell, in 1846, were found upon the most rigid tests to

---

* The word 'horizontal' is used ambiguously in reference to water-wheels, sometimes denoting a wheel revolving in a horizontal plane, and sometimes a wheel running on a horizontal shaft. Here it has the former meaning.

298

yield a useful effect of 88 per cent of the absolute power of the water, a result which, it is believed, has never been surpassed or equalled by any other form of wheel, though higher results are often claimed by inventors. In a wheel subsequently erected for the Atlantic Cotton Mills of Lawrence, where Mr. Boyden was not permitted to construct a weir for the measurement of the discharge, but arrived at it by computation, he confidently claimed an efficiency of 92 per cent.

Mr. James B. Francis, engineer of the Lowell Water-power, designed a wheel of this type which was erected in the Tremont Mills of Lowell. In 1851 he made a series of tests and measurements on it which, from their completeness and the care and painstaking accuracy with which they were conducted, have become classic in engineering literature. These are fully reported in Mr. Francis's book, "The Lowell Hydraulic Experiments," * comprising ninety-two experiments made under different conditions of velocity and quantity of water. These are accompanied by complete drawings and dimensions of the wheel, which make these results invaluable for testing theories of the action of water in the turbine.

This wheel is shown in Fig. 146, which represents it in plan. Fig. 147 is a vertical section, Fig. 147*a* an enlarged section through the rim of the wheel, showing the disk and gate. *W* is the wheel proper, consisting of an upper and a lower ring, united by the floats. These are forty-four in number, of Russia iron $\frac{9}{16}$ inch thick, inserted in grooves cut in the rings. Each float carries several tongues, which enter mortises cut through the rings and are secured by cold hammering. The wheel is fastened by countersunk screws to a broad web springing from the hub, which is secured to the shaft. The shaft rises through the penstock, and its weight with its attachments is sustained by a suspension-box above the crown gear which delivers the power. Surrounding the shaft is the disk-pipe *M* attached to the penstock at the top,

---

* Lowell Hydraulic Experiments. Van Nostrand & Co., 1868.

and at the lower part to a broad conoidal plate K called the disk, to which are attached the guides. These are thirty-three in number, of Russia iron $\frac{1}{16}$ inch thick, curved to give the proper direction in entering the wheel, attached to the disk by the same means as are used in the floats. They are also firmly united to the lower end of the supply-pipe or penstock. The regulating-gate G is a short cylinder, which moves telescope-wise on the lower part of the penstock, enters the annular space

Fig. 146.

between the guides and floats, and closes on the outer edge of the disk, thus controlling the flow of water to the wheel. The guides do not stand perpendicular to the disk, but are inclined backward in a direction counter to that of the wheel's motion.

This wheel did not give so high a result as the wheels designed by Mr. Boyden, the highest efficiency shown being 79 per cent, which occurred when the gate was fully raised and the velocity of the interior circumference of the wheel was 62 per cent of that due the head acting thereon, although the

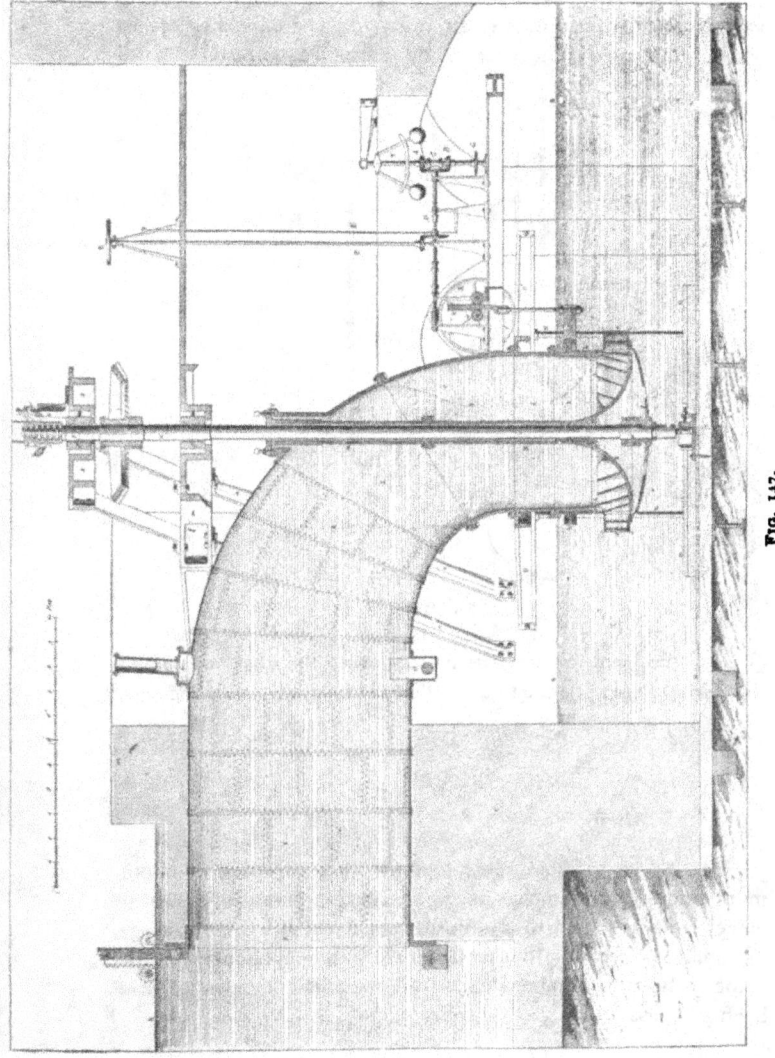

FIG. 147.

efficiency stood between 77 and 79, while the relative velocity varied between the limits of 47 and 70. The efficiency of this wheel diminished rapidly as the discharge was reduced. With

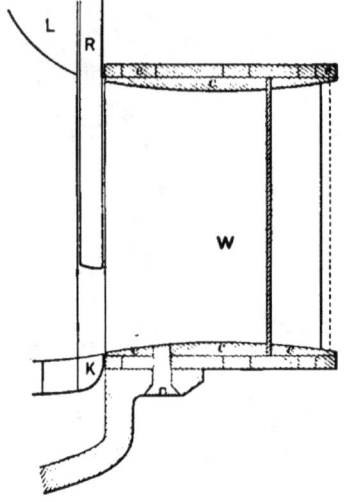

Fig. 147a.

$Q = 0.96$ of that in the experiment giving the best result, and the relative velocity 0.61, the efficiency was. 0.762

For $Q = 0.78$, relative velocity 0.61, efficiency 0.652

0.525	''	''	0.56	''	0.44
0.27	''	''	0.49	''	0.10
0.26	''	''	0.27	''	0.24

The above results appear to show that the speed of maximum efficiency diminishes as the discharge diminishes, but the wheel is usually held to a uniform speed whether the discharge be small or great. In the last case but one above, had the velocity been maintained at 0.62, the efficiency would have been still less.

A Method of Computing the Discharge of Water from the Boyden Wheel at Full Gate. —We will use the following notation:

$Q =$ quantity of water discharged by the wheel;

$h =$ head of water acting on the wheel;

$F =$ cross-section of guide-passages at exit;

$F_1 =$ cross-section of wheel-passages at entrance;

$F_2 =$ cross-section of wheel-passages at exit;

$f =$ cross-section of wheel-passages at any point whose radius is r;

$c =$ velocity of discharge from F, c_1 do. through F_1, c_2 do. from F_2;

$r_1 =$ radius of inner ends of floats, r_2 do. of outer ends;

$\omega =$ angular velocity of wheel;

$a =$ angle between tangent of guide and tangent of wheel at extremity of former;

$e =$ angle between direction of escaping water and tangent of wheel at extremity of float. The foot and second are the units.

The inner ends of the floats make substantially a right angle with the circumference of the wheel.

Aside from the centrifugal force, the general theory of the motion of water in the wheel may be stated very briefly. We suppose the velocity c to be decomposed into its radial and tangential components, viz., $c \sin a$ and $c \cos a$.

The tangential component $c \cos a$ is inoperative as regards the discharge. It simply puts the water in a position to enter the wheel without any effect in urging it through the wheel.

At the entrance to the wheel there is a loss of head, being the head due the difference between c_1 and the radial component of c, viz.,

$$\frac{1}{2g}(c \sin \alpha - c_1)^2 = \frac{1}{2g}\left(c \sin \alpha - c\frac{F}{F_1}\right)^2 = \frac{c^2}{2g}\left(\sin \alpha - \frac{F}{F_1}\right)^2. \quad (44)$$

There is a loss of head due the friction of the water in the supply-pipe and the guide- and bucket-passages, which might be arrived at by a laborious computation, but I adopt the value given by Weisbach,* viz., f, f_1 being numerical coefficients,

$$\text{Frictional head} = f\frac{c^2}{2g} + f_1\frac{c_2^2}{2g}.$$

Weisbach says, " we may take $f = f_1 = 0.05$ to 0.10." I put

$$\text{Frictional head} = 0.15\frac{c^2}{2g},$$

which is about the mean given by Weisbach. I also adopt Weisbach's coefficients for the discharge of the guide- and bucket-orifices. He finds that the cross-section of a stream issuing from a straight pipe is about 3 per cent less than that of the pipe, while a very slight convergence diminishes the cross-section by 5 per cent, which I adopt.

As soon as the water takes part in the rotary motion of the wheel, it is acted on by centrifugal force. Writers on the turbine have generally estimated this force upon the assumption that the water has the same angular velocity as the wheel, and have thus arrived at a very simple expression, though a decidedly erroneous one, for, while the bucket may be moving to the right with the angular velocity ω, a particle of water is moving to the left with a certain angular velocity which we will call ω_1, which is very small at the entrance to the wheel, but becomes considerable toward the efflux. The true angular velocity, therefore, is $\omega - \omega_1$, and the true centrifugal force acting on a mass M is $M(\omega - \omega_1)^2 r$.

We must next determine the head due the centrifugal force, by finding the work done by the same on a given weight of water while passing through the wheel, and dividing the result by the weight. To attempt this by the strict methods of the calculus would lead to a hopelessly complex expression. As in many hydraulic problems we shall find an advantage in approximating the truth by a rough graphical process of integration. At a point whose distance from the centre is r, let d represent the angle between the tangent of the float and

---
* Hydraulics. Du Bois's translation, p. 359.

radius of the wheel, f being the section of the bucket (space between two consecutive floats), by a cylindrical surface concentric with the wheel. Then $\omega_1 r = c \dfrac{F}{f} \tan d$, and the expression for centrifugal force becomes

$$M\left(\omega^2 - 2\omega c \frac{F}{f} \tan d + \frac{c^2}{r}\left(\frac{F}{f}\right)^2 \tan^2 d\right). \quad . \quad . \quad (45)$$

Dividing the bucket into any number of parts by concentric cylindrical surfaces, we can compute the coefficients of ω^2, ωc, and c^2 for each part. From the full-size drawings of the wheel, and from the data given by Mr. Francis in the Lowell Hydraulic Experiments, I obtain the following values: $r_1 = 3.375$ feet, $r_2 = 4.146$ feet, $F = 6.53$ square feet, $F_1 = 19.35$ square feet, $F_2 = 7.467$ square feet, $a = 19°\ 5'$. From these, together with the several values of d and f I construct the following table.*

TABLE 2.

Radius r, Inches.	Distance from Inner Edge of Crown, measured on Radius of Wheel, Inches.	f, Area of Float-passage or Bucket measured on Cylindrical Surface of Radius r_1 concentric with Wheel. Sq. Ft.	d = Angle between Tangent of Float and Radius of Wheel.	$\dfrac{F}{f}\tan d$ = Numerical Coefficient of c in the Expression for the Relative Tangential Velocity.	Numerical Coefficient in the Expression for the Centrifugal Force of		
					ω^2	ωc	c^2
41	0.5	0.4397	4° 27′	0.0263	3.417	0.0526	0.0002
42	1.5	0.4418	13 43	0.0820	3.500	0.1640	0.0019
43	2.5	0.4430	23 20	0.1445	3.583	0.2890	0.0058
44	3.5	0.4460	32 47	0.2143	3.667	0.4286	0.0125
45	4.5	0.4506	39 55	0.2756	3.750	0.5512	0.0202
46	5.5	0.4564	48 37	0.3691	3.833	0.7382	0.0355
47	6.5	0.4639	54 41	0.4515	3.917	0.9030	0.0520
48	7.5	0.4739	62 06	0.5915	4.000	1.0830	0.0875
49	8.5	0.4868	69 38	0.8232	4.083	1.6464	0.1660
49.625	9.125	0.5064	77 18	1.3006	4.135	2.6012	0.4091
49.75	9.25		Ext'mity 78° 27′				
Sum of the first 9 coefficients and ½ of the 10th					34.784	6.5063	0.4839
$\frac{1}{12}$ of do.					2.899	0.5422	0.0403

* This mode of determining the discharge of the turbine was published in the Journal of the Franklin Institute for July, 1884.

The centrifugal force is represented by the expression

$$\left(M = \frac{W}{g}\right)(34.784\omega^2 - 6.5063c\omega + 0.4839c^2), \quad . \quad (46)$$

in which M is the mass and W the weight of water included between two consecutive sections one inch apart; one-fourth the value of the last coefficient being taken because this applies only to $\frac{1}{4}$ inch. The expression may also be understood to represent, in inch-pounds, the work done by centrifugal force on the said mass while passing through the wheel; hence the division by 12 to reduce it to foot-pounds. The head due the centrifugal force is represented by

$$\frac{5.798\omega^2 - 1.0844\omega c + 0.0806c^2}{2g}. \quad . \quad . \quad (47)$$

This expression in the formula for discharge takes the place of

$$\frac{\omega^2(r_2^2 - r_1^2)}{2g},$$

given generally by writers on the turbine.

The principles stated in what precedes may be expressed algebraically as follows:

$$\frac{c_2^2}{2g} = h - \frac{c^2}{2g}\cos^2\alpha - \frac{c^2}{2g}\left(\sin\alpha - \frac{F}{F_1}\right)^2$$
$$- \frac{0.15c^2}{2g} + \frac{1}{2g}(5.798\omega^2 - 1.0844\omega c + 0.0806c^2). \quad (48)$$

Or, substituting numerical values, reducing, and observing that $c_2 = c\dfrac{F}{F_2}$, we get

$$1.7277c^2 = 2gh + 5.798\omega^2 - 1.044\omega c, \quad . \quad . \quad (49)$$

and $Q = 0.95Fc$.

The results given in Table 3 are computed by this formula. They are arranged according to the ascending values of ω, and embrace in reality the entire series with full gate, the experi-

ments omitted being substantially nothing more than repetitions
of those included.

TABLE 8.

Number of the Experiment in Mr. Francis's Series.	*n* Number of Revolutions of the Wheel per Second.	$\omega = 2\pi n$. Angular Velocity of the Wheel. Feet per Sec.	*h* Head Acting on the Wheel. Feet.	Q Discharge in Cubic Feet per Second.	
				By Experiment	By Computation.
43	0	0	12.797	135.65	135.41
42	0.45431	2.8534	12.948	133.43	134.57
41	0.53232	3.3447	12.977	133.75	135.19
40	0.60000	3.7699	12.973	134.80	135.76
39	0.64702	4.0653	12.963	135.34	136.20
36	0.69471	4.3650	12.944	136.49	136.69
35	0.74211	4.6629	12 939	137.71	137.31
34	0.78401	4.9261	12.941	138.09	137.98
32	0.83624	5.2542	12.915	138.27	138.68
29	0.86643	5.4439	12.906	138.51	139.18
21	0.90201	5.6675	12.899	139.90	139.80
18	0.94507	5.9380	12.880	140.47	140.56
16	0.99945	6.2797	12.890	141.98	141.76
15	1.02373	6.4323	12.888	142.04	142.28
14	1.06744	6.7069	12.856	142.52	143.15
11	1.12518	7.0697	12.819	143.91	144.38
10	1.18460	7.4431	12.800	144.87	145.83
9	1.24514	7.8234	12.777	146.02	147.38
8	1.30933	8.2268	12 720	147.29	148.98
7	1.38249	8.6864	12.696	149.47	151.08
6	1.46149	9.1828	12.653	152.27	153.41
5	1.53218	9.6270	12.611	154.39	155.57
4	1.59651	10.0313	12.554	156.65	157.57
13	1.78404	11.2095	12.510	163.43	164.26

The reduced efficiency at part gate, or, in other words, the
inability to use a small quantity of water with the same efficiency as a large one, is an inherent defect of the turbine and
is very strikingly apparent in the one under consideration, the
useful effect being not more than 40 per cent when the discharge is reduced one-half, and so small with a discharge of
one-fourth that it is hardly worth while to raise the gate. To
realize the cause of this difficulty we must consider how the
water traverses the guides. When the gate is fully raised, *R*,
Fig. 147*a*, the water which enters at the top of the wheel follows the curve of the garniture *L*, and takes the direction of the
extremity of the guide at its entrance to the wheel; but when

the gate is partly closed, as indicated by the dotted lines, a part of the water follows the inner surface of the gate, and reaches the opening in a vertical direction with no tangential velocity; tending to enter the wheel in a radial direction as if escaping from an orifice in a direction perpendicular to the plane thereof. The more the gate is closed down, the more the opening loses the character of a channel and acquires the character of an orifice, and the more the direction of the escaping water departs from that of the tangent and approaches that of the radius. The closing of the gate alters the direction of the water entering the wheel and impairs the efficiency with which it acts thereon. This, however, is not the only ill effect, and probably not the worst. The stream entering the bucket does not fill and pass smoothly through the same as it does at full gate, but, a part of the bucket being necessarily filled with dead water, the stream wastes its energy in commotion and eddies while traversing the wheel.

A great many devices have been introduced to obviate this difficulty, and it is to this object that the efforts of inventors have been mainly directed. A considerable measure of success has attended these efforts, but it has been attained at the sacrifice of some percentage of efficiency at full gate. Mr. Boyden's expedient for meeting the difficulty was by giving an inclined direction to the guides. The outer extremity of the guide was inclined in a direction reverse to that of the wheel's motion, making an angle of 60° or less with the horizontal. The water which passed down along the inside of the cylindrical gate pursued a spiral pathway and reached the opening with a considerable tangential element of velocity. The guides in the Tremont turbine had, as appears from the cuts, a slight inclination of this kind, but not enough to affect the efficiency. To realize the full benefit of this disposition the guides would require an inclination of less than 30° to the horizontal, an arrangement involving grave structural difficulties. I know of no experiments to show how much this disposition increased the efficiency.

Another expedient is by attaching to the gate a curved block or shell, Fig. 147*a*, in the form of the garniture or curb *L*. This contracts the channel of the guide as the gate descends, and thus forces the water to follow the direction of the guides.

The most promising method of meeting the difficulty incident to the efflux of water from the guides is by dividing the float-passages into several parts by horizontal diaphragms. This device was adopted in the earliest form of the Fourneyron turbine, and was used in many cheaper forms of wheel in this country, made wholly of cast iron. Its introduction in the higher grade of wheels with sheet-iron floats involved constructive difficulties. It has been recently adopted in the wheels of the Niagara Falls Power Company at Niagara Falls. The only perfect method of obviating the loss incident to part gate would consist in a device to contract simultaneously both the guide- and the float-passages according to the requirements of the power.

The improvements in the turbine which have been effected since Mr. Boyden's time are, in addition to an increased efficiency at part gate, a great diminution of the cost, the development of the draft-tube, and the adaptation of the wheel to a horizontal shaft. As to the wheel itself, meaning the revolving part carrying the organs on which the water acts, it has almost universally taken the form of Fig. 148, in which the water enters the wheel through openings having their longest dimensions parallel to the shaft, and leaves it through openings which have their longest dimensions radial. The differences between the wheels of different makers consist in the guides, gates, and mode of regulating the supply of water. It will be readily perceived that the form of float, Fig. 148, is not consistent with the highest efficiency, the imperative condition of which is that the water should escape from the float-passages with a velocity equal to that of the float and in a direction contrary to that of the float's motion, a condition which cannot be realized in this form of wheel. Except in so far as the motion

of the water is affected by centrifugal force, it has the same
velocity at the inner end of the opening as at the outer end,
while the velocity of the float at the latter point may be more
than twice as great as at the former.

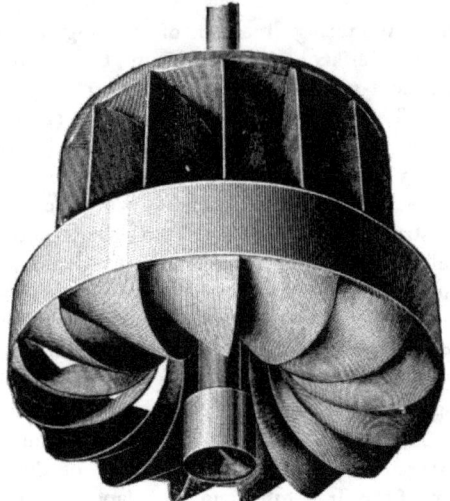

FIG. 148.

The Swain Wheel, shown in Figs. 149, 149*a*, and 149*b*,
is one of the most successful of existing types, especially as
regards the difficulties incident to part gate.   Fig. 149 is a
vertical section through the centre of the wheel, Fig. 149*a* a
plan of a portion of the wheel and gate, Fig. 149*b* a development of the outer surface of the wheel.

I describe it in the language of Mr. James B. Francis contained in his report of a test of a 72-inch wheel of this construction at the Booth Cotton Mills,* Lowell, in 1874.

The lower curb, *C*, is a strong disk of cast iron, with a short
cylinder upon which the gate moves, and an inner tube with

* Journal of the Franklin Institute, April, 1875.

diverging sides, through which the water leaving the wheel is
discharged into the pit.   There are three arms reaching from

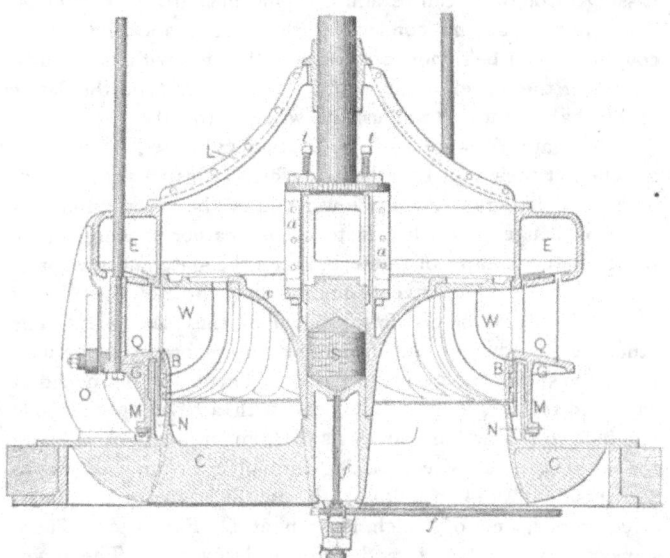

FIG. 149.

FIG. 149*a*.

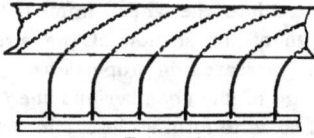

FIG. 149*b*.

the sides of this tube to the hub, which forms the pintle upon
which the wheel revolves.   The step S is a cylinder of white
oak, with conical ends, and is free to revolve with the wheel

or to remain stationary upon the pintle while the wheel revolves around it.   By means of pipe (f) water is supplied to the step, passing through its centre and escaping outward over its ends. The intermediate *aa* connecting the shaft *c* and the wheel-coupling *v* can be removed to replace the step without disturbing either the wheel or the shaft.   The screws *t t* in the flange of the shaft are used to adjust the wheel vertically.

The gate *G* was made with two cylinders, *N* and *M*, attached at their tops to a disk *Q*, which forms an angle of 80° with the cylinders.   At the lower end of the outer cylinder is a narrow flange to which is fastened the leather packing, which prevents the escape of water between the gate and the lower curb.   The gate has twenty-four guides, three of them being of cast iron and of the form shown in the plan at *e*.   The other guides, twenty-one in number, are of bronze, 0.23 inch in thickness and 18.94 inches long.   These are sharpened at each end to 0.04 inch in thickness, with a bevel on each side one inch long; and are so set as to form an angle of 14° with the tangent to the wheel passing through their inner edges.

Outside of, and in a line with, the thick guides are placed three stands, one of which is seen at *O*, Fig. 149.   These support the chamber *E* and the wheel-cover *L*.   The lower disk of this chamber is slotted, so that the guides may enter the chamber when the gate is raised, by means of the hoisting-rods which pass through the thick guides.   The gate is shown as fully opened.   The gate is opened by lowering, and closed by raising it, so that when the gate is first opened, the water is admitted into the wheel, immediately under the crown, and the depth of the section of the stream passing through the guides is increased in proportion as the gate is opened.   The lower edge of the chamber and the upper edge of the gate are finished so as to form a close joint when brought into contact. The inner edges of the guides are 1⅜ inches distant, radially, from the outer edges of the buckets.

The wheel *W* is 72 inches in diameter at the outer edges of the buckets, and 23.35 inches in depth from the under side of the crown to the lower edges of the band. It has twenty-five buckets of bronze, these being formed between dies in a press and having the crown-plate and the lower band cast upon them of iron.

Fig. 149*a* is a horizontal section just below the crown-plate, and represents the form of the bucket for the first six inches below the crown.

Fig. 149*b* is a development of a portion of the cylindrical surface of the wheel containing the outer edges of the buckets. The discharging edge of the bucket lies in a vertical plane passing through the axis of the wheel, and is parallel to this axis, from the under side of the crown, to a point about 8⅛ inches below it, and from this point is continued in the form of a quadrant having a radius equal to one-fifth of the diameter of the wheel, and having its centre in the cylinder forming the outer circumference of the wheel; thus forming, in connection with the surface of the adjoining bucket, an outlet which is in effect a union of two wheels, an inward discharge and a downward discharge.

The following measurements were made at the mill before the wheel was started:

Vertical distance from under side of crown to the
   lower edge of the buckets................. 23.35 inches.
Vertical distance from the under side of the
   crown to the top of band *B*............... 13.285 "
Total area of outlets of wheel (twenty-five in
   number)................................. 9.558 sq. ft.
Vertical movement of speed-gate............. 13.08 inches.
Mean shortest distance from the inner edge of
   one guide to side of adjacent guide (twenty-
   four in number)......................... 4.532 "
        Total area of inlet in speed-gate........ 9.880 sq. ft.

It will be perceived that centrifugal force, in this wheel, operates to diminish the discharge instead of to increase it as in the case of the Boyden wheel. The importance of this

TABLE 4.—TESTS OF DISCHARGE AND EFFICIENCY OF THE SWAIN TURBINE.

72-INCH WHEEL AT BOOTH COTTON-MILLS, LOWELL, AUGUST, 1874.						36-INCH WHEEL AT HOLYOKE TESTING-FLUME, JANUARY, 1897.						
Number of Experiment.	Height of Speed-gate.	Head Acting on the Wheel.	Discharge in Cubic Feet per Second.	Ratio of the Velocity of the Exterior Circumference to the Velocity due the Head.	Ratio of the Useful Effect to the Power Expended.	Number of the Experiment.	Height of Speed-gate, Full Height being 1.000.	Head Acting on the Wheel.	Discharge in Cubic Feet per Second.	Ratio of the Velocity of the Exterior Circumference to the Velocity due the Head.	Ratio of the Useful Effect to the Power Expended.	Percentage of Full Discharge.
	In.	Feet.						Feet.				
1	3.25	13.96	70.86	0.706	0.595	1	0.067	16.49	12.80	0.733	....	0.161
4		13.93	72.20	0.600	0.609	2	0.125	16.11	18.73	0.776	0.283	0.239
5		13.93	72.49	0.573	0.580	3		16.18	19.24	0.714	0.327	0.245
6		13.93	69.54	0.772	0.589	6		16.17	19.62	0.691	0.430	0.249
7	6.50	13.36	108.11	0.842	0.715	7		16.07	19.76	0.619	0.449	0.252
9		13.28	111.55	0.779	0.757	8		16.01	19.80	0.582	0.460	0.254
10		13.07	113.45	0.714	0.774	9		16.03	20.09	0.543	0.404	0.257
14		13.32	116.13	0.677	0.779	10	0.250	15.79	27.64	0.585	0.440	0.356
15		13.25	116.80	0.555	0.721	12		15.70	24.20	0.808	0.572	0.378
23	2.00	14.31	47.48	0.902	0.712	14		15.61	30.59	0.725	0.523	0.396
26		14.28	49.21	0.792	0.580	16		15.63	31.46	0.661	0.639	0.407
31		14.19	51.67	0.510	0.463	17		15.57	31.83	0.628	0.642	0.412
32	3.00	13.98	68.75	0.512	0.541	18		15.54	32.11	0.596	0.643	0.416
40		14.05	63.10	0.503	0.453	19		16.53	32.40	0.564	0.620	0.420
41	4.00	13.90	72.76	0.967	0.473	27	0.375	15.30	35.57	0.837	0.682	0.504
45		13.76	81.49	0.766	0.648	21		15.31	40.01	0.765	0.724	0.524
49		13.70	83.59	0.648	0.606	24		15.21	41.58	0.670	0.746	0.545
50	5.00	13.44	96.86	0.695	0.747	25		15.19	42.06	0.641	0.748	0.552
53		13.44	97.82	0.517	0.657	26		15.21	42.50	0.604	0.738	0.557
57		13.55	94.80	0.762	0.706	28	0.500	15.06	45.62	0.801	0.773	0.605
60		13.66	86.85	0.931	0.551	30		15.91	50.87	0.738	0.795	0.652
69	6.00	13.28	110.09	0.669	0.751	33		15.78	52.54	0.647	0.805	0.676
72	7.00	13.39	103.50	0.681	0.542	34		15.74	52.98	0.608	0.791	0.683
75		13.16	116.71	0.773	0.760	35	0.625	15.65	55.32	0.839	0.770	0.715
80		13.00	121.56	0.637	0.787	39		15.43	58.95	0.709	0.843	0.767
90	8.00	12.97	130.25	0.698	0.806	40		15.35	59.45	0.673	0.836	0.775
101	9.00	12.84	138.84	0.710	0.829	42	0.750	15.20	62.06	0.828	0.831	0.814
103	10.00	12.91	128.57	0.969	0.826	44		15.62	64.09	0.780	0.890	0.829
111		12.74	143.26	0.762	0.831	45		15.54	64.64	0.744	0.854	0.835
113		12.68	144.77	0.708	0.836	46		15.50	65.41	0.722	0.852	0.844
122	12.00	12.77	143.44	0.650	0.690	53	0.875	15.28	70.01	0.705	0.845	0.916
123		12.52	150.70	0.749	0.838	54		15.15	70.42	0.683	0.848	0.925
127		12.49	158.46	0.609	0.775	56	1.000	15.33	70.53	0.807	0.798	0.922
128	13.08	13.10	160.19	1.174	0.806	57		15.44	71.56	0.787	0.809	0.934
129	Full	12.88	137.57	1.008	0.490	58		15.47	72.73	0.767	0.821	0.945
130	Ht.	12.603	149.13	0.908	0.690	59		15.48	73.65	0.750	0.820	0.954
131		12.480	128.83	0.866	0.803	60		15.43	74.19	0.730	0.836	0.966
133		12.37	162.54	0.770	0.836	61		15.42	74.74	0.718	0.843	0.971
136		12.37	164.94	0.727	0.830	62		15.40	75.32	0.700	0.846	0.984
142		12.40	165.05	0.570	0.746	63		15.25	75.98	0.677	0.848	0.995
143		19.40	182.65	0.438	0.697	64		15.16	76.49	0.656	0.846	1.004
144		13.17	117.84	1.193	0.000							

action is apparent when we compare expt. 143 with 144, which were both made with the same height of gate. Had the head been 13 feet in each of these cases, other conditions remaining the same, the discharge would have been 166.75 cubic feet per second in 143, and 117.06 in 144. This difference of 49.69 cubic feet per second is due to an increase in the angular velocity ω from 4.12 in 143 to 11.56 in 144. We have already adverted to the apparently unscientific character of a wheel with radial orifices, page 309: The vertical edge of the float, in this wheel, moves with a velocity less than three-fourths that of the horizontal edge near its extremity, and both velocities could not bear the right proportion to the velocity of the water if the latter was uniform. We see how the centrifugal force goes far toward obviating the ill effect of this disposition. The water at the outer part of the orifice is actually discharged under a greater head than at the inner part, and so conforms more nearly to the theoretical conditions of efficiency than would appear without considering the action of centrifugal force. Nevertheless it may be safely affirmed that the highest efficiency can never be obtained from a wheel with radial orifices, or orifices lying in a plane perpendicular to the shaft. The highest efficiency yet obtained for the Swain wheel is 85.4, while the Boyden wheel, as stated, showed 88. This difference must be attributed to the difference in the character of the orifices of discharge.

The maximum efficiency in the trials of 1874 was 83.6 in expt. 113, with a relative velocity of 0.708, and a discharge of 144.77, the gate not being fully raised. Substantially the same efficiency was found in expt. 133, with the gate fully raised, a discharge of 162.54, and a relative velocity of 0.77, though the efficiency did not fall below 82, while the relative velocity varied from 0.71 to 0.81. As the head varied considerably during the trials, it is better, for purposes of comparison, to reduce the discharge to a uniform head. This was done by Mr. Francis in his report of the trials, in which he gave the discharge corresponding to a uniform head of 17 feet. This calculation is made upon the principle that the discharge is as the square root of the head. That is to say: If for a head of 12.373 feet the discharge is 162.54, then for a head of 17 feet the discharge will be $162.54\sqrt{\dfrac{17.00}{12.372}} = 190.53$. Three-fourths of this is 142.90. With this discharge and nearly the same relative velocity the efficiency was about 80.

With one-half of full discharge, viz., near 95 cubic feet per second, on 17 feet head and a relative velocity of 0.65 to 0.75, the efficiency was 65 to 67. With about one-fourth the full discharge and the same relative velocity we get an efficiency of something over 40. So far as this series of trials goes, therefore, we are entitled to say that the wheel gives an efficiency of near 84 at full discharge, 80 at three-fourths, 65 at one-half, and about 40 at one-fourth.

The tests of January, 1897, show some improvement in the wheel, as compared with those of 1874, the best efficiency in this series being a little over 85, which occurs with about five-sixths of the full discharge, and a velocity about three-fourths that due the head. It would appear from the table, however, that the proper velocity for the wheel would be about two-thirds that due the head. With this velocity the efficiency is substantially 85 at full discharge, 75 at one-half discharge, and 43 at one-fourth.

Fig. 150 is a section of the vertical Swain wheel in its present form. The only essential change consists in dispensing with the chamber into which the guides rise when the gate is closing. The guides are not attached to the gate but to the fixed curbs of the wheel, and pass through slots in the horizontal rim of the gate, which closes by rising as before. G is the gate, f the float, g the guide. In the former arrangement, where the guides were cast in one piece with the gate, they frequently broke at their junction with the latter in consequence of the vibrations. This accident does not occur in the present arrangement. Properly the edges of the floats should appear in elevation in the figure, but this feature is omitted. No improvement has been made in the step-bearing, which consists of a conical block of wood, thoroughly dried and soaked in hot tallow. A hole is bored through the centre of the step through

which water brought by the pipe *f* passes and spreads over the
conical surface. An improvement of some value has been
made in the means of replacing the step. To do this the
spool-shaped piece *T* is removed by taking out the screws.

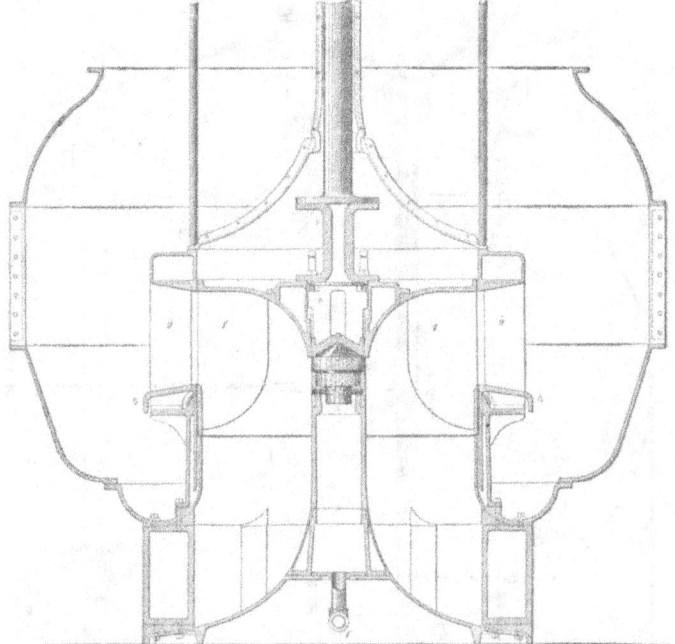

FIG. 150.

The wooden bearing-piece *P*, resting on the iron cap, is taken
out with the latter, the step is removed and a new one dropped
in its place.

Fig. 151 shows a pair of Swain wheels on a horizontal shaft.
This shaft passes through stuffing-boxes, and at the left end
has a thrust-bearing, to meet the considerable thrust that exists
when one wheel is out of use. At the opposite end the shaft
runs in a stuffing-box bearing. The guide, gate, and floats

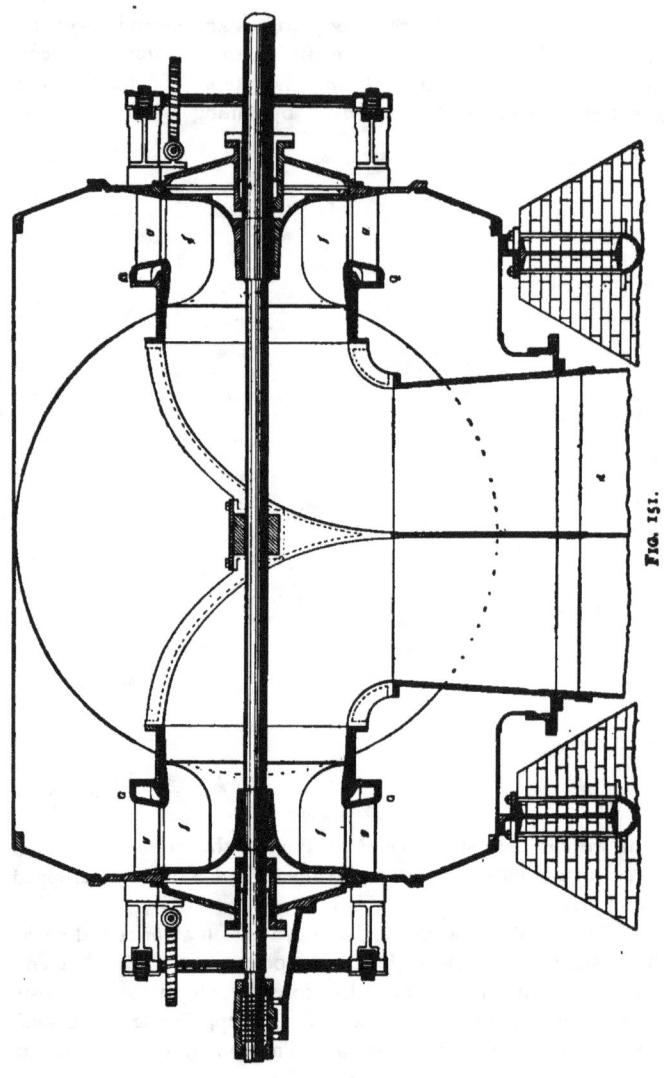

FIG. 151.

are indicated by the same letters as before. This arrangement is admirably adapted to a wide variation in the consumption of water. With both wheels running at full gate, the water is used with high efficiency, and the consumption is at its maximum. When the consumption falls to half the maximum, the water begins to be used with poor efficiency. One gate is then closed, and the other, under the action of the regulator, opens to full width, and the water continues to act with high efficiency till the consumption is reduced to about half the capacity of one wheel, or about one-fourth the maximum discharge. To avoid running the unused wheel in the water, the dividing partition between the two wheels is continued to low water, forming a separate draft-tube for each wheel. A small valve, not shown, is opened in the draft-tube pertaining to the unused wheel, admitting air and causing the water to disappear.

The American Turbine.—The Western States of the Union had use for water-power from their earliest settlement. Communication with the older parts of the country, at that time, was slow and expensive. For such heavy articles as water-wheels the cost of transportation was prohibitive, and the people were obliged to adopt such makeshifts as they could devise. Among the early settlers were many skilful mechanics, and necessity, the mother of invention, led to the development of turbines without much regard to the traditions of the engineering profession, the learning of the schools, or the laws of hydraulics; though, by a purely experimental process, they have, in some cases, been brought more or less into conformity with those laws. Of course many wheels of very crude design have been put on the market in that region, and have speedily disappeared, but certain types have persisted and not only held the market in that section, but have found extended application in the older parts of the country. Among these is the wheel we are now considering. In this case the revolving wheel is not essentially different in principle from the Swain wheel already described, though it differs in constructive details, being cast all in one piece, while in the Swain and Boyden wheels the floats are separate sheets of iron or steel, attached to the crowns and rims by mechanical means or set up in the moulds and incorporated with the wheel in casting.

Fig. 148 shows the wheel, i.e., the running part removed from the case. Fig. 152 is a horizontal section through the

FIG. 152.

wheel and case, showing the wheel as cast all in one piece. It also shows the chutes which control the admission of water and direct it upon the floats. The wheel is surrounded by an upper and a lower ring which are united by vertical plates, shown here as eight in number. Each chute consists of a pair of plates, one, called the fender or guard, rigidly joined to the upper and lower plate, the other, called the guide, turning on a joint near the periphery of the wheel. It is susceptible of a movement sufficient to bring it into contact with the adjoining plate and thus close the admission of water to the wheel.

Fig. 153 is a top view of the upper plate or dome, showing the mechanism for controlling the gates. A circular plate

FIG. 153.

rotates on a large hub on the dome and carries a strong arm to which is attached a toothed segment gearing with a pinion on the gate-shaft, which is controlled by the regulator. The circular plate carries also the arms attached to the rotating chutes, whereby the influx openings can be increased or diminished without altering the direction or velocity of the water to the disadvantage of the wheel.

This wheel is usually provided with a draft-tube and runs upon a wooden step resting on bridge-trees bolted to the interior of the tube. When placed in a wooden flume the weight is sustained by a broad flange resting on the floor of the flume (see Flume).

A "quarter-box" is fixed upon the dome of the wheel above the circular plate, forming a bearing for the shaft whereby the wheel is accurately adjusted in the case and prevented from binding. This box contains four blocks of wood, bearing against the shaft endwise of their fibres, and adjustable by means of set-screws.

Table 5 gives the results of certain tests of this form of wheel made at the Holyoke Testing-flume in July, 1894. They are given on the authority of Mr. E. S. Waters, engineer of the Holyoke Water-power Company.

Later tests of this wheel have shown a higher efficiency; viz., a test made in September, 1898, on a 36-inch wheel showed an efficiency

At full discharge of...............	82.57 per cent
0.92　　"	86.27　"
0.88　　"	86.69　"
0.83　　"	86.76　"
$\frac{3}{4}$　　"	85.1　"
$\frac{5}{8}$　　"	78.53　"
$\frac{1}{2}$　　"	72.09　"

These results are avouched by Mr. Waters.

The increased efficiency when the discharge is below the maximum is thought to be a peculiar advantage of this wheel. In its normal condition a wheel does not run with full discharge. It must have some margin, some reserve of power to meet contingencies, otherwise it would be liable to be stalled. It is a distinct advantage to be able to work with best efficiency in its normal running.

Fig. 154 shows an analogous mode of regulating the water-supply. The guides still serve as gates, but are hinged near the outer ends, the inner end being susceptible of motion till each guide touches its neighbor and so stops the influx entirely. This movement is effected by a toothed ring under the upper guide-plate, carrying a series of projections which act upon the outer extremities g of the guides. The ring is rotated by a pinion and shaft in the same manner as the toothed segment in Fig. 153. The cuts published by the maker are not clear

TABLE 5.— REPORT OF TESTS OF A 42-INCH RIGHT-HAND NEW AMERICAN TURBINE AT THE TESTING-FLUME OF THE HOLYOKE WATER-POWER COMPANY, HOLYOKE, MASS., JULY 14, 1894.

Number of the Experiment.	Proportional Part of		Head Acting on the Wheel in Feet.	Duration of the Experiment in Minutes.	Revolutions of the Wheel per Minute.	Quantity of Water Discharged by the Wheel. Cubic Feet per Second.	Power Developed by the Wheel. Horse-power.	Efficiency of the Wheel. Per cent.
	The Full Opening of the Speed-gate.	The Full Discharge of the Wheel, being the Discharge at Full Gate when giving the Best Efficiency.						
33	1.000	1.014	16.39	4	116.25	136.40	200.24	79.17
32		1.008	16.37	4	120.50	135.60	200.99	80.03
31		1.004	16.36	4	124.00	134.97	200.08	80.09
30		0.999	16.33	4	128.00	134.18	199.56	80.50
29		0.994	16.33	4	132.25	133.54	198.08	80.29
28		0.988	16.28	3	136.00	132.53	194.44	79.66
27	0.710	0.916	16.43	4	112.25	123.38	184.18	80.31
26		0.912	16.46	4	117.50	122.99	186.39	81.38
25		0.907	16.47	4	122.25	122.36	187.27	.82.13
24		0.900	16.51	4	128.25	121.60	187.73	82.65
23		0.893	16.56	5	134.80	120.85	188.14	83.09
22		0.877	16.55	4	144.00	118.57	181.37	81.69
21	0.504	0.798	16.87	3	110.00	108.99	164.76	79.20
20		0.795	16.74	3	113.33	108.15	164.34	80.24
19		0.793	16.53	5	117.00	107.18	163.29	81.47
18		0.785	16.56	4	124.00	106.25	164.62	82.70
17		0.774	16.59	3	129.33	104.85	162.89	82.77
16		0.764	16.67	4	133.75	103.67	159.35	81.50
15		0.736	16.77	4	141.75	100.20	149.58	78.68
14	0.389	0.700	17.10	4	106.00	96.17	140.73	75.64
13		0.697	17.04	3	113.33	95.58	144.28	78.30
12		0.685	17.11	4	120.25	94.22	144.91	79.45
11		0.674	17.13	4	125.25	92.76	142.40	79.21
10		0.660	17.15	4	130.00	90.85	138.96	78.83
9		0.646	17.20	4	136.00	89.10	134.26	77.43
8		0.633	17.25	4	144.50	87.44	127.89	74.94
7	0.230	0.530	17.42	4	102.62	73.51	100.61	69.44
6		0.527	17.44	4	108.50	73.10	101.20	70.16
5		0.520	17.48	4	113.25	72.27	100.23	70.13
4		0.509	17.48	4	121.00	70.80	98.85	70.60
3		0.501	17.48	4	128.75	69.66	96.42	69.99
2		0.494	17.51	5	136.20	68.73	92.73	68.10

enough to admit of a more detailed description. This wheel has met with some success. In fact it is named the "Success Wheel" by the maker. It has not yielded quite so high an efficiency as the Swain or the American, but it is probable that this does not result from the form of the influx orifices. The maker claims an efficiency of 80 per cent as the result of tests on this wheel.

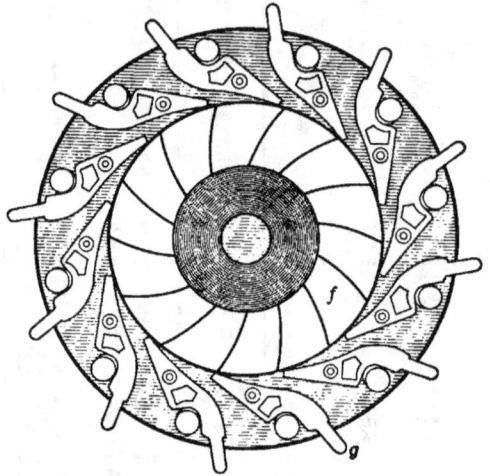

FIG. 154.

The Risdon Wheel is shown in vertical section with its attachments in Fig. 155, in exterior view in Fig. 156, and the wheel detached from the case appears in Fig. 157. *A* is the wheel, which is cast in a single piece; *B B* the guides, which are cast in one piece with the guide-plate *R*. Three of the guides are heavier than the others and support the pieces *F F* which sustain the crown-plate *E*, and have the slots *G* to admit of the movement of the gate. In other words, the supports *F F* rest on the thick guides, straddle the gate, and sustain the crown-plate. This latter is circular and makes a water-tight joint with the cylindrical gate *C* by means of the packing-ring *H*. From the crown-plate *E* rises the cylinder *I I* enclosing the shaft and forming, at the upper part, the bearing *K K*. Outside of *I I* is a second cylinder which fits *I I* closely at its lower end, and at the lower end is attached to the gate by the arms *L L*, and at the upper part is attached to the piston *oo*, which is fitted to the interior of the cylinder *P*. The space *oo* within the cylinder *P* communicates with the interior of the wheel through the orifices *Q Q*, while the lower side of the piston is exposed to

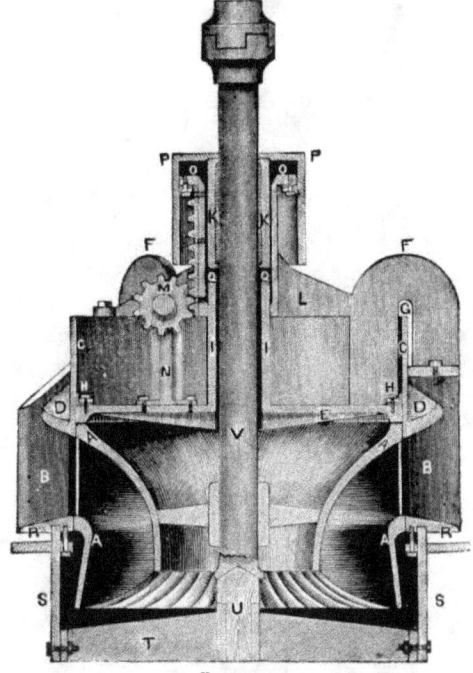

FIG. 155.

the pressure due the head, which balances the weight of the gate, the size of *P* being fixed with that view. The cylinder outside of *I I* carries a toothed rack, gearing with the pinion *M*, whereby the gate is raised and lowered. The shaft of the pinion *M* is supported by the standard *N* bolted to the crown-

plate. DD are projections or fingers attached to the gate
which fit between the guides and serve to contract the chutes
as the gate descends. V is the wheel-shaft, and W the shaft

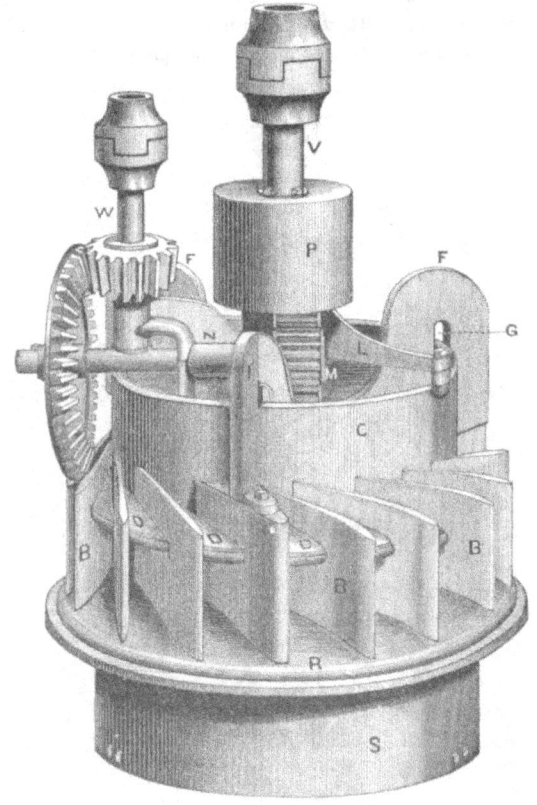

FIG. 156.

for operating the gate. U is the wooden step of the shaft
resting upon the cross-tree T. S is a short pipe discharging
the water below the level of the lower pool.   This wheel sits

inside a wooden or iron flume in which the water is under the pressure due the head.

From the shape of the buckets it will appear that the discharge of this wheel is mainly downward through orifices radial to the wheel, the inner end of the orifice of discharge moving not more than half as fast as the outer, a disposition clearly inconsistent with the highest efficiency. If we determine the speed of best efficiency of a wheel, and then reduce that speed

FIG. 157.

one-half, we shall reduce the efficiency as much as one-third or more. Making all due allowance for the effect of centrifugal force, it is plain that if the water near the outer part of the orifice acts with best efficiency, the water toward the inner end must act with diminished efficiency. The maker of this wheel claims an efficiency of over 91 per cent, and professes to have the certificate of an expert to verify this claim.

Fig. 158 is a section of the '' Victor wheel,'' made by Stillwell Bierce & Co. of Dayton, Ohio. W is the solid part of the wheel which revolves with the floats FF, the shaft S, and the

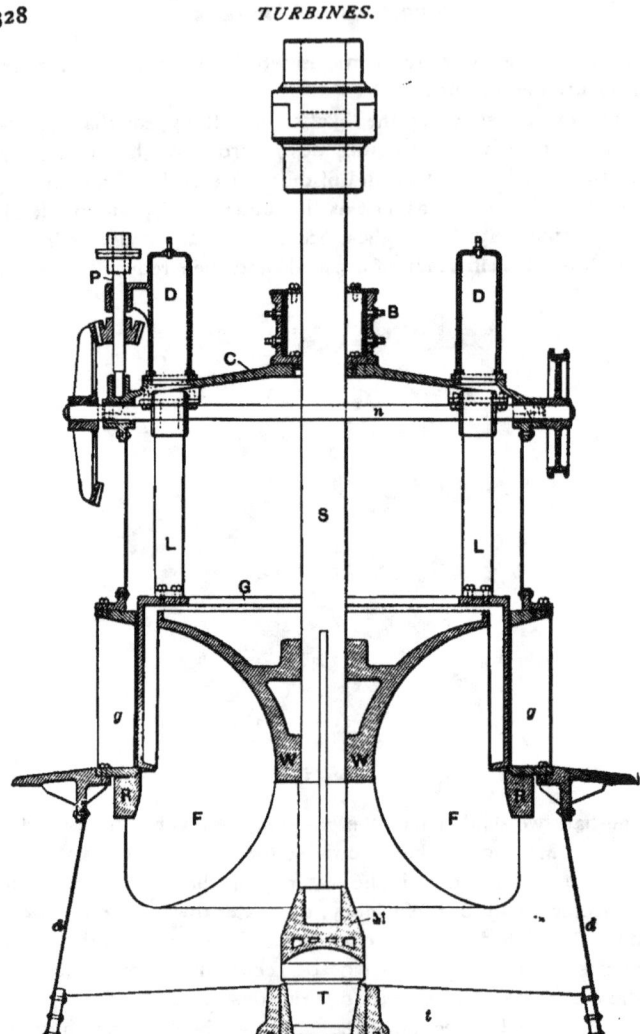

FIG. 158.

ring *R.* *g g* are the guides attached to fixed plates. *G* is the cylindrical gate to which are attached the toothed uprights *L L* operated by spur-gears on the shaft *n*, one end of which carries a bevel-gear for controlling the gate by means of a hand-wheel acting through the small shaft *P*, the other a wheel through which the regulator acts. The uprights *L L* rise through the crown-plate *C* into hollow chambers *D D.* The reader will notice the internally projecting lip on the bottom of the gate for giving the desired direction to the water entering the wheel. *M* is the foot on the bottom of the shaft, and *T* the wooden step on which it rests. *t* is a cross-tree fixed to the bottom of the draft-tube *d*, and sustaining the socket for the step *T.* Only the outlines of two buckets are shown in the drawing; the projections of the remainder are not shown. The discharge is all around the free edge of the bucket, so that this wheel has an inward, a downward, and an outward discharge.

Tests of this wheel at the Holyoke Company's flume have given an efficiency

For full discharge of............	80 to 82	per cent
$\frac{7}{8}$ "	78 to 81	"
$\frac{3}{4}$ "	75 to 77	"
$\frac{5}{8}$ "	70 to 72	"
$\frac{1}{2}$ "	63 to 64	"

Duplex Wheel.—Fig. 159 shows an arrangement which has met with some application, designed to use a varying quantity of water with good effect. It consists of a double wheel, that is, a wheel with two sets of floats and two sets of guides, each controlled by a separate gate, and susceptible of being used separately or in combination. *f f* are the floats, *g g* the guides, and *G G* the gate of the outer wheel *w*; *f'f'* the floats, *g'g'* the guides, and *G'G'* the gate of the inner wheel *w'*. The gate of the outer wheel closes by rising, that of the inner by descending. When the water is low, the gate *G* of the outer wheel is closed, and the water acts only on the floats *f'*; as the flow increases the inner gate is closed, the outer gate opened, and the water acts on the floats f. In high stages of the stream both gates are open. The shaft runs on a suspension bearing near the top and should have a guide-bearing or quarter-box near the wheel, above or below, to centre the

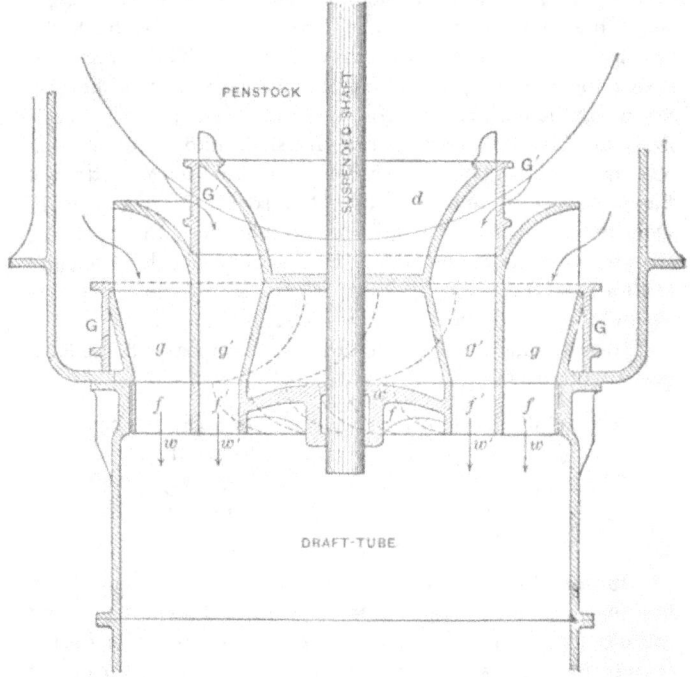

FIG. 159.

wheel in its case. The form of the buckets and guides appears in dotted lines.

This combination will, no doubt, use a quantity of water equal to the capacity of the inner wheel, with greater economy than the same quantity of water could be used by a single wheel of capacity equal to that of the two, and it would, no doubt, give a reasonably good result with the outer gate alone open. The combination presents in an exaggerated degree the defect already pointed out, viz.: if the velocity is such that the outer wheel gives its best effect, it will be much too slow for the inner wheel. Moreover, when the inner wheel is in use and the outer gate closed, the rotation of the outer wheel is an additional burden of useless work laid upon the inner wheel. The most serious defect of the combination, however, is this: in the arrangement shown, neither of the wheels acting by itself would give a good result on part gate. Wheels of this form are by no means new. They have given good results with full discharge, but their great defect has been the impossibility of a good result at part gate. The only successful mode of regulation has been by a separate gate for such guide opening, but no approach to economical regulation is possible with gates of the form shown. It must be observed that the gate is outside of both guides and floats, and that the water passes through a chamber of some size before reaching the guides. A serious loss of head always occurs when water passes through a narrow opening from one chamber to another. The defects of this arrangement are similar to what would be met with in attempting to regulate the supply of water to a turbine by a throttle-gate in the penstock.

There are other forms of turbine which the writer would be glad to describe and consider if he could obtain the necessary data. Prominent among these is the Leffel wheel, made at Springfield, Ohio, which has met with extended application in the Western States and is not unknown in the Eastern section, which facts entitle us to presume that it has points of merit. But nothing can be learned from the numerous publications of the makers as to the construction of the wheel or the manner in which the water acts to propel it. Equally reticent are the makers as to the efficiency of the wheel. They do not believe in tests and assume that their readers attach no importance to them. Their publications abound in vociferous assurances of the excellence of the wheel, but neither give the intending purchaser the means of judging for himself nor furnish him with the results of exact experiments as to its performance.

The Hercules Wheel, so called, made by the Holyoke Machine Company at Holyoke, Mass., is very extensively used in New England and the Middle States. The makers issue a very neat catalogue giving illustrations and description of all the appurtenances and attachments of the wheel, but the inquiring reader will search it in vain for any information as to the construction of the wheel, the forms of the floats and guides, the regulation of the water, or its mode of action. Results of experiments upon this wheel are published, which are of commercial interest, but such information has no scientific interest unless accompanied by a knowledge of the construction of the wheel and the manner in which the water acts thereon.

The Hurdy Gurdy Wheel.—We now come to a form of wheel which can hardly be classed as a turbine, since it cannot, generally, like the latter be placed below the level of the tailwater, neither can it act in connection with a draft-tube. It consists of a series of cup-shaped vanes attached to the periphery of a wheel and acted on by a jet of water. It is usually mounted on a horizontal shaft, which carries the organs for transmitting motion, and in that form is probably the simplest combination for the development of water-power that can be devised. It is an impulse-wheel in the most elementary form, neither reaction nor direct weight contributing anything to the result.

Fig. 160 is an elevation of the wheel in a plane transverse to the shaft, Fig. 161, parallel to the same. The wheel is here shown enclosed in a wooden case, to protect the surroundings from the spray which could not fail to be plentifully distributed by such a wheel. Fig. 160*a* is a section of the vane by a plane tangential to the wheel; Fig. 160*b*, a section perpendicular to *a*. It is shaped so as to divide the stream and throw the parts clear of the adjoining vane, and cannot, therefore, exactly reverse the direction of the stream, but it probably realizes this

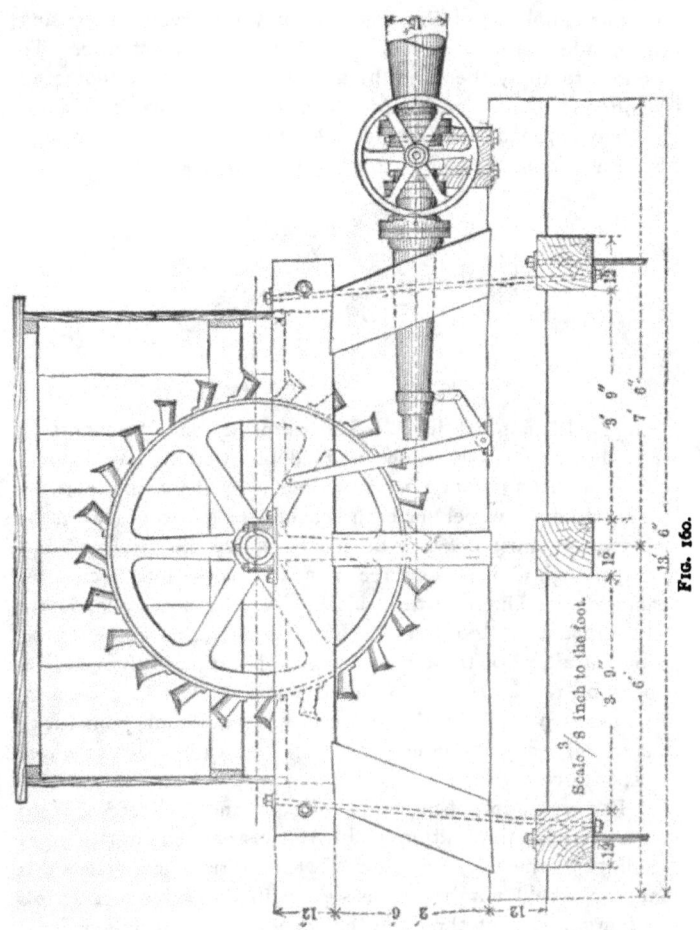

FIG. 160.

condition of perfect efficiency as closely as is possible in any wheel.

The simplicity of this wheel is only preserved by limiting the number of nozzles to two or three, or at most four. To attempt to apply the water by a series of nozzles extending all round the periphery would, with the necessary regulating mechanism, lead to as complex a form as in any existing wheel. For this reason the wheel cannot give a large amount of power

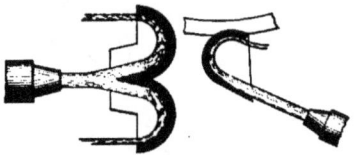

FIG. 160*a*.   FIG. 160*b*.

except with a great head. The simplicity and cheapness of the wheel, however, allow of its duplication to any extent, either by placing a number of wheels upon the same shaft, or by providing a wheel for each machine or group of machines. This wheel must not only be placed above the tailwater, but high enough above to be free from any risk of interference by backwater. The head thus sacrificed would, on any moderate fall, amount to too large a proportion of the whole to be economical. For these reasons the application of the wheel is not to be recommended for heads under 50 feet, nor even in that case when the head is liable to great variation from backwater. These conditions restrict the application of the wheel to special cases.

The regulation of the supply of water and the speed of these wheels is sometimes attempted by means of cocks in the pipes leading to the nozzles. For a wheel with four nozzles this method would not be inconsistent with a good result at full discharge, and at three-fourths, one-half, and one-fourth of same. But the water from the pipe which was undergoing contraction would act at a great disadvantage on account of the loss of head incident to the action of the cock. For a wheel propelled by a single nozzle as at Fig. 160 this mode of regulation would be wholly inapplicable. The speed of the vanes is fixed by the application or use made of the power and may be considered invariable. This velocity we may take as that of greatest efficiency, viz., one-half that due the full head

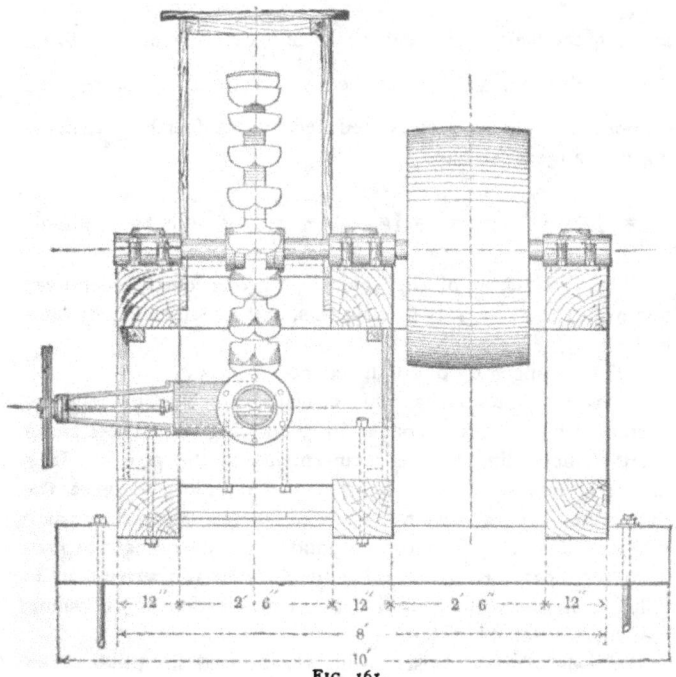

Fig. 161.

acting on the wheel. In the arrangement of Fig. 160, if the supply be controlled wholly by the gate there shown, the issuing stream will be of the full size of the nozzle efflux, irrespective of the extent to which the gate is closed. The gate acts to diminish the pressure in the space between the gate and efflux, and consequently the velocity of the issuing stream. Equation 39 gives for the energy exerted on a cup-shaped vane moving with the velocity u, under the action of a stream of water, of cross-section a, with the velocity v,

$$Pu = wav\,\frac{2u(v-u)}{g},$$

w being the weight of a cubic foot of water. Suppose u to be the velocity of maximum efficiency, viz., $u = \dfrac{v}{2}$. If now we suppose the velocity v to be reduced by one-fourth, u remaining unchanged, we have

$$Pu = \tfrac{3}{4}wav\,\frac{v(\tfrac{3}{4}v - \tfrac{1}{2}v)}{g} = \tfrac{3}{4}wav \times \frac{1}{2}\frac{v^2}{2g} = \tfrac{3}{4}Qw \times \tfrac{1}{2}h = \tfrac{3}{8}Qwh.$$

The diminution of the velocity by one-fourth, therefore, diminishes the power by five-eighths, and the efficiency by one-half.

Diminishing v by one-half reduces Pu to 0.

It is understood that this difficulty is, in some degree, overcome by a series of nozzle-tips of varying sizes, these being changed according to the requirements of the power. It is manifest, however, that no efficient regulation to meet the requirements of a varying load could be effected by such means. The only economical mode of controlling the flow from the nozzle is by a device to contract the stream at its efflux from the same; varying the cross-section of the issuing stream, but not its velocity.

Wheels of this construction were tested in 1896 at the Ohio State University, at Columbus, Ohio, and yielded under a head of 162 feet, according to certificates of officers of that institution, a maximum efficiency of 92 per cent, which statement is not beyond belief. Tests of these wheels under low or medium heads are liable to be misleading. The head on which the efficiency is computed is usually taken as limited by the height at which the stream strikes the vanes, a mode of computation which does not give a fair basis of comparison between this wheel and others. This wheel stands above the tailwater or lower pool, and often, when backwater is to be expected, many feet above the same; whereas another wheel would stand below the same level, and utilize the entire head at all times. Obviously the head to be used in these computations should be the entire head available between the upper and lower pools, with no deduction for that portion of the head which has to be sacrificed to the practical form and character of the wheel.

CHAPTER XV.

APPENDAGES AND ATTACHMENTS OF TURBINES.

The Case.—This term is usually applied to the fixed parts sustaining the guides and gate or gates, which the maker furnishes with the wheel and is regarded as a part of the same. It embraces the plate or disk which supports the guides and a plate which relieves the wheel of the pressure of the water. In the Boyden wheel, Fig. 147, these two plates are the same. This latter plate contains a stuffing-box through which the shaft passes, or, as in the Boyden wheel, a pipe reaching above the water. The case usually carries the mechanism for manœuvring the gate, and often the step on which the shaft runs. The case is usually enclosed in a flume in communication with the upper level, and in most modern wheels contains the guide-bearing which centres the wheel in the case. In every form of the turbine, properly so called, there are inevitably two annular openings between the wheel and case through which the water is free to pass under a pressure not greatly less than that of the full head acting on the wheel. With the most accurate adjustment of the guide-bearing this opening can hardly be less than $\frac{1}{8}$ inch in width. For a wheel 6 feet in diameter there is a space of $2 \times 6\pi \times \frac{1}{96} = 0.393$ square foot for the free escape of water, which under a head of 20 feet might occasion a loss of 5 or 6 cubic feet per second. There appears to be no means in use for obviating this waste, since, if the clearance were made narrower, maladjustment or derangement of the guide-bearing might cause the wheel to bind. From the rapid motion, anything in the nature of a

338

packing-ring would waste more power in friction than it could save in water.

A patent was issued to the writer under date of January 19, 1886, on the combination indicated in Fig. 162, in which W is supposed to pertain to the revolving wheel, C to the stationary case, with an interval between them as small as will permit the wheel to run. A ring R rests upon a seat formed on the case and encircles the wheel with a clearance much less than is necessary between the wheel and case. This ring is held upon its seat by the pressure of the water.

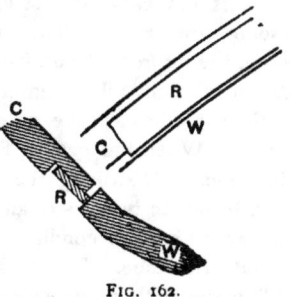

FIG. 162.

Any movement of the wheel by the derangement of the guide-bearing merely alters the position of the ring on its seat; and while a clearance of $\frac{1}{8}$ inch might be necessary between the wheel and its case, a clearance of $\frac{1}{32}$ would be ample between the wheel and ring.

The Draft-tube.—The head or fall acting on a wheel is determined by physical conditions. It is the elevation of the upper pool or upper level, otherwise called the headwater, above the lower pool or lower level, otherwise called the tail-water. A wheel set any number of feet above the latter and discharging into the air loses so many feet of the head; but if it discharges through a vertical pipe whose efflux is below the lower level, the water will free the pipe from air, and the pressure against which the water issues from the buckets will be reduced by the exact equivalent of that part of the head between the centre of the discharge-orifices and the tailwater. In other words, the draft-tube renders the entire head available, though the wheel may set considerably above the lower level. Of course this height cannot exceed that corresponding to the atmospheric pressure, which at the sea-level is about 34 feet as an average, otherwise there would be a vacuum immediately below the wheel, and a part of the head would be lost. In practice the draft-head never reaches this limit or closely approaches it; 27 feet may be considered the extreme height, and it is usually much less.

It is not easy, at first view, to understand how the water issuing from the discharge-orifices of the wheel operates to expel the air from the draft-tube. The process is very simple. When the draft-tube is empty, the surface of the water stands at the same level inside as outside, viz., the level of the tail-water. When the gate is raised, streams of water fall inside the tube, and, striking the surface, beat it into foam, i.e., fill it with minute bubbles of air and create a current which carries the water in that condition downward, and the bubbles rise outside the tube. The smallest bubbles of air do not rise in still water with a velocity of more than 9 to 12 inches per second; a velocity greater than the latter will carry them downward. In this manner the air is entirely expelled from the tube.

The pressure on the draft-tube is from without inward, and it must be formed to withstand such a pressure. At the foot of the tube, that is, at the surface of the tailwater, there is no pressure outward or inward. At any point above the tailwater the pressure is inward and equal to that due the height of the said point.

By means of the draft-tube, wheels on a vertical shaft may be placed so as to be accessible for examination and repairs without draining the pit. With this view it is advisable to fit the tube with a manhole for admission. A cock is sometimes inserted in order to fill the tube with air and free it of water before opening the manhole. As to wheels on horizontal shafts, it is only through the draft-tube that their use is possible. Such a shaft must carry the gear or pulley for transmitting motion, and these organs could not operate below the level of the tailwater. It is not too much to say that the adoption of the draft-tube marks a distinct era in the development of the appliances of water-power. By giving a "flare" to the draft-tube, causing it to expand downward, we may induce the same action as occurs in the diffuser (next paragraph) and thereby diminish to some extent the loss of head due to the velocity with which the water leaves the wheel.

The Diffuser is an expanding passage through which the water passes after leaving the wheel. As applied to an outward-flow turbine it takes the form of Fig. 163. To an inward-flow wheel it is not readily applicable unless the wheel works with a draft-tube, in which case the diffuser consists in a conical form of the latter, widening downward. The diffuser is the invention of Mr. U. A. Boyden, and is especially applicable to the Boyden wheel. Its purpose is to employ the momentum with which the water leaves the wheel, in diminishing the back pressure or the pressure acting against the discharge. Its practical effect is to slightly increase the head acting on the wheel. The extent to which it increases the head is represented by the head due the velocity with which the water leaves the wheel diminished by the head due the velocity with which it leaves the diffuser.

The effect of the diffuser can be best expressed in the language of Mr. J. B. Francis,* with reference to the case of Fig. 163: "On leaving the wheel, it [the water] necessarily has a considerable velocity, which would involve a corresponding loss of power but for the diffuser, which utilizes a portion of it. When operating under a fall of 33 feet and the speed-gate raised to its full height, this wheel discharges about 219 cubic feet of water per second. The area of the annular space o, o, o, o, Fig. 163, where the water enters the diffuser, is $0.802 \times 8.792\pi = 22.152$ square feet; and if the stream passes through this section radially, its mean velocity must be $\frac{219}{22.152} = 9.886$ feet per second, which is due to a head of 1.519 feet. The area of the annular space p, p, p, p, where the water leaves the diffuser, is $1.5 \times 15.333\pi = 72.255$ square

---

* Lowell Hydraulic Experiments, 1868, p. 231.

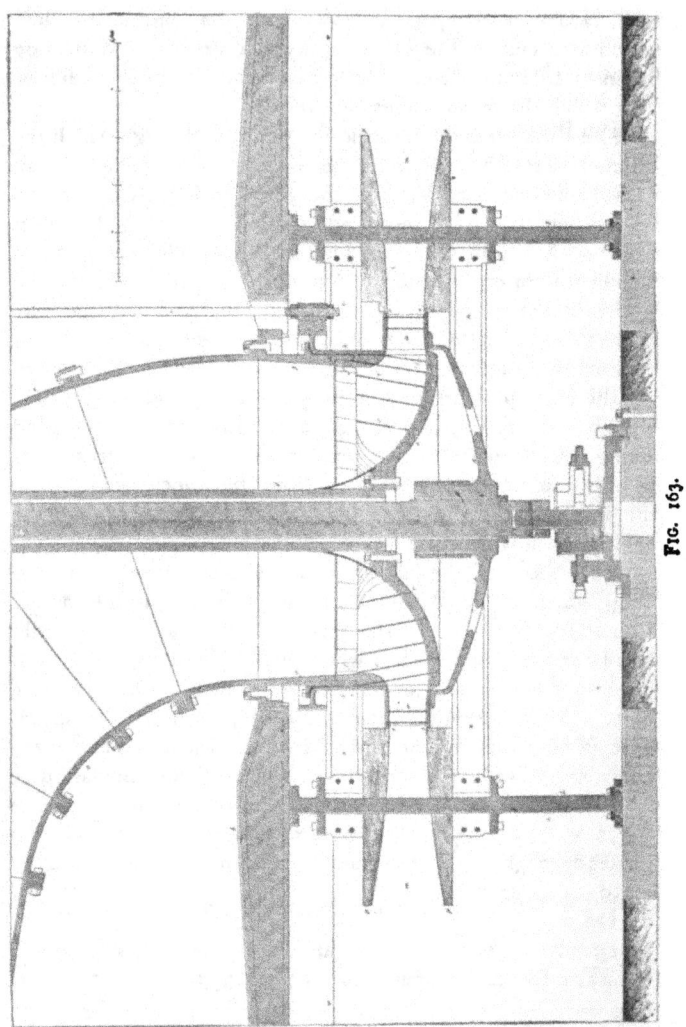

FIG. 163.

feet, and the mean velocity $\dfrac{219}{72.255} = 3.031$ feet per second, which is due to a head of 0.143 feet. According to this, the saving of head due to the diffuser is $1.519 - 0.143 = 1.376$ feet, being $\dfrac{1.376}{33 - 1.519}$, or about $4\frac{3}{8}$ per cent of the head available without the diffuser, which is equivalent to a gain in the coefficient of useful effect to the same extent. . . . Experiments on the same turbine with and without a diffuser have shown a gain due to the latter of about 3 per cent in the coefficient of useful effect. The diffuser adds to the coefficient of useful effect by increasing the velocity of the water passing through the wheel, and it must of course increase the quantity of water discharged in the same proportion. If it increases the available head 3 per cent, the velocity, which varies as the square root of the head, must be increased about $1\frac{1}{2}$ per cent, and the quantity discharged must be increased in the same proportion. The power of the wheel, which varies as the product of the head into the quantity of water discharged, must be increased about $4\frac{1}{2}$ per cent.''

The Flume.—In order to cause the water to act upon the wheel, the latter is usually enclosed in a chamber communicating with the upper level by a penstock or head-race, and with the lower level by a tail-race, which it enters by passing through the wheel. This chamber is called the flume. The outward-discharge turbine does not require a flume properly so called, but, as very few of that class of wheels are now built, we may regard the flume as a necessary appendage of the turbine. The flume may be of wood, iron, stone, or a combination of these materials, the choice depending very much upon the head. With a head not exceeding 15 feet, wood or a combination of wood and stone will generally be found preferable, and with the aid of draft-tubes the same construction may be employed up to a head of 30 feet. Above this limit iron is generally to be preferred.

Fig. 164 is a vertical cross-section of a flume for a double turbine on a horizontal shaft under a head of 15 to 30 feet. For a head less than 15 feet the arrangement would be in no way different except in the absence of the upper deck. The floor-timbers of the flume rest upon the walls of the race. Into

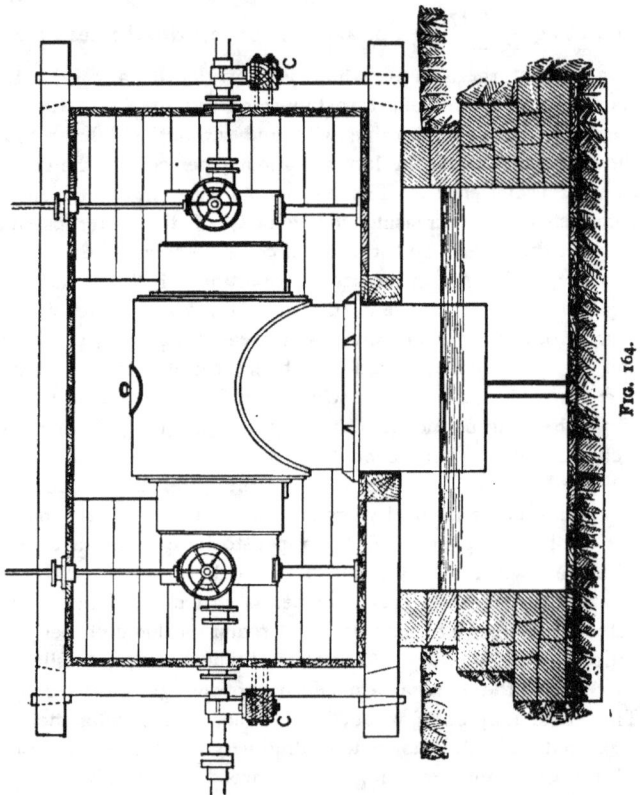

Fig. 164.

these are mortised the uprights, and the latter are joined at the top by the upper cross-timbers or ties which sustain the deck. These connections are made by dovetailed joints and wedges tomeet the pressure tending to lift the deck. The floor of the flume is represented at Fig. 165. The two central floor-timbers cannot extend across the flume, but are terminated by the timbers *E E*, which, together with the two cross-timbers into which they are mortised, form a square opening large enough to admit the draft-tube. This is reduced to an octagonal opening by the corner blocks *B B*, confined by bolts. The planking overlaps a little and is finished to a circle conforming

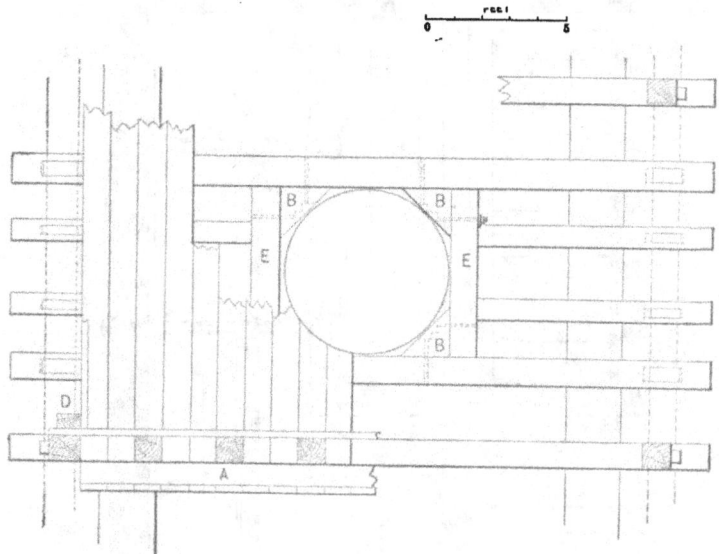

FIG. 165.

to the draft-tube, which rests thereon by a broad flange. The shaft passes through stuffing-boxes in the side of the flume and rests in bearings outside the flume, which are supported by timbers bolted to the uprights. The small shafts for manipulating the gates pass through stuffing-boxes in the top of the flume.

Fig. 166 is a longitudinal section showing the tail-race and penstock. The latter takes smaller dimensions both in height

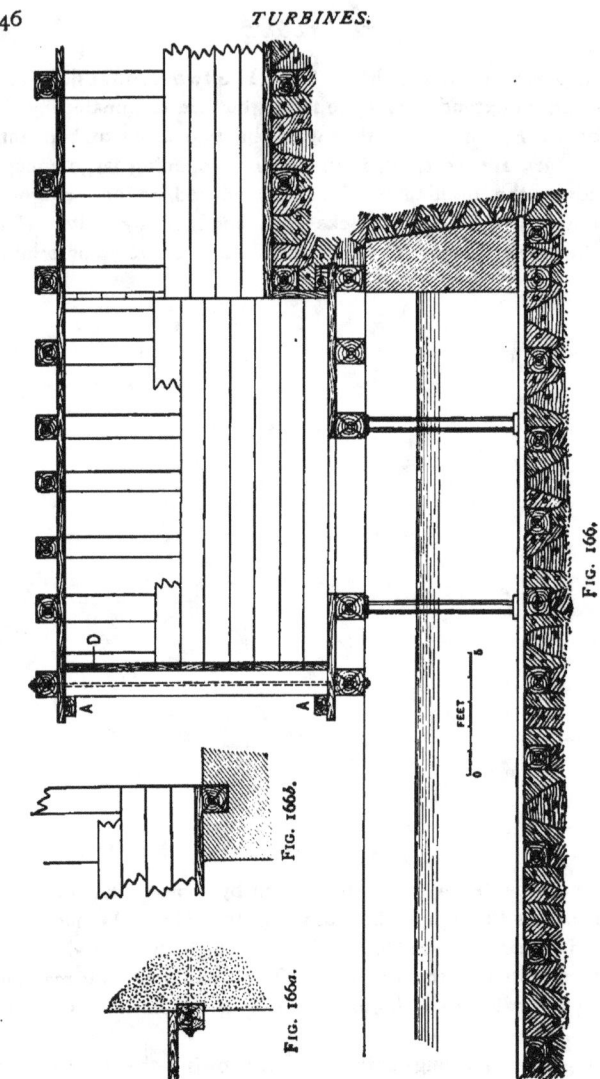

FIG. 166,

FIG. 166b.

FIG. 166a.

and width from the flume, to the source of supply. At the down-stream end of the flume the thrust tending to push the uprights out of place cannot be met in the same way as at the sides. The floor-plank are extended a couple of feet, and on the outer ends are bolted the timbers *A A* which sustain the pressure against the uprights. The pull upon the uprights is here met by iron rods joining the deck to the floor.

At the influx of the penstock it is usually necessary to make a water-tight connection with the masonry of the dam, or canal wall. This is shown at Fig. 166*b*, where the plank are spiked to timbers built into the masonry. The walls of the flume are sometimes of masonry, and the floor of timber. In this case the ends of the cross-timbers are firmly built into the masonry, and the joint between the plank and the latter may be calked with oakum and paid with tar. The joint at the ends of the plank is arranged as at Fig. 166*a*. The end cross-timber is partly imbedded in the masonry and confined thereto by bolts, the joint at the ends of the plank being calked as before.

The advantage of mounting two wheels on a horizontal shaft consists in mutually neutralizing the thrust. This, however, only occurs completely when both wheels are running at equal gate. A thrust-bearing cannot be dispensed with in this case, as one wheel is liable to run at full gate while the other has its gate wholly or partly closed.

Fig. 167 shows a very neat arrangement of a cast-iron quarter-turn flume for a high head. The flume, by means of a flange strengthened with gussets, rests on iron beams spanning the tail-race, or wheel-pit. The quarter-turn is made in two halves, and has a neck strengthened with gussets through which the shaft passes by a stuffing-box. This neck carries a broad flange to which is bolted the bridge-tree for sustaining the crown-gears, which give motion to the horizontal shaft carrying a pulley and belt for delivering the power. To the horizontal end of the quarter-turn is bolted a very short length of pipe or, more properly, a ring, pierced with rivet-holes for attaching the wrought-iron penstock. A manhole appears in

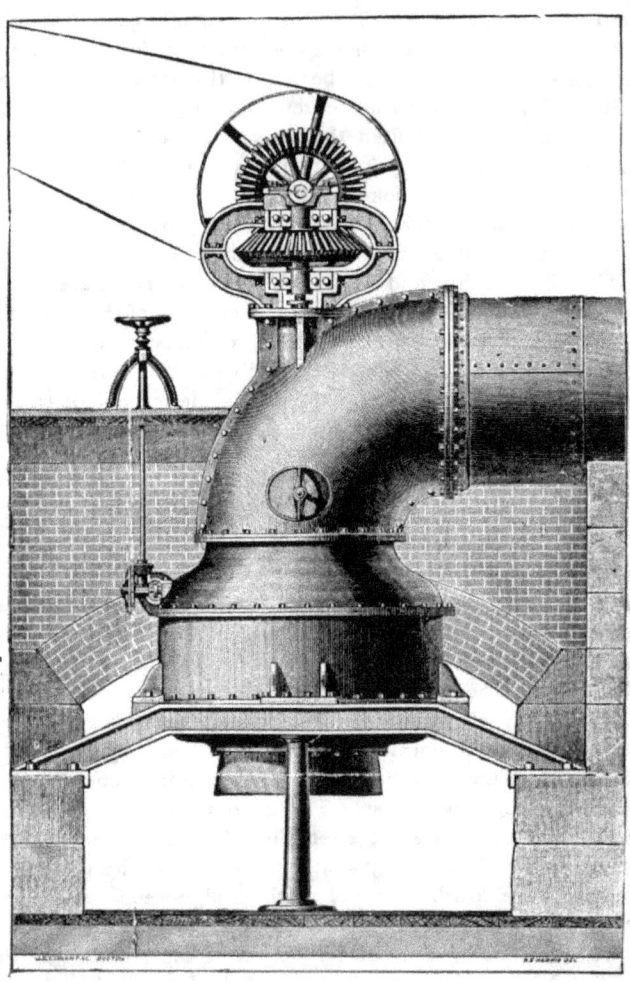

Fig. 167.

the side of the flume for admission to the interior. In tnis form the word "flume" is usually applied to the chamber enclosing the wheel, the remainder of the pipe being known as the penstock. This arrangement is presented by the Rodney Hunt Company of Orange, Mass.

In adopting this cut the writer feels obliged to disclaim the method therein indicated of constructing the arch over the tail-

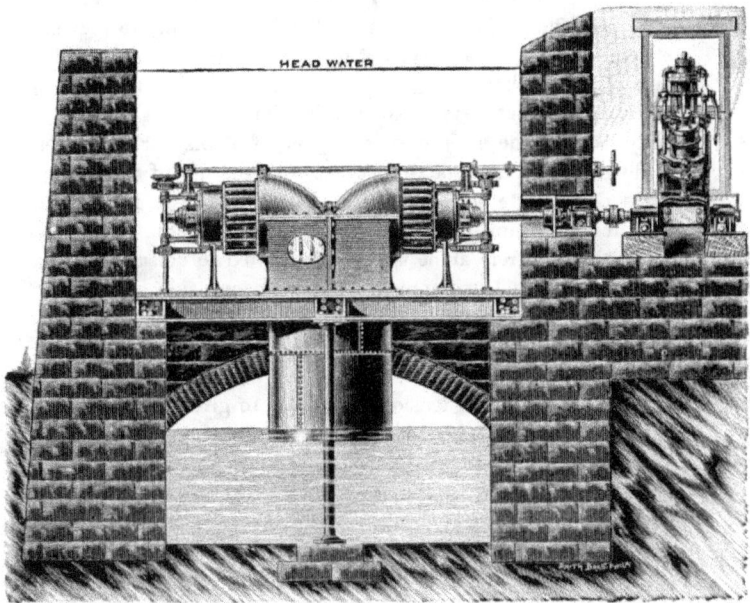

HEAD WATER

FIG. 168.

race. The bricks are shown as laid flatwise, which would be as unworkmanlike as putting bricks on end in a vertical wall. Bricks in an arch are always laid with their ends or their edges to the soffit, and so as to receive the pressure on their broadest faces.

Fig. 168 shows a flume of stone, iron, and wood, viz., stone side walls, iron beams, and wooden floor, for a pair of turbines to drive a pulp-grinder. The main shaft goes through a stuffing-box built into the wall and rests in a bearing in the same opening. This figure is taken from the catalogue of S. Morgan Smith of York, Pa. The hand-wheel shaft for manipulating the gates also passes through a stuffing-box; though, the flume being open at the top, there is nothing to hinder it from entering freely from above. When such a structure rests upon anything but good sound rock it should have a substantial flooring of timber and planking, not only to prevent settlement of the masonry, but to protect the bottom from the wash of the discharge.

Flumes of Concrete.—Noteworthy improvements have recently been made in the construction of floors for fire-proof buildings, in the attempt to dispense with the use of timber, by means of concrete strengthened with steel rods. A floor must either rest on beams or must act by itself as a beam. It is well known that concrete alone is very ill adapted to act as a beam, which is exposed to compressive strain on one side and tensile strain on the other. It resists compressive strain well, but offers comparatively slight resistance to a tensile strain. This defect is remedied by embedding steel rods or plates in the concrete on the side exposed to tension. In order to give the concrete a hold or grip on the rods, the latter are, in some of the systems of construction, corrugated, i.e., notched or ridged; in others they are twisted. It is claimed that the operation of twisting modifies, in some degree, the expansibility of the metal, so that it is affected by heat in almost precisely the same manner as the concrete itself. In a concrete beam reinforced in this manner the tension is thrown wholly upon the rods and the concrete acts entirely in compression. A beam of pure concrete is not susceptible of any deflection under a bending strain, but will break before the tension becomes great enough to cause any deflection. A beam strengthened by rods will deflect materially before rupture takes place, and will endure blows of considerable force. A concrete floor, resting on walls, resists as a beam both transversely and longitudinally, and is strengthened by rods embedded in the mass near the bottom, running in both directions and crossing each other. A floor of this construction, stated to be 9 inches thick, 17 feet long, and 11 feet wide, was loaded in the author's presence to the extent of 1200 pounds per square foot, equivalent to the pressure of 19 feet of water, without any sign of injury. This appears to indicate that a flume 20 feet square with floor and walls of this construction 18 inches thick would stand safely under a head of 20 feet. We may at least safely affirm that flumes could be built in this manner with dimensions which would not make them inadmissibly expensive.

Step and Suspension Bearing.—A bearing to support the shaft, wheel, crown-gear, and attachments while in rapid motion is no light problem in mechanics, though not the most difficult form of the problem. This occurs in the thrust-bearing of the steam-propeller which sustains the entire pressure that drives the vessel through the water, often amounting to 50 tons acting through a shaft in rapid rotation. This strain is successfully met by a thrust-bearing, Fig. 225, which consists of a series of deep grooves cut on the shaft near the end, forming a series of collars which bear against corresponding surfaces in a hollow box. Often the external collars are formed of soft metal, consisting mainly of tin, which is poured into the hollow box and forms a casting conforming to the grooves on the shaft. This method having been successfully employed for the thrust-bearings of screw steamers, there can of course be no doubt as to its applicability to the suspension bearings of turbines. The Boyden wheel was generally, at first, provided with a suspension box at the top of the shaft, made in the above-described manner, and a guide-bearing at the bottom, Fig. 147. This method has latterly fallen into disuse, and most vertical shafts of turbines now run on wooden step-bearings. This arrangement is seen in the Swain turbine, Figs. 149 and 150, and in the Victor turbine, Fig. 158. The wooden block rests in a socket which is usually supported by cross-trees in the draft-tube, but sometimes on the bottom of the pit.

Usually the bottom of the shaft is squared and enters a broad foot which rests on the wooden block, the lower surface of the foot being spherical or conical and bearing on the ends of the fibres of the block. In the Risdon turbine, Fig. 155, the foot is formed on the shaft. The block is sometimes free to revolve in its socket, so that if the upper surface becomes heated, so as to greatly increase the friction, the block will revolve on its lower surface and give the upper surface time to cool. The Geyelin wheel, Fig. 159, runs on a suspended shaft. When the step-bearing was introduced, it was thought that nothing but lignum vitæ or some of the very heavy, close-grained Southern woods would answer for steps, but maple and oak are found to work very well. It is a common practice, in preparing steps, to dry the wood thoroughly and then boil it in linseed oil, impregnating it so fully that it becomes in some measure self - lubricating. Such steps, although constantly immersed in water, sometimes become heated so as to char the wood and soften the iron.

Fig. 169 represents a type of step bearing which has found some application, though not in the precise form here shown. It usually takes the form of a chamber containing oil under high pressure, which the shaft enters through a stuffing-box, the pressure on the end of the shaft being sufficient to sustain the weight of the latter and its attachments. This requires the gland of the stuffing-box to be so strongly set up that the friction on the shaft occasions a serious waste of power. The arrangement of Fig. 169 is designed to avoid that difficulty. The shaft rests on the revolving plate a through the intervention of the four-armed cross n, which makes the pressure perfectly central and uniform on the plate a. The surfaces of contact between a and b are dressed to an exact fit, and the wear keeps them in that condition. The cylinder g is attached to b and provided with a stuffing-box embracing the shaft. Oil for lubrication comes through the pipe e, being forced in under great pressure by a small pump. It fills the space j, and exerts a pressure on the plate *a* nearly but not quite· equal to the weight of the shaft and its attachments, so that the contact-surfaces sustain very moderate friction. The oil passing these surfaces fills the space *c* and would come to great pressure there if confined. The pipe *d* communicates with this space and conveys the oil back to the tank from which the pump draws its supply. The pressure on the stuffing-box is very slight, and the latter need not be screwed tight enough to occasion any sensible loss of power.

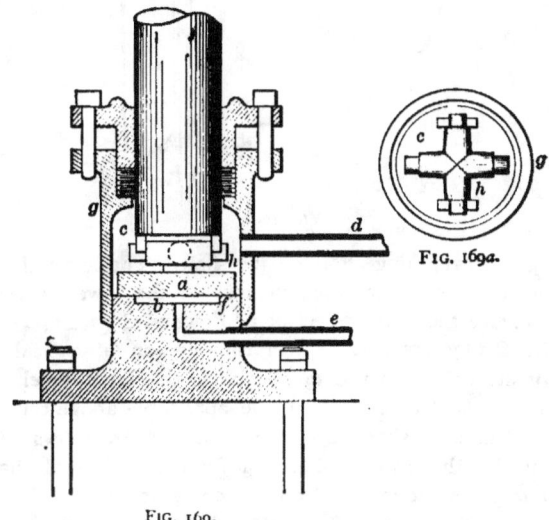

FIG. 169*a*.

FIG. 169.

Wheels Sustained by the Pressure of the Head.— Fig. 169*b* indicates the method adopted at Niagara Falls for sustaining the weight of the wheel, shaft and attachments, including the armature of the 5000 h.p. electric generator, without step or suspension bearing. *A* is the supply pipe or penstock delivering water under a head of 130 or 140 feet. There are two wheels, *W W*, on the same shaft *S*. This shaft passes through a disk-pipe like that of the Boyden turbine, Fig. 147. The disk-pipe widens into a disk both above and below, which bears guides

like the Boyden turbine and are thereby united to the casing.
It will be perceived that the upper disk has orifices through which
the full pressure of the head acts upon the upper wheel-web,

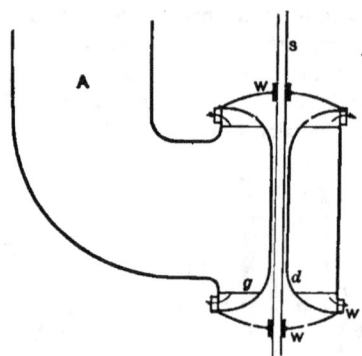

FIG. 169*b*.

causing a great lifting effort on the shaft. The lower disk has
no openings, but there are openings in the lower wheel-web
which relieve the lower wheel of all downward pressure.

The Bridge-tree.—A wheel running on a vertical shaft
usually sits below, or not much above, the lower level, while
the power usually has to be made applicable above the upper
level. This necessitates a support at a greater or less height,
not only for the upright shaft, but for the horizontal shaft to
which it gives motion, and so formed as not to interfere with
the motion of the gears which transmit the movement from one
to the other. This support is called the bridge tree. On a
moderate or low fall the bridge-tree can usually rest on the
flume; but a high fall requires such a support at a considerable
distance above the flume. A very neat example of a bridge-
tree resting on the flume is shown in Fig. 167. It is in two
halves, the lower of which is bolted to a flange on the neck
which projects from the iron flume. This carries a guide-bearing for the upright shaft. The upright half or yoke is bolted
to the lower by flanges and likewise carries a guide-bearing in

addition to the bearing for the horizontal shaft. The crown-gear thus works in the recess between the two halves of the bridge-tree and is very firmly supported by the two bearings.

Fig. 170 shows a bridge-tree in the form of a heavy casting resting on masonry walls. This occurs on a head of 50 or 60 feet, which brings the horizontal shaft at a considerable height above the flume.· The gears are proportioned so as not to change the relative velocity of the horizontal and vertical shafts. This design is given by the Dayton Globe and Iron Works, Dayton, Ohio.

Fig. 171 shows a bridge-tree designed to be supported upon rolled I beams. The beams are arranged in pairs, and the connection is made by means of cast-iron brackets bolted through the web of the beam where a bolt-hole does not sensibly weaken the beam. The tree carries the bearing for the vertical and also for the horizontal shaft, these bearings being to some slight extent adjustable laterally. It will be noticed that this arrangement contemplates an iron gear on the horizontal shaft, and a wooden mortise gear on the vertical. Such a pair of gears runs with much less noise and is less liable to breakage than a pair of iron gears. This design is given by the Dayton Globe Iron Works Company.

Fig. 172 is a simpler form of bridge-tree designed for attachment to a wooden beam, and suitable for a low head. The upright shaft here has two bearings, one attached to the beam, the other to the bridge-tree. It will be noticed that the bridge-tree is adjustable, as well as the bearings which are attached to it. The bolt-holes through the foot of the tree are somewhat elongated, so as to admit of a slight movement of the tree after the bolts are inserted. Before turning up the nuts the tree is brought into the right position by the wedges. Similar adjustments are seen on the bearings attached to the bridge-tree. This design is borrowed from S. Morgan Smith.

Disconnecting Mechanism.—The power of several turbines is often transmitted through the same shaft, either by bevel-

Fig. 170.

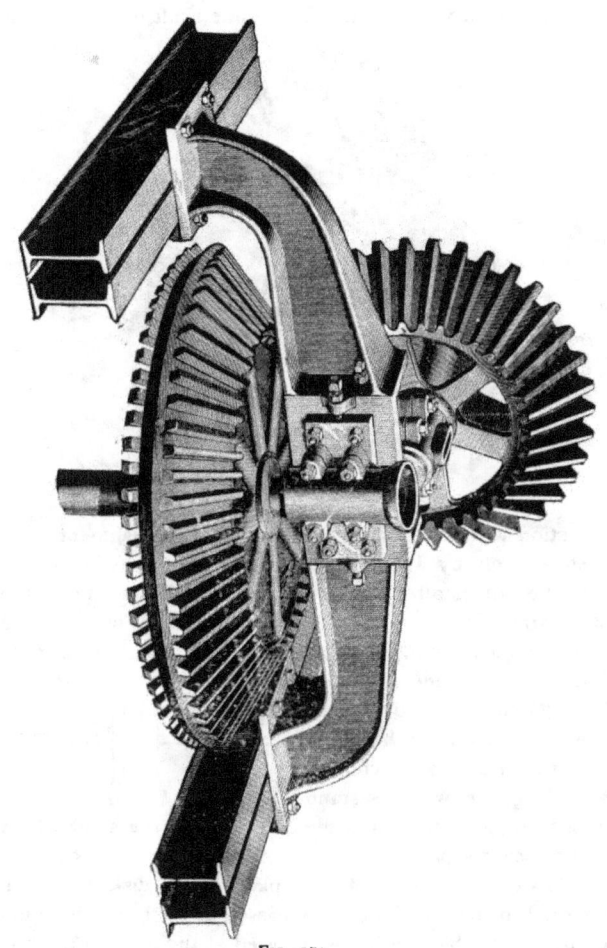

Fig. 171.

gears, when they run on vertical shafts, or directly, when all run on the same horizontal shaft. The same shaft may also have

FIG. 172.

a connection with the steam-engine. The establishment runs, at times, entirely by water; as the water diminishes in the dry season, one wheel after another is put out of use, and in the driest weather it may run wholly by steam. For this and other reasons it is often desirable to throw a wheel out of connection with the general system.

Fig. 173 shows the simplest device for disconnecting a horizontal shaft. It consists of a common face-coupling, the two faces stand-

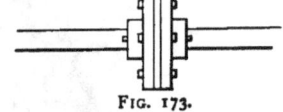

FIG. 173.

ing about an inch apart, and the two parts of the shaft which they unite being separated by the same distance. To connect the shafts we insert a disk of inch plank, both disk and faces being bored for bolts. These are passed through and the nuts turned up. A double bearing is required at the point of disconnection. This device is of limited application. In a series of turbines numbered 1, 2, 3, 4, etc., all transmitting through

the same shaft, it is possible by this means to cut off No. 1 or
1 and 2, or 1, 2, and 3, leaving the others running; but it
would not be possible to cut off No. 2 and leave No. 1 running, etc.

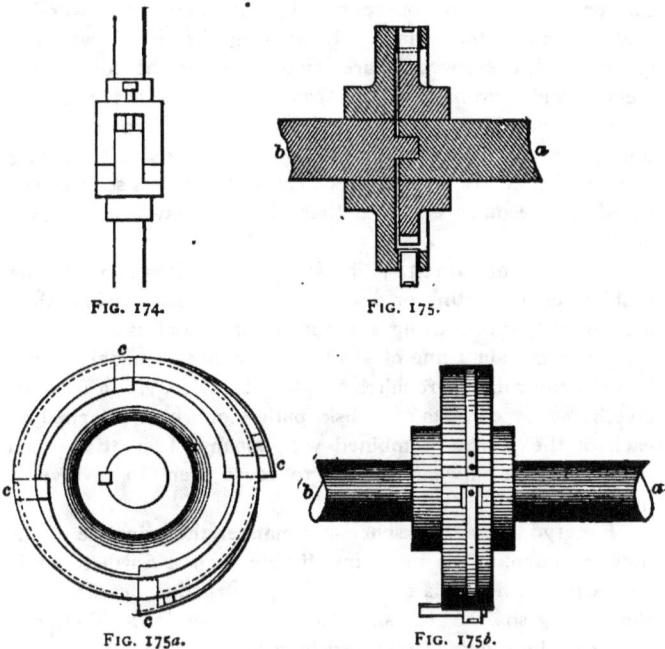

FIG. 174.   FIG. 175.

FIG. 175*a*.   FIG. 175*b*.

The disconnecting-coupling indicated by Fig. 173 has been
used on vertical shafts. By turning the screw we lower the
upper part of the shaft and draw the crown-gear out of connection. By this device, any wheel gearing into a horizontal line
. of shafting may be cut out of the system.

Figs. 175, 175*a*, and 175*b* show a disconnecting device
applicable to small shafts for which a patent was granted to
the writer from the United States under date of April 27, 1897.
Shafts *b* and *a* are supposed to be driven by separate motors,

and it may be required to throw on or off the line *b* without stopping the line *a*. *b* transmits power to *a* by means of the spring-pawls *c c*. When *b* is thrown off, the pawls begin to pound, and continue to do so till *b* comes to rest, when they can be confined out of contact by thrusting small wooden wedges under the springs. In starting the wheel, which is geared to *b*, these wedges are removed, when the noise commences and continues till the pawls come to a bearing. As the speed of *b* increases, the clicks diminish in frequency, and this guides the workman in bringing the pawls to a bearing without shock. In fact no shock can occur, for as soon as the speed of *b* becomes equal to that of *a* the pawls come to a bearing.

The Friction-clutch.—The foregoing devices relate to the problem of connecting or disconnecting two shafts while they are stopped, or of starting a motor to join one already running. To put in motion a line of shafting or a heavy machine by a line already running requires a different device, viz., a friction-clutch. This consists of a disk, pulley, or wheel attached to each of the shafts, combined with means of creating such mutual friction between them as to cause them to revolve in unison.

Fig. 176 shows in a schematic manner the principle of the friction-clutch. It is more intelligible than a finished cut in which the principle is obscured by constructive details. *S* is the moving shaft, *S'* the shaft to be set in motion. The shaft *S* carries the sleeve *A* attached by a loose-fitting spline which causes the sleeve to revolve with the shaft, but leaves it free to move longitudinally thereon. The collar *B* runs in a groove on the sleeve, which confines it laterally, but leaves it free to remain fixed while the collar revolves. This collar carries two gudgeons which are seized by the forked lever seen at Figs. 176 and 176*a*. To the shaft *S'* is affixed a hub carrying a pulley with a wide rim against which bear the friction-blocks *C C*. These latter are connected with the sleeve by the arms *R R*. A pull upon the lever, carrying the outer end to the left,

advances the sleeve and, through the levers *R R* which act as
a toggle, presses the friction-blocks against the rim of the pulley
with such force that the friction exceeds the resistance to

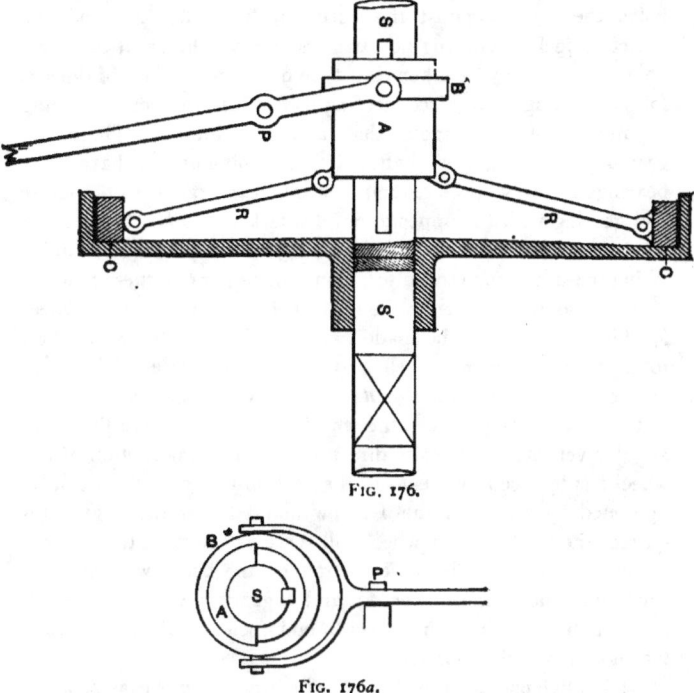

Fig. 176.

Fig. 176a.

motion and the shaft revolves. The arms are very strong and
strongly jointed to the sleeve to bear the transverse strain of
the friction. They may be two, four, or six in number. The
friction-blocks are elongated segments bearing against the inner
face of the pulley, and are formed so that the connections have
a slight degree of elasticity. They are also fitted with adjustments for taking up the wear. In the position of Fig. 176, if
we advance the sleeve so as to set *S'* in motion, and clamp the

lever in that position, there is an end-long strain on the shaft, requiring a thrust-bearing and occasioning friction. If, however, we advance the sleeve till the inner ends of the levers are in advance of the outer ends, the elastic strain on the levers holds the sleeve against the face of the pulley S', and the device is locked, causing no extra friction on the shaft-bearings.

* Fig. 177 may be taken as a type of a certain class of devices for connecting and disconnecting two shafts without stopping the motor. A is the motor-shaft and is continuous. The bevel-gear a is fixed on this shaft. The two pinions $d\ d$ have their bearings in the wheel $c\ c$, their journals being radial to the latter and gearing with the opposite bevel-wheels a and b. For a large amount of power it might be advisable, for the purpose of diminishing the strain upon the latter, to introduce 4 or 6 of these pinions. If, now, motion be given to the shaft $A\ A$, it is clear that the wheel b, which is loose, will be made to revolve in a contrary direction to the wheel a, which is fixed, by means of the carriers $d\ d$, but no motion of the wheel c, if *slightly opposed*, will ensue; and so long as this last remains at rest, the wheels a and b will have the same angular velocity in opposite directions. But if the motion of the wheel b is opposed by means of the friction-brake e, which can be tightened by means of the T-screw marked f to any degree required, the teeth of that wheel will serve as fulcra to the carrier-pinions $d\ d$, which will then carry round the wheel c with half the velocity of the prime mover A; and c, gearing with the wheel h on the main spindle of the wheel to be impelled, will transfer to it the motion which itself receives.

It is obvious that the shaft h will begin to move as soon as any friction is exerted on the wheel b, and will reach its full velocity when the latter is fully stopped, so that neither the starting nor stopping of n is attended with the slightest shock. It is obvious also that the shaft h can by this device be controlled to any desired velocity less than that corresponding to the speed of A, although this mode of running would be attended with great loss of power by friction.

* See Engineers and Machinists' Assistant. Blackie & Son, London, 1854.

Fig. 178 shows the most recent arrangement for throwing in and out a belt, shaft, or machine by means of a friction-

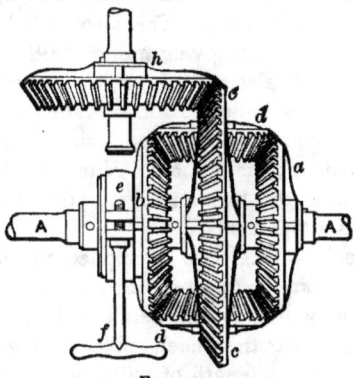

FIG. 177.

coupling. It shows a shaft, which we will call the main shaft, supported in bearings. Rigidly attached to this shaft is the

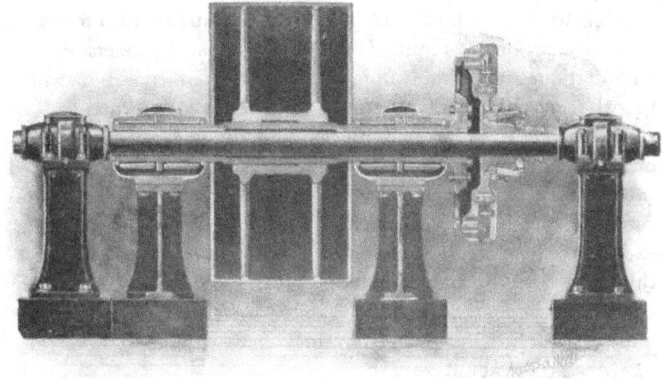

FIG. 178.

disk and the sliding sleeve of the friction-clutch. This is close to one of the bearings of the main shaft. The remainder of the space between two consecutive bearings is occupied by a hollow shaft outside the main shaft. This also rests in bearings and carries the pulley or gear giving motion to the disconnectible shaft or machine. It also carries the corresponding half of the friction-clutch. This hollow shaft is called a quill. When not in motion it has no connection with the main shaft and offers no impediment to its motion. When the clutch is thrown in gear by means of the handle, Figs. 176 and 176*a*, the quill revolves with its attachments.

The Penstock.—This term is applied to the pipe which brings the water from the canal or other source of supply to the flume. It is a very unimportant detail of the system when the source of supply is near the wheel. When, however, as sometimes occurs, it takes a length of miles, it becomes of primary importance, and phenomena are developed in it which it is very necessary to understand.

Such pipes are commonly made of riveted wrought-iron or steel plates. Modern rolling-mills produce single sheets large enough to make a length of pipe up to a diameter of 6 feet, and probably greater if required. The use of such sheets is very advantageous as diminishing the amount of riveting, which greatly increases the resistance to the flow of water. The running joints are formed by bending the sheet into the form of a cylinder and riveting the edges to a separate plate called a butt-strap. Sometimes two such plates are used, one inside, the other outside. The circular joints are formed by riveting each length to a ring outside the pipe. Necessarily a part of this latter riveting has to be done on the ground. To prevent corrosion and increase the durability of such pipes, they are immersed in a bath of coal-tar thickened with asphaltum, at a temperature of about 300° F.

In the case of a long line, nearly level but impracticable for a canal, large pipe made of wooden staves and banded with iron hoops has been successfully used. This pipe is not made

in uniform lengths and then put together, but built as a continuous structure, the staves breaking joints with each other.   Fig.

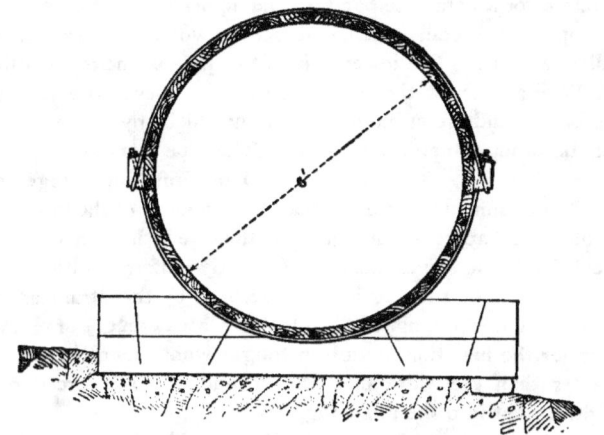

FIG. 179.

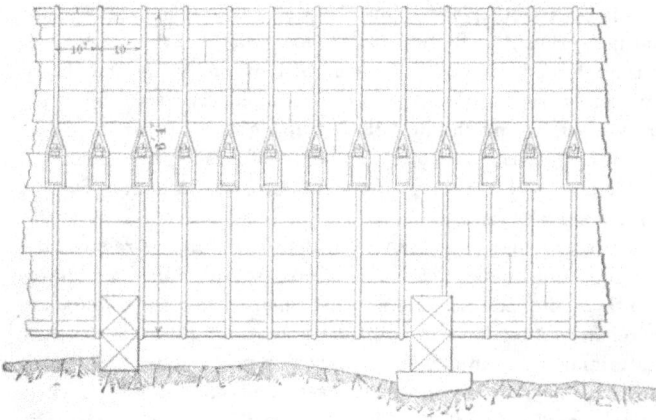

FIG. 180.

179 is a cross-section, Fig. 180 an elevation, of such a pipe
6 feet diameter.   It is composed of staves about 8 inches wide,

made of 2½-inch plank, and 12 to 20 feet in length. The bands are composed of ⅝- or ¾-inch round iron, two rods being required for a complete band. The upper one is formed with a loop at each end, which embraces a wooden or iron block called a shoe. The lower half of the pipe is encircled with a rod having a screw-thread at each end. These pass through the blocks and are secured by nuts, by which any desired strain can be brought upon the hoops. The pipe rests in cradles of 6″ × 8″ or 8″ × 8″ timbers spiked or drift-bolted together, of a length somewhat greater than the diameter of the pipe, and at distances apart about equal to their length. In order to break joints the staves must be of exactly uniform width. The tightness of the running joints is secured by the strain on the hoops. The abutting end-joints have a saw-cut ½ or ¾ inch deep for the insertion of an iron tongue whose length is a little greater than the width of the stave and which is forced to a slight extent into the adjoining staves.

Such a pipe can receive a considerable curvature either horizontal or vertical. To effect this the part under construction is put together and lightly banded so as to hold it in shape but not prevent a slight relative movement of the staves. It is then forced out of line by weights, by block and tackle, or by a screw-jack. When it has the desired curvature the bands are all put on and the nuts turned up tight. In this condition it will not return to its original form. If it is slightly deformed in the process of bending, the tightening of the bands tends to restore the circular form.

A long penstock presents peculiar difficulties as regards the regulation of the flow of water to correspond with the requirements of the power. In a 6-foot penstock a mile long, the water in motion has a weight of more than 4000 tons. Any material diminution of velocity in such a pipe implies the sudden slowing up of this enormous mass, and cannot fail to exert great pressure upon the pipe. We will endeavor to obtain some idea of the effect of changes of velocity in such a pipe.

Assume a long pipe filled with water, Fig. 181. Imagine

at AB a piston to start forward, and suppose that when it has moved the distance l it has set the water in motion as far as CD.

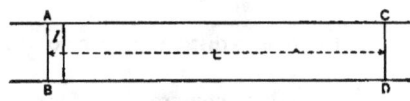

FIG. 181.

Let $r =$ radius of pipe in feet;

$w =$ weight of one cubic foot of water;

$m =$ modulus of elasticity of water, say 294 000 × 144 pounds per square foot;

$M =$ modulus of elasticity of metal of pipe, pounds per square foot;

$T =$ thickness of pipe, feet;

$t =$ time occupied by the piston in moving the distance l in seconds;

$v =$ velocity of piston in feet per second;

$f =$ force exerted upon the water by the piston in pounds per square foot;

$L =$ the distance from the piston to the particles of water that are in the act of commencing to move when the piston has moved a distance l in feet.

When the piston has moved a distance l, although it has only moved the centre of gravity of the mass L a distance $\frac{1}{2}l$, it has done work in compressing the water and distending the pipe equivalent to moving the mass a distance l and imparting to it the velocity $\frac{l}{t}$. This will appear on reflecting that if the forces opposing the movement of the mass were suddenly removed, the latter would, by release of pressure, acquire the velocity stated.

The bursting tension on a running foot of the pipe is rf.

The increase in the radius is $\dfrac{r^2 f}{MT}$. Increase in cross-section

$= 2\pi r \dfrac{r^2 f}{MT} = 2\pi \dfrac{r^2 f}{MT}$ Increase of volume $= Lf 2\pi \dfrac{r^3}{MT}$

Traverse of piston due to distension of pipe $= L\dfrac{2rf}{MT}$.

"   "   "   " " compression of water $= L\dfrac{f}{m}$.

Total traverse of piston $= Lf\left(\dfrac{2r}{MT}+\dfrac{1}{m}\right)=Lf\dfrac{2rm+MT}{m.MT}= l.$

Velocity of piston in feet per second $= v = \dfrac{l}{t}$

$$= \frac{Lf}{t}\frac{2rm + MT}{m.MT}. \quad \cdot \quad \cdot \quad \cdot \quad (50)$$

The total weight of water set in motion is $\pi r^2 Lw$.

The force acting to impart motion is $-r^2 f$.

Gravity, acting freely, would impart to the mass a velocity of gt feet per second in the time t, whence we have the proportion

$$\frac{l}{t} : gt = \pi r^2 f : \pi r^2 Lw,$$

whence

$$L = \frac{gt^2 f}{lw}; \quad \frac{f}{l} = \frac{mMT}{L(2rm+MT)}. \quad \therefore L = \frac{gt^2}{w}\frac{mMT}{L(2rm+MT)}.$$

Whence we find $\dfrac{L}{t} = \sqrt{\dfrac{g}{w}\dfrac{m.MT}{2rm + MT}}$ $\cdot \quad \cdot \quad \cdot \quad (51)$

Taking $M = 30\,000\,000 \times 144$, m as above, $T = \frac{1}{4}$ inch $= \frac{1}{48}$ foot, and $r = 2.5$ feet, we find $\dfrac{L}{t} = V =$ velocity of pulsation $= 2551$ feet per second.

So that, if we conceive of a pipe-line such as here contemplated reaching from Boston to Chicago, something over half an hour must elapse after the starting of the pumps at Boston before the water would commence to flow at Chicago. This would be true whether the pumps worked fast or slow. The entire pumpage of the interim would be absorbed by the distension of the pipe and the compression of the water.

Eq. (51) may be put under the form

$$\frac{L}{t} = \sqrt{\frac{g}{w} \frac{m}{\dfrac{2rm}{MT} + 1}}.$$

The assumption that there is no distension of the pipe implies an infinite value of M, which would make

$$\frac{L}{t} = \sqrt{m \frac{g}{w}} = 4672 \text{ feet per second}, \quad . \quad . \quad (52)$$

being substantially the velocity of sound in water, which is usually taken at about 4700 feet per second.

Instead of a piston suddenly starting forward, imagine water moving with the velocity v, to be suddenly arrested by the closing of a gate. f is then the force acting against the gate and walls of the pipe. To find the value of f, we reason as follows:

If O be the volume of a mass of water under atmospheric pressure, and O_1 its volume under the pressure f, then

$$\frac{O - O_1}{O} m = f.$$

Let L_1 be the value of L when $t = 1$. Then the volume of the mass of water that we are considering, under atmospheric pressure, is

$\pi r^2 (L_1 + v)$. Do. under pressure f, $\pi r^2 L_1 \left(1 + \dfrac{2rf}{MT} \right)$,

whence

$$f = m \frac{L_1 + v - L_1 \left(1 + \dfrac{2rf}{MT} \right)}{L_1 + v}. \quad . \quad . \quad . \quad 3)$$

If we make M infinite in this equation, it reduces to

$$f = \frac{v}{L_1 + v} m,$$

which for $v = 4$ feet would give $f = \dfrac{4m}{4676} = 251.5$ pounds per square inch.   Equation (53) takes the form

$$(L_1 + v)f = m\left\{L_1 + v - L_1\left(1 + \frac{2rf}{MT}\right)\right\},$$

or $MT(L_1 + v)f = m(L_1 + v)MT - mL_1MT - 2rmL_1f,$

whence

$$f = \frac{m(L_1 + v)MT - mL_1MT}{(L_1 + v)MT + 2rL_1m} = \frac{mvMT}{(L_1 + v)MT + 2rL_1m}. \quad (54)$$

Taking $v = 4$ and other symbols as before, we find $f = 75$ pounds per square inch.

As v is always very small compared with L_1, it will be seen that the pressure is sensibly proportional to the velocity destroyed.

The above equations indicate that f is theoretically independent of the length of the pipe, but this can only be true on the assumption that the stoppage is absolutely instantaneous. Under practical conditions, in which the stoppage occupies an appreciable time, the force developed is not independent of the length.   For the purpose of experiment a valve may be used causing an instantaneous stoppage, though such valves are ordinarily avoided, and upon the closure of such a valve the full pressure is instantly developed on it.   The section in which the water is coming to rest moves up-stream with the velocity indicated by equation (52).   The amplitude of this movement is only limited by the length of the pipe, above the valve, either to the reservoir or to the larger pipe or canal from which it branches.

Neglecting the elasticity of the pipe, the time occupied by its contents in coming to rest is $\dfrac{X}{4672} = t$, X being the total length.   The water has continued to flow into the pipe with the undiminished velocity v, for t seconds after the closure. The compression of the water in the pipe is represented by $\dfrac{vt}{X}$, and the force exerted on the interior by $m\dfrac{vt}{X} = m\dfrac{v}{L_1}$ pounds per square inch, m being the modulus per square inch. A pipe buried and firmly packed in the ground, as is the ordinary practice with penstocks, is probably in a condition to require the elasticity to be neglected in computing the pressure, and such a pipe might have parts not so protected on which the pressure would be liable to act dangerously.

The pressure on a suddenly closed valve is not released when the entire contents of the pipe have come to rest. The water at that instant is in a state of compression throughout the entire length X of the pipe. A release and reversal of motion takes place commencing at the origin of the pipe, and moves toward the valve with the velocity $V = L_1$ feet per second, so that the time from the stoppage to the release of pressure is $2t$. A series of blows or shocks on the valve will recur at intervals of $2t$ till the movement dies out. This is always observed in service-pipes where the abrupt closing of a valve is followed by a harsh grating sound. In a service-pipe 100 feet long, neglecting elasticity, these blows would occur at the rate of about twenty-three per second.

In confirmation of these views the following may be quoted from the *Engineer* (London) in reference to the power-house at Fresno, Cal., which is supplied by a pipe 4000 feet long under a head of 1400 feet, some 600 pounds per square inch: "The gates are controlled at the power-house by hydraulic rams. In opening the gates as at first designed a fluctuation of 170 pounds above and below the normal was brought about. That is to say: the pressure would first drop to 90 pounds below normal, then rise to 180 pounds above, then sink to 75 pounds below normal, and continue 'diminuendo' until the normal pressure was reached. A similar phenomenon, but reversed in order, occurred when the gates were closed," etc.

When a valve occupies an appreciable time in closing it is easy to understand that the resulting pressure will depend upon the length of the pipe. The closing may be supposed to take place by a great number of small steps each of which may be regarded as instantaneous. Each step occasions a certain diminution of velocity and is accompanied by an increment of pressure. Each increment runs to the head of the pipe, and the release returns with the same velocity. Every increment of pressure originating at the valve remains in force there till the pulsation has run to the head of the pipe and back to the valve. Therefore, whatever diminution of velocity may be effected during the interval $2t$ may, for purposes of computation, be regarded as instantaneous. This is true for any point of the pipe when we use the value of t corresponding to that point.

The important point in this inquiry is to determine the pressure generated in the ordinary operation of regulating the power. Consider a penstock in which the elasticity of the metal can be disregarded, and long enough so that $2t$ seconds will suffice for any required change in the power. Let $v =$ the normal velocity of the water, and $P =$ the normal pressure in the pipe. Suppose the demand for power to suddenly diminish one-fourth, and the power to automatically adjust itself to this requirement within the time $2t$. Let P_1 be the pressure, and v_1 the velocity, at the instant this adjustment is effected. We must have

$$P_1 v_1 = \tfrac{3}{4} Pv, \quad \text{also} \quad P_1 = P + m\frac{v - v_1}{L_1}.$$

From these two equations there will result

$$P_1^2 - \left(P + \frac{mv}{L}\right) P_1 = -\frac{3m}{4L_1} Pv$$

whence

$$P_1 = \tfrac{1}{2}\left(P + \frac{mv}{L_1}\right) + \sqrt{\tfrac{1}{4}\left(P + \frac{mv}{L_1}\right)^2 - \frac{3m}{4L_1} Pv}. \quad (55)$$

Taking the values of P and v as contemplated in the Pioneer Electric Plant* at Ogden, Utah, viz., $P = 216.67$, $v = 9$, we find $P_1 = 638.96$, which is the momentary pressure that the pipe would be subjected to in the case supposed if the velocity could be controlled in precise accordance with the requirements of the power.

We might make eq. (55) entirely general by putting a for the fraction that the reduced power is of the normal power. Then we should have

$$P_1 = \tfrac{1}{2}\left(P + \frac{mv}{L_1}\right) + \sqrt{\tfrac{1}{4}\left(P + \frac{mv}{L_1}\right)^2 - a\frac{mv}{L_1}P}. \quad (56)$$

Regulation.—Industrial operations require a uniform speed of shafting, although the quantity of work or the number of machines in operation may vary greatly from hour to hour or even from minute to minute. This condition necessitates an automatic device for controlling the admission of water to the wheel, diminishing the same where the velocity exceeds the normal rate and *vice versa*. The essential part of this device is an organ which moves in one direction and sets in motion the mechanism for closing the gate when the velocity exceeds the normal limit, and which moves in the opposite direction and sets in motion the mechanism for opening the gate when the velocity falls below that limit. Such a movement can only be secured through the agency of centrifugal force.

Many combinations may be imagined by which a revolving weight may be made to depart from its axis of rotation by the action of centrifugal force, and return by the action of a weight or spring; but after the most exhaustive study of the subject we should revert to the arrangement suggested by Fig. 182. A heavy weight W is carried by an arm which is pivoted to a vertical revolving spindle so as to be susceptible of a swinging movement in the plane of the picture. The arm and ball rotate around the spindle, and the centrifugal force causes the

---

* Trans. Am. Soc. C. E., vol. XXXVIII. p. 246.

arm to deviate from the vertical. For every velocity of rotation there is a corresponding angle of deviation which does not change while the velocity remains constant. This deviation of the arm and ball from the vertical constitutes the movement desired for the regulator. It increases or diminishes with the velocity, and by means of suitable linkages and connections can be transferred to any desired part of the mechanism.

Let ω represent the angular velocity of the spindle, Fig. 182 ;

l = the length of the arm from the centre of suspension to the centre of the ball or, if great accuracy is aimed at, the centre of gravity of arm and ball;

α = angle of deviation;

W = weight of ball, or of ball and arm.

The force tending to turn the ball around A to the right is $\frac{W}{g}\omega^2 l \sin \alpha$. The force tending to turn it in the opposite direction is W. Whence, by the law of moments,

$$\frac{W}{g}\omega^2 l \sin \alpha \,.\, l \cos \alpha = Wl \sin \alpha,$$

whence

$$\frac{\omega^2}{g} l \cos \alpha = 1, \quad . \quad . \quad . \quad . \quad . \quad (57)$$

whence $\cos \alpha = \frac{g}{l\omega^2}$, from which it appears that the angle of deviation does not depend at all upon the weight of the balls, but solely upon the velocity of rotation.

There is an apparent limitation to the application of equation (57). In the form $\cos \alpha = \frac{g}{l\omega^2}$, $\cos \alpha$ cannot exceed unity, while $\frac{g}{l\omega^2}$ can have any value from 0 to infinity. The value of ω from (57) is $\omega = \sqrt{\dfrac{g}{l \cos \alpha}}$. This equation can give no value of ω less than $\sqrt{\dfrac{g}{l}}$. For the arrangement of

Fig. 182 it would appear that values of ω less than $\sqrt{\dfrac{g}{l}}$ must

be excluded. This difficulty, however, has no practical con-

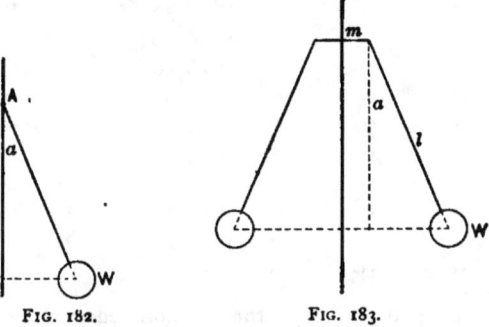

FIG. 182.   FIG. 183.

cern for us, since in practice the pivot of the arm never coincides with the axis of the spindle. The arrangement ordinarily
adopted is that of Fig. 183, in which the arm is pivoted to a
cross-bar fastened to the spindle. Let m represent the distance
of the point of suspension from the axis of rotation. In this
case

$$\frac{W}{g}\omega^2(m + l \sin \alpha)l \cos \alpha = Wl \sin \alpha,$$

whence

$$\omega = \sqrt{\frac{g \tan \alpha}{m + l \sin \alpha}}. \quad \cdot \quad \cdot \quad \cdot \quad \cdot \quad (58)$$

In this equation, which appears to be free from ambiguity,
it is not easy to assign a value to ω and compute the corresponding value of α, but it is easy to assign a value to α and
find the corresponding value of ω. ω represents the velocity
of a point at units distance from the axis of rotation, and $\dfrac{60\omega}{6.28}$

=: the number of revolutions per minute.　Making in equation (58) $l = 2$ and $m = 0.25$, we find that to give the balls a deviation from the vertical of

10 degrees requires a speed of 29.44 turns per minute.

15	"	"	. "	" 32.01	"	"	"
20	"	"	"	" 33.82	"	"	"
25	"	"	"	" 35.36	"	"	"
30	"	"	"	" 36.80	"	"	"
35	"	"	"	" 38.36	"	"	"
40	"	"	"	" 40.05	"	"	"
45	"	"	"	" 42.00	"	"	"
50	"	"	"	" 44.31	"	"	"
55	"	"	"	" 47.12	"	"	"
60	"	"	"	" 50.65	"	"	"
65	"	"	"	" 55.25	"	"	"
70	"	"	"	" 61.55	"	"	"

It will be noticed that as the rotation becomes more rapid it requires a greater increment of velocity to effect a given in-

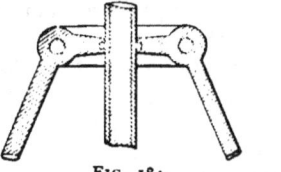

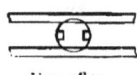

FIG. 184.　　　　　　FIG. 184a.

crement in the deviation of the balls.　The regulator is more sensitive to a change of speed at a low velocity than at a high one.　For this reason it is not customary to give the spindle a speed of more than 35 to 45 turns per minute.

Fig. 185 shows an old form of regulator, Y being the revolving spindle and balls.　The latter are connected with the collar C, which revolves with the spindle but is susceptible of a slight up-and-down movement.　The two bevel-gears A and B are loose on the spindle and mesh with the larger bevel-gear on the shaft D.　The connection of the arms with the

ɪollar *C* is shown at Figs. 184 and 184*a*. Two thin splines
run in grooves cut in the sides of the spindle and are attached
to the collar *C*. The arm forms an elbow at the suspension
joint, and terminates in a finger which seizes a projection on
the spline, raising the collar *C* when the balls fall and lowering

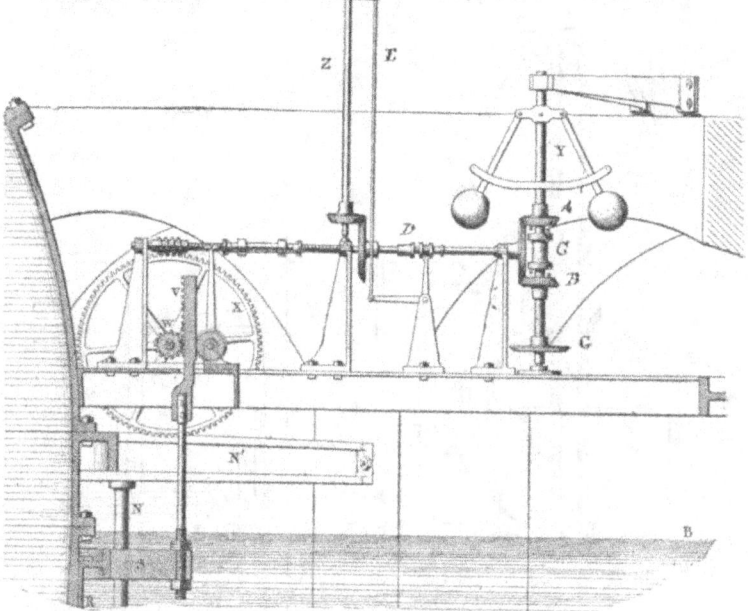

FIG. 185.

it when they rise. The collar carries a projecting stud on its
upper and another on its lower face, and a corresponding stud
is borne by each of the bevel-gears *A* and *B*. When the
velocity is normal, these studs clear each other; but when the
velocity diminishes, the balls fall, the collar rises and its stud
engages with the upper bevel-gear, setting it in motion and
turning the shaft *D* to open the gate. When the velocity rises
above the normal limit, the contrary action takes place and

closes the gate.   The shaft *D* carries a worm-gear which acts
on the wheel *X*, and, through the toothed pinion *W*, on the
toothed gate-stem *V*.   The spindle is revolved through the
lower bevel-gear, *G*, by a small line of shafting branching,
at any convenient point, from the main shafting.   The shaft *D*
carries a friction-coupling, which insures the mechanism against

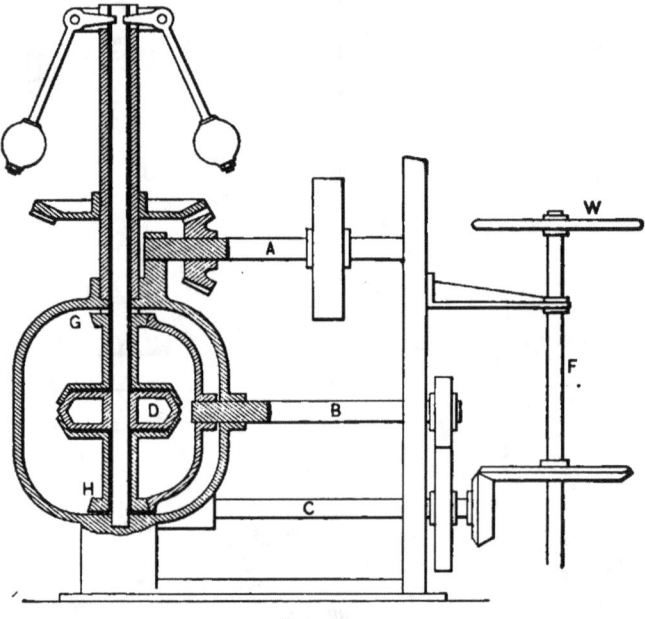

FIG. 186.

breakage when the gate reaches the end of its traverse.   The
regulator continues in action after the gate has reached its
upper limit, and if the mechanism were rigid it would inevitably
break.   When the wheel is coupled to others, or several wheels
are geared to the same line of shafting, the regulator may continue to act after the gate has reached its lower limit.   *Z* is a
rod reaching to the floor above and carrying a hand-wheel by

which the gate can be raised and lowered by hand; but before turning the hand-wheel, the workman places his foot on the head of the rod *E*, and operates a disconnecting-coupling which frees the gate from the action of the regulator.

Fig. 186 is a schematic representation of a governor which has received extended application. The upright spindle, which is hollow, receives motion from the shaft *A* driven by a belt and pulley. Within the hollow spindle is a rod firmly attached to the double friction-cone *D*. Two hollow friction-cones, *G* and *H*, stand one above and one below *D*. Connected with these hollow cones and forming a part of them are bevel-gears, which mesh with the large dish-shaped bevel-gear on the shaft *B*. At normal speed the double cone runs clear of both hollow cones; but a slight increase of speed forces it into the lower cone and turns the shaft *B*, to lower the gate, and *vice versa*. The disconnecting mechanism is not shown. In this form of regulator the cone will revolve against the friction, when the gate reaches the end of its traverse, before any breakage can occur, but it is usual to introduce a friction-connection at some other point in the machine.

Fig. 187 represents the leading element in the more modern mechanism of the regulator. It is shown as applied to the cylindrical gate of a turbine wheel, by means of three stems *A A A*. The regulator of Fig. 185 acts only on one stem, and would not work to raise a cylindrical gate unless the latter were accurately balanced by counterweights. The ring *F*, Fig. 187, is toothed on the outside and has a screw-thread on the inside, which acts, when the ring revolves, to raise the gate by means of toothed racks on the stems *A A A*. *S* is the shaft of the wheel. An arm *B* attached to a ring on the shaft carries the spring-pawls *D D*, and has a constant oscillatory motion of a few inches to right and left. A second arm, *C*, carries a shield *C*, which masks the teeth of the wheel from the action of the pawls. At normal speed the traverse of the pawls does not carry them beyond the limits of the shield, and the regulator does not act. The revolving balls act, through suitable linkages

and rods, upon the arm C, and, as the velocity varies, draw
the shield aside and allow the pawls to act on the teeth.
When, for instance, the speed falls below the normal, the shield
is drawn, say, to the right, and the left pawl acts to raise the
gate, and *vice versa*.   If the change of speed is great, the
movement of the shield is great, and the ring receives a large

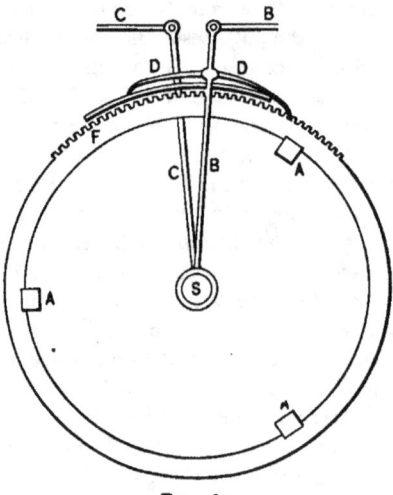

FIG. 187.

movement at each stroke of the pawls.   If the change is slight,
the movement imparted to the gate is slight, etc.   In addition
to the advantage of adapting the rapidity of its action to the
rapidity in the change of velocity, this mechanism may be
readily thrown out of connection with the wheel when the gate
reaches its limit.   A wheel running under control of a regulator
often runs near the limit of its capacity, and in that case a friction-coupling is liable to be constantly in action and to become
dangerously heated.   A device for throwing the regulator
wholly out of gear when the gate reaches its limit does not
meet the case.   It must be thrown out of gear with reference

to one movement of the gate, and remain in gear with reference to the contrary movement; otherwise it ceases to act as soon as the gate reaches its limit. Such a device would be complex when applied to the mechanism of Figs. 185 and 186, but is very simple in its application to Fig. 187. It is only necessary to arrange so that when the gate reaches its upper limit it shall throw the left pawl out of action, leaving the right ready to act as soon as the velocity increases. The left comes into action as soon as the gate lowers.

The preceding devices worked very well for the old applications of water-power in mills and factories, though in these cases there is constantly increasing necessity for closer regulation of speed. This necessity is still more imperative in systems of electric lighting, traction, and electric transmission, in which uniformity of movement is of the first importance, and which are liable to greater and more sudden variations of load than manufacturing establishments. The use of long penstocks, which is becoming more and more prevalent, tends still further to increase the difficulties of regulation. Especially is this necessity apparent in the steam-engines of great power systems, which have now attained enormous dimensions, and in which the bursting of heavy fly-wheels from uncontrolled velocity is no uncommon occurrence. As regards water-wheels, the chief defect of the old forms of governor is that they do not act with sufficient promptness. For instance, the device of Fig. 187 would probably occupy at least a minute in raising the gate an inch, a defect not noticeable in a system liable to slight variations of load, but which would be entirely inadmissible in a system liable to great and sudden fluctuations. The modern forms which have aimed to correct this difficulty have often incurred another quite as serious, viz., that while acting with promptness, they do not cease to act till too great a change has been effected in the position of the gate, and thus keep the speed in a continual state of fluctuation.

Figs. 188 and 189 represent the Lombard water-wheel governor, which is stated by users of water-power to have given good results both in promptness of action and in the avoidance of overaction. The first effect is secured by a reservoir of liquid under pressure, which, acting upon a piston in a cylinder, effects, through suitable gearing, a sufficiently rapid movement of the gate. Here, however, comes in the second difficulty. The fundamental element of the mechanism

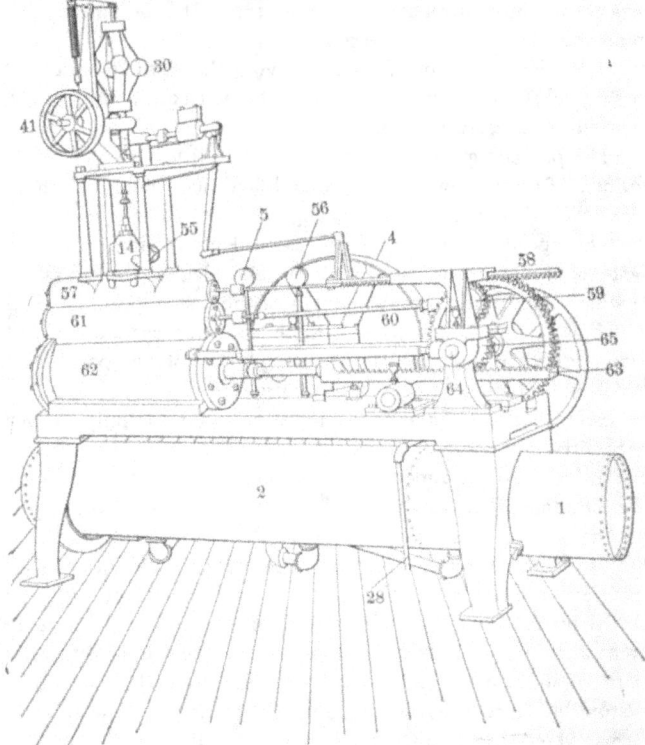

FIG. 188.

is, of course, the revolving balls. These act, when their amplitude exceeds the normal, to open a valve which lets liquid into one end of the cylinder and exhausts it from the other, and perform the reverse function when the amplitude falls below the normal. Now when a large part of the load is suddenly thrown off, the speed increases, the balls fly outward and open the valve to close the speed-gate. When the speed

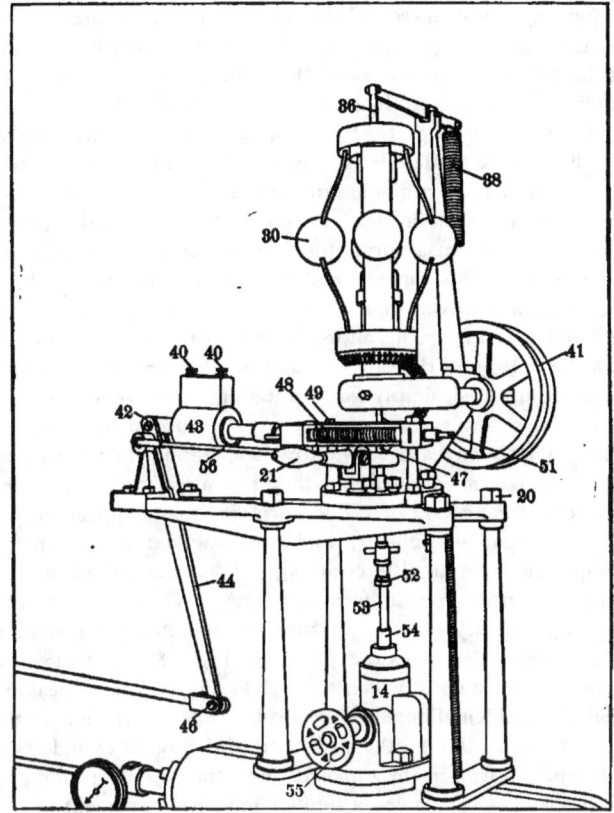

FIG. 189.

begins to slacken, or but little later, the closing movement of
the gate should cease, but it does not cease till the balls have
come back to their normal position, and then the gate is too
much closed, the speed falls below the normal, the gate opens
and the reverse movement ensues. In shop language the
machinery "races." Such, at least, would be the action were
there no devices introduced to correct it.

Moreover, the balls, unless very heavy, cannot be depended on for any great direct effort, as this prevents them from assuming their proper position. They, accordingly, do not act directly upon the valves of the main cylinder, but operate it indirectly through an intermediate cylinder.

Returning to Fig. 188, we notice below the machine a cylindrical tank of plate iron. It is divided into two compartments, 1 and 2, by a diaphragm indicated by a line of rivets. A vacuum is maintained in 1, and 2 is partly filled with oil over which is air, at a normal pressure of some 200 pounds per square inch. This vacuum and pressure are maintained by a pair of pumps driven by a slotted cross-head on the shaft of the band-wheel 4. The oil for operating the piston is drawn from the bottom of the tank, so that no air enters the cylinder, the air being used simply for maintaining the pressure on the oil and constituting a reserve of energy for use in handling the gate. The pumps are constantly in operation, but so long as the vacuum is maintained in 1 they run against no resistance and consume no power. The revolving balls appear at 30. They are four in number, and are mounted on·laminated springs, the centrifugal force acting to bend the springs instead of against gravity as in former figures. They revolve with very high velocity,—some 500 turns per minute. The bending of the springs depresses the plate 36, Fig. 189, and moves a valve, 14, the stem of which appears at 53. We are speaking now of a sudden diminution of load. The depression of the valve 14 lets oil into the right end of the small cylinder 57, and exhausts it from the opposite end, drawing its piston and rod inboard. This gives a limited inboard movement to the rod 60, which is attached to a piston-valve in the intermediate cylinder 61. The effect of this movement is to put the large cylinder 62 in communication with the reservoir and vacuum, and impart an inboard movement to the rack 63, rotating the shaft 64, which, through another gear and rack, gives a closing movement to the gate. It will be perceived at once that with no other attachments than those described the motor would inevitably race. We will now point out the devices for obviating that difficulty. It will be noticed that the piston-rod of 57 is attached to a toothed rack, 58. This is in mesh with a floating pinion, 59, attached to the rod 60 in the following manner: it is embraced by a fork in 60, and the pivot on which it rotates rests in elongated slots in the framework, which limit the in-and-out movement of the rod 60, although this movement is not intended to extend far enough to bring the pivot to a bearing. The floating pinion meshes with the rack 58 and with another spur-gear on the shaft 64. A stud will be noticed on 58, which acts through a system of jointed levers, 44, 46, etc., Fig. 189. Its effect when the rack is carried inward is to rotate the valve-stem 53 and, by a combination not necessary to describe, bring the valve back to its normal position. As soon as 60 has moved inward far enough to open the port, 63 starts inward, rotates the shaft 64, and gives the necessary closing movement to the gates. As soon as the rotation of the shaft commences, the inward movement of 60 ceases, and when it has proceeded far enough to bring 60 back to its normal position, the whole movement ceases by the action of the jointed levers in cutting off the supply of oil, leaving the balls to return to their normal position in their own time. The movement of 58 will be more rapid in proportion as the change of velocity is more sudden. The dash-pot 43, Fig. 189, is intended to prolong the movement of the gate in the latter case. The rod jointed at 42, which communicates the closing movement to the governing valve, is not continuous, but is interrupted by a piston attached to one part of the rod, working in a cylinder attached to the following part. This cylinder is filled with oil which must pass from one side of the piston to the other to admit of any movement of the latter. Under violent action the oil passes through the piston by valves controlled by stiff springs. Under gentle pressure it passes through small openings controlled by the thumb-screws 40. When the action is sudden and violent the valves open, and admit the movement of the gate, without instantly closing the governing valve, leaving that function to be performed by springs. The above is the theory of the machine as stated by the makers.

CHAPTER XVI.

CANALS. ·

A FALL available for water-power, whether created by a dam or resulting from the natural conformation of the ground, very rarely occurs in one perpendicular plunge. The more common case is a series of rapids below the dam or waterfall, necessitating a more or less extended canal in order to utilize the entire fall. Even when the entire fall occurs in one jump, a canal is commonly necessary to convey and distribute the water to the several wheels.

A canal in earth where the value of land is not a serious consideration is usually made with sloping sides. The slopes should be flatter than is admissible in dry embankments. The canal should admit of being frequently and rapidly emptied, and an abrupt lowering of the water tends strongly to the caving of the banks. For this reason it is not advisable to make the slopes steeper than 2 to 1. In populous districts, where land is of high value, it is often advisable upon economical and sometimes upon æsthetic grounds to flank the canal with vertical walls of masonry. Fig. 190 indicates a canal 12 feet deep with slopes of 2 to 1, the water-surface being 3 feet below the ground-level. In this case a wall placed so as not to change the cross-section of the canal would save about 18 square feet of ground per linear foot on each side the canal. A linear yard of this wall would contain about $7\frac{1}{2}$ cubic yards of masonry, and it would make available for other uses some 54 square feet of ground. The excavation would be greater in the case of the wall than in that of the slope. Allowing

386

3 dollars a cubic yard for the cost of the masonry, the land must be worth near 50 cents a square foot to justify the wall. Such walls are commonly laid without mortar, of quarried stone such as remain after the dimension stone have been culled out.

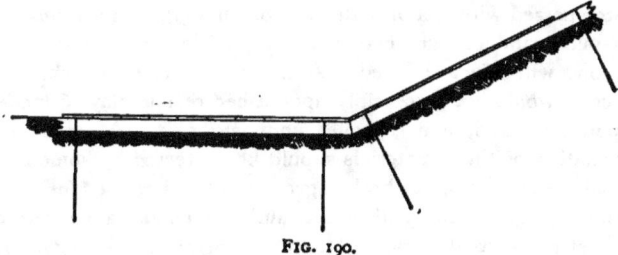

FIG. 190.

The thickness of such a wall should be at the bottom about two-fifths of the height, and about one-fifth at the top. It must, for the reason above given, be rather heavier than for a dry embankment of the same height. It should have extra strength when, as sometimes happens, a railroad runs along the bank of the canal, as the vibrations incident to the passage of trains and locomotives are very trying to such work. In such case the bottom width should be one-half the height.

When a canal is excavated in solid rock the modern channelling-machine may be used with great advantage. This is an automobile engine running on a track and carrying a wide flat tool working by percussion like a steam-drill, but capable of cutting a continuous channel some 3 inches wide and 10 feet or more in depth. This channel greatly facilitates the excavation of the rock and leaves a smooth vertical wall for the boundary of the canal, very favorable to the movement of the water. When the rock is excavated to the full depth of the channel, a second channel can be cut to an equal depth, leaving a horizontal offset of some 6 inches in the face of the wall. The machine can run on a curved track, to conform to the line of the canal, when required.

The Velocity in Canals is governed by two considerations, viz.: (1) the preservation of the bed and banks of the canal; (2) the preservation of the head. In a canal with side slopes the first consideration would limit the velocity to from 1 foot in sand and other material readily disturbed, to 3 feet in gravel interspersed with pebbles the size of an egg. In a formation of pure sand or pure clay a velocity of less than 1 foot per second will affect the bed. A mean velocity of 4 inches per second would act perceptibly upon a bed of fine clay, 6 inches upon fine sand, and 9 inches upon *buckshot*. A canal in a formation of these materials should be protected by some artificial means. Angular broken granite of the size used for road-coverings is not easily displaced and will endure a velocity of 5 feet per second, meaning thereby the mean velocity in the canal. This material is objectionable as forming a very rough surface and offering great obstruction to the flow of the stream. Coarse gravel with numerous pebbles, boulders, and cobbles soon comes to a permanent condition under any velocity attainable in a canal. The sand and clay is washed out, leaving the stones. The small stones fall into the cavities between the large ones and are not readily detached. In a walled canal with its bed in a refractory stratum almost any desired velocity can be used, the limiting consideration being the expense of removing the great quantities of silt and sediment brought in by an excessive velocity. As to the second consideration, it is a balancing of the value of the power sacrificed by loss of head against the expense of the dispositions necessary for preserving the head, viz., increased cross-section and smoother perimeter. In a canal carrying 1800 cubic feet per second, a loss of 1 foot head would involve a loss of 150 horse-power, having a value which would justify considerable expense in avoiding loss of head.

When, as is almost universally the case, the water-power is used in connection with steam, and the demand for power largely exceeds the low-water flow of the stream, the canal is required to carry a varying quantity of water. In the season of low water the velocity is moderate and the loss of head is slight. When water is abundant and running to waste, it carries the largest possible quantity of *power* without regard to loss of head. It would be easy to construct two canals of equal cross-section, one of which would, with equal loss of head, deliver twice as much water as the other. It would thus appear that loss of head is of primary importance in the design of water-power canals.

In considering the subject of Flowage, page 8, we have used the rougher and more approximate method of computation by the formula

$$v = c\sqrt{rs}, \qquad \cdots \qquad \cdots \qquad (59)$$

in which $v =$ the velocity in feet per second, s the slope of the water-surface in feet per foot, r the quotient of the cross-section divided by the wetted perimeter, and c a coefficient which is varied according to judgment to suit varying conditions. In computation of loss of head in canals we will use Kutter's formula, which is thought to be more exact. It is of the same form as (59), but c is a variable and may be computed by the relation

$$c = \frac{41.66 + \dfrac{1.811}{n} + \dfrac{.00281}{s}}{1 + \left(41.66 + \dfrac{.00281}{S}\right)\dfrac{n}{\sqrt{r}}}, \quad \cdots \quad (60)$$

where n is a numerical quantity depending on the nature of the surface in contact with the water. It is called the coefficient of roughness.

For straight and regular channels lined with planed boards carefully laid, or plastered with neat cement, we may put $n = $.................... 010

For similar channels lined with unplaned boards or plastered with mortar, or for the best and cleanest brickwork........ 012

Lining the boards with canvas raises the value of *n* to. . .015
Channels paved and walled, best work................017
    "   "   "   "   roughest work...........024
Small rivers and canals with fairly regular channels in
    earth 024
Channels in earth with masonry side walls, in best condi-
    tion.....................................020
    " " " " " " " in worst con-
    dition.................................050
Irregular channels in earth in best and straightest condi-
    tion.....................................025
Do. in worst condition, obstructed by detritus and vege-
    tation...................................050

In exceptional cases *n* rises even higher than the latter figure.

Channels in earth, however regular they may be on completion, will fall into an irregular condition in the course of a few years, with a corresponding increase in the value of *n*. It is the latter condition that must be assumed in computing the flow.

In applications (60) takes the form

$$v = c\sqrt{rs} = \frac{41.66 + \frac{1.811}{n} + \frac{.00281}{s}}{1 + \left(41.66 + \frac{.00281}{s}\right)\frac{n}{\sqrt{r}}}\sqrt{rs}, \quad . \quad (61)$$

from which *v* is found when the other elements are known. When *v* is known *c* is readily found, being $= \frac{v}{\sqrt{rs}}$. Two other cases may occur, viz.: (1) to find *n* when *r*, *s*, and *v* are known; (2) to find *s* when *r*, *n*, and *v* are known. We will take numerical values for the known quantities, which makes the equation less cumbrous, while the principle is no less obvious.

Suppose a canal in which $r = 12$, $s = 1$ foot per mile $= \frac{1}{5280}$, $v = 3.29$ feet per second, what would be the value of n? Eq. (61) becomes

$$3.29 = \frac{56.50 + \frac{1.811}{n}}{1 + 16.31n}\sqrt{\frac{r}{5280}}, \text{ or } 69.012 = \frac{56.50 + \frac{1.811}{n}}{1 + 16.31n},$$

or

$$69.012 = \frac{56.50n + 1.811}{n + 16.31n^2},$$

whence

$$69.012 \times 16.31n^2 + 12.512n = 1.811,$$

or

$$n^2 + .011116n = .0016089,$$

whence by the ordinary rule for quadratic equations

$$n = -.005558 + \sqrt{.0016089 + (.00556)^2} = .035.$$

Again, suppose $n = .035$, $r = 12$, $v = 3.29$, to find the value of s. Inserting these values in (61), it becomes

$$3.29 = \frac{323.55s^{\frac{1}{2}} + \frac{.0097341}{s^{\frac{1}{2}}}}{1.42092 + \frac{.000028391}{s}},$$

whence we obtain

$$323.55s^{\frac{1}{2}} - 4.6748s + .0097341s^{\frac{1}{2}} = .00009341.$$

This is a cubic equation from which the value of s may be obtained by the rules given in the books of algebra.

The engineer who deals with these problems has usually forgotten most of his algebra, in which case he will find it most convenient to obtain the value of s by a process of trial and error, i.e., he assumes a value for s and computes the several terms. If the first side of the equation exceeds the second, the assumed value of *s* is too great, and *vice versa*. He amends his supposition and tries again till he makes both sides of the equation equal. This is a laborious process, but is usually the one to be adopted.

Ganguillet and Kutter,* in the work referred to, give a diagram from which either of the quantities *r*, *s*, *c*, or *n* can be obtained when the others are known. Where many computations of this kind are required it might be advisable to construct such a diagram. It is seldom necessary to obtain *r* from the formula, but it is obvious that it can be obtained by a quadratic equation.

Slope of Bottom.—The bottom of a canal usually receives a slope based upon some assumed delivery of water. The slope *s*, determined from formula (61), is assumed to apply to both the surface and the bottom. If the slope of the surface differs materially from that of the bottom, the velocity and the value of *r* change from point to point. In fact the slope of the surface never is precisely the same as that of the bottom, though usually so near that the formula applies without material error. Problems often occur in which serious error would result from assuming the depth and velocity in a canal to be uniform, and it is necessary to take account of the variation in depth consequent on the loss of head. In this case the canal must be divided into a number of parts and the computation applied separately to each part. After computing the slope and loss of head in the first part, we find a revised cross-section, a revised velocity, and a revised value of *r* applicable to the second part, and so on.

Comparison of Smooth and Rough Canals.—Consider two canals of the same cross-section, the same mean radius $r = 12.25$, the same slope $s = .000256$, being about 16 inches in a mile. We have $\sqrt[4]{r} = 3.5$, $\sqrt{s} = .016$, $\sqrt{rs} = .056$. Suppose the bed and banks of the first to be in such a condition

---

* A General Formula for the Flow of Water, etc. New York, John Wiley & Sons, 1889.

of roughness that $n = .031$, which would be no uncommon case. Suppose the second to be straightened, trimmed, and lined with plank so that $n = .012$. Then in the first case

$$v = \frac{41.66 + 58.42 + 10.98}{1 + (41.66 + 10.98)\dfrac{.031}{3.5}}\sqrt{12.25 \times .000256}$$

$$= \frac{111.06}{1.4664} \times .056 = 4.24.$$

In the second case

$$v = \frac{41.66 + 150.92 + 10.98}{1 + (41.66 + 10.98)\dfrac{.012}{3.5}} \times \sqrt{12.25 \times .000256}$$

$$= \frac{203.56}{1.1805} \times .056 = 9.66.$$

That is, according to admitted principles, the second canal would carry, with the same loss of head, more than twice as much water as the first. Such a velocity, however, could only be maintained while the canal-surfaces remained perfectly clean. It would diminish as the latter became fouled by deposits. Moreover, in comparing the two canals, we must consider not only the loss of head incident to the flow of the water, but that necessary to impart the velocity. This would be for the first canal

$$H = \frac{(4.24)^2}{2g} = 0.28 \text{ ft.,}$$

and for the second

$$H = \frac{(9.66)^2}{2g} = 1.45,$$

so that the first canal would deliver the water 1.17 feet higher than the second and would have that much advantage in point of head.

To compare the two canals on the supposition that the total loss of head is the same in each, let us suppose each to be a mile in length. Then the total loss of head in the first case is .000256 × 5280 + 0.28 = 1.63 feet. By eq. (59) we have for the second canal $c = \dfrac{v}{\sqrt{rs}} = \dfrac{9.66}{.056} = 172.5$. c may be regarded as constant for any slight variation in depth and velocity. The head expended in imparting the velocity v is $\dfrac{v^2}{2g}$, and the descent in the canal may be represented by 5280s = $5280\dfrac{v^2}{c^2r}$, so that we have the total loss of head

$$= \left(\frac{1}{2g} + \frac{5280}{c^2r}\right)v^2 = 1.63, \text{ or } .0301v^2 = 1.63,$$

whence
$$v = \sqrt{\frac{1.63}{0.0301}} = 7.36.$$

In other words, the second canal could carry 74 per cent more water than the first with the same total loss of head.

Effect of an Ice-sheet.—The capacity of a canal to carry water is seriously impaired by the formation of a sheet of ice, which not only diminishes the cross-section, but increases the wetted perimeter. There is a lack of experimental data on this subject, and it is not known what value of n applies to the ice-surface. The under surface of the ice is usually corrugated and irregular. It is not so smooth as a surface of planking, though perhaps smoother than a wall of rubble masonry. In the absence of exact knowledge we will use the value of $n = .024$. Assume the same value to apply to the rest of the perimeter. Consider a channel 80 feet wide on bottom, 12 feet deep, side slopes 2 to 1, making surface-width 128 feet; slope 1 foot per mile. Find how much the flow would be diminished by a sheet of ice floating 20 inches deep.

In the open channel we have

Cross-section $= (80 + 2 \times 12) \times 12 = 1248.$

Perimeter $= 80 + 2 \times 12 \ \sqrt{5} = 133.66.$

Radius $= r = \dfrac{1248}{133.66} = 9.34.$

$$v = \frac{41.66 + 75.46 + 14.84}{1 + (41.66 + 14.84)\dfrac{.024}{\sqrt{9.34}}} \sqrt{9.34 \times \frac{1}{5280}}$$

$$= \frac{131.96}{1.4437} \sqrt{\frac{9.34}{5280}} = 3.84.$$

Quantity of water $= 1248 \times 3.84 = 4792$ cubic feet per sec.

Head required to impart a velocity of 3.84 feet per
second . . . . 0.23 feet
Loss of head in canal. 1   "

Total loss of head in canal free of ice . . . . . . $\overline{1.23}$ "

We have also

$$c = \frac{3.84}{\sqrt{9.34\dfrac{1}{5280}}} = 91.3.$$

In the case of the canal covered with an ice-sheet floating
20 inches deep, we have the cross-section

$$= (80 + 10.33 \times 2)10.33 = 1039.8 \text{ square feet,}$$

and the wetted perimeter

$$= 2 \times 80 + 4 \times 10.33 + 2 \times 10.33 \sqrt{5} = 247.51 \text{ feet.}$$

Whence we have for the mean radius $r = 4.20.$ If v
represent the velocity in the canal, we have $\dfrac{v^2}{2g} =$ loss of

head at entrance, also $s = \dfrac{v^2}{c^2 r}$, and the descent in the canal

$= \dfrac{v^2}{c^2 r} \times 5280$. These two losses must be equal to the loss in the open channel, whence

$$\left(\frac{5280}{c^2 r} + \frac{1}{2g}\right)v^2 = 1.23,$$

and

$$v = \sqrt{\frac{1.23}{0.16636}} = 2.71, \quad Q = 2.71 \times 1039.8 = 2827.3.$$

That is to say, the canal frozen over to the depth assumed would carry only 59 per cent as much water as the open canal with the same loss of head. Practically, the users of water, under such conditions, would not confine themselves to any prescribed loss of head, but would draw as much water as they required, or as much as they could, without regard to loss of head and consequent diminution of their fall. But the utmost that can be drawn is not over 59 per cent of what could be drawn in open channel with equal loss of head. Where there are several users on the same canal it often happens under these conditions that the water at some hours of the day is entirely drawn out of the canal at the lower end. In this case a singular and embarrassing phenomenon sometimes occurs. The ice-sheet becomes anchored by freezing solidly to the bottom, and so remains when the water comes to a level after the stopping of the mills at night. During the night a fresh coating of ice forms and is in its turn lowered and frozen to the anchored sheet. This goes on till, in protracted cold weather, a part of the canal is packed solidly with ice from top to bottom.

Lining or Sheathing Canals.—In the above comparison of velocity in a smooth and rough canal, page 393, suppose the cross-section of the canal to be 1500 square feet, other dimensions conformable to the above supposition. Suppose the canal to supply wheels on a clear head of 30 feet. We have found the velocity in the unlined canal 4.24 feet per second, corresponding to a flow of 6360 cubic feet per second, and in the lined canal a velocity of 7.36, corresponding to a flow of 11 040. Though, for reasons stated, we could hardly hope to reduce n to .012, there is little doubt that by reasonable expense in cleaning the canal it could be maintained practically at .015, corresponding to a velocity of 6.64 and a flow of 9960 cubic feet per second. Here is a gain, due to the improvement, of 3600 cubic feet per second in the delivery of the canal. In the case of an industry which has attained dimensions far in excess of the low-water flow of the stream, and where all the users are provided with steam-engines, this additional flow of water, though it may not be obtainable more than six or even four months in the year, has a value as suspending, for the time being, the consumption of coal and some other expenses of steam. Manufacturers in New England readily pay three and four dollars per diem per mill-power for water furnished in this manner. As the above-mentioned increase of 3600 cubic feet per second would amount, on a fall of 30 feet, to some 144 mill-powers, it is readily apparent that, under suitable conditions, a very large outlay would be justifiable in improving the carrying capacity of the canal. Improvements of this character are beginning to attract the attention of manufacturers. As this kind of work has not been practised long enough to establish any recognized method of conducting it, the writer deems it important to point out the several methods that may be followed.

1. *For Cement Work.*—The bed and slopes of the canal being finished to the desired grade, the slopes not steeper than 2 to 1, the whole is covered with a layer of broken stone 3 inches deep on the slopes, 4 inches on the bed. This is lightly rolled or rammed to settle it into the earth and bring it to a smooth surface. Cover this surface with a layer of mortar not so stiff as would be used for masonry, work it well into the crevices of the stone, and bring it to a smooth surface. Near

the water-surface, within the range of ice action, the protection must be heavier, consisting of good concrete as much as 9 inches deep. The bottom must be strong enough to stand the occasional passage of carts used for removing deposits. Near the foot of the slopes numerous weep-holes must be left to prevent any upward pressure when the canal is being emptied.

Where the canal has side walls, a great improvement can be made by simply pointing the joints of the masonry with mortar; but the cost of a facing of hard brick is not out of proportion to the benefit to be expected. This may be one-half brick thick, backed with sufficient mortar, which gets a good hold by being worked into the crevices of the wall. The rougher the wall, provided it is stable, the better it lends itself to this method. The utmost care is necessary to clean the surfaces of the stonework. Openings for the relief of pressure, near the foot of the wall, should not be forgotten. This method could only be adopted in the southern part of the country, where the action of frost is not severe.

2. For Timber Work.—The plank or board lining must, of course, rest on sills or sleepers imbedded in the ground.

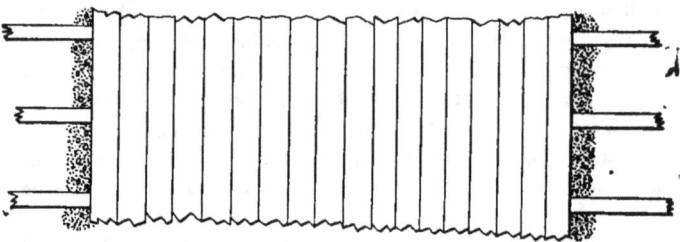

Fig. 191.

The difficulty is to prevent the work from rising or coming into disarrangement from its buoyancy. Figs. 190 and 191 represent a mode of confining the sills which, though somewhat expensive in material, is expeditious in performance; a point of no small importance in work of this kind, which has to be done during nights and holidays. The earthwork being finished, the timbers are imbedded in trenches cut therein, and are confined by long iron rods 1 inch diameter, driven through holes bored in the timbers. The length of these rods will depend on the nature of the material. If they are so long that the operation of driving will form a sufficient head on them, they will have abundant holding-power, which increases with time. The rods being driven and the earth packed around the sleepers and brought truly up to their surfaces, the planking is applied. This may be 2 inches thick at the bottom, which has occasionally to sustain wheeled vehicles, and 1 or $1\frac{1}{4}$ inches on the slopes. When the rod is put in position for driving, its head is far out of reach. Fig. **192** shows a tool for starting it down-

FIG. 192.

ward. To a piece of $1\frac{1}{2}$-inch round steel 18 inches long is welded a 2-foot length of pipe, forming a cap, which is slipped on the head of the rod before it is erected. The rod drives easily for the first 3 or 4 feet, and the cap is lifted and dropped by any convenient tool, as a shovel. When the pipe comes within reach of the hand it is worked more expeditiously till the head of the rod comes in range of the workman's hammer.

There is no doubt that in alluvial ground or ground favorable to driving, wooden stakes can be driven deep enough to hold the sleepers with safety. This can readily be determined by experiment, i.e., by driving stakes and observing the force required to draw them. The buoyancy of such a lining would in no case exceed 10 pounds per square foot, which for sleepers 3 feet apart and stakes 5 feet is about 150 pounds pull on each stake. This diminishes rapidly as the timber becomes water-soaked. In this method it is better to use 3-inch by 8-inch joists for the sleepers. These are laid flatwise and bored with 3-inch holes for the stakes, which are formed with heads that do not go through the holes.

Figs. 193 and 194 represent another method of lining which may, under some conditions, be used with advantage, especially when the ground is not adapted to driving, as, for instance, where the bed-rock is not far enough below bottom. The surface of the canal, bed, and slopes, being finished to subgrade, is covered with boards or planks running lengthwise of the canal. These are not necessarily of uniform length, width, or thickness. Provided they are sound, they may be

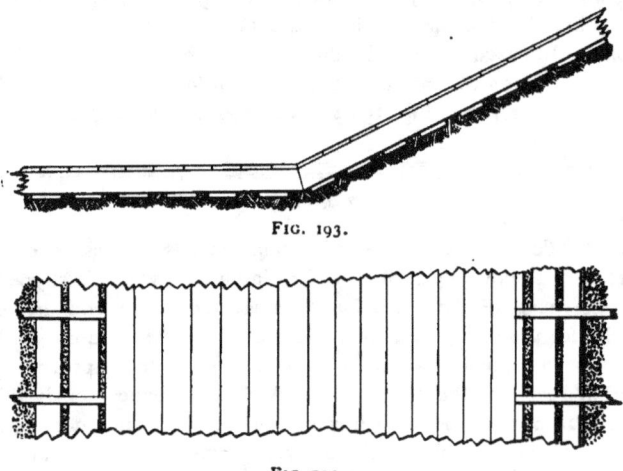

FIG. 193.

FIG. 194.

refuse lumber, culls, slabs, lumber taken from coffer-dams or old buildings. On this covering are laid the sleepers, consisting of 2×9 or 3×9 plank set edgewise and confined by toe-nailing, i.e., by nails driven obliquely through the lower corners into the covering. These nails, crossing each other, have a sufficient holding power. The spaces or pockets between the sleepers are then filled with gravel or whatever material occurs in the canal, which material is placed in position, tamped, and levelled even with the tops of the sleepers. Then the lining-plank are applied. In applying this method to an existing canal, the material excavated in front of the work is brought back and dumped into the pockets. Holes must be bored through the lining near the foot of the slopes for relief of pressure. In applying a timber lining to a canal with side walls, the bottom is treated by either of the methods pointed out, and light timbers are bolted to the side walls to receive the planking.

CHAPTER XVII.

DEVELOPMENT OF NATURAL WATER-POWERS.

Two general methods of applying water-power are now in use. The first, which is the method adopted in all the old manufacturing centres, consists in conducting the water by means of canals to the several establishments requiring power. In the second, the whole operation of converting the water into mechanical power and the latter into a form suitable for transmission is performed in a single establishment called a power-house, whence it is transmitted to the several users. The first requires the mills to be located in situations accessible to the water, and often calls for greater expense in their construction than would otherwise be necessary. The second carries the power to the machines, which are located according to convenience and, thereby adapting itself to the multifarious requirements of industry, it has found application in many if not most of the recent installations. Shafting, wire rope, air, and water under pressure have found wide application for conveying power from water-wheels to the machines, but since the great development of electricity dating from about 1880, this agent has taken the leading place as a means of transmitting power.

Fig. 195 represents a common arrangement for a single mill on a small stream. The dam forms the foundation for the upper end of the mill, and the mill forms one abutment of the wasteway, the opposite abutment being connected by an embankment with the high ground. At the upper end of the mill is a wooden platform raised above the level of high water. Here are located the gates for admitting water to the wheels.

This arrangement assumes that all the fall in the vicinity is utilized by the dam. In other words, that there is quiet water for some distance below the dam.

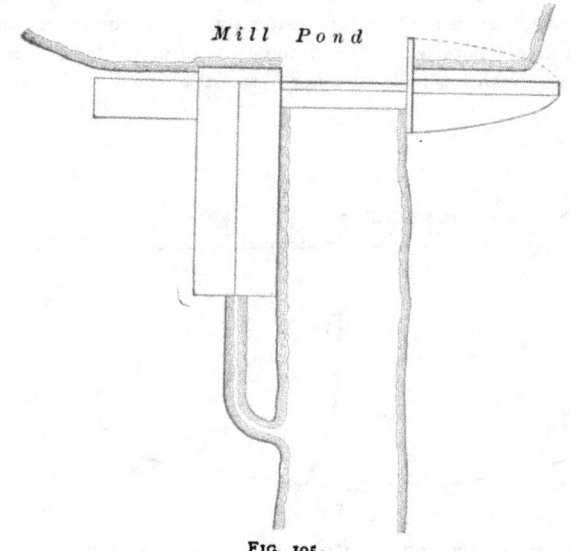

Mill Pond

FIG. 195.

Fig. 196 indicates the arrangement suited to the case of rapids below the dam, which it is desirable to include in the fall acting on the wheels, or the case of a number of mills occupying too much ground to admit of drawing directly from the pond. A canal leads from the mill-pond to a point below the rapids, or to ground favorable to the location of mills. Sometimes, instead of extending the canal, the river-bed can be deepened, or a separate tail-race constructed leading downstream far enough to comprehend the entire fall. It is desirable in that case to exclude the flood-waters from the tail-race, otherwise it is filled and rendered inoperative in time of high water. A canal such as is here contemplated generally has a wasteway at the lower end.   This is necessary when the canal has considerable length and supplies a number of mills, in order to prevent a dangerous surge of the water when the mills stop, as they often do, at the stroke of a bell.   The water in the canal continues to move after the gates are closed, and unless a vent is provided the momentum of the water causes it to overflow the banks.   In northern latitudes a wasteway is

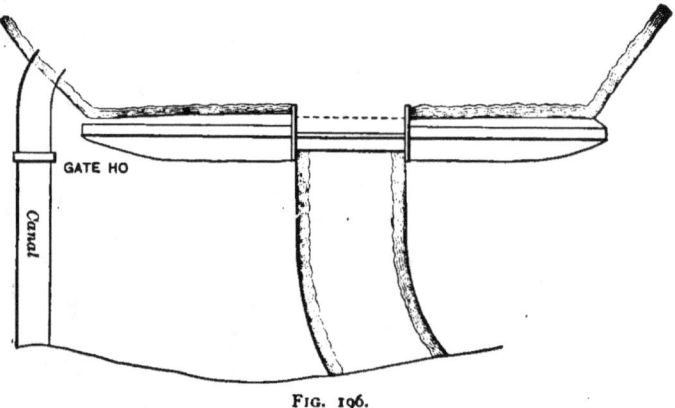

FIG. 196.

necessary to sluice off the ice and prevent its accumulation in the canal.

Fig. 196*a* represents the arrangement of, and mode of applying the water to, a modern cotton-mill.   *A* is a long, low building parallel to the canal and near it, containing the repair-shop, storerooms, offices, baling- and packing-rooms; *B*, cotton storehouse; *C*, one of the main mills, generally five or six stories high.   The small circles represent the wheel-pits.   The water leaves the canal at *D*, and, after passing through a covered chamber called the "feeder-head," is admitted to the penstocks through gates *E E*.   After passing the wheels, the water is discharged through arched culverts into the open race *G*.   A projection on the front face of the mill contains the entrance and stairways.   A projection on the rear contains the sinks and lavatories. When health laws do not intervene these discharge directly into the culverts. *H* is the engine-house, *K* the boiler-house, *L* the chimney. The wheels and engine give motion to a main shaft, running along the basement. Each of these motors can be thrown in or out of connection at will. The projection *F* is the belt-house, containing the belts which transmit motion from the main line of shafting to the several lines in the rooms above. In times of abundant water the engine does not run. It is coupled on

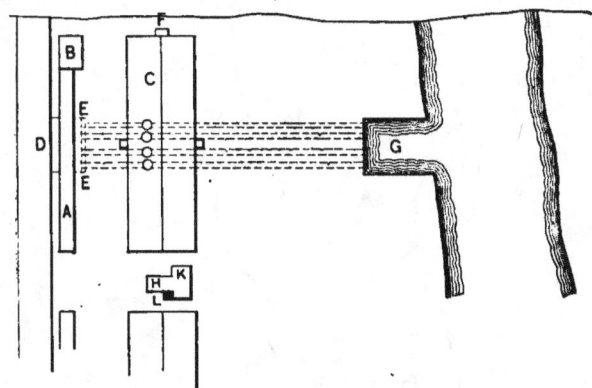

FIG. 196a.

when the flow of the stream falls short of the demand, and furnishes a larger and larger supply of power as the flow of the stream diminishes until, in the lowest stage, the mills often run almost wholly by steam. The water from the canal enters the feeder-head through a line of rack, and at *D* is a platform on which workmen can stand to rake out the trash which accumulates in the racks.

Fig. 197 shows a case in which the fall in a wide stretch of river can be utilized by a short canal. Such a situation occurs on the Cedar River, at Waterloo, Iowa, and works for development of power were undertaken there in 1883. The writer was consulted as to the general arrangement and scheme of these works. The promoters of the enterprise had built a dam as indicated, raising a head of about 6 feet, and had constructed head-gates at the entrance. They had excavated the canal, which was about 2 miles long, to a depth sufficient to allow the water, as raised by the dam, to flow through. There was a fall of about 2 feet 3 inches in the long bend of the river. They conceived the idea of excavating the canal by the scouring power of the water, the formation being of a character favorable to this action. Their intention was to use the canal

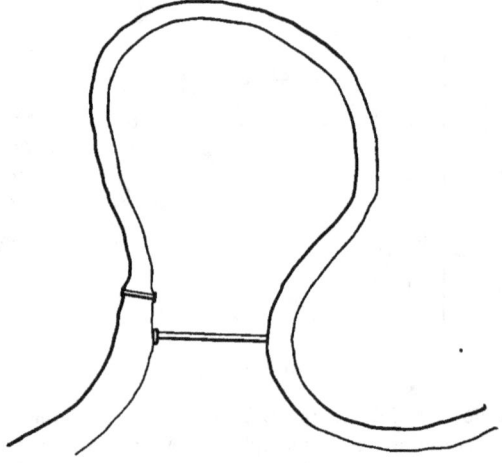

FIG. 197.

as a tail-race, placing the mills at the upper end. Their attention was strongly called to the greater probability of success if they placed the mills at the lower instead of the upper end of the canal, making the canal serve as a head-race instead of a tail-race, as, in the former case, it would not need to go so deep by something over 8 feet as in the latter. The desire to place the mills near the existing centre of population prevailed over the engineering considerations. They were unable to excavate the race to the required depth by the proposed method, and the insufficiency of the tail-race has made the water-power of little value.

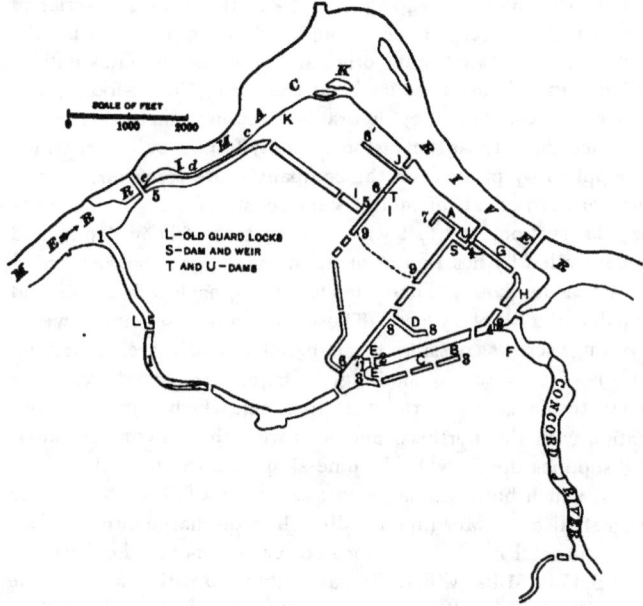

FIG. 198.

The recent development on the St. Lawrence at Messena Springs, New York, is of this character. Here the river plunges down the Long Sault Rapids, falling some 50 feet in a distance of 9 miles. Approximately parallel with the course of the St. Lawrence flows the Grasse River with a very gentle current, and at a point above the rapids the two channels are but 2 miles apart, with a fall of 42 feet between them. A canal some 3 miles long makes the water of the St. Lawrence available on a head of 32 feet at the power house.

Fig. 198 indicates the general arrangement of canals at Lowell,* Mass. The extreme fall here is now near 40 feet, having been recently increased several feet by deepening the Merrimack River below the mouth of the Concord. The dam crosses the river in a zigzag line at *e* at the head of a series of rapids which extend down to the bend in the river. I I I is the old Pawtucket canal, built originally for purposes of navigation. It has a gate-house near its head containing five 9-foot gates, which are controlled by hydraulic pressure, each gate being attached directly to the piston of a hydraulic cylinder, which is supplied by pipes from the company's fire-reservoir. 5 5 is the northern canal, of more recent construction. It is of rectangular section, 100 feet wide and 15 to 20 feet in depth, and is controlled by head-gates at the dam, the gate-house forming a part of the dam. There are ten gates, each 8 feet wide and capable of a lift of 15 feet. These gates are raised and lowered by long screws passing through nuts, which are turned by means of a system of shafting, belts, and pulleys driven by a small turbine. 5 6 is the western canal, which is in communication with the northern and also with the Pawtucket canal. 7 8 supplies the Lowell Machine-shop and the Lowell Carpet-mills, which both discharge into 2 2 with a fall of about 13 or 14 feet, also the Merrimack Mills, which discharge into the river with the total head. The northern canal supplies the Tremont and Suffolk Mills, which discharge into 6 6 with a fall of some 13 feet. 6 6' supplies the Lawrence Mills, which discharge into the river with the remaining fall. 3 3 supplies the Appleton and Hamilton Mills under a head of some 13 feet, discharging into 2 2. This latter canal supplies the Booth and Massachusetts Mills and, in part, the Middlesex, all discharging into the Merrimack or Concord rivers. The northern and western canals are connected with the Merrimack canal by an underground feeder on the line 9 9. Water falls from the basin 3 1 7 into the canal 2 2 over a wasteway controlled by flashboards.

---

* Tenth Census Report U. S., vol. XVI. Water-power of Eastern New England, p. 31.

Water is also discharged into the canal 6 6 over a wasteway at *T*, and similar wasteways discharge the surplus water from each of the lower-level canals into the river. These wasteways are essential in order to maintain the required level in the canals, since the discharge from the upper mills may greatly exceed or fall short of the draft of the mills on the lower level, and means must be provided for discharging the excess and supplying the deficiency. These wasteways are under the constant care of attendants, who remove and replace the flashboards according to the stage of water in the canal. The Merrimack Mills located at *A* use the water under the full head.

Fig. 199 shows the arrangement of canals at Holyoke, Mass. The total fall here, from the pond above the dam to quiet water below, is near 60 feet. The working fall is 50 feet and upward, according to the stage of the river. The fall is divided into three parts by the canals 1, 2, 3, No. 1 being the upper level, 2 the second, 3 the third. In describing these canals we cannot do better than use the language of Professor Swain in the Tenth Census Report, vol. XVI: "The system of canals at Holyoke comprises three levels from which water is drawn. The first or upper level strikes off across the bend which the river forms, and runs at a distance of from 2800 to 3400 feet from the latter; it has a length of 5700 feet at present" (1880, since extended some 250 feet), "and decreases in width from 150 feet near the bulkhead to, say, 105 feet at the lower end. The water-depth is about 20 feet near the bulkhead, but through the main portion of the canal is uniformly about 10 feet. The canal is walled throughout its length to a height generally of 2 or 3 feet above the water-surface, and is the only one of the three canals that has been completely walled. The fall from this level to the second is 20 feet. Near the upper end a few mills discharge from it directly into the river, using falls of 32 to 40 feet. The second level runs parallel to the first and 400 feet nearer the river, forming a straight reach of 6500 feet; continuing from the upper end it also sweeps around through a further distance

of 2600 feet at present, running parallel to the curving course
of the river and 500 feet distant from it.   Its width decreases
from 150 feet at the upper end to 90 feet at the lower end of
the straight reach, and in the curving portion lies mainly
between 140 and 150 feet.   The water-depth in this level is
uniformly about 8 feet.   The supply of water comes from the
first level, partly as tail-water from the mills and partly from
the waste-weir and gates between the two levels.   The fall to
the third level is $11\frac{1}{2}$ or 12 feet, and from the second level to
the river from 25 to 28 feet.   The third level, for a part of its
course, runs parallel to the river, at a distance from it of, say,
500 feet.   It has a total length of from 3500 to 4000 feet, a
width of about 100 feet, and a water-depth of 8 feet.   The fall
from this level to the river is substantially the same for all the
mills using it, but, according to the stage of the river, ranges
from 15 to 27 feet.   In addition to the wasteway, already
described, near the bulkhead, having a length of about 200
feet, and discharging into the river, there is another, of 40 feet,
over which water descends from the first to the second level.
Closely adjacent, the second level has a waste-weir 100 feet
long, toward the river, which the overflowing water reaches
through four arched openings underground.   At the lower end
of the second level there is another weir, 80 feet long, over
which water spills to the third level.   The latter has a similar
weir, 150 feet long, connecting with the river.   These various
weirs do not rise to the ordinary level of the water-surface, but
are surmounted by temporary flashboards, varying from 18
inches to 2 and even 3 feet in height, used for maintaining the
proper level.   At each weir there are also waste-gates at or
near the bottom, which may be used in connection with the
flashboards for regulating the level or for drawing it down
altogether.

   "On the left bank of the river is a short canal, 4, which
supplies three mills under a head of about 35 feet, discharging
directly into the river.   This is controlled by a small gate-
house.   6 is the gate-house on the right bank.   It is about 40

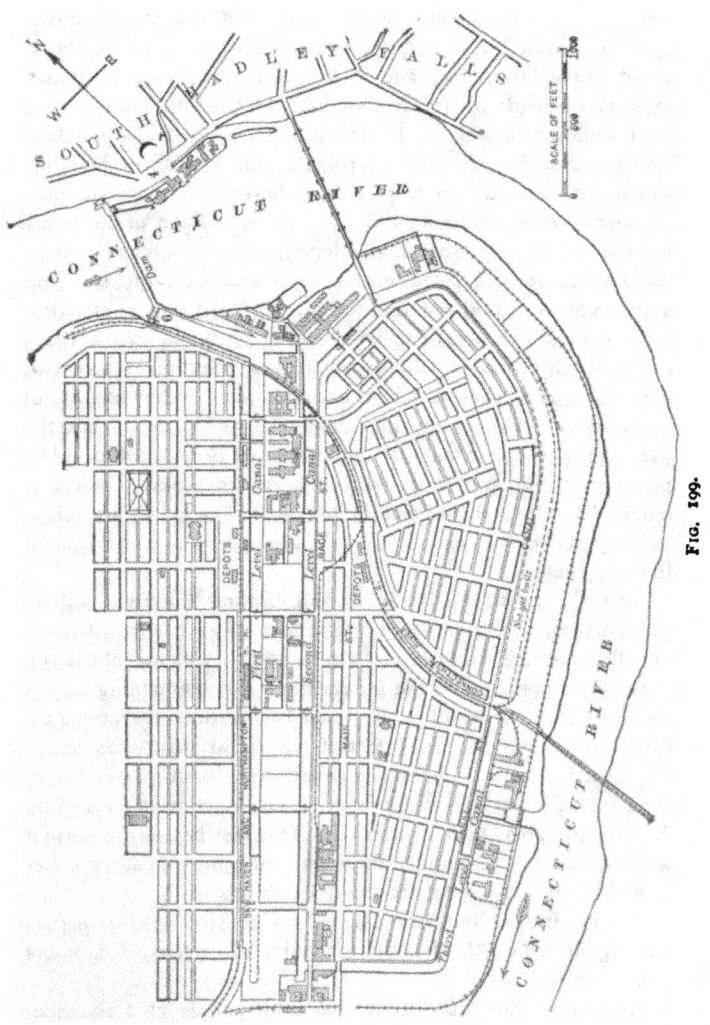

FIG. 199.

feet wide, measured with the current, and has twelve gate-openings, each 8 feet wide by 15 feet deep, separated by piers about 19 inches thick; and also two smaller gate-openings, each $4\frac{1}{2}$ feet wide by $10\frac{1}{2}$ feet deep.   This is surmounted by a brick building in which is the machinery for working the gates. The power is furnished by a turbine at one end of the building which acts directly to turn a long horizontal shaft running lengthwise of the building above the gates.   Each of the latter has two vertical wooden posts (stems), 10 by 13 inches in size, fastened to it, and faced on one side with iron racks.   The long shaft operated by the turbine, in turn, by connecting belts, causes a series of cog-wheels to revolve, the last of these engaging and moving the racks already mentioned and with them the gates, each of which can be moved independently of every other.   Commonly the water stands higher outside the gates than in the canal; at such times they are not raised to the full height, and in consequence of the pressure power is required to close them; but in low water the gates are raised entirely above the surface, and their weight alone is sufficient to close them.

"Adjacent to the bulkhead and forming the river-wall of the main canal is a wasteway $198\frac{1}{2}$ feet long; it is constructed of solid masonry, and rises to within 20 or 24 inches of the top of the gate-openings in the bulkhead; for the remaining height above its crest the water in the canal is controlled by temporary flashboards.   Of the $198\frac{1}{2}$ feet of length, 40 feet rises above the surface of the water in the canal, and about 5 feet above the crest of the waste-weir proper.   This portion is pierced by four waste-gates, their centres about 20 feet below the normal water-surface in the canal; these gate-openings measure 5 feet in width, and from 5 feet to 65 inches in depth."

The principal industry carried on at Holyoke is paper-making, in which the heavier machines run twenty-four hours a day.

Fig. 200 shows the arrangement of canals at Lawrence, Mass., where the plan of development was simpler than in the

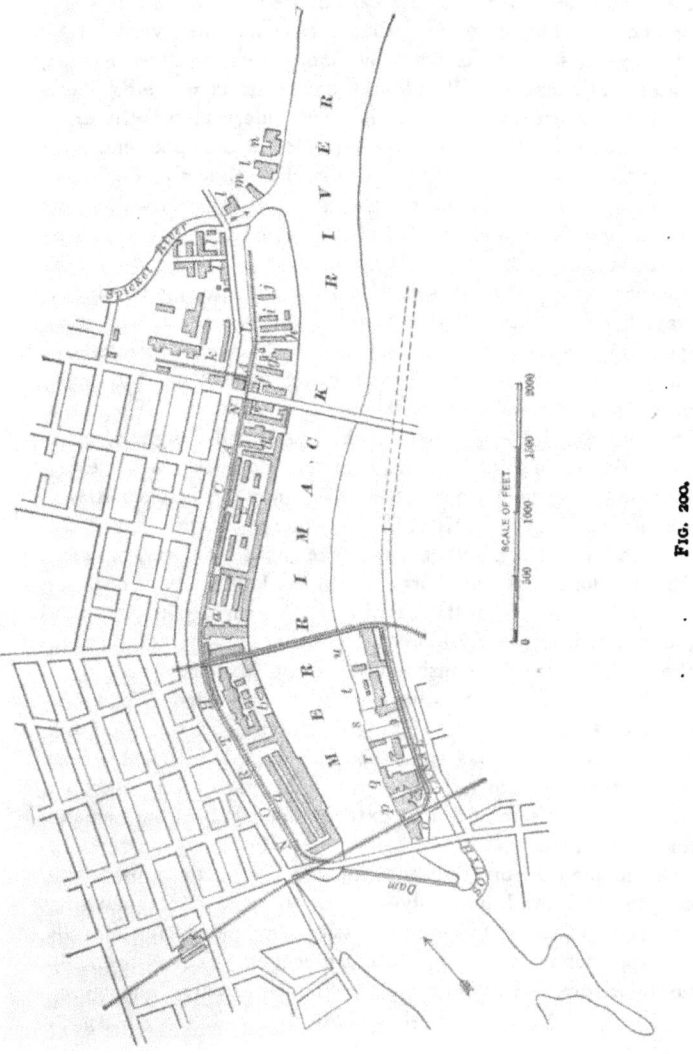

Fig. 200.

cases just described. The total fall here is 31 feet, and the practical working-head in ordinary stages of the river is 26 to 30 feet, subject to considerable diminution in times of high water. A canal on each side of the river starts at the dam and leads down-stream. That on the north side, which is the principal one, is 5330 feet long, 100 feet wide at the upper end, and 60 at the lower end, with the cross-section shown in Fig. 201. Its depth is 12 feet in the middle, 4 feet at the side walls, and the bottom is graded to a fall of 1 foot in 10 000, or 0.53 foot in the length of the canal. The head-gates are twenty-four in number, arranged in six sets of four each, or closing six sluice-ways between piers 12 feet deep and 9 feet wide, so that each gate is 3 feet high and 7 feet long. They are operated by hand. This improvement was made subject to the legal obligation of providing for the passage of boats. There is a lock of slight lift at the head of the canal, and a flight of three locks descending into the river at the foot; also a wasteway by which the canal may be emptied into the Spicket River, which here joins the Merrimack. There is no division of the fall; the mills are placed between the canal and river, drawing from the former and discharging into the latter. At a distance of about 80 feet from the canal a line of sheet-piling extends parallel to it for nearly half its length to prevent percolation, the bank being, to a considerable extent, artificial.

The south canal was built in 1866, and carried for a distance of 2000 feet with a rectangular section 60 feet wide and 10 feet deep. It has since been extended 750 feet. The head-gates are sixteen in number, in four sets, and are operated by hand. It is intended to extend this canal as the demand for power increases.

The plan of dividing the fall into two or three parts was adopted at Lowell and Holyoke with the view of commanding as large an area of ground as possible for the placing of mills and appurtenant buildings. It is a precedent that will probably not be followed in any future plan of development. The supply from one level cannot adjust itself to the demand of the next,

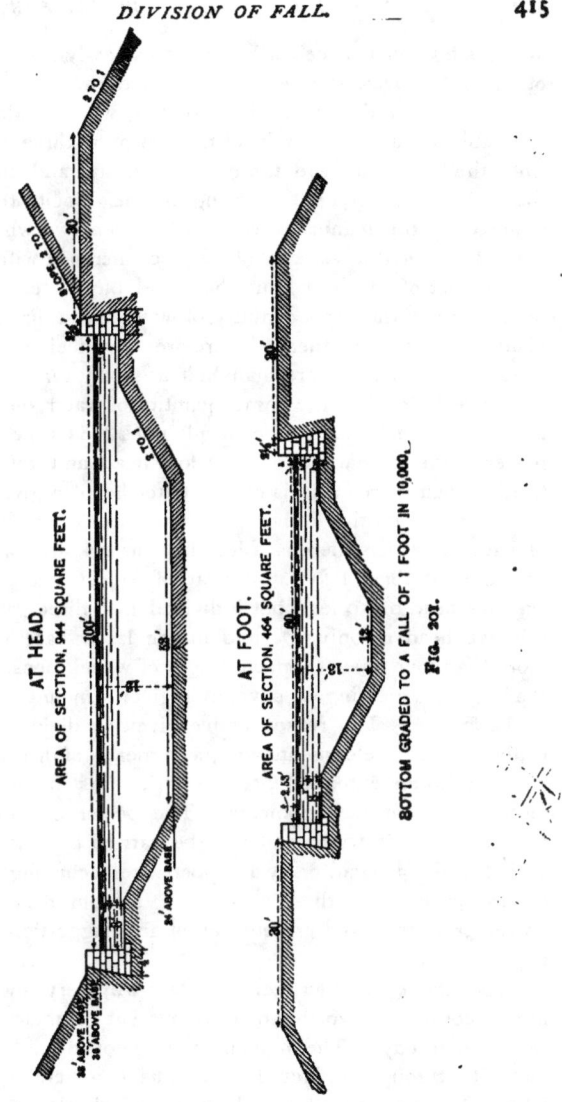

AT HEAD.
AREA OF SECTION, 944 SQUARE FEET.

AT FOOT.
AREA OF SECTION, 464 SQUARE FEET.

BOTTOM GRADED TO A FALL OF 1 FOOT IN 10,000.

FIG. 201.

and much water is necessarily wasted uselessly from one to the other. Moreover, the tendency of modern manufacturing is constantly toward more rapid velocities, which makes it more profitable to use a high head than two or three low heads, since the lower the head the more expensive and cumbrous is the mechanism required to bring up the velocity to the rate required by the manufacture. As between two wheels of the same diameter, the velocity of the circumference will be as the square root of the head, but the wheel on the lesser head, in order to draw the same quantity of water, must have a larger diameter than the other. Therefore, a wheel on a 12-foot head will not make more than half as many turns per minute as a wheel drawing the same quantity of water on a 24-foot head. A wheel on a 24-foot head gives 2.83 times as much power as the same wheel on a 12-foot head, and it would practically require three wheels on the latter head to give the same power as one on the former. Moreover, on every application of water to wheels there is a loss in penstock and race usually estimated at about 1 foot. A head of 60 feet used entire is an effective head of 59 feet, but if divided into three parts it is an effective head of only 57, and in the latter case we require more than four times as much weight of wheel, penstock, case, shafting, and gearing to develop a given amount of power as in the former. The reason above given for dividing the head in the earlier developments, viz., as a means of distributing the power over an extended factory district, has lost much of its force by recent developments. The power conveyed by a canal 100 feet in width can now be carried by a 30-inch pipe buried in the ground, or by a copper wire occupying no appreciable space. For these reasons we seldom hear in water-power projects of the present day of any suggestion for dividing the head.

The above considerations explain why very low falls are hardly considered worthy of development or maintenance at the present day. The numerous abandoned mill privileges met with throughout New England have generally very low falls. The expense of installing water-wheels on a fall of

6 feet is actually greater for the same quantity of water than on a fall of 24 feet, while the power is less than one-fourth as great. Such a fall can often be incorporated with the next fall on the stream above by removing the dam and deepening the channel, or with the next fall below by raising that dam and flowing it out. This is usually a more profitable use than any separate scheme of development.

Tidal Power.—Twice a day, on all the coasts of the world, the sea rises and falls under the influence of lunar and solar attraction. Speaking precisely, the phenomenon does not occur twice a day, but the time of high tide falls each day 40 or 50 minutes later each day than on the preceding day. At any given point high water occurs at the same hour at the end of each lunation. When the sun and moon act in conjunction, and also when they act in opposition, the tidal oscillation reaches its maximum. When their action is at right angles to each other, tide is at its minimum. Otherwise stated: at the time of new and full moon we have high tides, called spring-tides. When the moon is at its first and last quarter we have low tides, called neap-tides.

The amplitude of the tidal oscillation is very various.[*] On the Atlantic coast of the United States it has its maximum at the north part, viz., at Eastport, Me., where the range is 15 to 22 feet. It diminishes toward the south, being 7 to 12 feet at Portland; a little greater at Boston; 3 to 5 at New York; 1 to $1\frac{1}{2}$ at Baltimore; 4 to 7 at Fernandina, Fla., and 12 to 18 inches in the Gulf of Mexico. The tides of the Pacific Coast are low and erratic and not to be considered in reference to power. It would appear, therefore, that there are few localities on the coasts of the United States at which development of tidal power is a matter for practical consideration. The highest tides occur on the coasts of England and France, and here projects for making tidal power available for industrial purposes are often considered from a financial point of view. The range of spring-tides at different points on these coasts is given below.[†]

[*] See Tide Tables, U. S. Coast and Geodetic Survey.
[†] Hagen, Seeufer und Hafenbau, 1863, 1. 127.

ENGLAND.	Ft.	In.
Devonport	15	4
Dover	18	9
London	19	6
Hull	20	10
Leith	16	4
Liverpool	26	0
Holyhead	16	0
Pembroke	21	0
Weston-on-the-Sea	37	2

FRANCE.	Ft.	In.
Cordonan	15	6
La Rochelle	17	6
St. Nazaire (river Loire)	17	7
Brest	21	0
Isle Bréhat	32	7
St. Malo	37	3
Granville	40	5
Les Ecrehont	33	4
Cherbourg	18	6
Barfleur	18	6
La Hongue	19	11
Port-en-Bassin	21	0
Mouth of the Orne	24	1
La Havre	23	5
Fécamp	25	4
Dieppe	29	10
Cayeux (Somme)	30	1
Boulogne	26	0
Calais	20	6
Dunkerque	17	7

Fig. 54, on p. 111, indicates an arrangement for utilizing tidal power. The dimensions of the dikes are such as would be required for the highest tides on the coast of England or France, but the method is applicable to all tides. *BB* is the main dike, uniting two headlands and cutting off a basin of some hundreds or thousands of acres. It is shown as having an exterior slope of 4 to 1 and an interior of 2 to 1. *A* is a partition-dike with twofold slopes, separating the basin into two parts, *C* and *D*. Turbines are located in the pits *E*, *E'*, the former discharging into the basin *C*, the latter supplied from the basin *D*. Each basin is provided with sluices, as indicated, communicating with the sea, and controlled by automatic tide-gates. Water cannot enter the basin *C* through the sluices, but can freely escape thereby when the level of the sea is below the level of the basin. No water can escape from *D* through the sluices, but can freely enter thereby when the relative height permits it.

During the stand of low tide the level in the basin *C* comes very close to the external level. Water is passing the turbines *E'* from the basin *D* under a head which is diminishing both from the rise of the tail-water and the fall of the head-water. When the tide has risen so much as practically to obliterate the head on the turbines *E'*, it is several feet above the level in the basin *C*. The turbines in *E'* are now shut off and those in *E* started, the tide being now the head-water and the basin C the tail-water. When the sea reaches the level of D, water begins to flow into that basin, and at high tide it is filled. The tide falls and reaches a level at which there is no longer a working head with reference to C. The turbines in E are now shut down and those in E' started, and so on.

This method gives a continuous development of power. Other arrangements may be adopted giving a greater head at times, but involving an entire cessation of power at other times. It is probable that the hydraulic compression of air (p. 518) might be advantageously adopted in the utilization of tidal power.

For any given conditions of tidal development a diagram can readily be constructed to show the average power, and the power to be expected at different stages of the tide, when the following data are known:

1. The extent and contours of the tidal basins.

2. The discharge of the turbines under any given head.

3. The hydrograph of the tide; that is, the height of the tide at any given time, most conveniently reckoned from low water of each tide.

Development by Means of Tunnels. — Situations often present themselves in which a subterranean channel becomes feasible for the passage of water drawn for power. This is more commonly used for the discharge of the tail-water than for the access of the supply, but the latter arrangement is by no means unknown. This case occurs where the stream, after plunging down a high fall or rapid, changes direction and approaches closely to its channel above the fall. In such case, a tunnel through the intervening ridge becomes the most suitable means of making the fall available. The development at Messena Springs on the St. Lawrence is a case of this kind, except that the ridge intervening between the two streams was not high enough to call for a tunnel. To justify from a financial point of view the construction of a tunnel of the necessary dimensions for such a distance would usually require a fall reaching into the hundreds of feet.

Commonly the tunn l is introduced to discharge the water into the lower level of the stream after passing the wheels. Prominent among installations of this character is the system of the Niagara Falls Power Company at Niagara Falls, N. Y., where a tunnel some 7000 feet in length conveys the water from the wheel-pit to the river. Thi tunnel is of egg shape, about 19 feet extreme width and 21 feet extreme height, its cross ection being something like 320 square feet. It has a declivity of 7 feet in 1000, intended to give the water a velocity of 25 feet per second, and a capacity of 8000 cubic feet per second to the tunnel. Previously to the inauguration of this work a velocity of 25 feet per second in a water-power channel was a thing utterly unheard of. It traverses a formation of limestone very liable to disintegration upon exposure, and is for that reason lined with four half-brick courses of brickwork. That brickwork will endure su h a velocity cannot be regarded as an established fact.* It was adopted upon a somewhat short view of the subject, saving the cost of a sufficient tunnel, furnishing a sufficient amount of power for present uses, establishing the company's right to the full quantity of water discharged, and leaving them at liberty to make a more economical use of the water at a later date. The adoption of this arrangement restricts the fall utilized by the wheels to about 136 feet, whi e the total fall, from the i take of the canal to the outfall of the tunnel, is 216 feet. Taking 20 feet as a reason ble allowance for loss of head in channels of not over 3 miles length, this leaves 60 feet, implying some 40 000 effective horse-power, as sacrificed on account of this insufficient tunnel.

The law department of the State of New York has repeatedly declared that further diversions of water from Niagara Falls are unlawful; that, the cataract with its surround-

---

* A letter to the author from Coleman Sellers, Chief Engineer of the Niagara Falls Power Co., under date of Jan. 16, 1901, says that "the last examination of the tunnel made after a large amount of power was being developed showed no signs whatever of wear.

ings having been dedicated by the State to public uses, the grandeur of this majestic spectacle cannot be further impaired for private purposes. The cost of the tunnel is stated to have been about a round million of dollars. Under these conditions, when the time comes that the company has use for all the power obtainable and could find use for more, it will find itself confronted by this fact: That by the expenditure of three millions an addition of 40 000 horse-power could be obtained. Power, then, could not probably be considered worth less at that place than $25 per annum per horse-power, and the securing of an income of one million per annum by the expenditure of three millions is an opportunity not to be disregarded. The construction of three such tunnels would reduce the velocity to some 8 feet per second, and the loss of head there into one-ninth of what occurs at present. This change would be effected by carrying the wheel-pit down, say, 20 feet below the surface of the river at the outfall, and leading out the tunnels on a level or with an inclination of about one-tenth that of the present tunnel. But here comes in a point that does not appear to have been foreseen: the existing tunnel can form no part of the amended system, since its influx will be full 50 feet above the water-level in the wheel-pit. Had the tunnel been placed at the above-indicated level, it would have carried the same quantity of water as at present, with the loss of the same head, and would have been available in an amended system designed to use the water with reasonable economy.

Snowqualmie Falls on the river of that name, in the State of Washington, is a very conspicuous instance of the use of a tunnel in the development of water-power.* The Snowqualmie empties into Puget Sound on the Pacific coast, in latitude 47° 30″ N., though the climate is much milder than in the same latitude on the Atlantic coast. At a point above the falls the river is some 600 feet above the sea-level, with a drainage-area of some 500 square miles in extent, reaching to the summit of the Cascade Range of mountains, an elevation

---

* See *Engineering News*, 1900, p. 398.

of some 8000 feet; being, in part, above the perpetual snow-line, so that the stream never gets very low in hot weather. The total precipitation at the falls is said to be 90 inches per annum, and it has reached 150 inches in parts of the drainage-area, though the rainfall on the borders of Puget Sound is generally not more than 35. A low-water flow of 1000 cubic feet per second is expected, and it is said that opportunities exist for the construction of storage-reservoirs to greatly increase this flow. The cataract is one of the most remarkable in the world, being a sheer plunge of 270 feet. The formation here is a basaltic rock of a very firm texture, very impervious to water, and with no tendency to disintegration on exposure. These conditions led to the project of placing all the machinery in a subterranean chamber, receiving the water through a vertical shaft and discharging it through a tunnel. A low dam was built across the river, sufficient to turn the water into the penstock. This may be raised, as occasion requires: not to increase the fall, but to create a pondage-basin sufficient to meet the hourly variations in the demand for power. The water passes down the shaft in a $7\frac{1}{2}$-foot penstock and there enters a horizontal pipe or receiver, about 10 feet in diameter, extending along the side of the chamber, which is 200 feet long, 40 feet wide, and 30 feet high, and stands directly over the discharge-tunnel, separated from it by a 5-foot concrete floor. The arrangement, though differing in detail, is precisely the same in principle as indicated by Fig. 226. The chamber contains at present (1901) four generators of 2000 H.P. (1500 K.W.) each. The turbines are of the impulse type as indicated, Fig. 226, arranged in groups of six, each group driving a generator. The chamber is sufficiently above the tunnel to be free of water in the highest stage of the stream, an arrangement involving some waste of head at lower stages. The total head utilized is about 250 feet. The tunnel has a cross-section which enables it to carry off the water with a moderate velocity, and has a fall of but 2 feet in its entire length of some 650 feet.

. The arrangement at Snowqualmie is one that could not be universally adopted. It was rendered admissible here by the firm and impervious character of the rock. In a softer and more permeable formation there would have been continual danger of damage to the machinery by the falling of masses of rock from the roof, and the continual dripping of water from the roof and sides would have made the chamber too damp. Of course it is within the resources of engineering to obviate these difficulties, but the expense of constructing a suitable chamber in such a formation would probably have exceeded the cost of transmitting the power to a suitable power-house at the surface, together with the construction of such a power-house.

The Falls of St. Anthony, at Minneapolis, on the upper Mississippi, have been made available for power mainly by the aid of tunnels. The river has here a drainage-area of about 20 000 square miles. The formation at the falls is a series of strata of Trenton limestone so called, some 30 feet in thickness, overlaid by soil and gravel, and underlaid by an indefinite depth of friable material called St. Peter sandstone by the geologists, though it can hardly be classed as stone in an engineering sense, being in some places hardly firm enough to stand at a vertical face, and nowhere, in the vicinity of the falls, too hard to be excavated with the pick. The limestone is of a tolerably firm texture, some of the strata being extensively used for building, while others are liable to entire disintegration upon exposure to frost. At the crest of the falls the limestone has been worn down by the river to a thickness of 10 or 12 feet, and some 1200 feet further up-stream it has disappeared entirely from the bed of the stream. A dam, above the falls, holds the water within a few feet of the surface of the limestone. The first establishments located here, about 1855, were sawmills, drawing their supplies of timber from the vast pine-forests to the northward; but these have latterly given place to flour-mills, to which the power is now mainly devoted. The first establishments used only the head created by the dam,

allowing their discharge to run to waste over the falls. The
peculiar facilities offered by the situation for the use of tunnels
were soon realized, and the existing method came gradually
into use. Canals and races in the limestone lead the water to
the mills. The wheels are placed in vertical shafts sunk
through the limestone. A floor in this shaft near the bottom
of the limestone forms a flume from which a draft-tube descends
into a tunnel excavated in the sand rock and leading to the
river below the falls. A vertical shaft, using the word in a
different sense, transmits the power from the wheel to the
several floors of the mill. The above is the most common
arrangement, though erratic methods of course find occasional
application. The great advantage of this system over the
common arrangement will be apparent. In the ordinary
method the mill must stand between the canal and the river.
In this system its location is subject to no such restriction. It
can stand on the opposite side of the canal, and at a considerable distance therefrom if necessary, the discharging-tunnel
passing under the canal on its way to the river, the limestone
stratum forming a convenient roof for the tunnel. The first
tunnels were only made deep enough below the limestone to
make them accessible. They did not require any lining, but
left a considerable part of the head unutilized. Where they
are carried deep enough to utilize the entire head, they are not
safe without side walls reaching up to the limestone roof. The
head utilized by tunnels at Minneapolis is about 50 feet, though
the entire head appertinent to the water-power is 75 or 80 feet.

A reckless tunnel project undertaken in 1870 opened a
channel for the river to pass under the limestone stratum and
discharge below the falls, threatening the destruction of the
water-power, a result which was with great difficulty averted.
To guard against a repetition of this disaster, the U. S. Government has constructed a dike under the limestone, reaching
entirely across the river and some 40 feet deep into the sand
rock. This is constructed of concrete, with a general thickness
of about 4 feet and has fulfilled its purpose well.

CHAPTER XVIII.

TRANSMISSION OF POWER.

THE points at which water-power can be most economically developed are often more or less remote from those at which it is required. In the plans of development thus far described, the mills are forced into locations accessible to the water flowing by gravity, and their construction is thereby often rendered much more expensive than it would be if their location could be chosen upon other considerations. Many large water-powers lie undeveloped within a few miles of centres of industry and population having unlimited use for power. The transmission of power, therefore, has an intimate connection with the subject of water-power.

Shafting is used to transmit power from one part of a mill or shop to another, and is sometimes employed for external transmission to distances of a few hundred feet. Beyond 300 or 400 feet it becomes too expensive as compared with wire rope. Shafting consists of a line of round iron or steel bars rigidly fastened together, resting in bearings which permit it to rotate, and carrying the organs for delivery of power. Shafting is made in lengths of 12 to 24 feet, the latter being as long as can be conveniently handled and transported; but the largest mills can produce lengths up to 40 feet if desired. It is now almost universally made of steel, by rolling, and hardened on the exterior by compression. It is usually turned to an exact cylindrical surface for the convenient application of its attachments, and in situations exposed to view is smoothed by polishing. Coupling-plates, gears, and pulleys are fixed by keys, which are short prismatic pieces of hardened

425

steel driven into grooves, called seats or splines, cut partly in
the shaft and partly in the attachment.   The separate pieces
of shafting are united into one continuous line of shafting by
couplings of various forms, a standard form consisting of two
circular plates keyed to the ends that are to be united and
fastened to each other by bolts.   James B. Francis, from an
experimental inquiry made in 1867, deduced the following table
(Table 6) of the diameters of shafting.

The table gives the power which can be safely carried by
shafts making 100 turns per minute.   The power which can
be carried by the same shaft at any other velocity may be
found by the following simple rule:

. *Multiply the power given in the table by the number of
revolutions of the shaft per minute, and divide the product by
100; the quotient will be the power which can be safely
carried.*

The table for steel shafts contemplates the best quality of
steel.   Makers of the present day, using the metal that passes
in the trade for steel, make the diameters about 6 per cent
greater.   For wrought or cast iron the table gives ample
dimensions.   In shafting which both transmits power and distributes it along the line the losses are considerable.   In many
cotton-mills not more than 75 per cent of the power furnished
by the motors reaches the machines, the remainder being
absorbed in resistance of shafting, belts, and gearing.   In a
line used wholly for transmission without any intermediate
delivery the loss should not exceed 3 per cent in 100 yards.

In modern mills and factories the lines of shafting are supported by bearings, usually about 10 feet apart.   In a line
used wholly for transmission, the intervals may be much
greater.   Mr. Francis gives the following as the greatest
admissible distance between the bearings of a continuous shaft,
which carries no attachments other than the couplings.   The
figures relate to steel shafting, but they would be substantially
the same for wrought iron.   The two end intervals should be
considerably less than those given.

TABLE 6.

Diameter of Shaft in Inches.	Horse-power which can be safely carried by shafts for prime movers and gears, well supported by bearings and making 100 revolutions per minute, if of			Horse-power which can be safely transmitted by shafts making 100 revolutions per minute, in which the transverse strain, if any, need not be considered, if of		
	Wrought Iron.	Steel.	Cast Iron.	Wrought Iron.	Steel.	Cast Iron.
1.00	1.0	1.6	0.6	2.0	3.2	1.2
1.25	2.0	3.1	1.2	3.9	6.2	2.3
1.50	3.4	5.4	2.0	6.7	10.8	4.1
1.75	5.4	8.6	3.2	10.7	17.2	6.4
2.00	8.0	12.8	4.8	16.0	25.6	9.6
2.25	11.4	18.2	6.8	22.8	36.4	13.7
2.50	15.6	25.0	9.4	31.2	50.0	18.7
2.75	21	33.3	12.5	41.6	66.6	25.0
3.00	27	43	16	54	86	32
3.25	34	55	21	69	110	41
3.50	43	69	26	86	137	51
3.75	53	84	32	105	169	63
4.00	64	102	38	128	205	77
4.25	77	123	46	154	246	92
4.50	91	146	55	182	292	109
4.75	107	171	64	214	343	129
5.00	125	200	75	250	400	150
5.25	145	232	87	289	463	174
5.50	166	266	100	333	532	200
5.75	190	304	114	380	608	228
6.00	216	346	130	432	691	259
6.25	244	391	146	488	781	293
6.50	275	439	165	549	879	330
6.75	308	492	185	615	984	369
7.00	343	549	206	686	1098	412
7.25	381	610	229	762	1219	457
7.50	422	675	253	844	1350	506
7.75	465	745	279	931	1490	559
8.00	512	819	307	1024	1638	614
8.25	562	898	337	1123	1697	674
8.50	614	983	368	1228	1965	737
8.75	670	1072	402	1340	2144	804
9.00	729	1166	437	1458	2333	875
9.25	791	1266	475	1583	2533	950
9.50	857	1372	514	1715	2744	1029
9.75	927	1483	556	1854	2966	1112
10.00	1000	1600	600	2000	3200	1200

For a shaft 1 in. diam. greatest distance between bearings 1 2.6 ft.

"	"	2	"	"	"	"	"	16	"
"	"	3	"	"	"	"	"	18	"
"	"	4	"	"	"	"	"	20	"
"	"	5	"	"	"	"	"	22	"
"	"	6	"	"	"	"	"	23	"
"	"	7	"	"	"	"	"	24	"
"	"	8	"	"	"	"	"	25	"
"	"	9	"	"	"	"	"	26	"
"	"	10	"	"	"	"	"	27	"
"	"	11	"	"	"	"	"	28	"
"	"	12	"	"	"	"	"	29	"

Telodynamic or Wire-rope Transmission.—When we consider that a 4-inch line of shafting transmitting 200 horse-power at 200 revolutions per minute would require, with its couplings, some 50 pounds of metal per linear foot, exclusive of the bearings, while the same power could be transmitted by a wire rope weighing not more than 1¼ pounds per foot, we realize the immense advantage of the latter mode of transmission where applicable. Wire rope for transmission is made of steel or wrought-iron wire twisted into strands, which in turn are twisted around a hemp centre to form a rope. The hemp centre makes the rope more flexible and diminishes the internal wear. The flexibility is also greater as the number of wires is greater and their diameter less. The common American practice is

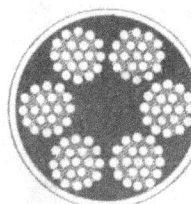

to put 19 wires in a strand and 6 strands in a rope, Fig. 202. Some makers put a hempen core in each strand, but this is not the common practice. The twist of the rope is usually in the opposite direction to that of the separate strands. One important application of wire-rope transmission is in the cables * of street railways, and these are said to last longer and wear

FIG. 202.

---

* See Fairchild on Street Railroads, 1892, p. 102.

better when the twist of the rope is in the same direction as that of the strands.

Wire rope running on a pulley and enveloping one-half the circumference undergoes a contraction of the inner wires and an extension of the outer. For the part in contact with the pulley, the outer wires exceed the inner by about 1½ times the diameter of the rope. In a 12-foot pulley this difference is distributed over about 19 feet of rope. In a 24-inch pulley it takes place in a length of about 3 feet, and the buckling of the inner fibres and stretching of the outer lead to a very rapid destruction of the rope. On this account, the pulleys for wire-rope transmission should be as large as practicable. Diameters of 12 to 15 feet are usually adopted; smaller sizes can be used when other considerations control, but the smaller the size the more rapid is the deterioration of the rope.

TABLE 7.—CRUCIBLE-STEEL WIRE ROPE, 6 STRANDS.
19 WIRES PER STRAND. HEMP CENTRE.

Diameter in Inches.	Breaking Strain in Tons of 2000 Pounds.	Working Load in Tons of 2000 Pounds.	Weight per 100 Feet, Pounds.
⅜	4.5	0.75	26
0.5	7.5	1	35
0.62	14	2	63
0.75	18	3	88
0.87	25	5	120
1	33	6	158
1.12	42	8	200
1.25	52	10	250
1.37	63	11	300
1.50	77	12	365

The Velocity of Wire Rope is of course limited by the consideration that no dangerous strain shall occur in the rim of the pulley. This, however, is of but slight importance, because, long before the velocity reaches a point that can set up any dangerous strain in the rim, the centrifugal force in that part of the rope which is in contact with the pulley so loosens the grip of the rope that it slips. In a 15-foot pulley with its rim

running 100 feet per second, each pound of the rim would tend
to fly off radially with a force of 41.66 pounds, which would
occasion a bursting strain in the material of about 1041 pounds
per square inch—a matter of no consequence.   The centrifugal
force acting upon the rope, assuming the latter to weigh
1 pound per foot, diminishes its pressure upon the pulley by
nearly 1000 pounds, and to that extent diminishes its adhesion
to the pulley and the power that can be transmitted by it.
The centrifugal force not only diminishes the pressure of the
rope on the pulley, but diminishes the arc of the pulley

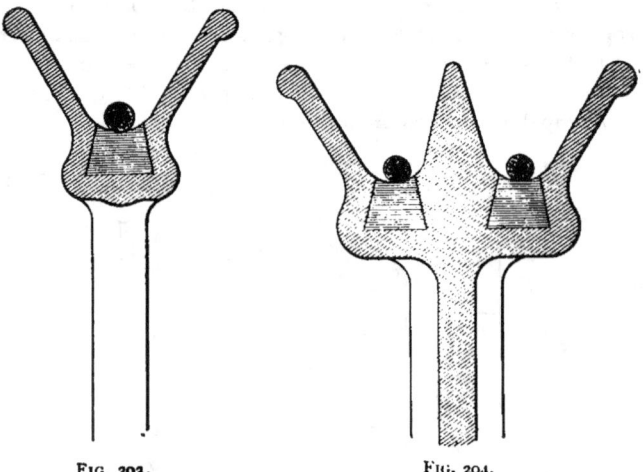

FIG. 203.   FIG. 204.

embraced by the rope.   For these reasons 100 feet per second
is thought to be the extreme limit of the velocity, and in
practice it is usually put considerably below that figure, often
not more than 75.
   Fig. 205 shows the general arrangement of turbines on
vertical shafts for a system of wire-rope transmission.   The
turbines are set with their centres in the same straight line, and
give motion, by means of bevel-gears, to a horizontal shaft

elevated so that the rope will clear the ground and intervening objects. As many wheels and as many pulleys can be used as the extent of the development calls for, and pulleys can be placed on each end of the line of shafting if desired. A transmission-line sometimes consists of a single rope reaching the

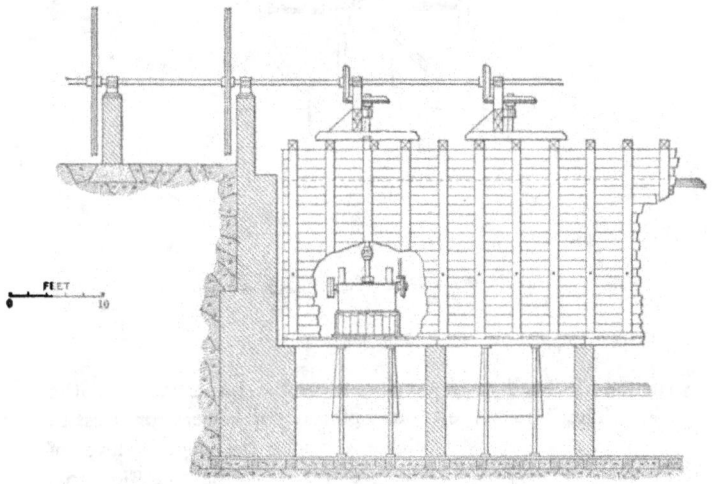

FIG. 205.

entire distance and resting on intermediate carrier-pulleys. This arrangement is unavoidable in street-railway cables. The best practice consists in dividing the distance into a number of spans and using a separate rope for each span with pulleys, as indicated by Fig. 204. The rim of this pulley has two grooves, one for the incoming and one for the outgoing rope. Fig. 203 is a section of the rim of a pulley for a single rope, such as would be used to deliver power from a shaft to a rope, or to drive a shaft by means of a rope. The bottom of the groove in which the rope runs is deeply widened into a channel of trapezoidal cross-section reaching around the pulley.

This channel is packed solidly with pieces of leather inserted radially and presenting their edges to the rope.   This con-

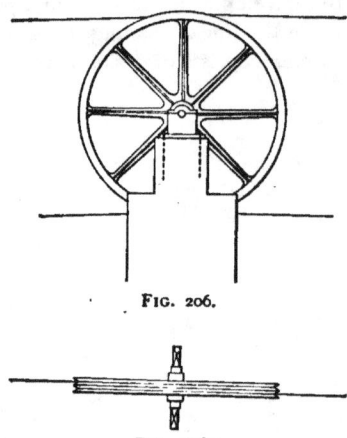

FIG. 206.

FIG. 206a.

struction is found very conducive to the preservation of the rope.   The line pulleys are placed on towers or trestles elevated above the general level of the ground.   The pulley, Fig. 204, can only deliver power in the same line as it is received.   When it is necessary to change the direction of the rope the disposition of Fig. 207 is used, in which the shafts of the pulleys are connected by bevel-gears, made to an angle equal to the angle of deviation of the line, i.e., the angle between the pitch-circles of the gears is equal to the angle of deviation.

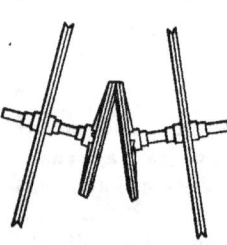

FIG. 207.

The Horizontal Turbine, placed above tail-water and discharging through a draft-tube, lends itself, under certain limitations, very readily to wire-rope transmission.

In the arrangement of Fig. 164 it is obvious that a wire-rope pulley might be applied to either end of the horizontal shaft, or one to each end.

There is no insuperable mechanical difficulty in driving a rope directly from a vertical shaft, although the arrangement is somewhat complex, as appears from Figs. 208 and 209.

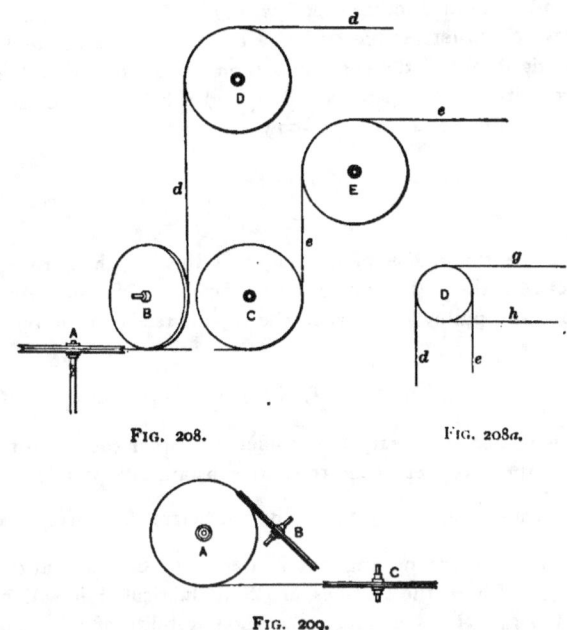

FIG. 208.   FIG. 208a.

FIG. 209.

The wire-rope pulley A is fixed upon the head of the water-wheel shaft, receiving and delivering the rope horizontally. The rope envelops considerably more than half the circumference of the pulley, so that the two plies come into the same vertical plane at a short distance therefrom. Here each of the two plies passes a pulley on a horizontal axis B and C and takes a vertical direction. At a sufficient elevation, the two plies again change direction by the two pulleys D and E which run in the same vertical plane, and thence the rope runs horizontally, forming a part of the main line. It is manifest that the pulley E could be left out of the system, the two plies d and e passing over the pulley D and the rope which envelops the four pulleys A, B, C, and D, forming a complete circuit. D in that case is a double pulley of the form indicated at Fig. 204, and sends off another rope, gh, Fig. 208a, to the main line.

Power Transmissible by Wire Rope.—If $t_1 =$ the tension on the tight ply of the rope, and t_2 on the loose ply, v being the velocity in feet per second, t_1 and t_2 being expressed in pounds, the power P in horse-power is

$$P = \frac{t_1 - t_2}{550} v. \quad . \quad . \quad . \quad . \quad . \quad (62)$$

Let $r =$ the radius of the pulley, s the length of rope in contact with the pulley, and f the coefficient of friction between the rope and pulley; then when the rope is ready to slip on the pulley

$$t_1 = t_2 \times (2.718)^{\frac{s}{r}f}. \quad . \quad . \quad . \quad . \quad (63)$$

It is seldom necessary to consider any other case than that in which the two plies of the rope are substantially parallel, and in that case $\frac{s}{r} = \pi = 3.14$. The coefficient f, between iron and leather, where no unguent is used, may be taken at 0.31 to 0.34. Where the surfaces are fully lubricated it will not exceed 0.14. It is conducive to the durability of wire ropes to keep the hempen core saturated with oil, and in hoisting-ropes, which run on small pulleys, this lubrication cannot be dispensed with. This necessity diminishes as the diameters of the pulleys are greater. In pulleys of the size contemplated here, we may, most commonly, regard the exterior of the rope as dry. Adopting $f = 0.3$, we have $\frac{s}{r}f = 0.94$, and for the extreme value of t_1 we may put $t_1 = 2.56t_2$. On the other hand, if we regard the surfaces as fully lubricated, $f = 0.14$,

$$\frac{s}{r}f = 0.44, \text{ and } t_1 = 1.55t_2.$$

Tension and Deflection.—A rope or chain suspended from two fixed points and hanging freely under the action of gravity assumes a curve called the catenary, whose equation, referred to horizontal and vertical lines passing through its lowest point, is

$$y = \tfrac{1}{4}l\left(E^{\frac{x}{2l}} - E^{-\frac{x}{2l}}\right)^{2*}, \quad \cdots \quad (64)$$

in which x and y are the coordinates of the curve, $l =$ a length of the rope equal in weight to the horizontal tension on the lowest point of the curve, $E =$ the base of the Naperian system of logarithms $= 2.7183$.

When $x = a =$ the half-span, y represents the deflection or sag of the rope.

Also, if s represent the half-length of the rope,

$$s = \tfrac{1}{2}l\left(E^{\frac{a}{l}} - E^{-\frac{a}{l}}\right). \quad \cdots \quad (65)$$

Table 8 will be found useful in any calculations relative to the deflection, tension, or length of wire ropes used in transmission. It contains the values of the quantity within the parenthesis for all practical values of $\frac{a}{l}$ or $\frac{x}{2l}$. Suppose, for instance, it be required to find the sag of a rope weighing 1.2 pounds per linear foot, with a span of 600 feet and a tension of 6000 pounds on the lowest point. Here $l = \dfrac{6000}{1.2} = 5000$ ft., $a = 300$, $\frac{a}{2l} = .03$. Looking in col. 5 of the table, opposite 0.030 in col. 1 we find 0.0036, which is the value of $\left(E^{\frac{a}{2l}} - E^{-\frac{a}{2l}}\right)^2$ in eq. (64). This multiplied by $\tfrac{1}{4}l$ gives $y = 2500 \times .0036 = 9.00$.

---

* See Moseley's Mechanics, 1866, p. 506.

TABLE 8.—TENSION AND DEFLECTION OF WIRE ROPE.

1 $\frac{a}{l}$ or $\frac{x}{2l}$.	2 $E^{\frac{a}{l}}$.	3 $E^{-\frac{a}{l}}$.	4 $E^{\frac{a}{l}} - E^{-\frac{a}{l}}$.	5 $\left(E^{\frac{a}{l}} - E^{-\frac{a}{l}}\right)^2$
0.005	1.00501	0.99501	0.01000	0.00010
0.010	1.01005	0.99005	0.02000	0.00040
0.015	1.01511	0.98511	0.03000	0.00090
0.020	1.02020	0.98020	0.04000	0.00160
0.025	1.02530	0.97531	0.05000	0.00250
0.030	1.03045	0.97045	0.06000	0.00360
0.035	1.03562	0.96561	0.07001	0.00490
0.040	1.04081	0.96079	0.08002	0.00640
0.045	1.04603	0.95600	0.09003	0.00811
0.050	1.05127	0.95123	0.10004	0.01001
0.055	1.0565	0.9464	0.1101	0.0121
0.060	1.0618	0.9418	0.1200	0.0144
0.065	1.0671	0.9371	0.1300	0.0169
0.070	1.0725	0.9324	0.1401	0.0196
0.075	1.0779	0.9275	0.1504	0.0226
0.080	1.0833	0.9231	0.1602	0.0257
0 085	1.0887	0.9185	0.1702	0.0290
0.090	1.0942	0.9139	0.1803	0.0325
0.095	1.0997	0.9094	0.1903	0.0362
0.100	1.1051	0.9048	0.2003	0.0401
0.105	1.1107	0.9003	0.2104	0.0443
0.110	1.1163	0.8958	0.2205	0.0486
0.115	1.1219	0.8914	0.2305	0.0531
0.120	1.1275	0.8869	0.2406	0.0578
0.125	1.1331	0.8825	0.2506	0.0628
0.130	1.1388	0.8781	0.2607	0.0680
0.135	1.1445	0.8737	0.2708	0.0733
0.140	1.1503	0.8694	0.2809	0.0789
0.145	1.1560	0.8650	0.2910	0.0847
0.150	1.1618	0.8607	0.3011	0.0911
0.155	1.1677	0.8564	0.3113	0.0969
0.160	1.1735	0.8521	0.3214	0.1033
0.165	1.1794	0.8479	0.3315	0.1099
0.170	1.1853	0.8437	0.3416	0.1167
0.175	1.1912	0.8395	0.3517	0.1237
0.180	1.1972	0.8353	0.3619	0.1310
0.185	1.2032	0.8311	0.3721	0.1384
0.190	1.2092	0.8270	0.3822	0.1461
0.195	1.2153	0.8228	0.3925	0.1541
0.200	1.2214	0.8187	0.4027	0.1622
0.205	1.2275	0.8146	0.4129	0.1705
0.210	1.2337	0.8106	0.4231	0.1790
0.215	1.2399	0.8066	0.4333	0.1877
0.220	1.2461	0.8025	0.4436	0.1968
0.225	1.2523	0.7985	0.4538	0.2059
0.230	1.2583	0.7945	0.4641	0.2154
0.235	1.2649	0.7906	0.4743	0.2250
0.240	1.2713	0.7866	0.4847	0.2349
0.245	1.2776	0.7827	0.4949	0.2449
0.250	1.2840	0.7788	0.5052	0.2552

Suppose we wish to transmit 400 horse-power by a rope of the above weight connecting two 15-foot pulleys, 540 feet centre to centre, running 80 feet per second, and suppose both plies of the rope to be under a tension of 3180 pounds when not running.   To find the sag.

When not running

$$l = \frac{3180}{1.2} = 2650, \quad \frac{a}{2l} = \frac{270}{5300} = 0.051,$$

for which col. 5 gives .01043, and we have

$$y = \frac{2650}{2} \times .01043 = 14.34 \text{ ft.}$$

When the rope is running we have

$$t_2 - t_1 = \frac{400 \times 550}{80} = 2750 \text{ lbs.}, \quad t_2 + t_1 = 6360,$$

whence $t_2 = 4555, \quad t_1 = 1805.$

For the tight side

$$l = 3796, \quad \frac{a}{2l} = .0356,$$

from which we find

$$y = 1898 \times .00508 = 9.64,$$

And for the loose side

$$l = 1504, \quad \frac{a}{2l} = 0.0897, \quad y = 752 \times 0.0322 = 24.21.$$

To find the total length of the rope, consider it when not running. $\frac{a}{l} = 0.102$, for which col. 4 gives 0.2043, and by eq. (65) $2s$ = one ply = 2650 × 0.2043 =........ 541.14
For the other ply............................. 541.14
One entire circumference =....................   47.12
                                              ─────────
     Total length of rope.................... 1129.40

The length of the rope does not change when running; for though the tight side stretches slightly under the increased tension, the loose side contracts to substantially the same extent under the diminished tension.

There is an advantage in making the lower ply of the rope the tight side, as we can then reckon the sag of the loose side from the top of the pulleys. This arrangement has another slight advantage in that the rope envelops a trifle more than half the circumference of the pulley, whereas, with the loose ply below, it envelops a trifle less than half. The practical limit to the length of span that can be employed, subject to this method, is the condition that the loose ply shall not sag below the tight one and occasion mutual injury by chafing. The above supposition of a span of 540 feet comes sufficiently near to this limit, as the loose ply would hang 24.21 feet below the top of the pulleys, and the tight one $15 + 9.64 = 24.64$. It would no doubt be practicable to run a rope on pulleys a thousand feet apart, but, in that case, the loose ply would have a sag of more than 60 feet, and this must be reckoned from the lowest point of the pulley, as otherwise the loose ply would hang far below the tight ply.

The critical reader will notice that these results, though sufficiently accurate for any practical purpose, are not consistent with strict mathematical nicety, as we assume the upper ply to hang from the highest point of the pulley, and the lower ply from the lowest. This is not strictly correct, owing to the slight inclination of the rope. It would be easy to correct this trifling error if it were worth while.

Case of One Pulley Higher than the Other.—The foregoing assumes that both pulleys are on the same level, in which case the relations of sag and tension are comparatively simple; but these methods become inapplicable when, as most commonly occurs, the two pulleys are not in the same horizontal line. In this case we must regard the curve as forming a part of a more extended catenary whose suspension points are on the same level, viz., the level of the higher point. In Fig.

210, d and e are two pulleys at different elevations, connected by a wire rope with lower side tight. This side is to be regarded as a part of the catenary ACB, and the elements of this catenary are to be arrived at by a laborious process of trial and error. We will state without demonstration some properties of the curve, referring the reader who is desirous of fuller information to the discussion of the catenary in treatises on the calculus. Let CH be a horizontal line through the lowest point of the catenary; let x and y be the coordinates of any point r of the curve. When $x = a$, the semispan, $y = h$, the

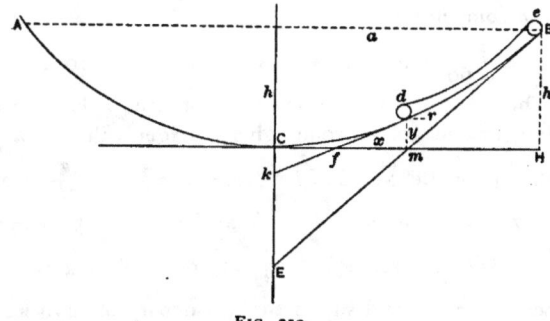

FIG. 210.

sag of the rope. At the point r, if a tangent be drawn to the curve, it forms with the horizontal and vertical through C a triangle such that if Cf be taken to represent the tension on the lowest part of the curve, fk will represent the tension at r, and Ck the weight of the rope Cr. Expressing these several tensions in equivalent lengths of rope, we put l_0 for the tension at C, l_1 for that at r, and l_2 for that at B. The quantity in column 4 of the table, corresponding to $\frac{x}{l_0}$ in the first, is twice the tangent of the angle Cfk. As a basis of computation, suppose the pulleys to be 300 feet apart horizontally, 50 feet vertically. We know within narrow limits what tension l_1 we require at r, say $l_1 = 3600$. We know that l_0 is but little less than l_1, say $l_0 = 3480$. To find a we proceed thus: Assume a value of a, and find by equation (61) the corresponding value of h. Thus $a - 300 = x$, $h - 50 = y$. See if these values will satisfy equation (64). If not, amend the supposition and try again.

First, take $a = 700$. $\dfrac{a}{2l_0} = \dfrac{700}{6960} = 0.1006$. Find in the

fifth column of the table the quantity corresponding to 0.1006 in col. 1, viz., 0.0406. Then $h = 1740 \times .0406 = 70.64$.

$\therefore$ $x = 700 - 300 = 400$, $y = 70.64 - 50 = 20.64$.

To compute y:

$$\frac{x}{2l_0} = \frac{400}{6960} = .0575. \quad \therefore \quad y = .01325 \times 1740 = 23.05.$$

This shows that we have not taken a large enough. A larger value of a would give a larger value of h and a

relatively smaller value of y. Take $a = 725$. $\dfrac{a}{2l_0} = 0.1042$,

$h = 1740 \times .0436 = 75.86$. Whence $x = 425$, $y = 25.86$,

$\dfrac{x}{2l_0} = .061$, $y = 1740 \times .01493 = 25.98$. This agreement is

sufficiently close and shows that a catenary of 1450 feet span and a tension of $l_0 = 3480$ at the lowest point will go through the centres of the pulleys. We want it, however, merely to touch the circumference of the pulleys. Call A the angle which a tangent to the curve makes with the horizontal. Then for $x = 425$ and $y = 25.86$ we have

$$\frac{x}{l_0} = \frac{425}{3480} = 0.1211.$$

The quantity in the fourth column of the table corresponding to 0.1211 in the first, is 0.2428, and 0.1214 is the tangent of A for $r =$ tangent cfk. $\therefore$ $cfk = 6° 54'$. In like manner $CmE = 11° 48'$. Therefore for the practically correct value of l_0 we

have $l_0 = 3600 \cos 6° 54' = 3574$, expressed in feet of rope.
And for the tension at B we have $l_2 = \dfrac{l_0}{\cos 11° 48'}$

We can now correct the horizontal and vertical distances of the points of suspension. Call R the radius of the pulley; the radius through the point of tangency of the rope makes an angle A with the vertical through the centre. The horizontal distance of these tangent points is

$$300 + R \sin 11° 48' - R \sin 6° 54',$$

and the vertical distance

$$50 + R \cos 6° 54' - R \cos 11° 48'.$$

With these revised values we repeat the trial and error process and find a with sufficient accuracy. The same process can be applied to the upper ply of the rope if desired. The main object of these computations is to obtain the required difference of tension without too great a strain on the axis of the pulley and without so great a sag in the loose ply as will allow it to interfere with the tight one. Much information as to existing wire-rope installations is contained in Mr. W. C. Unwin's book, "Development and Transmission of Power," London, 1894.

The Efficiency of Wire-rope Transmission for short distances is probably greater than for any other mode. The principal losses are the friction of the pulleys on their journals and the bending of the rope, which latter loss diminishes greatly as the diameter of the pulley increases. For the end pulleys of a transmission system the tension of the rope acts on the journals, but the intermediate pulleys bring no pressure on their journals but their own weight and the weight of the rope. It is the disadvantage of the system that these losses are the same whether the power transmitted be great or small.

Zeigler, at Oberursel, found a loss of 13.5 per cent in a line of seven spans carrying 104 h.p. at the sending end. This would indicate a loss of 2 per cent in each span; that is to say, the power received at any station is 98 per cent of that transmitted from the next preceding station. According to this, at the end of a mile, allowing ten stations of 528 feet each, the efficiency would be $0.98^{10} = 0.817$, and at the end of 2 miles $0.98^{20} = 0.668$. At half-load the relative loss would be twice as great. The efficiency in one span would be 0.96; at the end of a mile 0.664, 2 miles 0.442. Wire rope, though not suitable for long distances, has advantages for distances between 100 yards and 1 mile not possessed by any other method, especially as the power does not require to be changed into any other form of energy before transmission.

CHAPTER XIX.

HYDRAULIC TRANSMISSION.

THE use of water under high pressure very naturally suggests itself as a means of transmitting energy to distant points. Pipes buried underground are preferable on many accounts to towers, ropes, and their accessories, especially in populous districts. Systems of water-distribution for domestic purposes often involve arrangements for the use of the water for power. The difficulties met with in this application are: (1) Under the moderate pressure required for domestic uses, seldom exceeding 200 feet, or 86.5 pounds on a square inch, a comparatively large quantity of water is required, and as this must all be of the quality required for domestic consumption, which costs something for collection, purification, and storage, the water itself is expensive. (2) The points where power is required are commonly much higher than the source from which it is pumped or the watercourse to which it returns after use; so that a large percentage of the power represented by the water is wasted. In a city like Chicago, which has the inexhaustible volume of Lake Michigan to draw from and lies on a level plain only a few feet above the lake-surface, these difficulties do not appear, and it is accordingly not strange that large amounts of power are drawn from the mains of this city especially for operating passenger-elevators. The above objections, however, are so serious, in the ordinary case, that the plan of combining domestic water-supply with power-distribution, though it once looked inviting and had zealous advocates, has never met with extended application. Where the latter

443

has been applied, it has usually been by a separate system of pipes, and under pressures greatly in excess of those required for domestic uses.

This mode of transmission was first applied to the operation of cranes and lifts in the docks and warehouses of London, where the only mechanical movement required was the hauling in and letting out of a rope. This was accomplished by means of a piston moving in a cylinder and carrying a series of pulleys or sheaves attached to the free end of the piston-rod. Around these pulleys and a contiguous series of fixed pulleys the rope or chain was rove, so that a small movement of the piston caused a great movement of the rope or chain. The unavoidably intermittent operation of these machines necessitated means for storing the energy of the motor during those times when no demands were made on it. A reservoir was out of the question because in such situations no sufficient height could be obtained. An air-chamber is inadmissible for the reason, first, that it would deliver the water at a varying pressure, and, second, because water at such high pressures speedily absorbs the air and necessitates special means for its renewal. Out of these conditions grew the

Accumulator, which is a large cylinder with an enormously heavy plunger. The surplus water not required by the machines passes into this cylinder, raises the plunger, and forms a reserve of power to be drawn when the demand of the machines exceeds the capacity of the motor. The absolute amount of power stored in this fixture is very small, not often more than one horse-power for a couple of hours, but it serves a very useful purpose in regulating the system. Of course when a reservoir at a sufficient height is obtainable it is to be preferred. Such a system, which hardly ever commands a head of more than 500 feet or a pressure of, say, 220 pounds per square inch, may for distinction be called a low-pressure system, while a system of hydraulic transmission involving the use of accumulators, where the pressure is usually 700 or 800 pounds per square inch, may be called a high-pressure system.

Mr. Unwin gives the following table of the gross horse-power transmissible by different mains at a velocity of 3 feet per second:

GROSS H.P. TRANSMITTED BY DIFFERENT MAINS.

Low-pressure System, Head 500 Feet.		High-pressure System, 750 Pounds per Square Inch.	
Diameter of Main in Inches.	Gross H.P. Transmitted.	Diameter of Main in Inches.	Gross H.P. Transmitted.
9	75	3	29
12	133	6	116
18	300	9	260
24	533	12	463

Loss of Head in Pipes.—In a mile of 6-inch pipe with a velocity of 3 feet per second the loss of pressure will be some 15 pounds per square inch when new, and this will be increased to 31 when the pipe is rusted and covered with incrustations, as it usually is after the lapse of time. The following formula is convenient for application to pipes of 6 inches diameter and under:

$$h = 148\frac{lq^2}{d^5};$$

l being the length in feet, q the quantity of water flowing in cubic feet per second, d the diameter in inches, and h the loss of pressure represented by feet of head. Mr. Unwin gives the formula

$$h = k\frac{l}{d}v^2,$$

in which h and l are as before, $d =$ diameter in feet, $v =$ velocity in feet per second, and $k =$ a coefficient which for $d = 0.5$ may be taken as 0.000375.

For $d = 1$, $k = 0.000344$.
" 2 " 0.000312.

The following tables are also taken from Mr. Unwin's book:

Diameter of Main in Inches.	Loss Due to Friction per Mile.			
	In Feet of Head.		In Pounds per Sq. In.	
	Clean.	Incrusted.	Clean.	Incrusted.
6	35.5	70.9	15 37	30.70
12	16.3	32.5	7.06	14.07
24	7.4	14.8	3.20	6.41

Diameter of Main in Inches	Loss per Mile in Per Cent of Total Head.				
	For Pressures in Feet of				
	100	250	500	1000	1600
6	35.5	14.2	7.1	3.5	2.2
12	16.3	6.6	3.3	1.6	1.0
24	7.4	2.8	1.4	0.7	0.5

The above is for new and clean pipes. For pipes fully incrusted the percentage loss will be twice as great. This table also contemplates a velocity of 3 feet per second. It shows clearly the great advantage of high pressures in a transmission system. With such pressures as are usual in water-supply systems, the loss of head in a mile of pipe amounts to a very material percentage of the entire power; but where we deal with pressures of 700 or 800 pounds per square inch, the loss cuts so small a figure that it may, in preliminary computations, be neglected.

Systems of hydraulic distribution for general industrial purposes exist in London, Liverpool, Hull, Birmingham, and Manchester, England; in Antwerp, Holland; and in many other places. The installation recently completed at Manchester is understood to use water under a pressure of 1600 pounds per square inch. All the above-named systems derive their power primarily from steam, and do not directly concern a treatise on water-power.

At Zurich in Switzerland, power derived from a small fall in the river Limmat is distributed to a great number of small industries by hydraulic pressure. Water is pumped into mains communicating with a reservoir some 6000 feet from the pumping-station, at an elevation of over 500 feet. The effective pressure at the motors is 475 feet, and the distributing mains have an aggregate length of 15 000 feet. The charge to consumers who use less than 20 000 h.p. hours per annum is 2.5 cents per h.p. hour. To those who use 50 000 and more, about 1.25 cents per h.p. hour. This amounts to from 40 to 80 dollars per horse-power for a year of 3000 working-hours.

At Geneva, on the river Rhone, near its issue from Lake Leman, is a considerable amount of water-power, transmitted and distributed by hydraulic pressure. The fall here varies from 12 feet in low water to 5.5 in flood. Turbines are used similar to the Geyelin wheel, Fig. 159, which adapt themselves to the varying conditions of flow. A 16-inch main leads from the pump-station to the reservoir, which is some 2¼ miles distant, at an elevation of 390 feet above Lake Leman, with some half million cubic feet capacity. This reservoir is of great importance to the system, as it enables the water-wheels to run twenty-four hours a day and store up their energy when not wanted; whereas, without it, they could only run during the continuance of work, and would run at considerable disadvantage, being obliged to pump more than the requirement of the motors and wasting the excess through a relief-valve. Previous to the construction of the reservoir, large air-chambers were used to regulate the pressure, and these were kept charged with air by a compressor.

The Motors originally used at Geneva were "Schmid" pressure-engines, and these are still used for small powers.

They use a quantity of water which depends on the speed only, and not on the quantity of work done; hence are uneconomical with light loads. They are convenient and cheap, can be run at any speed, and act as meters of the quantity of water used. A counter on the pressure-engine, recording the number of revolutions, gives the means of ascertaining accurately the quantity of water consumed. At full load their efficiency is 80 per cent.

The rotary pressure-engine is not necessarily subject to the disadvantage of requiring the same quantity of feed-water for little work as for much. A device can be introduced whereby the governor acts to vary the position of the crank-pin according to the load, so that for a light load the traverse of the piston and consequent consumption of water is small, the piston having a large "clearance" at every stroke. This large clearance leads to no waste in an engine worked by an inelastic fluid, though it would be ruinously wasteful in a steam-engine.

For all large motors, at Geneva, impulse-turbines are used of the general type of the Pelton wheel, Figs. 160, 161. These are peculiarly suitable for high pressures, and give, as we have seen (page 336), an efficiency fully equal to the best, an efficiency which is not much impaired by a diminished load. They are susceptible of close regulation, which is important at Geneva, the industries connected with watch-making requiring uniform speed.

A difficulty occurs here which is inseparable from any system having a reservoir at a considerable distance from the pumps and not in line with the supply-pipe. Water flows toward the reservoir while the delivery of the pumps exceeds the consumption, and from the reservoir to the supply-mains in the contrary case. The result is that the water is delivered to the small motors at an excessive pressure in the former case, and a deficient pressure in the latter, the difference being the loss of head in 5 miles of main, with a flow of water representing the excess or deficiency of the pumpage. Such variations are inadmissible here, because the water is paid for, not by meter measurement, but by computation based on the orifices of discharge, which assumes a constant pressure. This diffi-culty is met by a centrifugal pump actuated by a separate motor, which expends a part of the water passing the main in increasing the pressure of the remainder—a wasteful device, but necessitated by the conditions of the case.

The Pipes for a pressure system require greater strength than ordinary water-pipes, and the joints require more attention, as a small leak under such great pressure represents a large loss of power. The tensile strain per running inch on one side of the pipe is represented by $\frac{1}{2}PD$, P being the pressure per square inch, and D the diameter in inches. Thus, for a pressure of 750 pounds per square inch and a diameter of 8 inches the tensile strain is $4 \times 750 = 3000$ pounds per running inch, and if we assume cast iron with a safe tensile strength of 6000 pounds per square inch, this would require the pipes to be $\frac{1}{2}$ inch thick. They are usually made considerably thicker than this consideration would dictate. The following rule or some-thing very similar is usually applied, t being the thickness in inches:

$$t = .000178DP + \tfrac{1}{4}.$$

For $P = 750$ and $D = 4$ this gives $t = 0.78$ inch, say $\frac{3}{4}$ inch.
"   6 "   "   " 1.05 "   " 1 "
"   8 "   "   " 1.32 "   " $1\frac{1}{4}$ "
" 10 "   "   " 1.57 "   " $1\frac{1}{2}$ "
" 12 "   "   " 1.85 "   " $1\frac{7}{8}$ "

For such pressures the pipes are put together with flanged joints and bolts, a ring of gutta percha or lead being interposed between the abutting surfaces. For pressures up to 500 feet head, such as usually occur in water-power systems, the ordi-nary bell-and-spigot joint packed with lead is sufficient. As regards weight of pipes, there is but slight economy in high pressure and small volume of water as compared with low pressure and large volume, the increased diameter of the pipe consequent on increased volume being substantially offset by the diminished thickness consequent on diminished pressure.

The thickness above given refers to cast-iron pipes. It is very probable that recent improvements in metallurgy would now warrant the use of steel pipes, drawn, forged, or welded, with a safe tensile strength of 15 000 pounds per square inch —an improvement calculated to greatly extend the hydraulic transmission of power.

A pipe of novel character has recently been used to conduct water under a pressure of 750 pounds per square inch for operating the movable span of a bridge at the mouth of the river Dee in England. It is a continuous drawn lead pipe such as can be procured in great lengths, without joints. To give it the necessary strength it is wound externally with copper wire. It rests on the bed of the river and by its flexibility adjusts itself to the inequalities of the bottom. A copper pipe previously used put together with ordinary joints had failed. The lead pipe has worked successfully.*

Pumps.—The installations at Zurich and at Geneva use the Girard pump, which consists of a somewhat long cylinder divided into two cylinders by a central diaphragm. A solid plunger works into each end of the cylinder. These plungers carry cross-heads at their outer ends which are united by rods, forming a large frame to which a reciprocating movement is communicated by a crank, one plunger moving ''outboard,'' while the other moves ''inboard.'' The distinctive advantage of this arrangement is that it has no packings which require the dismemberment of the machine for their adjustment.

In the Zurich system a group of turbines drives a horizontal shaft by means of bevel-gears. This, says Mr. Unwin, runs at a speed of 50 turns per minute and actuates a second shaft at a speed of 100 turns per minute. To the latter the pumps are coupled, it is presumed through the intervention of spur-gearing, as a crank running 100 turns would give them too high a velocity.

---

* *Engineering News*, vol. XXXVII. p. 17.

At Geneva each turbine drives a pair of Girard pumps, by means of a crank on the head of the vertical turbine-shaft. These pumps are set in lines at right angles to each other, and the plungers move 188 feet per minute, the stroke being 3.61 feet. Another form and arrangement of pump will be described presently.

It has always appeared to the writer that there must be some more rational method of pumping water by means of water-power than by using the water-power to drive a wheel and the wheel to drive a pump. This method involves two transformations and two sources of waste, when there should be but one. He would suggest to inventors to turn their attention to a method dispensing wholly with the water-wheel. Let the power-water act in a very large cylinder to give motion to a piston and rod, the latter carrying at its opposite end a small piston acting in a cylinder and forcing the water into the pressure-pipes.

Hydraulic Transmission Applied to Locks. — Modern navigation locks have attained dimensions which call for considerable power in moving the gates. Not only the main gates which control the channel, ordinarily called the mitre gates, but the sluice-gates or valves for the admission and discharge of water. * The lock recently completed at St. Mary's Falls, Mich., has a length of 800 feet, a width of 100, and a maximum lift of 20 feet, requiring the admission and discharge of 1 600 000 cubic feet of water at every lockage. Such dimensions call, during the season of active navigation, for the frequent and rapid movement of massive gates. As the purpose of a large lock is usually to pass vessels around a fall or rapid in a running stream, it is almost invariably associated with water-power, and this is usually the most economical source of energy available for manipulating the gates. Moreover, the power being required at a number of points some hundreds of feet distant from the source, transmission is neces-

* Report of Chief of Engineers, U. S. A., 1898, p. 2554.

sary, and water under pressure is usually the most economical means. The writer gives the following description of a system designed by himself and applied to a lock on the Muscle Shoals Canal, on the Tennessee River, in Alabama. This lock was not of the largest class, being about 300 feet long from hollow groin to hollow groin and 60 feet wide, with a lift of 8 or 10 feet. There are two pairs of mitre-gates swinging horizontally and several sluice-gates moving up and down for filling and emptying the lock-chamber, 150 000 to 180 000 cubic feet of water being required for each lockage.

A parallel-flow wheel some 40 inches external diameter was inserted in a pit formed in one of the approaches of the lock. The head of the vertical shaft was disposed as indicated in Fig. 211. It had a square neck which entered a sort of hub

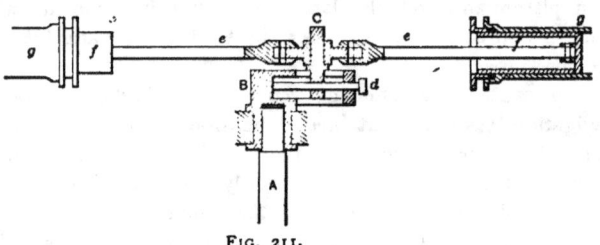

FIG. 211.

B revolving in a fixed bearing. This arrangement allowed the wheel to sink by the wearing of the step without bringing any strain on its upper supports. This hub forms a part of the crank, which revolves with the wheel, and carries the crank-pin *C*. This crank-pin carries a circular disk to which are attached the pump-rods *e e* of six pumps arranged in a circle, with the wheel-shaft in the centre. These pumps are worked by hollow plungers *f*, to the bottom of which the pump-rod *e* is jointed, an arrangement which gives sufficient play to the pump-rod within the hollow plunger and dispenses with a connecting-rod. The crank-pin *C* is adjustable by means of a screw *d* giving the pump-plungers a greater or less stroke

according to the fall available. The wheel is only intended
to run while the lock is in use, and the pipes are provided with
a relief-valve, adjusted to the desired pressure, for wasting the
superfluous water. One of the rods e is rigidly fastened to the
disk, the rest are jointed. The discerning reader will see that
the device would not operate if all the rods were jointed, neither

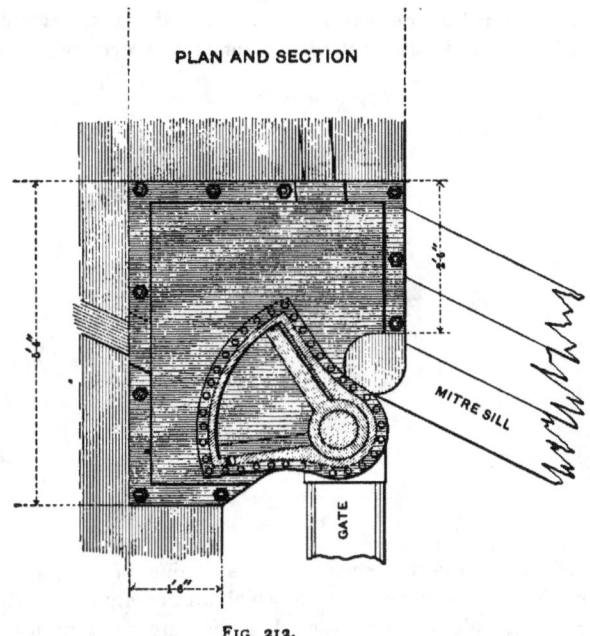

FIG. 212.

would it do so if more than one were rigidly attached. Small
pipes conduct the water from the pumps to the several points
of use. The sluice-gates are raised and lowered by hydraulic
cylinders with piston-rods attached to the gates. The mitre-
gates are swung by a hydraulic engine of a peculiar form,
Figs. 212 and 213, called a crab. This is a quadrantal case

with a rectangular piston having an angular traverse of 60 or 70 degrees, equal to the swing of the gate in opening or closing. The piston consists of a hollow cylinder with a wing attached. Passing through the cylindrical part is a short shaft placed directly over the heel-post and united thereto by an Oldham coupling, i.e., a coupling which admits of a slight deviation from a straight line in the centres of the shafts united. A good deal of leakage is to be expected in this arrangement, but this is not material, as there is abundant power and ques-

SECTION

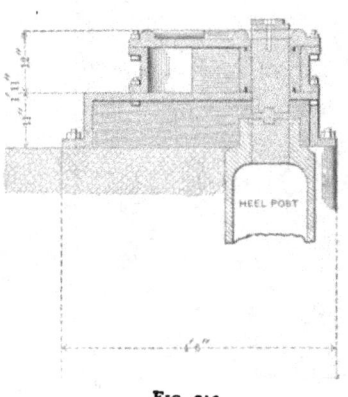

HEEL POST

FIG. 213.

tions of economy do not enter. The swinging of the gate is accomplished by a four-way valve which alternately admits the water to one side of the piston while discharging it from the other.*

Efficiency of Hydraulic Transmission.—We may generally count upon an efficiency of 80 per cent in the hydraulic motor, i.e., the power communicated to the pump is 80 per cent of the

---

* For more detailed illustrations of this mechanism, see Report of Chief of Engineers, U.S.A., 1890, p. 2111.

water-power expended. This is only true when the head is measured close to the wheel. When, as is more commonly the case, the head appertinent to the water-power represents the descent from the pond or canal to smooth water below the wheel, not more than 75 per cent can be expected. A loss of as much as 12 per cent must be expected in pumping, and ordinarily as much as 5 per cent in transmission through the pipes. An efficiency of 80 per cent may be assumed in the motors through which the water acts to drive machines. Thus, the percentage of the gross power which can be imparted to machinery at a distance of a mile or more from the source is not more than $0.75 \times 0.88 \times 0.95 \times 0.80 = 0.50$.

Storage of Energy by Pumping.—In the ordinary case of a water-power on a running stream, there is a great super-abundance of water at certain seasons of the year, and a deficiency at others. It is a very natural suggestion that the surplus energy of wet months could be employed in pumping water into a high reservoir for use during the dry months. At most water-powers there is a surplus of power six months in a year, and the case is very rare when water does not run to waste fully four months in the year. A pumping-plant could run during these periods night and day, and could accumulate a large store for use during the ensuing time of scarcity. The chief difficulty in such a project is to find a suitable site for the reservoir, and when a sufficient elevation exists it is usually on the summit of a hill where the construction would be inordinately expensive. Another difficulty is that in using the variable flow of the stream it would be necessary to install two or three horse-power of plant in order to obtain an average effect of one horse-power. The interest and depreciation on this plant will usually be found to more than offset the advantage to be expected from it.

Nevertheless, in textile mills running 10 hours or less per day, where a suitable reservoir-site exists, the following project will often be found well worthy of consideration: To adapt the wheels to the driving of pumps, so that on the cessation of each day's work they can readily be changed to that duty, and to keep them employed in non-working hours, during the season of abundant water, in raising water to the reservoir for use in the dry period. A wheel of 200 h.p. working twelve hours would raise water sufficient for $6 \times 200 = 1200$ h.p. hours, and during six months of abundant water would accumulate enough to maintain 200 h.p. for nearly four months of deficient flow, and under ordinary conditions a wheel could raise enough during the period of abundance to carry the work through the period of scarcity. It is true that the power obtained is only half that expended, but the power obtained is useful power, and that expended is of no value.

A desideratum in this mode of running would be a hydraulic engine capable of acting as a pump propelled by a shaft, to raise water to the reservoir during the season of abundance, and as a motor to drive or aid in driving the same shaft during the season of scarcity. It is true that, in order to do a given amount of work on the shaft, a larger quantity of water must pass the engine when running as a motor than when running as a pump. This condition can be met by the adjustable crank-pin.

CHAPTER XX.

PNEUMATIC TRANSMISSION OR TRANSMISSION BY COMPRESSED AIR.

AIR, like water, may be transmitted under a high pressure, and render up, at the receiving end of the line, a considerable fraction of the power expended in compressing it. The difference in the methods of transmission by air and by water results wholly from the difference in weight and compressibility of these two media. Water is practically incompressible, and the piston employed in forcing it works under a uniform pressure during the entire stroke. At an elevation corresponding to the pressure it may be confined in an open reservoir. Air, for practical purposes, may be regarded as compressible without limit. The piston employed in compressing it undergoes, during a single stroke, all gradations of pressure from that of the atmosphere to that of the pressure-pipe. It requires closed vessels for its retention. Its volume must be greatly reduced in putting it under the required pressure, and the pressure exerted during the resumption of its original volume, called the expansive pressure, must be utilized in the development of the power, if the latter is to approach in any reasonable degree the power expended in compressing the air.

Compressed air is much employed in mining, quarrying, and tunnel-work. In the construction of great locks, dams, reservoirs, bridges, and other important works where a great many isolated machines, such as power-drills, stone crushing and dressing machines, grindstones, mortar- and concrete-mixers, repair-works, and hoists, have to be operated at different points, it is coming into general use as a means of

457

distributing small quantities of power from a central station. As examples we may mention the reservoir for the New York water-works now under construction at Jerome Park, and the lock at Meeker's Island on the Mississippi between St. Paul and Minneapolis. A system of pumping sewage at a number of isolated stations by compressed air supplied from a central station has recently been perfected by Mr. Isaac Shone and is finding some application. In London compressed air carried in a strong vessel has been extensively used for the propulsion of street-cars.

"Compressed-air transmission," says Mr. Unwin, "is a perfectly general method of distributing power for all purposes. Whether in any given case it is the most advantageous, the least wasteful of power, or the cheapest in working cost depends on various circumstances. M. Hanarte believes that it is and will contiune to be the most economical method of transmission to considerable distances. The loss in the air-mains is very small. The motors worked expansively are efficient. The mains can be carried by any path, and the differences of elevation between the compressing and working points do not sensibly affect the results. In hydraulic transmission the water must be collected, stored, and in some cases filtered, and having actuated a motor, means must be found for removing it. But air is everywhere available and can be discharged anywhere without causing trouble. Compressed air has peculiar advantages in the case of underground transmissions. It has been used to replace manual labor in situations where hardly any other motive power could have been employed."

Compressed air is susceptible of indefinite subdivision and of measurement of the quantity supplied to each consumer. The latter is not so simple as in the case of pressure-water, but presents no mechanical difficulty. Meters similar to gas-meters are used which are not costly and which register the quantity of air with sufficient accuracy for commercial purposes. The pipes are less costly than for water-pressure and liable to less loss from leakage. We have seen that a 12-inch pipe would transmit 463 h.p. with water at 750 pounds per square inch. An air-pipe of equal size at 7 atmospheres would deliver considerably more, and need not be a third part as heavy as the water-pipe. Moreover, the air-pipes do not need the same precautions for protection against frost.

The disastrous consequences that sometimes result from the sudden stoppage of the current in a water-pipe or from the breach of the pipe have no parallel in the air system. The bursting of a receiver is a case parallel to the breach of an elevated reservoir, but the consequences only reach a very limited area.

Compressed air is susceptible of useful applications other than the development of power—applications which do not pertain to any other known form of energy. In foundations and tunnels water is excluded by an atmosphere of compressed air. It can generate light by driving a dynamo. The exhaust of the air-engine can be used for refrigerating purposes, for cooling and ventilating rooms, for forced draft of furnaces.

Relations of Volume and Pressure in Air.—In considering questions of this kind it is convenient to fix attention upon a definite quantity of air, say one pound.

Suppose P_1 to represent the pressure and v_1 the volume of compression; P_2 the pressure and v_2 the volume of release, which is usually at atmospheric pressure; P any pressure, and v the corresponding volume. It is sometimes convenient to consider P_1, P_2, P as the pressures upon a square foot instead of a square inch as is usual. The standard atmospheric pressure per square foot is 2116.4 pounds.

For Isothermal Change of Volume, i.e., when air during change of volume is maintained at a constant temperature, $P_1 v_1 = P_2 v_2 = Pv$. For a temperature of 60° F. $Pv = 27\,709$ foot-pounds.

For convenience, imagine the change of volume to take place in a cylinder of one square foot cross-section, then v_2

represents the total stroke, v_1 the portion of stroke performed under full pressure.

The work done during an indefinitely small portion of the stroke dv is $dw = Pdv = P_1 v_1 \dfrac{dv}{v}$. The total work between the limits $v = v_1$ and $v = v_2$ is

$$P_1 v_1 \text{ hyperbolic log. } \frac{v_2}{v_1} \quad . \quad . \quad . \quad (66)$$

This represents the work required to compress a pound of air from the volume v_2 to the volume v_1. It also represents the work yielded by a pound of air in an engine, entering at the pressure P_1 and expanding to the pressure P_2. The expression does not specifically include the work of admission nor exclude the work done against the pressure of the atmosphere. The former is $P_1 v_1$, and the latter $P_2 v_2$, and as these are equal they mutually offset each other.

For Adiabatic Change of Volume, i.e., when the air neither receives nor parts with heat during the process, we have

$$\frac{P_1}{P_2} = \text{not } \frac{v_2}{v_1}, \text{ but } \left(\frac{v_2}{v_1}\right)^{1.408},$$

so that

$$P = P_1\left(\frac{v_1}{v}\right)^{1.408},$$

and the work due to an elementary change of volume is

$$dw = Pdv = P_1\left(\frac{v_1}{v}\right)^{1.408} dv = P_1 v_1{}^{1.408} v^{-1.408} dv.$$

The integral of this expression between the limits $v = v_1$ and $v = v_2$ is

$$w = \frac{1}{0.408} P_1 v_1{}^{1.408}\left\{ \left(\frac{1}{v_1}\right)^{.408} - \left(\frac{1}{v_2}\right)^{.408} \right\},$$

or putting R to represent the ratio of expansion $= \dfrac{v_2}{v_1}$,

$$w = 2.451 P_1 v_1\left(1 - \left(\frac{1}{R}\right)^{0.408}\right). \quad . \quad . \quad (67)$$

This expression represents only the work of compression or expansion, not the total work. In developing power, the work done by the air includes admission P_1v_1, and excludes the work P_2v_2 consumed in atmospheric resistance. These are not, in this case, equal to each other. In compression, the work expended includes the work of forcing the air into the receiver or pipes after it has reached the pressure P_1, and excludes the work done by atmospheric pressure acting on one side of the compressing piston. The expression representing the net work done in compressing a pound of air to the pressure P_1, or the net work yielded by the same after compression, is

$$w = 2.451 P_1 v_1 \left(1 - \left(\frac{1}{R}\right)^{.408}\right) + P_1 v_1 - P_2 v_2. \quad . \quad (68)$$

The process of compression is attended by a great development of heat, which increases the tension of the air and consequently the work of compression. Were the stroke of compression followed immediately by the stroke of expansion, the power developed would be exactly equal to the power expended. This, however, never occurs. The air after compression, is transmitted through pipes to a distant point, and unavoidably cools to the temperature of the external air or earth, so that the power obtainable from it, without an addition of heat, falls greatly below that expended in compressing it. This is the inherent defect of the system of transmission by compressed air. As to the loss in compression, this can be diminished, but not wholly avoided, by the use of means to absorb the heat. As to expansion, a very inconsiderable addition of heat will not only obviate the loss but transform it into a gain.

The full theory of this subject involves the science of heat, which cannot be gone into in this treatise. Those who wish to pursue it should study Rankine on the Steam-engine, Weisbach (" Heat and Steam "), Zeuner, Maxwell, Tait, and other writers who have exhaustively treated the subject.

The accompanying tables give a practical idea of the relations of volume, pressure, and temperature in compressed air. Table 9 gives the power required to compress a pound of air from atmospheric pressure to different pressures up to 100 pounds per square inch. The pressure P_1 is the absolute pressure of the air, which is supposed to be maintained constantly at a temperature of 60 degrees. Omitting unavoidable losses from friction and resistance to movement, the power

TABLE 9.—ISOTHERMAL CHANGE OF VOLUME; Pv. CONSTANT = 27 709.

1	2	3	4	5
Pressure of Compression, Pounds per Square Inch. P_1	Volume of Compressed Air, Cubic Feet. v_1	Ratio of Volumes and Pressures, $\frac{P_1}{P_2} = \frac{v_2}{v_1}$	Power required to Compress 1 Pound of Air from P_2 to P_1. Foot-pounds.	Power required to Compress 1 Pound of Air per Second. Horse-power.
100	1.925	6.80	53 118	96.5
95	2.026	6.46	51 705	94
90	2.139	6.12	50 208	91
85	2.265	5.78	48 629	88
80	2.406	5.44	46 946	85
75	2.567	5.10	45 154	82
70	2.750	4.76	43 245	79
65	2.962	4.42	41 189	75
60	3.208	4.08	38 972	71
55	3.500	3.74	36 562	66.5
50	3.850	3.40	33 921	62
45	4.279	3.06	31 001	56
40	4.809	2.72	27 737	50
35	5.500	2.38	24 038	44
30	6.417	2.04	19 706	36

expended in compressing the air is the same as that obtainable from it in an engine which expands it to atmospheric pressure. That is to say: If air could be maintained at uniform temperature during compression and expansion, there would be no loss inherent in this method and peculiar to it. The fifth column gives with the same omission the number of horsepower expended in compressing a pound of air per second or furnished by the same when used in an engine admitting of its expansion to atmospheric pressure. In practice the power expended would be somewhat in excess of these figures, and that obtained would fall short of them. In addition there would be a certain loss in the transmission-pipes.

Tables 10 and 11 exhibit the conditions incident to the compression and expansion of air without gain or loss of heat. To compress a pound of air from atmospheric pressure and 60 degrees temperature to 100 pounds requires 70 987 foot-pounds

TABLE 10.—POWER REQUIRED TO COMPRESS 1 POUND OF AIR FROM ATMOSPHERIC PRESSURE AND 60 DEGREES TEMPERATURE WITHOUT GAIN OR LOSS OF HEAT.

1	2	3	4	5	6	7	8
After Compression.			Ratio of Pressure.	Cut-off.	Total Power Expended.	Power Expended excluding Atmospheric Pressure.	Power to Compress 1 Pound of Air per Second.
Pressure in Pounds per Sq. In. P_2	Volume in Cu. Ft. v_1	Temperature in Degrees, F.	$\frac{P_3}{P_1}$	$\frac{1}{K}$	Foot-lbs.	Foot-lbs.	Horse-power.
100	3.3510	448	0.147	0.256	98 696	70 987	129
95	3.4767	434	0.155	0.266	96 358	68 649	125
90	3.6141	421	0.163	0.277	93 950	66 241	120
85	3.7637	406	0.173	0.287	90 957	63 248	115
80	3.9283	390	0.184	0.300	88 277	60 568	110
75	4.1129	378	0.196	0.314	85 360	57 651	105
70	4.3210	358	0.210	0.330	82 375	54 666	99
65	4.5553	341	0.226	0.347	79 052	51 343	93
60	4.8197	322	0.245	0.368	75 740	48 031	87
55	5.1286	303	0.267	0.392	71 234	43 525	79
50	5.4860	282	0.294	0.419	68 414	40 705	74
45	5.9140	259	0.327	0.452	64 347	36 638	67
40	6.4285	235	0.367	0.491	59 881	32 172	58
35	7.0686	209	0.420	0.540	55 035	27 326	50
30	7.8867	180	0.490	0.602	49 650	21 941	40

of energy. The air so compressed has a volume of 3.35 cubic feet and a temperature of 448 degrees F. In the course of transmission to a distance of several miles the air unavoidably cools to the surrounding temperature, say 60 degrees. If we suppose the volume of the air to remain unchanged, it would at the latter temperature have a pressure of only 57.34 pounds. In this state, led into an engine and expanded without addition of heat to atmospheric pressure, it has a temperature of $-110°$, and a volume of 8.82 cubic feet, and yields only 31 121 foot-pounds of energy, viz., 44 per cent of the power expended. Even from this low percentage must be deducted the losses due to friction and resistance to motion.

TABLE 11.—POWER OBTAINABLE FROM 1 POUND OF COMPRESSED AIR EXPANDED WITHOUT GAIN OR LOSS OF HEAT TO ATMOSPHERIC PRESSURE, AFTER BEING COOLED TO 60 DEGREES.

1	2	3	4	5	6	7	8	9	10
Pressure in Pounds per Square Inch.		After Expansion to Atmospheric Pressure.		Ratio of Pressure	Cut-off.	Gross Power Obtainable.	Power due to Atmospheric Resistance.	Net Power.	Power Furnished by 1 lb. of Air per Sec.
Of Compression. P_1	After Cooling to 60 Degrees P_2	Temperature, Degrees, F.	Volume, Cubic Feet. V_2	$\frac{P_2}{P_1}$	$\frac{1}{R}$	Foot-lbs.	Foot-lbs.	Ft.-lbs.	H.-P.
100	57.34	−110	8.82	0.256	0.380	49 791	18 670	31 121	56.6
95	55.30	−106	8.92	0.266	0.390	49 333	18 882	30 451	55.4
90	53.17	−102	9 02	0.276	0.401	48 779	19 094	29 685	54 0
85	51.11	− 98	9.12	0.288	0.413	48 265	19 305	28 960	52.7
80	48.96	− 93	9.24	0.300	0.425	47 700	19 559	28 141	51.2
75	46 59	− 88	9.37	0.315	0.439	46 889	19 834	27 055	49.2
70	44.53	− 83	9.50	0.330	0.455	46 369	20 110	26 259	47.7
65	42.24	− 77	9.65	0.348	0.472	45 624	20 427	25 197	45.8
60	39.92	− 71	9.80	0.368	0.492	44 770	20 745	24 025	43.7
55	37.52	− 64	9.97	0.392	0.514	43 858	21 104	22 754	41.4
50	35.06	− 56	10.17	0.419	0.530	42 828	21 528	21 300	38.7
45	32.58	− 47	10.40	0.451	0.560	41 721	22 015	19 706	35.8
40	29.94	− 37	10.65	0.491	0.604	40 343	22 544	17 799	32.4
35	27.21	− 25	10.95	0.540	0.646	38 781	23 179	15 602	28.4
30	24.39	− 11	11.30	0.602	0.698	36 959	23 920	13 039	23.7

It is obvious from the foregoing that the success of any system of air-transmission depends very largely upon the efficiency of the means employed for absorbing the heat of compression and supplying the heat of expansion. If the former be thoroughly done, and the air before admission to the engine be sufficiently heated, the power obtained may not only equal but even exceed that expended in compression. Without such heating the use of air at high pressures is impracticable, owing to the formation of ice on the valves of the engine due to the extremely low temperature of release. Power developed by compressed air is substantially the mechanical equivalent of the heat which disappears in expansion. Coal used in heating the air, if we credit the coal with the entire power developed by the air, furnishes power at the rate of 3 or 4 horse-power per pound of coal per hour; whereas the most efficient steam-engines do not in practical working furnish more than 1 brake horse-power for every $2\frac{1}{2}$ pounds of coal per hour. A pound of coal per hour to every 3 or 4 horse-power is so trifling an expense as to be hardly worth considering. Theoretically it should take a little more than 1 pound of coal to 6 horse-power.

The Air-compressor. — Air is compressed by a piston moving in a cylinder. The pressure opposing the piston is very slight, in fact nothing, at the commencement of the stoke, and becomes great toward the end — a condition which introduces a serious difficulty into the design and operation of the machine. In the case we are considering, viz., where the motive power is furnished by a water-wheel, the latter works most efficiently with a uniform resistance at all points of its revolution. When applied to an air-compressor making a forward and return stroke at each revolution, the power of the wheel is greatly in excess of the resistance at the commencement of the stroke, and only becomes equal to it when the air has reached the required pressure and is being forced into the receiver. The wheel takes an irregular velocity, starting forward at the commencement of the stroke and slowing up at the end. The power of the wheel in excess of the resistance of the air is lost, and its working is very wasteful. When the wheel runs on a horizontal shaft this difficulty is remedied by a fly-wheel, but the latter is not applicable on a vertical shaft. It is sometimes avoided in a measure by arranging a number of single-acting compressors in a circle, all driven from a crank on the vertical shaft, as in the case of the force-pumps described page 451. When the power is derived from a steam-engine, the compressing piston being on the same rod as the steam-piston, the condition is still more uneconomical. In fact it is precisely the reverse of economical. The pressure on the steam-piston is great at the commencement of the stroke and diminishes to nearly nothing at the end, whereas the pressure on the air-piston is extremely slight at the commencement and becomes great toward the end. This difficulty is very effectually met by a heavy fly-wheel.

To absorb the heat generated by compression and prevent the air from exerting a greater pressure during compression than during expansion, the most effective means is, first, to surround the compressing cylinder with cold water in constant circulation, and, second, to inject cold water into the cylinder in the form of a fine spray. This latter expedient is very effective, but is attended with constructive difficulties. It is important that the clearance-space between the piston and the cylinder-head should be as small as possible, because the air contained therein represents so much waste of power—air compressed, but not secured in the receiver. When, however, this space is too much reduced there is danger that the injection-water will more than fill the clearance and cause a rupture of the cylinder-head, as often occurs in steam-engines from the accumulation of condensed water.

Engines Worked by Compressed Air are in no essential respect different from non-condensing steam-engines, and require no additional appurtenances or attachments except a cock for injecting a spray of water into the cylinder. Even when the air is heated before admission this spray of water is found very advantageous. The process of heating consists in transferring to the air the heat of the gases arising from the combustion of the fuel. This process is very simple in the case of air compared with the generation of steam, where the gases after leaving the boilers must retain heat enough to create a draft up the chimney. The exhaust air from the engine slightly raises the pressure in the engine-room and creates a current through any aperture or passage leading to the open air. There is no necessity for a chimney to create a draft. The air may approach the engine-room through a pipe of such size as to give it a low velocity. Enclosed in this pipe is another which conducts the furnace-gases under the pressure of the exhaust. The current of air to be heated moves in the opposite direction to the heating current. At all points of this pipe the gases are warmer than the air and in a condition to impart heat to the same. The air-pipe has a non-conducting coating, and all the heat generated by the fuel can thus be transferred to the air.

The air can also be heated by petroleum burned within the air-pipe or by illuminating-gas burned within the inner flue. There has recently been a marked cheapening of illuminating-gas and it is now extensively used for heating. The ideal system of transmission and distribution of power by compressed air in point of efficiency, cleanliness, and convenience would be a system of air-pipes combined with a system of pipes for the distribution of gas.

In southern latitudes the system of air-transmission recommends itself strongly on account of its applicability to purposes of cooling and ventilation. By controlling the heat imparted to the air the exhaust can be delivered at any desired temperature and serve either for warming or cooling the work-rooms. Where required for refrigeration only, a small motor can be run to drive a compressor, forcing air into the pipes, thus partly compensating for the air drawn, and delivering the exhaust at a very low temperature.

Raising Liquids.—Beer and other liquids are raised from cellars to the upper parts of buildings by connecting the tanks or barrels containing the liquid with the air-pipes. Compressed air is employed to raise water from non-flowing artesian wells, without pumps or mechanism of any kind. The depth of water in the well should be four or five times the lift. A small pipe is carried down to near the bottom of the well and delivers air into the water. The air, diffusing itself through the water of the well, lightens the aggregate mass so that it exerts less pressure at the bottom than the source from which it is supplied. The water consequently rises and overflows. The pressure of the air in this application should be but little greater than that due the depth of water at the point of delivery. The air expands isothermally in rising, the heat for maintaining the temperature being derived from the water.

Compressed-air Installation of Paris. — About 1870 a system of small pipes was laid down in Paris for the purpose of actuating clocks throughout the city by means of compressed air. Some eight thousand clocks, public and private, came into the system, all moved isochronously by impulses transmitted through the air-pipes. This installation gradually grew into the furnishing of air from a central station for purposes of motive power. New and larger compressing stations were erected and larger pipes laid, till in 1889 the system required a compressing plant of some 8000 horse-power. About 1891 a new station was erected at the Quai de la Gare contemplating the ultimate use of some 24 000 horse-power. There were at that time something over 10 miles of pipe for the transmission of air. In the early history of the enterprise the pressure was about 45 pounds per square inch absolute; later a pressure of 90 has been adopted. It is found that a velocity of 26 feet per second in the pipes is permissible and occasions no serious loss of head, though in 1891 the velocity did not generally reach that figure. At this velocity and pressure a 12-inch pipe would carry very nearly 1000 h.p. About 35 pounds of air per hour is required for a horse-power when heated to 300° F. before admission. The heating consumes about 1 pound of coke per hour for 5 h.p.

The most important detail of the whole system is the joints of the pipe. Air-pipes are liable to greater variation of temperature than others, and rigid joints have always given trouble. Cast-iron pipes have been mainly used at Paris. The most satisfactory joint is the one shown at Fig. 214. The pipes are perfectly plain, having no bells, spigots, flanges, or other means of attachment. In joining two pipes a heavy collar provided with bolt-holes is slipped on the end of each piece. Then a rubber ring is slipped on; then a sleeve is applied embracing both pieces of pipe, and the latter are brought

nearly together but not into contact. Bolts are then inserted
and the nuts turned up strongly, compressing the rubber rings
against the ends of the sleeve. This joint admits of free
expansion and contraction as well as some deviation from a
straight line, which in a line of pipe is unavoidable. The

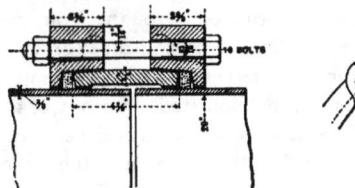 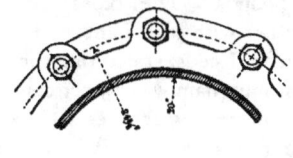

FIG. 214.   FIG. 214a.

leakage of these joints is very slight. Prof. Reidler * gives results of experiments on a system of 9000 yards of pipe at Paris,
and found the loss from leakage 3 per cent. The system of
air-pipes at Offenbach,† near Frankfort-on-the-Main, when
under a pressure of 6¾ atmospheres (about 100 pounds) showed
a leakage of about 1.7 cubic feet air per hour for each mile of
pipe. This consisted of 13 000 feet 4-inch, 5000 feet 3-inch,
and 5000 feet 12 inch pipe. Prof. Reidler in 1891 was of
opinion that, under conditions then existing in Paris, motors of
ordinary construction could utilize 80 per cent, and with some
simple and obvious improvements 100 per cent of the power
expended by the compressors, always assuming the expenditure of a small quantity of fuel in heating the air.

The Birmingham Compressed-air System was commenced
about 1886, and it was complete and in operation in 1890.
Preliminary to this installation the conditions at Birmingham
affecting such an enterprise were examined by Sir F. Bramwell
and Mr. Percy and found very favorable. This city contains
an immense number of small establishments run by steam.
The small engines in general use were found to consume coal

* London *Engineering*, March 13, 1891.
† *Ibid.*, October, 1891.

at a most extravagant rate—in no case less than $8\frac{1}{2}$ pounds, and in some cases as high as 36 pounds per horse-power per hour. It was fairly reasoned that such users would find great advantage in taking their power from the air-mains at rates remunerative to the company. Compressing engines to the extent of 3000 h.p. were installed. The pressure adopted was 45 pounds. The pipes were of riveted wrought iron, put together with lead joints. Prof. A. Lupton * of the Yorkshire College, Leeds, stated that the indicated efficiency of the motors at Birmingham was 73 per cent of the indicated efficiency of the air-compressing engines. Great trouble was always experienced here from leakage. It is stated that the pressure in the mains was but 20 pounds, instead of 45 as intended. This would indicate leakage of enormous proportions. This company † failed, and its affairs were wound up in 1891. If the statements as to leakage and loss of pressure are true, this would sufficiently account for the failure. However that may be, it cannot be denied that the Paris enterprise is a conspicuous success and demonstrates the commercial feasibility of such projects when justified by existing conditions and properly conducted. Experience at Paris and Offenbach justify the statement that air may be transmitted through pipes, excluding leakage, at a velocity of 25 feet per second, with a loss of not more than 1.5 pounds pressure per mile. It must be borne in mind that a loss of pressure in air does not imply a proportional loss of power, as is the fact in case of water. Loss of pressure is accompanied by an increase of volume. Table 9 shows that reducing the pressure from 100 to 95, i.e., 5 per cent, reduces the power only 2.7 per cent. By Table 11 it would be under 2 per cent. Both these results are theoretical; the actual loss would be between the two, say something under 2.5 per cent. We may say, therefore, that the loss of power with the above velocity need not exceed 1 per cent per mile. Another point of importance is this: For the proper regulation of the

---

* Meeting of British Assoc., September, 1890.
† London *Engineering*, January 1, 1892.

pressure in such a system a considerable receiver capacity is necessary. This receiver capacity costs no more in the form of enlarged pipes than in the form of separate vessels.

The longest air transmission in America is at the Chapin Mine, Iron Mountain, Michigan. The source of power is a waterfall. There is one pair 36 × 60 and three pairs 32 × 60 Rand compressors. The air is sent 3 miles through a 24-inch pipe. The pressure at the compressors is 60 pounds, at the motors $57\frac{1}{2}$ to 58, implying a loss of power of about 2 per cent. The air being used mainly underground, it is not practicable to reheat it. The pipe is 24 inches diameter, of $\frac{1}{4}$-inch sheet iron, each sheet wide enough to make a 24-inch pipe, and 8 feet long, single-riveted on all joints. The pipe is put together in lengths of 58 feet with riveted flange-joints, a paper gasket being interposed at each joint. There are 29 expansion-joints in the course of the pipe, each admitting a movement of 18 inches. The pipe rests on bents or trestles 58 feet apart, and is provided with rollers or rocker feet so that it accommodates itself readily to changes of temperature. These facts were communicated by Mr. James MacNaughton, superintendent of the mine.

As already mentioned, grave mechanical difficulties attend the application of the turbine to the duty of compressing air. Unless a number of compressors are driven by the same wheel, which is not always desirable, a heavy fly-wheel is imperatively essential to economical working. This on a vertical shaft is generally out of the question; and on a horizontal shaft, which must run in stuffing-boxes, is attended with difficulties. Moreover, on a high head the motion of the wheel becomes too rapid for the economical working of the air-compressor without the intervention of gearing. For these reasons the attention of engineers is often directed to devices for compressing air by the direct action of water, without the intervention of water-wheels.

The Tronc, formerly used for creating a blast for iron-furnaces, is well known to mechanicians. A column of water descending through a vertical pipe carries air with it and deposits the same in a receiver, under a pressure somewhat less than that due the head. It can under no condition bring the air to a pressure greater than that due the head.

The Ram.—By alternately establishing and interrupting the movement of water in a long pipe, the momentum of the latter, acting upon a confined mass of air, will bring it to a pressure far exceeding that due the head acting on the pipe. A pipe 1 square foot in cross-section and a mile in length contains 330 000 pounds of water. Such a column being in motion with a velocity of 4 feet per second would in coming to rest give out energy to the amount of some 80 000 foot-pounds—sufficient, according to Table 9, to bring 1.5 pounds of air to a tension of 100 pounds. The head acting on the pipe need not be more than 20 or 30 feet.

The simplest application of this principle is in the well-known hydraulic ram of Montgolfier, which at each stoppage compresses air in a chamber, and the latter by its expansion forces water to a height above the source of supply. Of this character was the first air-compressor used at the tunnel of Mont Cenis, designed by M. Sommeiller, to furnish air for the drills. This, on account of the high velocity required and the friction in the line of pipe, was found to be wasteful and, from the violence of its action, difficult and expensive to keep in repair.. It was discarded and replaced by a turbine driving piston compressors.

An English engineer, Mr. E. D. Pearsall, has devoted himself to the improvement of the Sommeiller compressor. He avoids the violence of the former action by a cylinder-valve for arresting the motion. This is operated by a small air-motor. This modification is said to work very well, having shown on a small scale, as is claimed, an efficiency of 80 per cent.

Fig. 228 shows a method of compressing air* by the direct

* A committee of the Franklin Institute lately gave their attention to this method, and, in a formal report, pronounced it identical in principle with

action of water without the use of mechanism of any kind. This method was patented by the writer in 1878, and has received some application. It is founded upon the familiar observation that water thrown into commotion breaks into foam and becomes impregnated with minute air-bubbles, and upon the further fact that these bubbles rise through water with a very moderate velocity and, in a descending current, are carried downward and subjected to a pressure conformable to the depth. Figs. 228 and 228*a* show the arrangement for applying these principles on a practical scale. It assumes a fall created by a dam on a running stream. It assumes, what is generally the fact at a fall or rapid, a rock formation at no great depth. On the up-stream side of the dam a vertical shaft descends into the ground to a depth corresponding to the required pressure. At that depth its direction changes to horizontal, and it pursues this course to a point below the dam, where it joins a second vertical shaft opening into the river. Around the entrance to the shaft a masonry structure rises to the elevation of the dam or something more, and is provided with means for controlling the influx. Over the horizontal passage a capacious chamber is formed, provided with an impervious lining. The transmission-pipe leads from this chamber. Water entering the descending shaft from different directions meets with a high relative velocity, and is thrown into violent commotion which impregnates it with air-bubbles of a minute and very uniform size. Such bubbles tend to rise with a velocity of not more than 1 foot per second, and the descending current being 4 feet or more, they are carried downward. In the horizontal part of the passage the air is free to rise and accordingly enters the air-chamber and there accumulates under a pressure corresponding to the depth of the surface of water in the chamber below

---

the ancient trompe, meaning the tronc above described. This statement is entirely incorrect, and reflects little credit upon that time-honored institution. The conclusive and indisputable criterion between the two methods is this: The tronc cannot produce a pressure equal to the working head or closely approaching the same. The device under consideration can produce a pressure many times the working head.

the surface in the stream.   The water divested of its burden of
air pursues its course and, rising through the ascending shaft,
joins the river.

The compression in this method takes place isothermally
and avoids one principal source of loss in machine compressors.
The losses inherent in the method are:

1. The head expended in impregnating the water with air,
which includes the head due the initial velocity.   A fall of
1 foot in entering the shaft brings in the water from both sides
with a velocity of 8 feet per second, and these two currents
meet with a relative velocity of 16 feet per second, which
creates a sufficient commotion.

2. A loss which may be called the *slip* due to the velocity
with which the bubbles tend to rise.   It is obvious that the rise
of the bubbles during the descent of the water is a lost motion,
to be deducted from the efficiency of the system.

In addition there is the head consumed by the friction in
the channels.   An extended series of experiments* on this
method was made by the author at Minneapolis, Minn., in
1880, the fair deduction from which was that on a practical
scale and in a plant intelligently designed and carefully
executed as much as 75 per cent of the absolute power of the
water would be represented in the air compressed.

A method of compressing air on these principles has been
in use for several years at the outlet of Magog Lake in the
province of Quebec (Fig. 229).   In this case a combination of
pipes has been used for introducing the air.†   Some 62 per cent
was the greatest efficiency found in this apparatus.   Even this
low result compares favorably with what could be expected
from water-wheels, where, reckoned on the total fall, we could
not count on an efficiency of more than 75 in the water-wheel,
and say at best 80 in the compressor, giving for the energy
represented by the compressed air $.75 \times .80 = 60$ per cent of
the total absolute power of the water.   This method is more
fully considered under the head of Power-houses.

---

* See Journal of the Franklin Institute, September, 1880.
† London Society of Engineers, June, 1897.

CHAPTER XXI.

TRANSMISSION BY ELECTRIC CURRENT

I$\scriptstyle{\text{T}}$ was discovered by Faraday in 1831 that an electric conductor crossing a magnetic field, i.e., the space in the immediate vicinity of a powerful magnet, undergoes a certain attraction or repulsion; that when moved against the repulsive force it generates a current of electricity; and that when supplied with electric current and allowed to move in obedience to the force, it may be made to generate mechanical energy. This principle has received immense application in the last twenty years in the form of electric generators and motors. An enormous development of the science of electricity has taken place within that period. It forms the subject of lectures and periodical literature, of special courses in industrial schools, and of books that cannot be numbered for multitude. It is not intended in this work to enter into the science of electricity or the construction of electric machines, beyond a brief reference to such general principles as relate to the transmission of power.

Analogy between Electric and Other Modes of Transmission.—In transmission by fluids under pressure, the elements of the power are the pressure and the rate of flow, i.e., 1 cubic foot of water per second under a head of 9 feet is very closely a theoretical horse-power, namely, a horse-power neglecting unavoidable losses in application. Practically 1 cubic foot per second under a head of 11 or 12 feet is a horse-power, and if h represent the head in feet, and q the flow in cubic feet per second, the power may be represented by the product hq. In electric transmission we deal with two analogous elements, viz., the electromotive force or potential E in volts, and the current I expressed in amperes, the power being proportional to the product EI. Thus, neglecting losses in application, a flow of 20 cubic feet of water per second under a head of 18 feet would represent about $\dfrac{20 \times 18}{9} = 40$ h.p. A current of $I = 40$ amperes under a potential of $E = 500$ volts would represent $\dfrac{EI}{746} = \dfrac{500 \times 40}{746} = 26.8$ h.p. The product EI represents the power of a current of electricity in units called watts, a watt being 1 ampere of current under 1 volt of potential, and 746 watts being equal to a horse-power. An analogous term is found in leases of water-power, viz., the mill-power, which is such a quantity of water-power that the product of the quantity of water in cubic feet per second by the fall in feet shall equal a certain fixed sum. Oliver Evans,* one of the earliest water-power engineers in the United States, introduced the term *cubock* in the sense of the product of cubic feet per second by feet of head, i.e., 10 cubic feet per second under 12 feet head = 120 cubocks. This term was the exact analogue of the watt. In fluid transmission we sometimes use the symbols h and P to mean, not the absolute head or pressure in a pipe or channel, but the loss of head, i.e., the head or pressure expended in maintaining the flow in a certain length of the pipe or channel. Likewise in electricity, the symbol E is sometimes used to designate, not the absolute potential of the current, but the loss of potential in a foot, a mile, etc., of the conductor. In each case the proper use of the symbol is always clear from the context, and the apparent confusion occasions no misapprehension to those familiar with such computations.

The power transmissible in a fluid form depends upon the nature of the channel of transmission, and the more difficult the passage the greater the quantity of power consumed in

---

* Millwright's Guide. Philadelphia, 1848.

transmission. Electric transmission is perfectly analogous. The channels of transmission are metallic wires. The pressure or voltage consumed in transmission is directly as the length and inversely as the cross-section. The lost power is in both cases transformed into heat, though in the case of electricity the heat becomes very sensible on account of the comparatively small size of the conductor, while in air it is but slightly sensible, and in water entirely insensible.

A quantity q of compressed air under pressure P may be changed into a greater volume under a less pressure or a smaller volume under a greater pressure, representing the same amount of power in each case.* A quantity q of water under the head h may, neglecting losses, be employed to pump a greater quantity aq to a height $\dfrac{h}{a}$ or a smaller quantity $\dfrac{q}{a}$ to a height ah, the result representing, in each case, the same power. In like manner an electric current I at potential E may, by certain devices called transformers, be changed into a current $\dfrac{I}{a}$ at potential aE or a current aI at potential $\dfrac{E}{a}$; in either case representing, unavoidable losses excepted, the same amount of power. The device which diminishes the potential and increases the current is called a step-down transformer; that giving the reverse effect, a step-up transformer. It is through these devices that electric transmission to long distances is possible.

Expansion. — Unlike elastic fluids, electricity does not develop power in passing from high to low potential, or absorb power in the reverse transformation. The energy derived from expansion of air has no analogue in electricity.

* This statement might appear contradictory, as a quantity q of air at volume v_1 and pressure P_1 would develop power in passing to volume v_2 and pressure P_2; but if air passes freely from P_1 to P_2, it will develop heat in v_2, which preserves the power unaltered. Likewise if the air passes from v_1 to the lesser volume v_2 and greater pressure P_2, the power of P_2v_2, omitting losses and dissipation of heat, is exactly equal to that of P_1v_1 plus the power expended in compressing the air from P_1 to P_2.

Resistance to the passage of the current follows a more simple law in electric conductors than in pipes and channels. It is directly as the length of the conductor traversed, ordinarily called the circuit, and inversely as the cross-section. Resistance is expressed in ohms, the relation between the current, the potential, and the resistance being represented by the equation $E = RI$, whence $R = \dfrac{E}{I}$, which means that a difference of potential of 1 volt, in a circuit whose resistance is 1 ohm, will cause a current of 1 ampere. Resistance varies greatly in different substances. In some it is so great that the passage of the current under any attainable difference of potential is practically null. Such substances are used as insulators to prevent the escape of electricity from conductors at their resting-points, and for other purposes. Conductivity is the reciprocal of resistance, high conductivity meaning low resistance. The following table, borrowed from Unwin, gives the relative conductivity of metals susceptible of use as conductors.

TABLE 12.—COMPARISON OF ELECTRIC CONDUCTORS.

Material.	Conductivity.	Density.	a Conductivity of Equal Weights.	b Tenacity, Tons per Square Inch.	ab Product of a and b.
Pure copper................	100	8.9	100	17	1700
Soft copper................	98	8.9	98	13	1280
Hard copper...............	97	8.9	97	29	2810
Swedish iron	16.5	7.8	18.8	22.5	423
Galvanized iron............	14	7.7	16.1	25	400
Cast steel.................	10.5	8.0	11.7	58	680
Aluminum..................	55	2.6	188	11	2380
Silicon-bronze.	97	8.9	97	28	2720
" "	80	8.9	80	34	2720
" "	45	8.9	45	49	2200
Phosphor-bronze...........	26	8.9	26	45	1170

The tenacity of the material is important as regards the supports of the wires; the stronger the material the greater the distances allowable between the supports. The tension on a wire suspended from two fixed points is arrived at with sufficient accuracy for this purpose in the following manner. Let $AB = l$ be the distance between the supports, Fig. 215. The

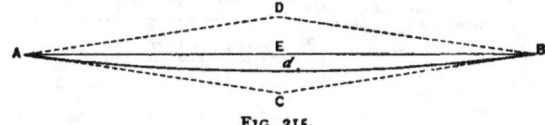

Fig. 215.

deflection d is small compared with l, and the curve is not essentially different from a parabola, in which the intersection of the tangents C is at the distance d below the curve. w being the weight of the wire per foot, the total weight wl may be supposed concentrated at C, causing a tension E on the tangents equal to that on the wire at A and B. Completing the parallelogram of forces we have, neglecting the difference in length between BC and BE, $t : \dfrac{wl}{2} = \dfrac{l}{2} : 2d$. $\therefore t = \dfrac{wl^2}{8d}$. Dividing t by the cross-section of the wire gives the tension per square inch of metal.

The sixth column of the table shows the two qualities of conductivity and tenacity in combination, in which respect hard copper heads the list, though silicon bronze follows very closely. As regards conductivity per unit of weight nothing approaches aluminum. The ores of this metal in the form of oxides exist in great abundance in all parts of the world, and modern methods of metallurgy are constantly tending toward its cheaper production. It is very probable that it will eventually become the most available material for electric conductors.

Table 13 gives numbers, sizes resistances, etc., for copper wire. It can be used for wire of other metals by the aid of Table 12. The first column gives the designation of the wire, following the system of Brown & Sharpe of Providence, R. I. After the numbers and sizes were established it became necessary to find designations for sizes larger than No 1, which led

TABLE 18.—DIMENSIONS AND RESISTANCES OF PURE COPPER WIRE.

1 American Gauge, Brown & Sharpe's Numbers.	2 Diameter in Mills.	3 Area of Cross-section Circular Mills = Diameter².	4 Weight and Length. Pounds per 1000 Feet.	5 Weight and Length. Feet per Pound.	6 Resistance at 75° F. Ohms per 1000 Feet.	7 Resistance at 75° F. Feet per Ohm
0000	460	211 600	639.33	1.56	.04906	20 383
000	409.64	167 805	507.01	1.97	.06186	16 165
00	364.80	133 079.40	402.09	2.49	.07801	12 820
0	324.05	105 592.50	319.04	3.13	.09831	10 409
1	289.30	83 694.20	252.88	3.95	.12404	8 062
2	257.63	66 373.00	200.54	4.99	.15640	6 394
3	229.42	52 634.00	159.03	6.29	.19723	5 070
4	204.31	41 742.00	126.12	7.93	.24869	4 021
5	181.94	33 102.00	100.01	10.00	.31361	3 189
6	162.02	26 250.50	79.32	12.61	.39546	2 529
7	144.28	20 816.00	62.90	15.90	.49871	2 005
8	128.49	16 509.00	49.88	20.05	.62881	1 590
9	114.43	13 594.00	39.56	25.28	.79281	1 261
10	101.89	10 381.00	31.37	31.38	1.00000	1 000
11	90.742	8 234.00	24.88	40.20	1.2607	793.2
12	80.808	6 529.90	19.73	50.69	1.5898	629.0
13	71.961	5 178.40	15.65	63.91	2.0047	498.8
14	64.084	4 106.80	12.41	80.59	2.5908	386.0
15	57.068	3 256.70	9.84	101.63	3.1150	321.0
16	50.820	2 582.90	7.81	128.14	4.0191	248.8
17	45.257	2 048.20	6.19	161.59	5.0683	197.3
18	40.303	1 624.30	4.91	203.76	6.3911	156.5
19	35.390	1 252.40	3.78	264.26	8.2889	120.6
20	31.961	1 021.50	3.09	324.00	10.163	98.40
21	28.462	810.10	2.45	408.56	12.815	78.04
22	25.347	642.70	1.94	515.15	16.152	61.91
23	22.571	509.45	1.54	649.66	20.377	49.09
24	20.100	404.01	1.22	819.21	25.695	39.92
25	17.900	320.40	0.97	1 032.96	32.400	30.86
26	15.940	254.01	.77	1 302.61	40.868	24.47
27	14.195	201.50	.61	1 642.55	51.519	19.41
28	12.641	159.79	.48	2 071.22	64.966	15.39
29	11.257	126.72	.38	2 611.82	81.921	12.21
30	10.025	100.50	.30	3 293.97	103.30	9.68
31	8.928	79.71	.24	4 152.22	127.27	7.86
32	7.950	63.20	.19	5 236.66	164.26	6.09
33	7.080	50.13	.15	6 602.71	207.08	4.83
34	6.304	39.74	.12	8 328.30	261.23	3.83
35	5.614	31.52	.10	10 501.35	329.35	3.04
36	5.000	25.00	.08	13 238.83	415.24	2.41
37	4.453	19.83	.06	16 691.06	523.76	1.91
38	3.965	15.72	.05	20 854.65	660.37	1.52
39	3.531	12.47	.04	26 302.23	832.48	1.20
40	3.144	9.89	.03	33 175.94	1049.7	0.95

to the symbols 0, 00, etc. The second column gives the diameter in mills, i.e., in thousandths of an inch. The third gives the equivalent number of wires of 1 mill diameter, i.e., a No. 1 wire is equivalent to 83 694 wires of 1 mill diameter. The remaining columns explain themselves. The resistance varies materially with the temperature. It is given here for a temperature of 75° F.

Potential in Transmission.—We are now in a position to understand the enormous advantage of high tensions in electric transmission. Let it be required to find the loss of potential in sending 20 h.p. over a line of No. 1 copper wire 10 miles long, with a potential of 550 volts at the receiving end, which is about the voltage required for street-cars. Such a line has, by the table, a resistance of $0.12404 \times 5.280 \times 10 = 6.55$ ohms. The current at the receiving end must be such that

$$EI = 20 \times 746 = 14\,920 \text{ watts.} \quad \text{Whence } I = \frac{14\,920}{550} = 27.1$$

amperes. To prevent confusion let e represent the electro-motive force required to move the current. Then

$$e = 6.55 \times 27.1 = 177.5.$$

The loss in the wire therefore is

$$177.5 \text{ volts} = \frac{177.5}{550 + 177.5} = 24.0 \text{ per cent of the power}$$

delivered to the wire.

Again, let it be required to send 10 h.p. $= 7460$ watts over the same line at a potential of 120, which is suitable for electric lighting. Here we must have $I = \frac{7460}{120} = 62.2$ amperes, and $e = 62.2 \times 6.55 = 407$. That is to say, not more than 23 per cent of the power imparted to the wire would be delivered at the receiving end.

Now assume the transmission of 1000 h.p. over the same line at a potential of 15 000 volts. In this case $EI = 746\,000$, $I = \frac{746\,000}{15\,000} = 49.7$ amperes, $e = RI = 6.55 \times 49.7 = 325.5$, and the loss in the wire is $\dfrac{325.5}{15\ 325.5} = $ a little over 2 per cent. Of course this is not the only loss in such a transmission; there are liable to be losses from defective insulation, and unavoidable losses in the transformers which wholly overshadow the above. Moreover, the above resistance does not include the return wire. The aggregate loss is less influenced by the length of line than by other factors.

Temperature.—As already mentioned, the resistance of electric conductors is affected by temperature. Table 14, given by Houston on the authority of Latimer Clark, shows the effect

TABLE 14.—RESISTANCE AND CONDUCTIVITY OF PURE COPPER AT DIFFERENT TEMPERATURES.

Temperature Centigrade.	Resistance.	Conductivity.	Temperature Centigrade.	Resistance.	Conductivity.
0°	1.00000	1.00000	16°	1.06168	.94190
1	1.00381	.99624	17	1.06563	.93841
2	1.00756	.99250	18	1.06959	.93494
3	1.01135	.98878	19	1.07356	.93148
4	1.01515	.98508	20	1.07742	.92814
5	1.01896	.98139	21	1.08164	.92452
6	1.02280	.97771	22	1.08553	.92121
7	1.02663	.97406	23	1.08954	.91782
8	1.03048	.97042	24	1.09365	.91445
9	1.03435	.96679	25	1.09763	.91110
10	1.03822	.96319	26	1.10161	.90776
11	1.04199	.95970	27	1.10567	.90443
12	1.04599	.95603	28	1.11972	.90113
13	1.04990	.95247	29	1.11382	.89784
14	1.05406	.94893	30	1.11782	.89457
15	1.05774	.94541			

of rise of temperature upon resistance and conductivity of pure copper, up to 30° C. = 86° F. It will be noticed that at the latter temperature the resistance is about 12 per cent greater than at 0° C. = 32° F., so that cold weather is more favorable to transmission than warm. A current traversing a conductor imparts heat to it at the rate of 1 British thermal unit for every 1058 watts * of energy consumed in resistance. The perma-

---

* This is S. P. Thompson's figure. *Electric Machinery*, 1892, p. 427. I make it 1057.—J. P. F.

nent temperature of the wire will depend upon the rapidity with which the heat is dissipated. In computations relative to transmission strict accuracy would require us to take account of the effect of this heating upon the resistance; but we cannot here go into these refinements.

A soft copper wire 1 mill diameter, 1 foot long, called 1 "mill-foot," at 10.22° C. = 50.4° F., has the standard resistance of exactly 10 legal ohms. At 15°.56 C. or 59°.9 F. it has a resistance of 10.20 legal ohms, and at 23°.9 C. or 75° F. 10.53 legal ohms.

Efficiency of Electric Transmission.—One of the earliest attempts at long-distance transmission was made under direction of Marcel Duprez in 1882. This was from Meisbach to Munich, a distance of 34 miles, on the occasion of an industrial exhibition at the latter place. A double line of telegraph-wire was used having a resistance of 950 ohms. The potential was 2700 volts, and the net efficiency of the transmission, as stated by Professor Von Beetz, president of the exhibition, was 32 per cent. This is, no doubt, as high as could have been expected in the state of the art then existing. In a later experiment by the same electrician, viz., in 1886, power was transmitted between Creil and Paris,[*] a distance of 36 miles, with a voltage of 6000 and an efficiency of 45 per cent. M. Fontaine, about 1886, transmitted some 50 h.p. under a voltage of 6000 with an efficiency of 52 per cent. These successive results very clearly show the progressive development and perfecting of methods. In 1887 power was transmitted between Kreigstetten and Solothurn,[†] through a conductor of 9.23 ohms resistance, with a net efficiency of 74.7 per cent.

At Steyermuhle [‡] in Tyrol, in 1890, some 8 h.p. was transmitted a short distance—less than half a mile—with an efficiency of 80.6 per cent. In this and all the preceding cases, the power applied to the line as well as that delivered

---

[*] *Electrician*, 1886, XVII. 318.
[†] *Journal Soc. Telegraph Engineers*, 1888, XVII. 337.
[‡] *Electrotechnische Zeitschrift*, 1890, XI. 11.

by it was measured by brake or dynamometer. 78 per cent is the efficiency claimed for the transmission from the turbines to the factories at Schaffhausen, a distance of some 750 yards. This is about 500 h.p. sent over the wires at 624 volts.

In 1891 the principles of electric transmission were fully understood and the methods developed to a degree not materially exceeded since that date. In this year, on the occasion of an electric exhibition at Frankfort, Germany, power was transmitted from a turbine located at Lauffen, on the Neckar, to Frankfort, and there applied to electric lighting and other purposes. This distance is stated to be 175 kilometers, about 109 miles. A line consisting of three bare copper wires, about No. 6 B. & S. gauge, was carried on tall poles from Lauffen to Frankfort. It is stated that 10 000 porcelain insulators were used, indicating some 3300 poles, being at the rate of 30 to the mile. The turbine at Lauffen drove a 3-phase alternator capable of giving three currents of about 1400 amperes at 50 volts. These were delivered to a step-up transformer of a ratio of 1 to 160. The current went over the line at from 12 000 to 25 000 volts. At Frankfort it was delivered to a step-down transformer and converted to 60 volts to supply either lamps or 3-phase motors. Table 15 * gives the power taken from the turbine and the losses in the several elements of the system. The average efficiency of the system is 73.3 per cent according to these figures, but it is given by S. P. Thompson as 72, possibly on more reliable data. The latter is a very gratifying result and probably could not be exceeded in practice. Supposing the turbine to give 75 per cent of the gross power of the water, this would indicate $.72 \times .75 = 54$ per cent of the water-power transmitted a distance of 109 miles.

These tests were made by a jury of experts under direction of Prof. H. F. Weber, and may be presumed correct. They are especially valuable for the reason that disinterested statements of efficiency are seldom met with. Men assuming to

---

* These figures are taken from the *Electric World* of July 2, 1892.

TABLE 15.—TRANSMISSION FROM LAUFFEN TO FRANKFORT, POWER IN H.P. DELIVERED AT DIFFERENT POINTS OF THE LINE.

Number.	By Turbine.	By Dynamo.	By Step-up Transformer.	At Step-down Transformer.	To Lamps.	Net Efficiency.
1	78.2	66.1	61.1	58.0	53.5	0.684
2	99.3	86.8	81.5	76.5	71.4	0.719
3	105.9	93.3	87.7	81.7	76.3	0.720
4	105.9	93.3	87.7	81.8	76.4	0.722
5	112.7	100.1	94.5	87.6	82.2	0.729
6	117.6	104.9	99.2	91.7	86.2	0.733
7	120.9	108.1	102.4	91.5	89.5	0.740
8	121.1	108.3	102.6	95.0	89.4	0.738
9	127.0	114.4	108.7	100.7	95.1	0.749
10	127.5	114.8	109.0	100.9	95.3	0.747
11	151.8	139.1	132.8	120.0	114.0	0.751
12	151.7	139 0	132.7	120.2	114.2	0.753
13	189.2	177.0	169.9	145 3	138.9	0.732
14	190.0	177.3	170.3	145.3	138.9	0.731
15	190.7	177.9	170.8	145.3	138.9	0.728
16	194.7	182.2	175.1	150.7	144.2	0.741
17	197.4	184.8	177.6	152.4	145.8	0.739

be electricians often state the efficiency of transformers at 97 and even 99 per cent. The *Engineering News*, vol. XXXIV. p. 256, estimates the loss in transmission of power generated by a steam-engine and sent over a line some 20 miles in length as follows, the indicated power of the engine being taken as unity:

Efficiency of engine.................. 0.92	Efficiency remaining..... 0.920
" " belting and jack-shaft. 0.90	" " 0.828
" " dynamo................. 0.92	" " 0.762
" " step-up transformer... 0.93	" " 0.709
" " line... .... 0.88	" " 0.624
" " step-down transformer. 0.93	" " 0.581
" " rotary converter....... 0.84	" " 0.488
" " railway circuit......... 0.90	" " 0.438
" " car motors............ 0.85	" " 0.372

That is to say, only 37 per cent of the indicated power of the engine becomes available in propelling cars. In the case of a water-wheel whose efficiency could not ordinarily be taken over .75, but in favorable cases .80, the percentage of power made available would not be over 32, understanding that this figure refers to the absolute power of the water and not to the power rendered by the wheel.

On the other hand an experiment made at Ogden, Utah, in 1898, under the auspices of the General Electric Company, and conducted by its agents, in which current was transmitted 73 miles at a tension of 30 000 volts, is stated to have shown a loss of only 9 per cent, including 4 per cent in transformers.

Systems of Electric Transmission in America. — The mountainous mining regions of the Pacific slope have usually a considerable rainfall, and the streams are fed till late in the season by melting snows. They abound in steep declivities, where great heads can be obtained, and the evaporation is not excessive. There is great demand for power in extracting the precious metals from auriferous and argentiferous rock, as well as for electric lighting and other purposes, and coal is very expensive. These regions have offered an attractive field for electric transmission.

The concentrating works of the Silverton mines, 4 miles southeast from Silverton,* Colorado, were, previous to 1895, run by steam. The works being 12 300 feet above sea-level and at a distance from railroad communication, accessible only by wagon-roads through a very rough country, the coal cost, delivered to furnace, $8.75 per ton, entailing an expense for this item of about $1000 per month. As a relief from this expense, a water-power and electric-transmission plant was adopted. Water was conducted from a point on the Animas River, above Silverton, in a 3 × 4-foot flume about 9750 feet to a point where a fall of 180 feet was obtained. The flume rested largely on trestles, and at some points was 50 feet or more above the ground. It carried some 40 cubic feet per second, yielding, on the above-mentioned fall, something over 600 h.p. It was used on two double-nozzle, 4-foot Pelton wheels, belted to two 150-K.W. General Electric Company's

---

* *Engineering News*, vol. XXXIV. p. 114.

3-phase generators, producing a current of 2500 volts. The current is transmitted 3 miles over an exceedingly rugged country by bare copper wires, No. 3 B. & S., one for each of the 3-phase circuits.

At Fresno,* Cal., power is received from the north fork of the San Joaquin River, over 40 miles distant. The water is conducted in a canal about 7 miles to a reservoir. Thence it is conveyed in a steel pipe 20 to 24 inches diameter to the power-house, a distance of over 4000 feet, where it is used on Pelton wheels under the enormous head of 1400 feet. These wheels are 57 inches diameter and make 600 revolutions per minute, developing 500 h.p. each. A separate wheel is used for the exciters of the dynamos. The current is generated at 700 volts and raised by transformers to 11 200 volts for transmission to Fresno, where it arrives at a voltage of 10 000, the distance being 35 miles. It is said that 75 per cent of the current generated by the wheels reaches the switchboard at Fresno. The pipe leading from the reservoir to the power-house is $\frac{1}{4}$ inch thick at the former point and $\frac{3}{8}$ at the latter. In case of a breach of the pipe a vacuum would be created near the influx, and for this reason a series of valves are inserted opening inwards to let in the air, in such an event, and prevent the collapse of the pipe.

Power † is transmitted from the American River near Folsom, Cal., to Sacramento, a distance of 24 miles. The current is generated at 800 volts and raised for transmission to 11 000. The line is in duplicate. The loss in the line is stated as 7.5 per cent with 3000 h.p.

The longest transmission in America and probably the longest in the world now in practical operation, the Frankfort-Lauffen system being merely an experiment, is the installation of the Southern California Power Company, who develop some 4000 h.p. at Santa Ana Canyon, near Redlands, and transmit it to Los Angeles,‡ a distance of 80 miles. A head of 750

* *Engineering News*, vol. XXXVI. pp. 12, 225.

† *Ibid.*, vol. XXXIII. p. 243.

‡ Communicated by General Electric Co.

feet is obtained at the canyon, which is used on Pelton wheels. These give motion to four 750-K.W. dynamos, coupled direct to the wheel-shafts and running 300 revolutions per minute. The current is delivered at 750 volts and transformed to 33 000 for transmission. The voltage is stepped down by rotary transformers for railways, and by static transformers for light and power.

The highest voltage known to be used in transmission is in the line of the Telluride Power Transmission Company at Provo,* Utah, which transmits 2000 h.p. a distance of 55 miles at 40 000 volts.

The above was written in 1900. The present (3d) edition of this book is able to chronicle a much more extended system of transmission, and one which leaves the celebrated Frankfort-Lauffen installation far behind. This is the system recently completed by the Bay Counties Power Company of California. From power stations on the Yuba River in the Sierra Nevada Mountains the current goes through two parallel lines of wire to Oakland, a distance of 142 miles. Two lines are used to meet the possibility of accident to one of them. One line is of No oo medium hard-drawn copper wire throughout; the other of aluminum, except in the vicinity of the salt water, where it is of copper; it being apprehended that the salt air from the sea would act injuriously on the aluminum. The two lines are 25 feet apart. They are supported on poles of Oregon red cedar, 14 inches diameter at butt, 8 at top, 35 feet long, set 6 feet deep in the ground, 132 feet apart. The system is not yet worked to its full capacity. The power-houses are now said to be capable of furnishing 17 000 h.p., and it is thought that there is water for 13 000 more. A tension of 60 000 volts is contemplated on this line. From Oakland, power is transmitted south to San José, making a total distance from the power-houses of 184 miles.

In later installations of electric transmission a strong tendency

---

* *Idem.*

is manifest to dispense entirely with the use of shafting in establishments driven by electric power. Each separate machine has its own motor and its own wires, dispensing not only with shafting and its accompaniments of bearings, hangers, couplings, pulleys, belts, etc., but giving a much neater and lighter appearance to the shop, and saving the very considerable loss of power incident to the use of shafting. In the old-fashioned cotton-mills the power consumed by the shafting was often as much as 25 per cent of the total power required by the mill, and it probably reaches 15 in mills of the most modern construction, which is much in excess of that consumed by the motors. The extensive ship and engine works recently erected at Weymouth Fore River, near Boston, is a conspicuous example of this mode of operation. Each lathe, drill, boring machine, riveting machine, crane, etc., has its own separate motor and runs independently.

This system has not yet been applied to cotton-mills to the extent of introducing a motor for each separate machine; but cotton-mills driven by transmitted power have often been arranged with a separate motor for the main shaft of each room, thus dispensing with large belts, which are very troublesome organs of transmission, besides readily admitting of the stopping of one room without interfering with the others. In mills driven by water-power, this saving goes far to neutralize the losses incident to the several electric conversions.

· In concluding the subject of transmission this remark appears pertinent. The older manufacturing cities are intersected by canals, which, at the date of their construction, were the only means of distributing power to different mills. In the course of time these canals have become lined with buildings, and the land has acquired a value of 3, 5, 10 dollars or more per square foot. A canal 100 feet wide might carry 10 000 h.p. on a 30-foot head. Such a canal would usually have marginal reservations of 15 feet or thereabouts, and the land occupied by it represents a value of $300 to $1200 per linear foot.   This property is maintained as a conductor of power to perform á function which might be equally well fulfilled by copper wires of half a square inch cross-section or by a 36-inch pipe.   It appears probable that these considerations will, before long, lead to important modifications in existing arrangements of water-powers.

CHAPTER XXII.

THE POWER-HOUSE.

THIS establishment contains the wheels and generators with their connections and appurtenances for developing the power, transforming it into electric current, and controlling the delivery of the latter to the wires. A central establishment for developing power and transmitting it to a distance is wholly a creation of the past twenty years. As early as forty years ago it was found convenient in some large manufacturing establishments to group the turbines in one building or one room, called the wheel-house or wheel-room, and couple them all to one shaft which supplied power to the entire mill and often to several mills. This establishment, through recent developments in electricity, has grown into the modern hydraulic power-house.

To treat the subject of power-houses in detail would require a large folio volume of drawings and descriptions, and these would become mainly obsolete in ten years. All that can be attempted here is a very general view of the subject, aided by such sketches as can be introduced in these pages. We will consider the subject according to the head acting on the wheels, commencing with the lowest heads.

On such a head as is ordinarily created by a dam it is convenient to put the power-house in a line with the dam and forming a part of the same, being adapted to sustain the pressure of floods. The height of floods should be moderated by as great a length of overflow as possible. This is the best arrangement when the dam commands the entire fall. When it is at the head of a rapid and it is desired to make the entire fall

available by means of a canal or race, it will usually be found
cheaper to extend the high level down to the foot of the rapid
by a canal than to extend the low level up to the dam by a
race.   This consideration would dictate placing the power-
house at the foot of the rapids, guarding against floods either
by raising the river-bank of the canal above flood-level or by
a bulkhead and gates at the influx of the canal.

Fig. 216 shows a power-house on the lowest head that
would be worth utilizing, viz., 6 feet.   This design was made
by the writer to show the use that could be made of an
abandoned mill privilege by transmitting the power to other
mills owned by the same corporation about a mile distant, in
a case where the water of the stream was to be diverted for the
supply of the Metropolitan District of Massachusetts.   This
may be taken as a type of a low-head hydraulic power-house
for electric transmission.   The stream furnishes water to drive
the four wheels for three or four months in the year.   At other
times there will be water for three, then for two, then for one,
and at times a single wheel would have to run at part gate.
The wheels are disconnectible, and are stopped one after another
according to the stage of water, commencing with the wheel
most remote from the dynamo.

Fig. 218 shows the general situation, the canal and power-
house being indicated by dotted lines.   The formation is
gravel, with no rock near the surface, and the following mode
of construction is contemplated in the design.   The entire site
is excavated to a depth of 8 feet below low water, after excluding the water of the river by a coffer-dam.   A substantial bulkhead of masonry is laid across the canal, and joined to the
bottom by a row of sheet-piling which extends well into the
banks on each side.   On the down-stream side of the bulkhead
a platform of heavy timbers covered with two thicknesses of
plank is laid.   On this are laid the walls of the power-house,
and the piers which separate the wheels.   The down-stream
wall has four arched openings, through which water escapes to
the river after passing the wheels.   The piers are built up to

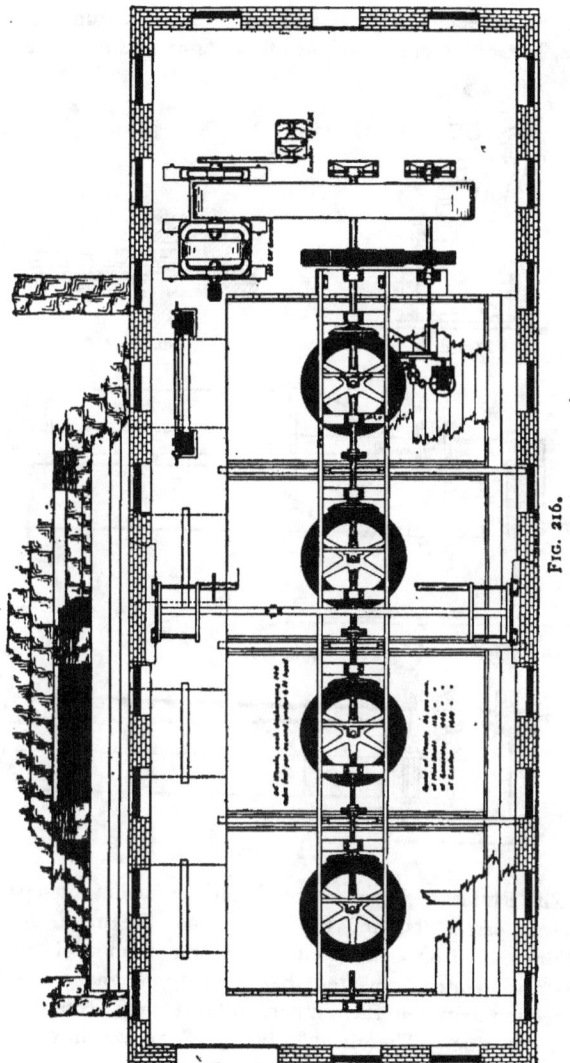

Fig. 216.

about the level of low water and sustain the floor-timbers of the flume, in fact the entire weight of the flume and its contents.

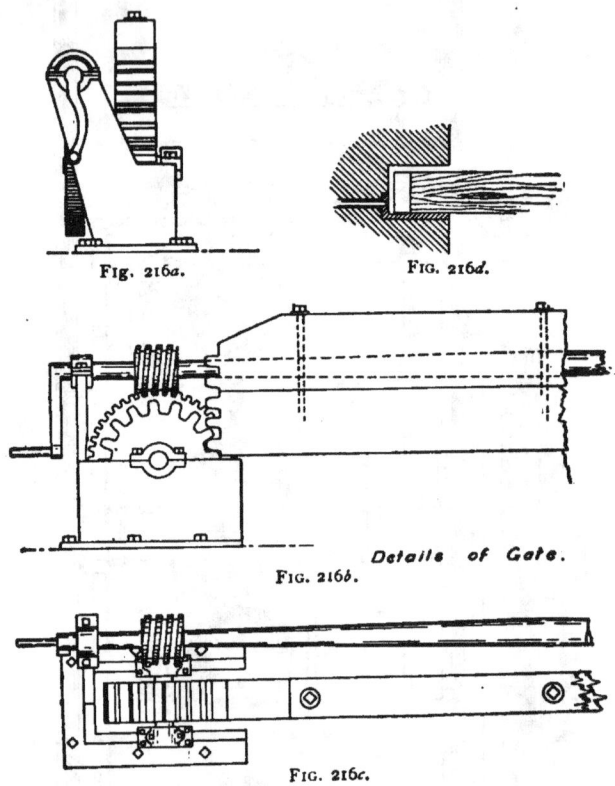

Fig. 216*a*.

Fig. 216*d*.

Details of Gate.

Fig. 216*b*.

Fig. 216*c*.

The Flume is shown as one continuous basin, not separated into compartments, and the bearings of the main shaft are sustained by iron beams running across from wall to wall. This arrangement obviates the necessity of any water-tight connection between the timber and masonry, except on the up-stream side.   There would be an advantage in separating

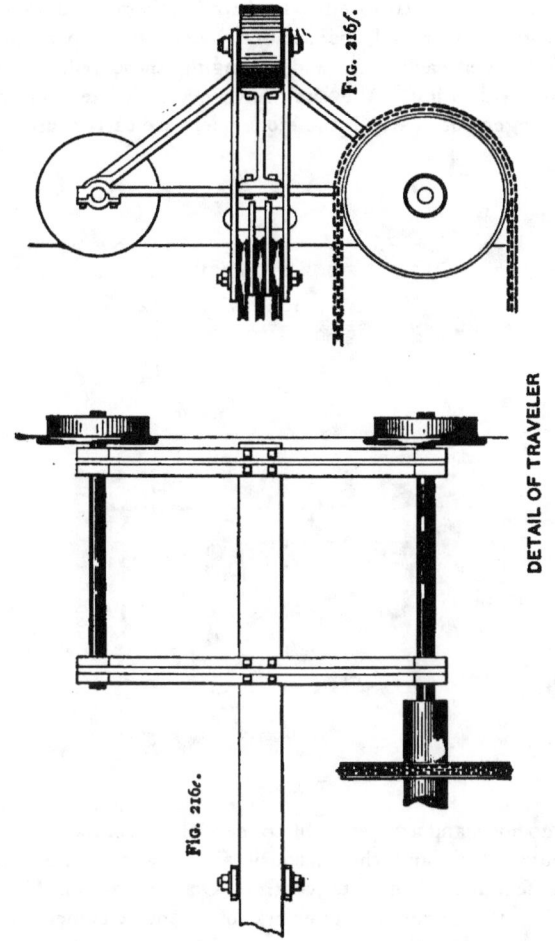

FIG. 216*f.*

DETAIL OF TRAVELER

FIG. 216*e.*

the flume into compartments, one for each wheel, in case of repairs being required to one wheel while others are in motion. This could very readily be secured by carrying up a timber bulkhead over each pier, and arranging these bulkheads to sustain the shafting. We shall have occasion later to discuss the arrangement in which the stone piers are carried up to the

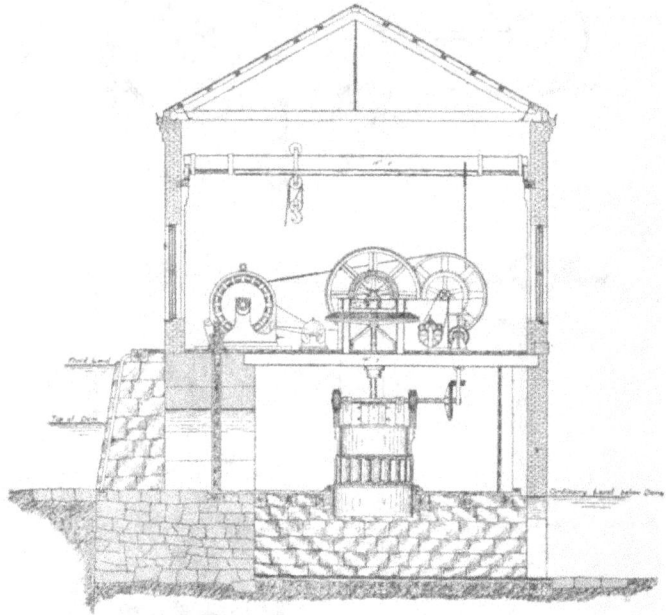

FIG. 217.

dynamo floor, and a water-tight connection is necessary between the flume floor and the masonry all around. The wheels are 14 feet apart centre to centre. On the 12-inch I beams spanning the flume a framework of I and channel bars is erected to sustain the bridge-trees and bearings. The crown-gears have wooden teeth, the jack-gears are all iron; these gears being proportioned to give the required velocity to the

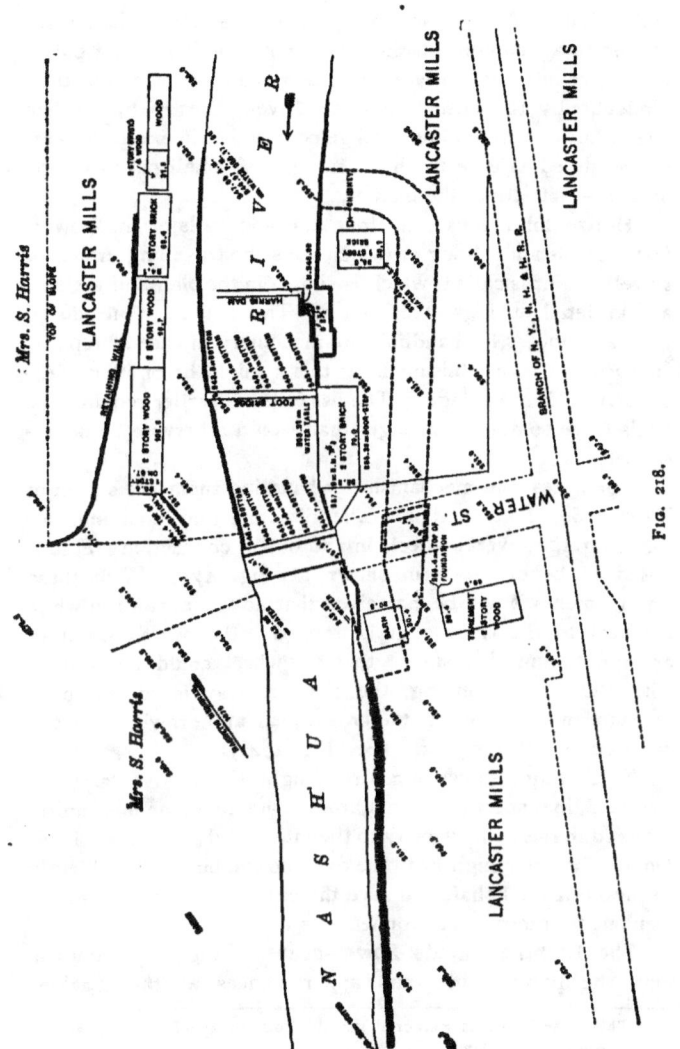

Fig. 218.

main shaft. A mortise spur-gear on the main shaft gives motion to a countershaft through an iron pinion, and a pulley on the countershaft drives the generator by means of a belt. Undoubtedly the latter might be driven directly by toothed gears, but this would involve a velocity rather in excess of what is considered safe for toothed gearing. A small pulley on the dynamo-shaft drives the exciter.

Horizontal ledges run along the side walls of the power-house to sustain a track on which runs the travelling crane, or traveller as it is called, which is shown in the plan and section, and in detail at Figs. 216f and e. This is usually introduced in power houses for handling the machinery in case of repairs. In so small an establishment as this it might be omitted without serious disadvantage. The use of the traveller requires the walls to be somewhat stronger than would otherwise be necessary.

The gates and mechanism for handling them are shown at Figs. 216a, b, c, and d. The disconnecting couplings for throwing the several wheels in and out of connection are supposed to be of the form shown at Fig. 175. With these arrangements it will be perceived that no wheel can run while a wheel to the right of it is stopped. If it were desired to arrange the mechanism so that *any* wheel could be stopped with the others running, the simplest way is to introduce mechanism for lowering the crown-gear and drawing it out of connection with the jack. (See Fig. 174.)

This design shows in a strong light the disadvantage of a low head, in respect of the cumbrous character of the mechanism required to raise the velocity to the rate called for by electricity. On a sufficiently high head the dynamo can be coupled directly to the horizontal shaft and save the expense and loss of power incident to intermediate connections.

The Lachine Rapids Power-house.*—Fig. 219 shows in plan the power-house and appurtenances at the Lachine

---

* This description is gathered mainly from an article in *Engineering News*, February 18, 1897.

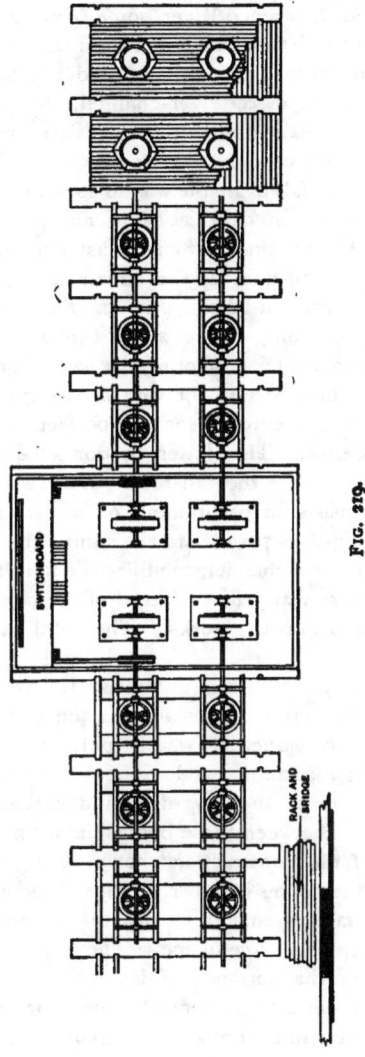

FIG. 219.

Rapids on the St. Lawrence River some 6 miles above Montreal. This is one of the numerous rapids which occur on the St. Lawrence between Lake Ontario and tide-water. They occupy some 5 miles of the river-channel, with a total fall of about 30 feet. A dam across the river here would not be permissible even were such a work practicable. The only mode of making the fall available was to construct a spur-dam, which we call the main dam, at right angles to the shore, extending into the stream a sufficient distance, and from that point to extend a wing-dam up the stream to a distance sufficient to raise the required head. Fig. 220*f* shows the general situation and surroundings. Fig. 220*g* shows the main dam, wing-dam, and booms for protecting the work from ice. Fig. 220*h* is a longitudinal section showing the wing-dam in elevation. The wing-dam extends some 3700 feet above the main dam and 1200 below. This latter portion adds to the fall by excluding the river from the tail-race. Otherwise stated, the improvement consists in partitioning off a part of the river-channel by a wing-dam parallel to the shore, and constructing a dam across this new channel, to utilize the fall therein. The wing-dam reaches nearly to the head of the rapids, the main part of which is below the works. The total fall utilized is about 12 feet.

From the inception of this enterprise it was apprehended that ice would be a serious obstacle to the operation of the works, and the construction was largely modified by this expectation. At Fig. 220*g* two long booms are shown, reaching from the shore to the wing-dam and sustained by cribs filled with stone. Between these booms the wing-dam is over-flowable, and of the construction shown at Fig. 220*c*. The construction of the booms is shown at Figs. 220*d* and *e*. They have mass and stability enough to sustain any number of men necessary to keep them free of ice. The upper boom is supposed to prevent the entrance of ice into the water-power channel. The lower one is intended to exclude any ice which passes the upper boom or forms in the channel, and throw the

same over the wing-dam. In addition there is a system of
booms attached to the main dam for sluicing ice through waste-

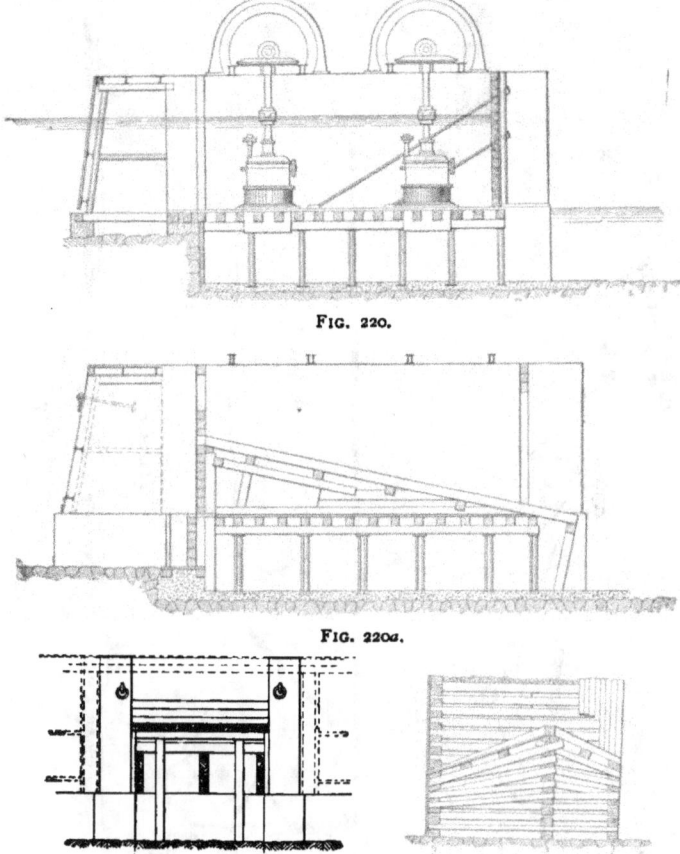

Fɪɢ. 220.

Fɪɢ. 220*a*.

Fɪɢ. 220*b*.   Fɪɢ. 220*c*.

ways in the latter. The construction of these wasteways is
shown at Figs. 220*a* and *b*.

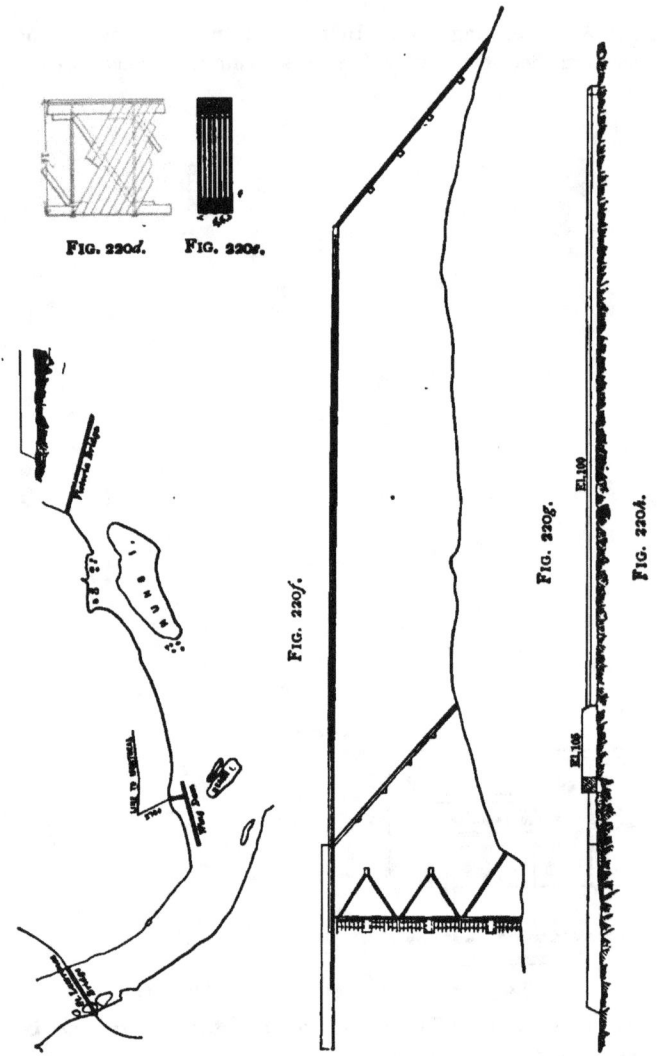

Fig. 220d.   Fig. 220e.

Fig. 220f.

Fig. 220g.

Fig. 220h.

The ice to be guarded against is of two kinds: anchor-ice, or frazil-ice as it is here called, which forms on the surface of rapidly running waters and ceases to run when the stream freezes over, and the heavy ice brought down at the annual breaking up. The St. Lawrence flowing northward, and commencing to break up at the head, the ice-jams are very formidable, and it remains to be seen how these works will stand against them.

The main dam is merely a series of flumes, power-houses, and wasteways. A general idea of it is given in Fig. 220g. It consists of a row of piers extending entirely across the channel and sustaining the machinery and buildings. Except at the power-houses these piers are 48 feet long up and down stream, and about 21 feet apart centre to centre. Up to the floors of the flumes these piers are of concrete 5 feet thick; above this level they are of cut-stone masonry 4 feet thick. The piers rise about 6 feet above ordinary water-level, which is thought to be sufficient, as the fluctuations of the St. Lawrence are not great. A rise in the river here does not imply a corresponding rise immediately above the main dam. A rise of 5 feet at the latter point would imply some 140 000 cubic feet per second going over the wing-dam, which would greatly lower the level in the channel.

The floors of the flumes are supported by timbers 12 inches wide, 15 inches deep, imbedded at their ends in the concrete masonry, and these are also supported by cast-iron columns. Near the upper end of the pier is a check 12 inches wide, 8 inches deep for the gate, and near the lower end is a similar check for a barrier of stop-logs. This latter feature was, it is presumed, introduced on the ground of economy. The substantial construction would have consisted in throwing an arch over the lower end of the wheel-pit, with its soffit at low-water level, and resting a heavy wall thereon united with the piers. The stop-logs are strengthened by the inclined iron tie-rods appearing in Fig. 220, which are attached to a channel-bar placed vertically on the down-stream side of the stop-logs, and at their up-stream ends to a long iron plate bolted to a number of the floor-beams. The gates are only inserted in emergencies. They are handled by the travelling crane and are ordinarily kept stored under water. The piers support steel-frame sheds erected over the wheels and shafting, and substantial buildings of brickwork and skeleton steel cover the electric machinery, constituting the power-houses proper, the whole forming a continuous building nearly 1000 feet in length, through which a 25-ton hand-power travelling crane runs from end to end. Two of the wheel-pit spaces are occupied by wasteways for sluicing off the ice diverted into them by the booms. The heavy rings anchored in the walls, as shown at Figs. 220*b* and *a*, are for the attachment of the booms. When racks are exposed unprotected to heavy flake and anchor ice, the latter is liable to pack solidly against them from top to bottom. The water in the flumes escapes through the wheels, leaving the pressure of the entire head acting on the racks, which are then very liable to break down.

Wheels.—Each flume contains two special 54-inch Victor turbines, a total of 72, which are connected in sets of 6 turbines to each of 12 generators, 4 in each power-house. These wheels are set vertically on 3-inch plank flooring and will develop 200 h.p. each under a head of 11 feet, a total of 14 400 h.p. Under this head they run at 64 revolutions per minute, discharging 200 cubic feet of water each per second, being a total discharge of 14 400 cubic feet per second. This is but an insignificant fraction of the total flow of the stream, which carries here not less than a quarter of a million cubic feet per second, including the channel to the northward of the island of Montreal.

Each set of 6 wheels is connected by bevel-core gears to a common jack-shaft which transmits their power, at a speed of 175 revolutions per minute, to a 3-phase generator of 750 K.W. guaranteed to stand 25 per cent overload. The speed is regulated by a governor guaranteed to control it within 2 per cent from 0 to full load. The gears and shafting are sustained

by bridge-trees resting on 12-inch I beams which extend from
pier to pier.

Mechanicsville on the Hudson.—Figs. 221 and 222 relate
to the power-house* at Mechanicsville on the Hudson River
about 18 miles above Albany, which city, together with Troy,
some 11 miles distant, has large use for power.   At Schenec-
tady, 17 miles away, are the great works of the General Electric
Company, covering some 130 acres of ground, and it was chiefly
with a view to supplying these works that this development
was undertaken.

The river at this point is about 1200 feet wide.   It is
separated by Bluff Island into two channels, the westerly about
400, the easterly some 800 feet in width.   The ordinary
summer flow of the Hudson at this point may be placed at
4000 or 5000 cubic feet per second, though often falling much
below the lower figure.   The available fall is 18 feet.   The
westerly channel is occupied by the power-house, supplemented
by a bulkhead rising above high water, and is entirely closed
to the passage of floods.   The easterly channel is occupied by
the overflow-dam already noticed, page 137.   A floating
boom, anchored to timber cribs filled with stone, extends
obliquely across the westerly channel for the exclusion of drift.
The lower part of the power-house is of concrete resting on
rock.   The floor is formed by arches springing from steel box
girders which are supported by steel I-beam columns.   A concrete head-wall 6 feet thick divides the floor into two parts.
The up-stream portion is the flume, containing the wheels and
draft-tubes.   The down-stream part is the generator-room,
which the horizontal shafts enter through stuffing-boxes in
plates built into the concrete wall.   Along the up-stream face
of the power-house runs a rack formed of flat steel rods, sustained by a frame of steel beams and pillars, the whole surmounted by a bridge for the convenience of workmen in
cleaning the racks.   There are ten pairs of 42-inch horizontal

---

* The drawings and description of this plant are from *Engineering
News*, vol. XL. p. 130, and *Engineering Record*, vol. XXXVIII. p. 299.

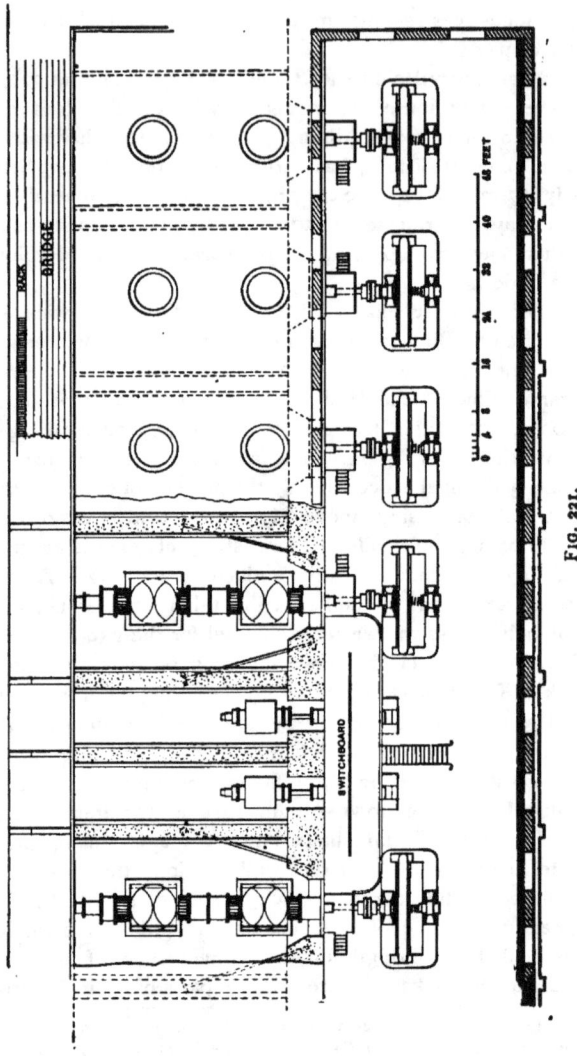

FIG. 221.

Victor turbines, two pairs on each shaft. On the obtainable
head of 18 feet these run at 114 turns per minute, each wheel
being rated at 250 h.p. and each set of four driving a 750-

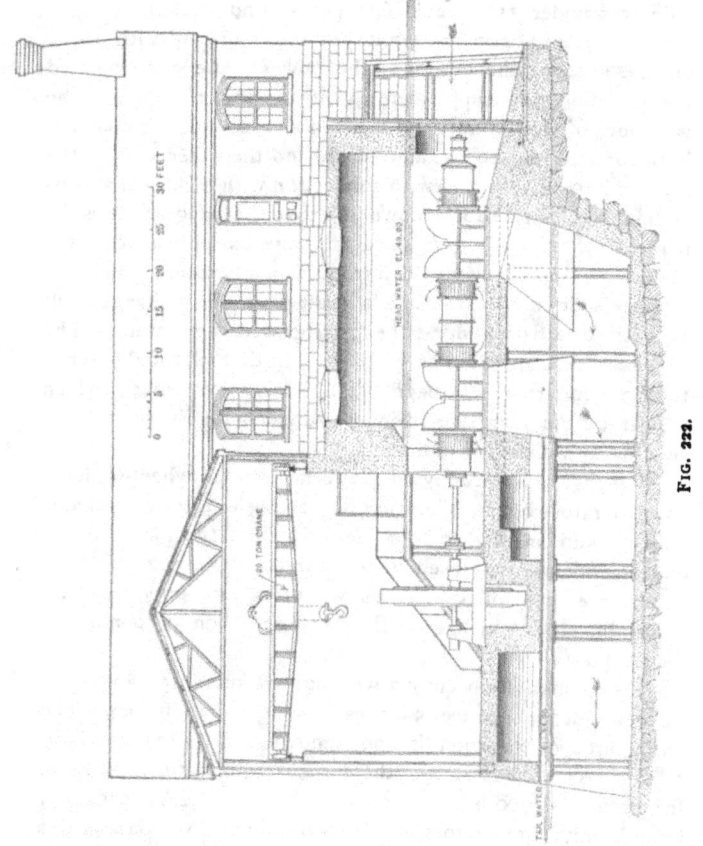

FIG. 221.

K.W. generator. Two additional sets of wheels are contemplated for the future, making a total development of 7000 h.p.
A development of 7000 or even 5000 h.p. would in the normal

state of the river be subject to serious interruptions. Data obtained by Mr. Geo. W. Rafter in the course of examinations relative to the system of reservoirs proposed by the State of New York, showed that the flow of the Hudson at Mechanicsville was under 1500 cubic feet per second (less than 3000 h.p.) for 124 days in the eight years terminating with 1895. On this account the design of the power-house contemplated the possibility of supplementary steam-power to meet the variations of the stream. To this end a disconnecting coupling is introduced on the dynamo-shaft, and the other end of the shaft is formed with a view to connection with a vertical steam-engine. It is probable, however, that the progress thus far made in the execution of the State's system of reservoirs will obviate any necessity for the introduction of steam. Each pair of wheels discharges through a draft-tube, which has, as will be noticed, a flaring shape, being largest at the bottom. The effect of this disposition is analogous to that of the diffuser of the Boyden turbine (see page 341). The exciter-wheels, which appear on the plan, are 18-inch cylinder-gate Victor turbines running 259 turns per minute.

The regulators occupy the platforms over the wheel-shafts in the generator-room. These are of the Geisler electro-mechanical type and are said to be so sensitive that the gates can be entirely opened or closed in 6 seconds. The exciter-wheel gates are controlled by Snow governors provided with adjustable stops which limit the hoisting action as soon as the gate is fully open.

The generators occupy a well-lighted room 255 feet long, 34 feet wide, and 30 feet 5 inches from floor to roof-truss. The entire area of the room is commanded by a 20-ton travelling crane. As already stated, the ultimate generator capacity of the station is 7000 h.p. in 7 generators of 750 K.W. capacity, though only 5 generators had been installed at the date of this information (1898). These are uni-tooth, 3-phase, 40-pole, 750-K.W., 114-revolution alternating machines, having revolving fields and stationary armatures. They are wound to deliver to the transmission-lines 36 amperes at a periodicity of 38 cycles and a pressure of 12 000 volts, and are arranged for operation in parallel at constant voltage. By using the revolving-field type of generator it is possible to secure this high voltage directly from the machine without the use of transformers. As the current is to drive synchronous and induction motors, to operate lights, and to be converted into direct current through rotary converters, a frequency of about 40 cycles was selected as most suitable.

The exciters for magnetizing the fields of the generators are placed one on each side of the stairway leading to the switchboard gallery. They are 6-pole, 100-K.W., 125-volt standard General Electric machines with ribbed field-frame and iron-clad armatures.

Fig. 223 with subsketches 223a and b shows a general idea of the proposed power-house of the Michigan Lake Superior Power Company on the Canadian side of Sault Ste. Marie. These are from blue-prints kindly furnished in advance of the entire maturity of the plans by Mr. H. von Schon, the engineer of this work. Between Lakes Superior and Huron there is a fall of 19 to 20 feet, which is substantially all concentrated at a single rapid. By means of a canal something over 2 miles long, this fall is made available at the power-house. This is in one sense the largest water-power development yet undertaken, not in the sense of developing the largest amount of power, but in the sense of using the largest quantity of water, viz., 30 000 cubic feet per second. The canal is constructed with a view to the attainment of a high velocity without undue loss of head. It is partly in rock formation, partly in earth. The rock is to be cut vertically with channelling-machines; the bottom is to be made smooth with hydraulic cement mortar. The earth part is to have a smooth and even lining of timber and plank. By these means it is expected to secure a velocity of 7 feet per second and a head of 16 feet at the power-house. Assuming a loss of 1 foot in penstocks and wheel-pits, and an efficiency of 80 per cent in the wheels, this would realize some

41 000 h.p.   The flow of Sault Ste. Marie ranges from 60 000 to 100 000 cubic feet per second according to the stand of Lake Superior.

Fig. **223** is a general section of the power-house, which has **a** length of about a quarter of a mile.   Fig. **223a** shows, on a

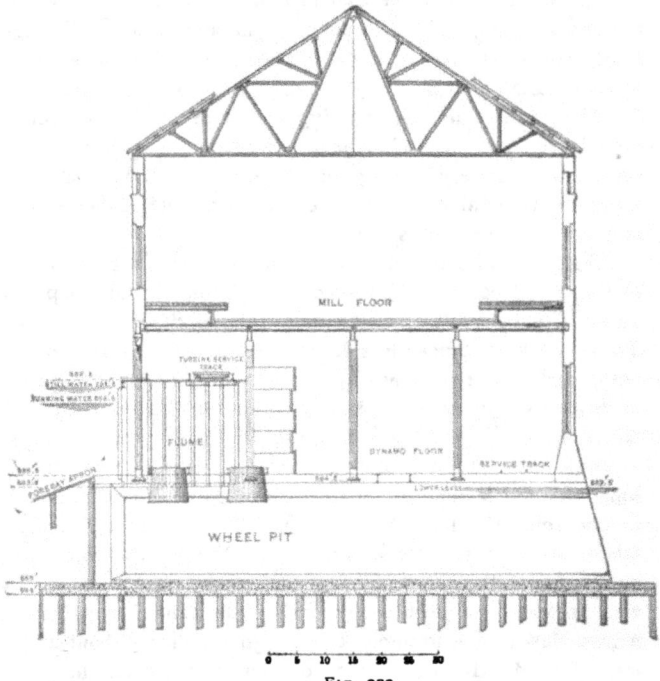

FIG. **223**.

small scale, the general arrangement of the wheel-pits, eighty in number, each containing two wheels or two pairs of wheels. The upper part of the building is reserved for industries dependent on the water-power, but the greater part of the latter will be applied to the production of electric current. This locality being remote from any large centre of industry, it is not known to what purpose the electric current will be applied.

The building differs from the types hitherto considered, in not being provided with a travelling crane. The unusual width of the building, and the intermediate floor, forbid this appurte-

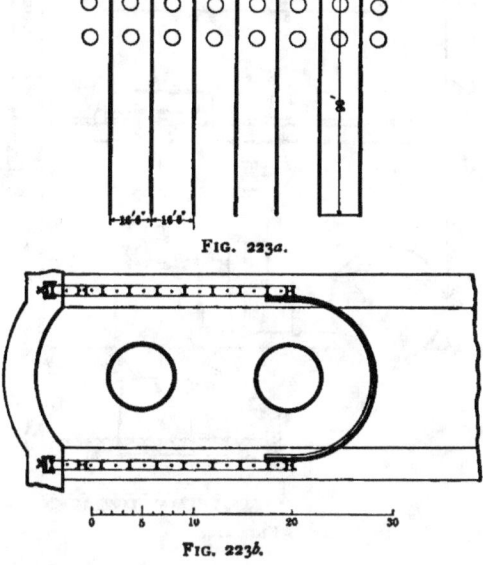

FIG. 223*a*.

FIG. 223*b*.

nance. Instead of the crane, light cars running on service-tracks are to be used for the moving of machinery. Along the front of the building arrangements are made for the attach-ment and movement of a small boom-derrick to handle the penstock-gates. Fig. 223*b* shows one of the flumes or pen-stocks in plan. These are constructed of steel plates supported by a framework of steel beams. Nothing of the wheels appears except the draft-tubes. It is understood that horizontal wheels are to be used, the shafts passing through stuffing-boxes in the flumes and extending to the dynamo floor.

For a Power-house Supplied by a Pipe under a head of 200 feet or less Fig. 224 indicates a suitable arrangement. It is suposed to be located near the watercourse which receives

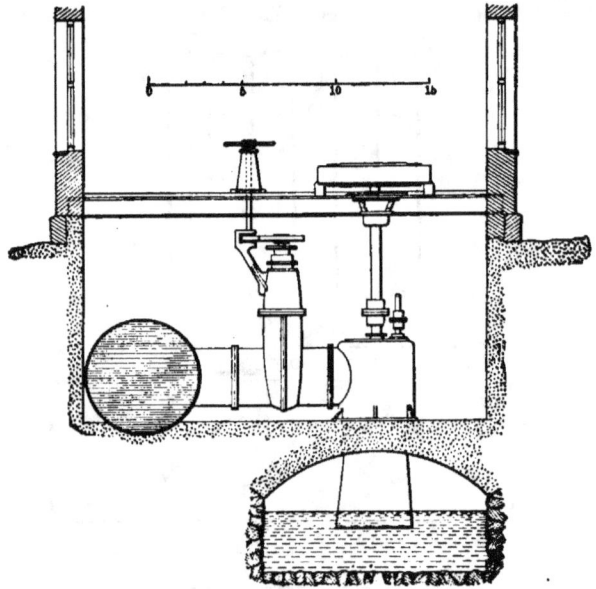

FIG. 224.

the discharge, and the sketch assumes a rock formation. A tail-race is cut leading into the watercourse. In a soft formation this would be provided with side walls and a plank or concrete bottom. An arch is thrown over this channel and levelled up with concrete, which extends the whole width of the building, forming the basement floor. The basement contains the supply-pipe and wheels, the former running lengthwise of the building and sending off branches at suitable intervals to drive the wheels. The sketch assumes a head of 200 feet and a velocity of some 7 feet per second in the supply-pipe. With such great heads it is not customary to count the loss of head so closely as is proper with ordinary heads. The wheel is supposed to draw about 50 cubic feet per second, giving something over 800 h.p. A wheel of ordinary proportions under this head would have too rapid a velocity, as it would not have a diameter greater than 15 inches and would revolve as much as 1000 times a minute. To limit the velocity to 500 turns a minute, its diameter must be as much as 30 inches and the discharge must be restricted by limiting the height of the orifices. The wheels discharge through draft-tubes passing through the arch and dipping into the tail-race. The generator floor is sustained by steel beams extending from wall to wall, and may consist of arches resting on the lower flange of the beam or of thick planking. Planking may be laid on steel I beams by attaching a timber to the latter confined by bolts through the web. The generator is directly coupled to the shaft. It is represented with a revolving armature and stationary field, though the reverse arrangement is more common. The weight of the revolving part comes upon the shaft and is borne by the thrust-bearing, Fig. 225. The

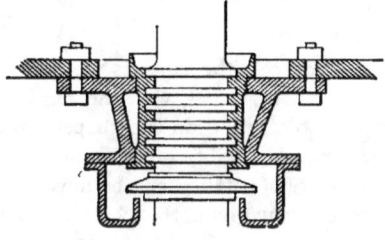

FIG. 225.

weight of the wheel and other revolving parts may be thrown upon the same bearing if desired, and this is generally the more judicious arrangement, as it admits of more efficient lubrication than the ordinary bearing at the bottom of the shaft.

For heads exceeding 200 feet, and even for heads exceeding 100 where the velocity of the generator is limited, the disposition of Fig. 226 may be adopted.   The general arrange-

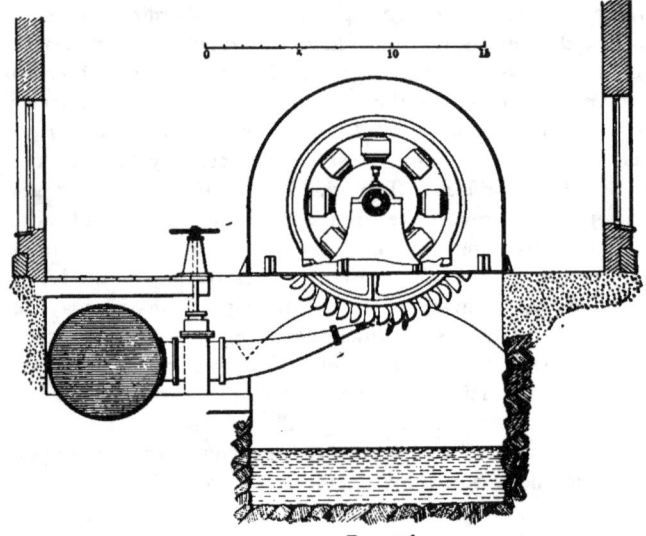

FIG. 226.

ment of the power-house is the same as the preceding, but we adopt a different form of wheel, viz., the hurdy-gurdy wheel driven by a jet of water acting on cup-shaped vanes (see page 332).   The peculiar adaptability of this wheel consists in the fact that a low velocity of rotation can be imparted to the shaft under the highest head by giving the wheel a suitable diameter. The head, in this case, is assumed to be 1000 feet, implying a velocity of 254 feet per second.   The diameter of the wheel is 12 feet measured to the centre of the vanes, and, as it gives its best effect with a velocity of vane equal to half that of jet, the speed will be $127 \times 60/12 \times \pi = 200$ per minute.   A 3-inch stream would carry 12.4 cubic feet of water per second and would furnish over 1000 h.p.   As in the former case, the tail-

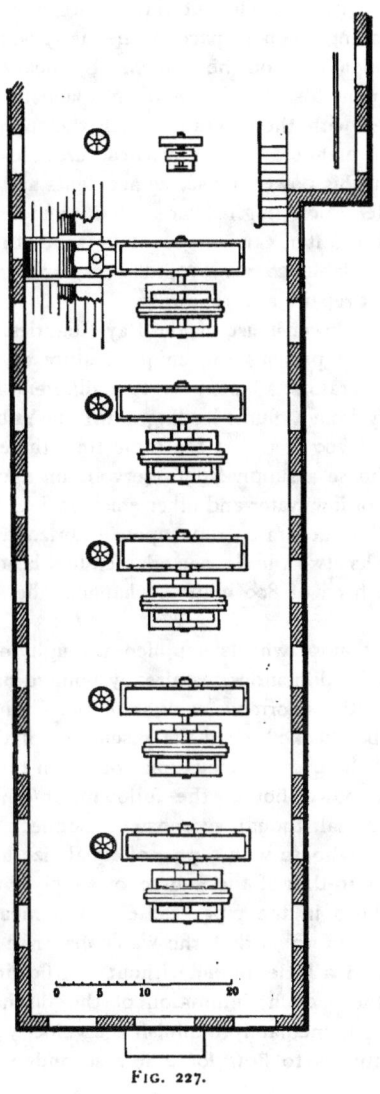

FIG. 227.

race is covered by an arch, but this is interrupted at each of the wheels, leaving an open space for the insertion of the wheel, the case of which rests on the masonry by means of a flange. This arrangement also leaves an open space for the nozzle, which, together with the recess in which the supply-pipe lies, is covered with planking.   The overhead crane cannot be dispensed with in this power-house, as accidents are more liable to occur under such high heads than under low heads. Especially if the water carries gravel or sand, the buckets of the wheel are liable to very rapid wear, and the wheel to require frequent repairs and renewals.

The Nevada Power-house of the Bay Counties Power Company in California presents the unique feature of a system of wheels and generators adapted to two different heads.   The ordinary supply from a flume leading down the Yuba River acts under a head of 200 feet.   At the same time there is available at this power-house a supply from reservoirs on another stream, for use in case of low water and other emergencies, under a head of 800 feet.   The generator runs upon a horizontal shaft, and each generator has two wheels, one adapted to a head of 200 feet, the other to a head of 800 without changing the speed of the generator.

The regulation of wheels supplied through long pipes is subject to peculiar difficulties, as already pointed out (see page 366), owing to the enormous weight of the column of water which has to be checked when the discharge of the wheels is diminished.   Where great elevations occur in the immediate vicinity of the power-house, the following method is practicable and effectual, though expensive: Connect the supply-pipe at the power-house with a pipe of equal size leading to an elevation equal to that of the source of supply diminished by the frictional head in the pipe, these communicating with a small reservoir so formed that the water may rise to the level of the source and a little higher without overflowing.   In this arrangement the sudden diminution of the discharge of the wheels does not immediately diminish the velocity in the pipe. The water continues to flow for a few seconds with sensibly unabated velocity, but discharges into the reservoir instead of through the wheels. Of course the pipe leading to the reservoir may be made smaller than the supply-pipe without being valueless, but the diminution of its size will be accompanied by a more than proportional diminution of its efficiency.

An Air-chamber in lieu of the preceding arrangement for moderating the fluctuations of pressure is a very natural suggestion. Some advantage may be derived from this attachment, but it is liable to more serious objections than appear at first sight. It must have great size in order to be of any value, and with any rational size the variations of pressure are greater than desirable. To form some idea of the size, suppose a pipe 1 mile in length with a head of 200 feet on the wheels. Suppose a velocity in the pipe of 6 feet per second and impose the condition that it can suddenly be reduced to 3 feet without changing the pressure more than 20 per cent. Consider only a portion of the pipe 1 square foot in cross-section. A velocity of 6 feet per second implies an amount of energy equal to $36/2g \times 5280 = 9/16 \times 5280$ cubic feet of water raised 1 foot. One of 3 feet represents $9/64 \times 5280$. A change from 6 to 3 implies the absorption of energy equal to $27/64 \times 5280 = 2227.5$ cubic feet water raised 1 foot, or, what is the same thing, 111 cubic feet raised 20 feet. The increase of head is 0 at the instant of shutting off, and is limited to 40 feet at the instant the velocity reaches 3 feet in the supply-pipe, so that the average head which opposes the momentum of the water is 20 feet, and under this excess of head 111 cubic feet of water will enter the chamber before the movement is checked to the assumed extent. Let v be the volume of air at the normal pressure. We have the relation $200v^* = 240(v - 111)$, whence $v = 666$. That is to say: to meet the above conditions, the chamber should have a capacity of 666 cubic feet for each square foot of cross-section of the supply-pipe and each mile of its length, requiring for a pipe 5 miles long and 6 feet

---

* When a given quantity of air changes its pressure at constant temperature the product of the volume by the pressure remains constant. See p. 459.

diameter a capacity of 94 156 cubic feet, being equivalent to 3330 linear feet of 6-foot pipe. The compression of the air and consequent slowing up of the velocity occupies about $\frac{111}{6} = 18$ seconds. Then the air expands and the pressure diminishes till it reaches near 40 feet head below the normal, and a series of long oscillations occur. To obviate this result it has been suggested that the communication between the pipe and air-chamber should be nearly, but not quite, closed by a flap-valve opening toward the chamber, allowing the water to flow freely into the latter, but making its return very slow.

The most perfect remedy for the difficulties of regulation in long pipe systems is one which dispenses with any necessity for regulation, viz., a sufficient storage-battery, which absorbs the surplus power when in excess of requirements and gives out power when the demand exceeds the supply, thus allowing the wheels to run with absolutely uniform discharge.

Hydraulic Compressed-air Power-house.—The method of compressing air by the direct action of water already summarily described (page 472) has been applied at Magog, P. Q., and is now in course of installation at Norwich, Conn. (1900). It is therefore entitled to a place in a description of methods of developing water-power. Being, moreover, adapted for the generation of power on a large scale, it may properly be considered under the head of Power-houses.

Figs. 228 and 228*a* show the application of the method as proposed by the writer. Fig. 229 is a section of the apparatus used at Magog, P. Q. The water is here supplied through a penstock and enters a circular plate-iron tank, which we may, by analogy, call the flume. From this flume a circular pipe 3 feet in diameter descends vertically into a pit, to a depth implying a pressure of 52 pounds per square inch. This pipe discharges into another tank, larger than the first, covered on top and open at the bottom, which is fixed some 2 feet above the floor of the pit. This we call the air-chamber. It has a diaphragm, not shown, the purpose of which is to detain the water in the chamber and facilitate the elimination of the air.

In the flume there is a separate pipe some 10 feet long, with a wide flaring mouth at top, telescoping into the descending pipe, and adjustable vertically by a screw and hand-wheel seen above the flume.   Above the mouth of the telescoping pipe is

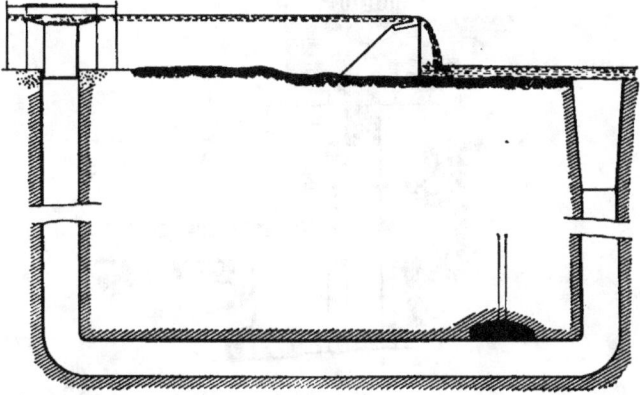

Fig. 228.

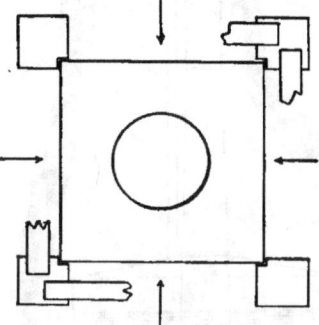

Fig. 228a.

a circular frame bearing a system of short vertical 2-inch tubes, the upper ends of which are open and the lower ends dip into the water.   The part of the tube immersed in water is provided with a number of $\frac{3}{8}$-inch orifices, in which small pipes 6 or 8

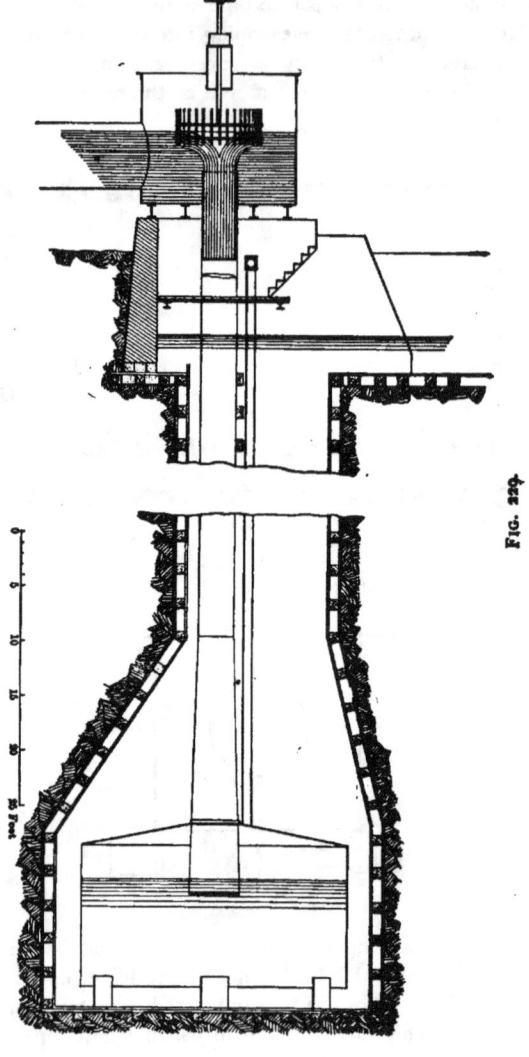

FIG. 229.

inches long are inserted all horizontal and pointing inward, i.e., toward the centre of the 3-foot pipe. The water passing the system of pipes in rapid movement, the pressure at the free ends of the small pipes is less than that of the atmosphere, and air in small bubbles enters the water and is carried downward. The pit toward the bottom is enlarged for the accommodation of the air-chamber, and from the summit of the latter a 7-inch pipe leads upward and, after passing the level of the tail-race, turns horizontally toward the mill requiring power.

Operation.—The water is set in motion by lowering the telescopic pipe, the position of the mouthpiece determining the quantity of water flowing through the system. Air enters through the small pipes and is carried downward into the air-chamber. The water, supposed to be freed from air, passes under the rim of the air-chamber, rises through the pit, and flows away through the tail-race. The air accumulates in the chamber under a pressure represented (when the water is not moving) by the depth of the surface in the air-chamber below the surface in the tail-race.

Figs. 228 and 228*a* show the system already partly described (page 472). We may add that these figures contemplate a telescoping pipe entering the descending shaft, carrying a hopper-shaped mouthpiece, which we will call the *hopper*, and adjustable vertically by mechanism not shown. The hopper is guided in its vertical movement by four piers shown in Fig. 228*a*. These form four channels of approach through which four currents of water enter the hopper and in meeting cause the necessary commotion. It was clearly shown by the writer's experiments at Minneapolis, in 1880, that throwing the water into commotion is a perfectly effective mode of impregnating it with air. Any arrangement of pipes for this purpose appears to the writer entirely superfluous, as well as very objectionable. The pipes could not fail at times to become clogged with floating débris to the extent of wholly arresting the flow of water. In the arrangement of Fig. 228, trash of every kind, blocks of ice, and even logs could go through the system without embarrassment or interruption, not even a rack being required. In this disposition the position of the hopper determines the quantity of water drawn. The quantity of air regulates itself and is neither more nor less than the given quantity of water can carry. If the descending column is so loaded with air that it does not preponderate sufficiently over the ascending column, the water in the former will rise, the commotion will diminish, and less air will enter. In the contrary case the water falls, commotion increases, and more air is taken in.

It may be useful to attempt an approximate estimate of the efficiency of this method upon admitted principles of hydraulics. Instead of attempting a general discussion of the matter we will assume specific data as follows: Circular channels lined with brickwork with uniform diameter of 10 feet; a velocity of 8 feet per second; a head of 24 feet; a pressure of 100 pounds net, i.e., 100 pounds above the atmosphere. We assume the ascending shaft to widen so as to reduce the velocity of efflux to 4 feet.

1. Fall at influx, or that required to impregnate the water with air. The arrangement of Fig. 228 is more efficient in this regard than that used at Minneapolis, in which this loss did not exceed.................................... .... 1.00

2. The head due the velocity of 8 feet would be 1.00 foot, but we need only consider the head due the velocity of efflux, which is............................... 0.25

3. Loss due to friction in the channels $= h$. We will use the formula $v = c \sqrt{rs}$. We will call the total length of the channel 650 feet, so that $s = h/650$. We have $r = 2.50$. Experiments on the Sudbury River conduit pertaining to the Boston Water-works show that for a surface of brickwork carefully laid c may be taken $= 140$. We have therefore $v^2 = 64 = 19\,600 \times 2.5 \times \dfrac{h}{650}$, whence $h = \dfrac{64 \times 650}{19\,600 \times 2.50} = $ 0.85

4. Loss from bends. The descending passage joins the horizontal one by a curve whose radius of curvature

cannot much exceed the radius of the circular passage. In this case we may, according to Weisbach,* put the loss of head = head due velocity = ... 1.00

The other bend may have a longer radius, and the loss need not exceed 0.50

5. Loss from the slip of the air-bubbles, i.e., from their velocity of descent falling short of the velocity of the water. It was a fair inference from the experiments at Minneapolis, in 1880, that bubbles of air, such as result from commotion, tend to rise with a velocity not exceeding 0.80 foot per second. In fact this figure is probably in excess, since their velocity is as the square root of their diameter, and their diameter diminishes as they descend. The percentage of loss, therefore, from this source is 0.8/8 = 10 per cent, which represents a head of........ .... 2.40

We have therefore a total loss thus far of....... ... 6.00

which is 25 per cent of the total head, but this is not the whole loss.

Loss due to Solution of Air.—We must now consider another loss of a more complex character resulting from the physical relations of air and water. At a temperaure of 32° F. a cubic foot of water in contact with air, at any pressure whatever, dissolves .049 of the oxygen and 0.0235 of the nitrogen contained in a cubic foot of the air. These gases so dissolved appear to assume the liquid form and remain inseparable from the water so long as the pressure and temperature continue. A change of temperature, within the natural range does not affect the phenomenon materially. A diminution of pressure leads to immediate elimination of air in the form of bubbles so minute that they are only discernible as a milky hue in the water. The oxygen forms substantially 0.21 and the nitrogen 0.79 of the atmosphere, so that each cubic foot of water takes up and holds in solution 0.049 × 0.21 + 0.0235 × 0.79 =

---

* Mechanics, Coxe's Translation, 1870, vol. I. p. 898.

0.0288 cubic foot of air under whatever pressure the air may be. This law,* it may be remarked, has not been verified by experiment at pressures above two atmospheres, but all analogy would lead us to suppose that it holds good for higher pressures, and we shall consider the subject on that assumption.

The water carrying the air down the shaft is in a condition eminently favorable to the absorption of the latter, and we must assume that it passes the air-chamber with all the air in solution that it can hold at that pressure. Under a pressure of 100 pounds net and temperature of 32° F., air weighs at the rate of 1.59 cubic feet per pound, and a cubic foot of water would contain 0.0288/1.59 = 0.0181 pound of air. We have seen (Table 9) that 1 pound of air per second at 100 pounds corresponds to 96.5 h.p., therefore each cubic foot per second carries past the air-chamber air representing $96.5 \times 0.0181 =$ 1.75 h.p. At one-half the pressure the quantity of air by weight would be but half as great, and being under but half the pressure would represent but one-fourth the power. We may therefore assert that the power represented by the air carried off uselessly is as the square of the pressure and is independent of the working-head.

But the air thus compressed without result is by no means to be accounted as a loss. Air is carried down the descending shaft at the expense of the working-head until it dissolves in the water. It passes the air-chamber, and in the ascending shaft is eliminated in the same order as it dissolves in the descending shaft, and it aids the movement in ascending as it retards it in going down. Its effect would be null if the energy exerted by it in rising were equal to that expended on it while going down. To the extent that the former falls short of the latter, the action involves a loss.

When the air starts on its descent, that portion destined for solution in common with the rest, undergoes a loss at the rate of 10 per cent. The loss on this portion diminishes as the air

---

* Roscoe and Schorlemmer's Chemistry, 1894, vol. I. p. 284.

enters into solution, and at the bottom of the shaft ceases entirely. We may therefore put the average loss during the descent of the air at 5 per cent. During the ascent of the water the loss is represented by the excess of the velocity of the air over that of the water; that is, by the ratio of the velocity with which the eliminated air-bubbles rise in still water to the velocity of the water. The velocity with which air-bubbles rise is approximately as the square root of their diameter. Bubbles resulting from commotion are perhaps $\frac{1}{4}$ inch in diameter. Those resulting from elimination are immeasurably small—less than $\frac{1}{100}$ of an inch at any stage of their ascent. Therefore, if the former rise 0.8 foot per second, the latter cannot rise more than 2 inches. These considerations would indicate a loss near the efflux of $\frac{2}{16}$, say 2 per cent. The loss being 0 at the bottom, the average is 1 per cent. We are not far wrong, therefore, in putting the total loss due to solution of air at 6 per cent of the power represented by the air which enters into solution. In the case supposed the dissolved air represents 1.75 h.p. per cubic foot of water per second, and the loss is $0.06 \times 1.75 = 0.105$ h.p. With an efficiency of 0.75, 1 cubic foot per second on a head of 12 feet is about 1 h.p. The loss therefore represents a head of $0.105 \times 12 = 1.26$ feet of head, and the total efficiency of the system is $(24 - 7.26)/24 = 69.75$ per cent.

The power represented by the dissolved air being as the square of the pressure, the loss due to solution follows the same proportion. For a pressure one-half that assumed above, other elements being unchanged, the loss would be 0.315 foot; for double pressure over 5 feet. Should pressures of 400 or 500 pounds be attempted this loss would assume overwhelming proportions.

The statement that the power represented by the dissolved air is as the square of the pressure, though sufficiently near the truth for our present purpose, is not strictly correct, because the power represented by a given weight of air at 100 pounds is not twice as great as at 50.

CHAPTER XXIII.

MEASUREMENT OF WATER.

In the execution of projects for development of water-power, and in the control and management of water-powers, the necessity is constantly arising for determining quantities of water. The measurement of the absolute volume of water in a reservoir of any kind is a matter of little difficulty, being a simple geometrical computation. The question is usually as to the flow of water, i.e., the quantity of water passing a fixed point in a given time, involving the unit of time as well as the unit of volume. The second is usually taken as the unit of time, and the cubic foot as the unit of volume, in all questions relating to water-power, the problem being to determine the number of cubic feet of water flowing in one second. Two general classes of methods are available: 1. The quantity of water flowing in a channel is determined when the cross-section of the channel and the average velocity are known, the quantity being the product of these two. In operations of this class the important point is the accurate measurement of velocities. 2. When water can be discharged through an orifice either closed on all sides or open at the top, in which latter case it is called a weir, the size of the orifice, the head acting thereon, and certain modifying conditions determine the quantity of water. The critical part of such measurements is the accurate determination of the head which generates the velocity.

Measurement of Velocities. Floats.—An approximate idea of the flow of a natural stream may be obtained by surface-floats, that is, by placing in the water chips of wood or any

526

objects that will float, and observing the time occupied in traversing a given distance. These observations, combined with soundings and measurements for determining the cross-section of the stream at the place, will give the means of ascertaining, with an approach to accuracy, the quantity of water flowing.

Surface-floats give the velocity at the surface of the stream. In point of scientific nicety the float takes a velocity a little greater than that of the water at the surface. To understand how this may be, suppose a stream with an inclination of 1 foot in a mile. If we could suppose the water to stand at this inclination without motion, we can readily see that a float placed in it would move down-stream. The inclination of the surface is a force tending to produce motion, and a floating body obeys any force that acts on it, however slight. Now the forces acting to give motion to the float would not be affected by the movement of the water. Therefore the float will take a velocity somewhat in excess of that of the water in which it floats. Practically it is not necessary to take account of this refinement, and we may regard the velocity of the float as representing that of the water; but it is very necessary to remember that the surface velocity is considerably in excess of the average velocity of the stream. The average velocity in any vertical line is generally from 80 to 90 per cent of the surface velocity. The surface velocity in midstream is also considerably in excess of that nearer the shores.

Submerged Floats or Double Floats.—To determine the velocity at depths below the surface, observers have often used a float susceptible of sinking, but maintained at the desired depth by a cord attached to a surface-float, the velocity of which is assumed to represent that of the submerged float. Within moderate limits this assumption can, no doubt, be admitted, but it is sometimes carried to an unwarrantable length. For instance, in the early surveys of the Mississippi, where the cord uniting the submerged with the surface float presented more surface to the action of the current than the float itself, the assumption becomes entirely untenable. Such are the resources of modern engineering that, except for very approximate operations, floats whether double or single need not be used beyond very moderate depths. There is a certain advantage in mid-depth floats, for the reason that the ratio of the mean to the mid-depth velocity is more free from uncertainty than its ratio to the surface velocity. The mean velocity in any vertical may be taken at 95 per cent of the mid-depth velocity. The mid-depth velocity moreover is entirely free from the influence of wind, whereas the surface velocity is much affected by that cause. In measurements on such small streams as usually come under the notice of the engineer in questions of water-power and water-supply, the writer finds nothing better for the submerged float than a potato united by a fine cord with a block of dry pine wood $1\frac{1}{2}$ inches square and $\frac{1}{4}$ inch thick. The former is heavier than water, but not enough so to require a large block of wood to sustain it. Where the distance becomes great enough to make the surface-float indistinct, it can be made conspicuous by inserting a peg carrying a bit of red cloth or paper.

Mode of Measuring a Natural Stream by Floats.—Find a straight and regular reach of the stream with a gentle uniform current, not obstructed by rocks, weeds, drift, or overhanging trees. Measure a base line, parallel to the course of the stream and as near the same as convenient. This need not exceed the width of the stream in any case, and for large streams is much less than the width. For the sake of a definite rule, make it, say, five times the square root of the width. Mark the extremities, and lay off at each a line at right angles to the base reaching across the stream (Fig. 230). These are called the upper and lower transit lines. On these lines stretch cords across the stream, and affix tags of tin or pasteboard to the latter at convenient intervals, each tag bearing a number showing its distance from the left bank,* the base being prefer-

---

* In speaking of the right or left bank of a stream the observer is supposed to be facing down-stream.

ably on the right bank. Measure the depth at each tag.
Take the average of each two corresponding depths and therewith draw a profile or cross-section of the bed, as in Fig. 231.

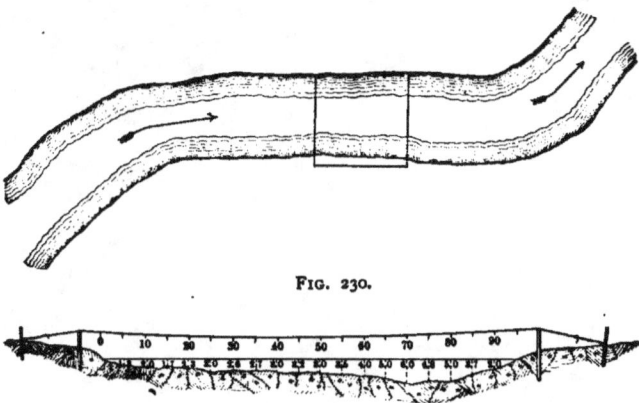

FIG. 230.

FIG. 231.

The assistant who launches the floats, from a boat or wading,
has these soundings and adjusts each float to mid-depth.   He
also observes the distances at which each float passes the
transit lines.   It is not worth while to take the trouble of
recovering these floats after passing the lower transit.   The
assistant who keeps the record notes the time occupied by the
float in passing the transit interval, and when this is recorded
he receives from the other assistant the distances and records
the same.   The recording assistant is preferably provided with
a stop-watch, but an expert observer can do very well with a
common watch beating four times a second, by keeping the
watch to his ear and counting the beats.   The following table
in connection with Figs. 231 and 232 shows the record and
computation of such a measurement.   The diagram Fig. 232
is plotted with the distances in col. 6 as abscissas and the
velocities in col. 8 as ordinates.   Col. 10 contains the velocities
read from the diagram for each 5 feet of width, though for

TABLE 16.

Number of Float.	Depth of Submerged Float.	Distance from Left Bank.				Time occupied by Float in moving 50 Feet. Sec.	Velocity Feet per Sec.	Partial Area for Each 5 Feet Width. Sq. ft.	Velocity for Each 5 Feet Width. Fig. 154 b. Feet per Sec.	Discharge for Each 5 ft. Width. Cu. ft. per Sec.
1	2	Intended.	At Upper Transit.	At Lower Transit.	Mean.	7	8	9	10	11
1	0.6	5.0	3.8	4.8	4.3	36.4	1.37	3.25	1.30	4.22
								8.25	1.43	11.80
2	0.8	15	16	13.3	14.6	33.7	1.48	9.25	1.47	13.60
								10.00	1.49	14.90
3	1.5	25	27.1	22.6	24.8	34.0	1.47	13.25	1.47	19.48
								14.00	1.44	20.16
4	1.3	35	34.2	36.3	35.2	32.9	1.52	13.25	1.46	19.34
								14.25	1.60	22.80
5	1.6	45	42.2	49.6	45.9	27.5	1.82	15.50	1.74	26.97
								15.50	1.86	28.83
6	1.7	55	52.3	55.1	53.7	25.1	1.99	16.25	1.97	32.01
								18.75	2.07	38.81
7	2.5	65	67.1	62.3	64.7	22.4	2.23	22.50	2.21	49.72
								27.50	2.23	61.32
8	2.4	75	72.9	78.3	75.6	24.6	2.03	27.00	2.13	57.51
								24.50	1.97	48.26
9	1.8	85	85.0	87.6	86.3	29.1	1.72	21.75	1.86	40.45
								14.25	1.66	23.65
								5.00	1.30	6.50

Aggregate uncorrected discharge............... ....|540.33

True discharge of stream 540.33 × 0.95 = 513.31 cubic feet per second.

greater exactness they might be read to each single foot. Col. 9 contains the area of cross-section to which each velocity

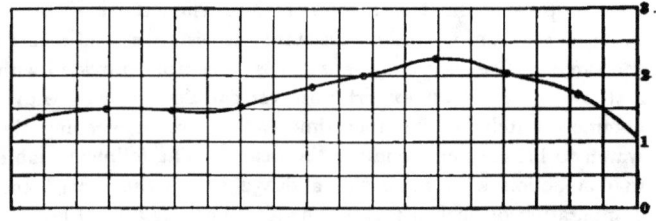

FIG. 232.

in col. 10 may be supposed to apply, and col. 11 the product of the corresponding quantities in 9 and 10, being the uncorrected quantity of water passing the partial areas. The aggregate of col. II is multiplied by 0.95 for the true discharge of the stream. In the application of floats to deep and wide rivers, too wide to distinguish the float with the naked eye, and especially to navigable rivers, the transit lines cannot be marked by ropes or cords. In lieu of these a theodolite is positioned at each end of the base line and is used to mark the transits and positions of the floats. The base line here has a length of 100, 200, or 300 feet. The floats are launched from a boat anchored in the stream, and, if they are too elaborate and expensive to be sacrificed, a second boat is kept in waiting below to pick them up. When the float is launched and approaches the transit line, the up-stream observer has his instrument clamped on that line; the down-stream observer keeps his instrument focussed on the float. When the float crosses the line the up-stream observer utters a shout. Thereupon the down-stream observer reads the angle which determines the position of the float at its upper transit, and the time-keeper starts his stop-watch. The down-stream observer then clamps his instrument on the lower transit line, and the up-stream observer focusses his instrument on the float. At the down-stream transit the down-stream observer utters a shout, the up-stream observer reads his angle and the time-keeper stops his watch, etc. The time-keeper may be charged with the duty of recording all the observations, though this is sometimes made the duty of a separate assistant. A system of signals is used for communicating with the boats. The submerged is connected with the surface float, preferably by a copper wire, presenting but little surface to the current, and special devices are necessary in launching the float to prevent the latter from becoming kinked and snarled. In the Mississippi River Survey of 1850–60,[*] the submerged floats were kegs without top or bottom, 6 to 12 inches diameter and 9 to 15 inches high, weighted with hoops of lead. The surface-

* Report of Humphreys and Abbott, 1861, p. 224.

floats were 6 or 8 inches diameter, 3 inches deep, of cork or pine wood. Sometimes tin vessels were used of ellipsoidal shape. They were connected with the submerged float by cords one to two tenths of an inch diameter.

Floats reaching nearly to the bottom of the channel are used in artificial channels of uniform depth. They consist of loaded poles or of closed hollow tubes carrying heavy weights at the lower end. Preferably they are stout tubes of tin 2 inches diameter, loaded with lead to give them the desired immersion. The upper end projects 6 or 8 inches above the water, and is closed with a cork; this part is painted red to make it conspicuous, and bears figures indicating its effective length, i.e., its depth of flotation. This method is used where water-measuring is carried on in a regular and systematic manner as incidental to the control and distribution of water-power. A portion of the channel through which the water reaches its destination is fitted with smooth vertical and parallel sides, and level bottom of planking. This is called a measuring-flume. The transit lines are marked by square timbers laid across the flume, the up-stream face vertical and marked with conspicuous figures showing distances from left side of flume. To facilitate computations of velocity the interval between the up-stream faces is accurately adjusted to a round distance of 50, 80, or 100 feet. A foot-bridge consisting of two square timbers is laid some 15 feet above the upper transit, and another the same distance below the lower one. The upper bridge also carries figures indicating distance from left side. To conduct such a measurement requires four persons: 1. The person in charge of the work, who is supposed to be a trained assistant; 2. An intelligent workman who launches the floats; 3. A laborer who catches the floats and removes them from the water; 4. A laborer who takes the floats from No. 3 and carries them to No. 2. A gauge at the flume shows the height of water and indicates the length of tube required. Assistant No. 1 places himself at the up-stream transit on the right bank. Stop-watch and note-book in hand, pencil in mouth; the note-book ruled with appropriate columns, the date, time of commencement, height of water, and length of tube used entered therein. Assistant No. 2 on the upper bridge launches the floats at regular intervals, guided by the figures on the bridge. This is a movement requiring some skill. Facing down-stream and resting on his knees, he passes the tube into the water in a slanting direction, throwing the bottom end well up-stream under the bridge, holds it by the top till the current brings it into an erect position, then abandons it, being careful to release it at its proper depth of immersion, otherwise it would oscillate and perhaps strike the bottom. No. 1 starts his watch at the crossing of the upper transit, records the distance of the tube from the left side, which is called to him by No. 2, and walks briskly to the lower transit station. He stops his watch when the float crosses the lower transit, and, after he has recorded the time interval, receives from No. 2, and records, the distance from the left side at the lower transit; then returns to the upper station. Where the velocity exceeds 5 feet per second this method becomes too trying for No. 1, and it is better to have a fifth man to keep the record.

Sources of Error in Measuring with Loaded Tubes.—Mr. James B. Francis,* who developed and applied this method in the management of the Lowell Water-power, examined carefully the several circumstances which tend to make the velocity of the tube different from that of the water in which it floats. These are:

1. In the most regular channel the length of the tube is less than the depth of the water. In an irregular channel or one liable to have deposits and obstructions this difference is considerable. The slowest-moving water does not have its full effect upon the velocity of the tube, and the latter is somewhat greater than that of the water. As the result of a long series of experiments, Mr. Francis determined the following correction for this source of error:

---

* Lowell Hydraulic Experiments, 1867.

Let d be the depth of water and d_1 the immersion of the tube;

q_1 the quantity of water as determined by the floats;

q the corrected quantity.   Then

$$q = q_1 \left\{ 1 - 0.116 \left(\sqrt{\frac{d - d_1}{d}} - 0.1 \right) \right\}. \quad . \quad (66)$$

This formula was later revised by Mr. Francis, and appears in the fourth edition of his work under the following form:

$$q = q_1 \left(1.012 - 0.116 \sqrt{\frac{d - d_1}{d}} \right). \quad . \quad . \quad (67)$$

2. The velocity of the filaments of water below the surface does not diminish uniformly with the depth.   The pressure exerted by moving water on a solid body is not proportional to the velocity, but rather to the square of the velocity.   It results from these laws that, while the float takes such a velocity as equalizes the pressure on opposite sides of it, it does not take the mean velocity of the water in which it is immersed. Mr. Francis finds from a mathematical investigation that, when the mean velocity of the water is 2.6 feet per second, the velocity of the tube would fall short of that of the water by about the $\frac{1}{85}$ part.

3. For the reason already alluded to (page 527) the velocity of the float would tend to exceed that of the water. The labors of physicists who have investigated this question show that in the case supposed this tendency would be equal to the $\frac{1}{85}$ part of the mean velocity of the water.   That is, Nos. 2 and 3 would practically offset each other.

4. A heavy body suddenly immersed in water and floating therein does not instantly acquire the velocity of the latter. Its inertia must be overcome by the pressure which the water exerts on it in virtue of its relative velocity, and this pressure becomes less and less as the velocity of the body approaches that of the water.   In fact, mathematical analysis indicates that

the float never acquires the velocity of the water, though it continually approaches the same.

The result of Mr. Francis's inquiry is that if we suppose the tube to be placed in moving water and held at rest, then suddenly released, it will, after moving 20 feet, have a velocity $\frac{1}{74}$ part less than that of the water, the latter being as formerly supposed. The supposition, however, of starting from rest has no application. The tube is placed in the water in a much inclined position, generally nearer horizontal than vertical. It is held by the top till it comes into a vertical position. When it comes into that position the centre of gravity of the tube, which is near the lower end, is moving faster than the water. We may therefore neglect this source of retardation and conclude that No. 1 is the only source of inaccuracy that need be taken account of.

It is only in certain situations that floats can be employed. In a closed conduit they are obviously inapplicable, and in an open channel they require, for the accurate determination of the velocity, a considerable length of channel, which it is often difficult to obtain. In any case they require a degree of regularity in the channel not always to be found. It is often necessary, therefore, to infer the velocity from effects other than the transportation of floats. Among these effects are: 1. The pressure of the current on a square or round disk placed normal to the direction of the current. This pressure is supported by a spring and is measured by the extent to which the latter is deflected. 2. A round metallic ball, supported by a wire, is immersed in the current, and the velocity is ascertained from the angle of inclination which the wire takes under the combined action of gravity and pressure of water. If α be the angle between the wire and the vertical, and w the weight of the ball, the pressure of the water will be represented by $w \sin \alpha$. No practical method of water-measurement has been developed on the line of 1 or 2, but 2 would probably be found to present as few difficulties of application and as few sources of error as any method in use. 3. Vanes fixed upon a rotating

axis cause it to revolve under the action of the current with a
velocity bearing to that of the current a certain fixed relation,
which must be determined by experiment.   This principle has
received wide application in the current-meter, which has
latterly been more and more taking the place of other modes
of velocity measurement.   4. A vertical tube bent to a right
angle at its lower end, and immersed in the water, will, when
the horizontal extremity is turned .toward the current, be filled
to a height above the hydrostatic level.   When turned in the
reverse direction the water in the tube sinks below the same
level.   From the elevation or the depression, or from a combination of the two, the velocity is determinable.   An instrument founded on this principle, called the Pitot tube, has been
much employed in the measurement of velocities.   Passing over
methods 1 and 2, as mainly speculative, we will first consider

The Current-meter.—This instrument takes two general
forms, viz., cup vanes and helical vanes.   In the first, four
cup-shaped vanes, Fig. 233, are arranged around an axis.
The current acts upon the convex side of the vane during one-
half its rotation, on the concave side during the other half.
The pressure exerted on the concave side greatly exceeds that
on the convex side.   The system of floats will continue to
rotate with the convex side forward.   Fig. 233 shows a meter
of this form as made by Buff and Berger of Boston.   It is
shown as arranged for deep-water observations, being adjusted
upon a rope suspended from an anchored boat, and carrying a
heavy weight which rests upon the bottom and keeps a strain
upon the rope.   For shallower channels the meter is mounted
on a rod consisting preferably of several pieces of brass tubing
screwed together and capable of being separated for convenient
packing in a box.   The instrument is shown as rigged for
registering the revolutions by electricity, the clockwork and
dial as well as the voltaic cell which generates the electric
current being in the boat or on the bank.   The electric current
goes through the rotating axis, one of the wires being held in
contact with it.   A bit of non-conducting material is inserted

in the axis, and comes in contact with the wire at each revolution, breaking the circuit and advancing the prime-mover wheel of the clockwork by one tooth. The makers say: " This form was used upon the gauging of the Connecticut River by General Ellis, and was designed particularly to avoid the catching of floating substances, such as leaves and grass, upon either the

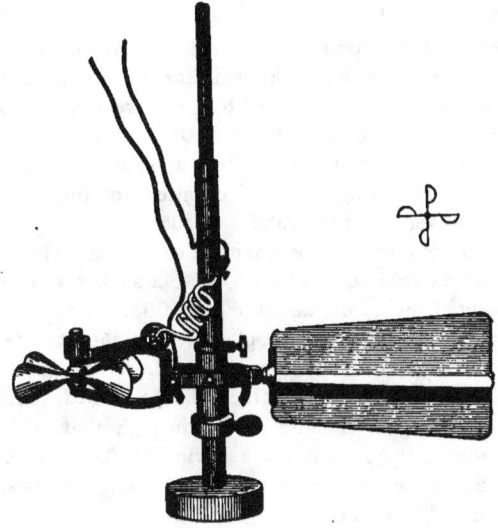

FIG. 233.

vanes or the axis, and to render the record of the instrument independent of the position of its axis with respect to the line of the current; also, to get less friction upon the axis so as to measure low velocities accurately.

" This current-meter is constructed upon the principle of Robinson's anemometer, turning by the difference of pressures upon opposite vanes of the wheel. The vanes of this meter, however, instead of being hemispherical cups with a straight stem, are made conical at the ends, and are made hollow and taper to the central hub, so as to offer no obstruction to the slipping off of straws, leaves, or grass as the wheel revolves. The central hub is made tapering so that any object can slide off easily, and it extends over the joints at the ends of the axis, so as to enclose and protect them from floating substances. The axis runs in iridium bearings. The forward end of the frame which carries the wheel can be turned and secured in any position, so that the wheel can be horizontal, vertical, or at any desired angle.

" The electrical connection is made by carrying an insulated wire from near the centre of the instrument, where the insulated wire from the battery is attached to it by a binding-screw when in use, out to the end of one arm of the wheel-frame, where it ends in a fine platinum wire resting upon a ring in the hub of the wheel. This ring is made of alternate interchangeable sections of silver and hard rubber, secured in place by screws, so that their position can be changed to register whole or part revolutions as desired. There is also a socket and set-screw in the body of the frame, near the centre, for the return current, which can be carried through a plain wire slightly twisted round the insulated wire so as to form one cord. If the instrument is run upon a wire or has a metallic connection with the surface, the return current can be made through that. A better method now in vogue is to use a 'twin' insulated wire."

Fig. 234 shows a current-wheel with helical blades, giving motion to a horizontal shaft. The blades join a hub on the shaft, and are limited by an exterior rim. In some cases this rim is made double and encloses a vacant space giving buoyancy to the wheel and relieving the bearings of pressure except the thrust. The wheel is encircled by a stout frame to protect it from injury. A small bevel-gear on the shaft gives motion to a train of wheelwork enclosed in a box *A* with a glass cover. The ratchet-wheel shown is moved, one tooth at a time, by a spring-catch to which is attached a cord reaching to the surface of the water. A pull on the cord throws the clockwork out of gear, and the next pull throws it in. The meter is shown as attached to a rod. The observer stands on a bridge or temporary platform over the watercourse. After noting the reading of the dial he places the meter in the water at the required depth, guided by a graduation on the rod, with the clockwork out of gear. An assistant holds the meter in

FIG. 234.   FIG. 234a.

place. A stop-watch is commonly used for noting the time, but an ordinary watch does very well. At the commencement of an even minute the observer pulls the string to throw the wheelwork in gear. After the lapse of a suitable number of seconds he throws it out, raises the meter, and notes the dial.

As indicated by dotted lines the instrument admits the attachment of electric wires when required.

An Acoustic Meter.—The writer, while in the service of the U. S. Engineer Department, had occasion to observe the working of current-meters. He became convinced that their chief source of difficulty and error lay in the means used for conveying their indications to the observer. Especially is this the fact when the revolving shaft operates a train of wheel-work from which sediment cannot be excluded. The attempt to exclude it involves the use of a stuffing-box, which is in itself a source of uncertain and varying resistance. The electric meter, also, owes its chief difficulties to the same cause. The necessity of being accompanied by attachments of wire, battery, and dial is a serious limitation on its usefulness, and the maintenance of the current, unless in the hands of trained electricians, is a source of continual trouble, vexation, and delay. It appeared, therefore, that the meter most suitable for ordinary uses, and to be entrusted to ordinary assistants, would be the one most free from attachment and appendage. The ear appeared to be just as accurate a receiver of these indications as the eye, and the metallic rod sustaining the meter a perfectly satisfactory channel for their conveyance. If a sensitive ear be applied to a hollow brass rod 15 feet long, it is hardly possible to touch the other end with a pin so lightly that it cannot be heard. These considerations determined the form of the meter to be used on the headwaters of the Mississippi. It consisted simply of the wheel, Fig. 234, with its frame and rod, divested of gear, ratchet, dial, wheelwork, string, and wire. The shoulder-bearing of the shaft was formed as in the magnified sketch, Fig. 234a, in a helical form, so that the shaft carrying the meter advanced a little during its revolution and at the completion of a revolution dropped back, under the pressure of the water, with a click. The advance need not be more than $\frac{1}{64}$ of an inch. The click was more pronounced than necessary, being audible 50 feet away. There was no difficulty in counting as high as six turns per second or as low as one turn in 10 seconds, the velocity ordinarily met with in running streams being from one to three turns per second. Several meters of this form were made and were used exclusively in the measurements of the Mississippi, Crow Wing, and St. Croix rivers, which continued daily from January 1 to December 31, 1882. Hardly any delay was experienced from the derangement of the meters, a fact which will be the more readily believed when we remember that there was nothing about them susceptible of derangement.

Rating of Current-meters.—For ordinary purposes the inclination of the screw-blades is made such that a turn of the wheel implies a movement of 1 foot of the water. The maker usually verifies this adjustment, but it cannot be depended on as permanent, and the observer cannot feel assured that wear, sediment, and the action of water will not change it. It is imperatively necessary to verify or revise the adjustment from time to time. This is usually done by moving the meter at an accurately determined velocity through still water.

Fig. 235 indicates the arrangement used for rating the acoustic meter just described. A round float was made, 12 feet in diameter, and capable of sustaining 1000 pounds or more. Across this was laid a stout timber of greater breadth than depth, reaching a few feet beyond the float on one side and to a distance of about 16 feet from the centre on the other. The projecting part tapered to small dimensions at the outer end. The float was anchored in the vicinity of a high railroad trestle crossing a stagnant bayou of the Mississippi, by driving a stout iron rod through a hole in the timber at the centre of the float, deep into the bottom. The rod rose several feet above the water, and a hollow wooden tube was slipped over this part of the rod. The rod rose considerably higher than shown, and was braced from the trestlework. The bottom of the tube was made fast to the timber, which was fastened to the float, and the upper end carried a round drum or pulley on which a cord was wound. This cord extended horizontally to the trestlework, passed over a pulley, and up to the bridge floor, there over another pulley, and sustained a weight as shown, the action of which caused the float to revolve. By changing the weight the velocity could be varied in any desired ratio. The observer sat on the float near the middle, counted

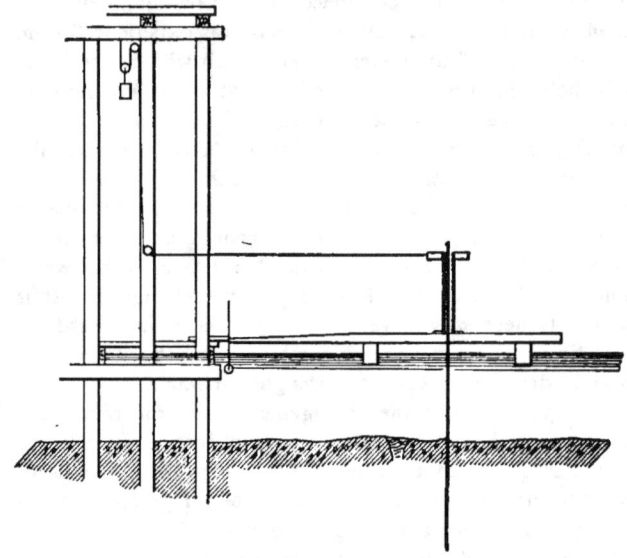

FIG. 235.

the clicks of the instrument, and noted the time of a revolution. The distance of the centre of the meter from the centre of the float was such that the former described exactly 100 feet at each revolution. The float was not revolved more than one turn at a time, as the rotatory movement imparted to the water was found to affect the velocity of the meter on the second and subsequent turns. The extremity of the timber carrying the meter stopped against a small wooden bar. It started when this bar was momentarily removed, and brought up short against it at the end of a revolution. 100 divided by the number of clicks in a revolution of the float represented the distance traversed in one turn of the wheel. Except for very low velocities it was found that the number of clicks in a turn of the float was the same whether the velocity was fast or slow. When the weight ran down it was raised by revolving the float in the reverse direction, which was easily done by the attendant grasping a rope attached to a timber of the bridge and walking on the circular float.

The Pitot Tube.—If a tube open at both ends be held vertically in a running stream, the water will stand within the tube a little lower than outside. If the lower end be bent to a right angle and the horizontal part, called the ajutage, be directed up-stream, the water will stand considerably higher inside than outside. If the ajutage be turned at right angles to the current or down-stream, the water will stand lower inside than outside. In each case the difference in level between the water inside and outside the tube has a definite relation to the velocity of the current at the ajutage. As it is very difficult to measure the level of water in a running stream with sufficient accuracy for this purpose, the instrument usually consists of two tubes, the first opening up-stream, the second straight and opening downward or with an ajutage opening at right angles to the current or down-stream. The difference of level in these tubes is a correct indication of the velocity of the water. The most important part of the instrument consists in arrangements for the accurate measurement of this difference of level. These tubes are of glass above, joined to copper below, the former being fixed upon an accurate scale. The glass tubes open into a bulb or vessel communicating with the atmosphere through a cock, and a cock controls the access of water to the copper tubes. To admit of accurate measurement the level of the water in the glass tubes is raised by suction, the simplest means being a nipple or mouthpiece on the bulb before mentioned, to which the mouth is applied. The lungs are readily capable of creating a vacuum of 3 feet. This being done and the bulb excluded from the air by the cock, the two columns stand at a height convenient for measurement, and their relative level is in no way affected since the pressure of the air is the same in both. The measurement is still further facilitated by closing the lower cock at the proper moment, after which the two columns are cut off from external influence and are perfectly still. Accurate observations may be taken with the glass tubes and bulb wholly immersed in the water, in which case the air above the water is condensed without affecting the relative height.

In this instrument, as in the current-meter, it is the rating that gives validity to the indications. A very slight difference in the ajutage makes a material difference in the rating, so that the rating of one instrument is of no value for another. The most convenient mode of rating is by comparing the surface velocity of a stream as determined with the instrument with the same as determined by surface-floats, taking due account of the source of error in the velocity of surface-floats already pointed out (page 527). The method of rating by causing the instrument to move in still water is more accurate in theory, but beset with greater practical difficulties.

This instrument was used by the French experimenters Darcy and Bazin * in determining the laws of fluid motion, and is well adapted to uses requiring extreme nicety. It is probably not destined to extensive application in the practical work of water-measurement.

Measurement of Heights of Water.—No operation in water-power engineering is more common or more necessary than the accurate observation of heights of water. In measurement of water by means of weirs and orifices the height is the main element. A record extending over a period of years, of the height of water in a stream, gives the means of ascertaining the flow with an approximation to accuracy. In legal questions which are constantly arising as to the flow of streams, amplitude of floods, value of water-power, etc., no data obtainable at equal expense and trouble are so valuable as accurate gauge-records.

* Recherches Hydrauliques. Paris, 1865.

A gauge designed to be permanent should be of cast iron, bolted solidly to masonry or natural rock. It should be 5 or 6 inches wide and an inch thick. As such gauges are intended to be read by unskilled persons, they are usually graduated in feet and inches as indicated in Fig. 236. The gauge is shown

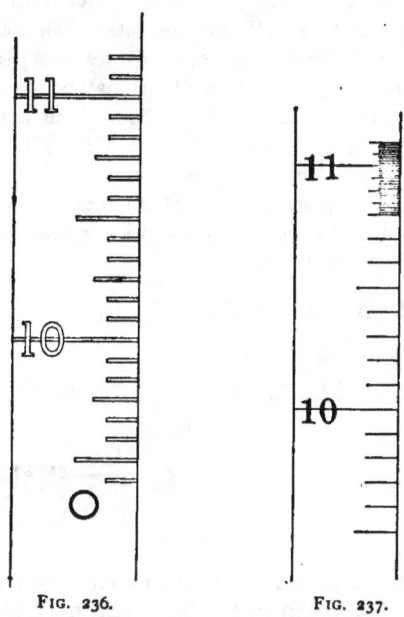

FIG. 236.   FIG. 237.

6 inches wide. The inch-marks are $1\frac{1}{2}$ inches long, the quarter-foot marks 2, the half-foot 3. The foot-marks extend across the face of the gauge. It is thus not necessary to attach figures to any but the foot-marks, the inches being readily discerned by the length of the marks. Fig. 237 is a wooden gauge graduated in tenths and hundredths. In marking the graduation a line is drawn 1 inch from the edge, a second $1\frac{1}{4}$ inches, a third $1\frac{1}{2}$, a fourth 2. The hundredth marks extend to the first line, the half-tenths to the second, the tenths to the

third, the half-foot marks to the fourth, and the foot-marks
entirely across the face. The gauge is generally painted white,
and the marks and figures put on in black with a brush, the
latter with stencil plates. A more permanent method is; first,
to oil the wood, then cut the marks with a sharp instrument,
then rub the face with powdered charcoal mixed with oil,
working the mixture well into the cuts; then clear the surface
with a plane. For temporary purposes the graduation (not
finer than inches or tenths of a foot) may be marked with a hot
iron, using Roman numerals for figures. This makes the marks
quite conspicuous, and they are not readily obliterated.

These forms of gauges presuppose solid artificial structures
for their attachment, as dam abutments, bridge piers, river
walls, etc., for the iron, timber-work for the wooden gauges.

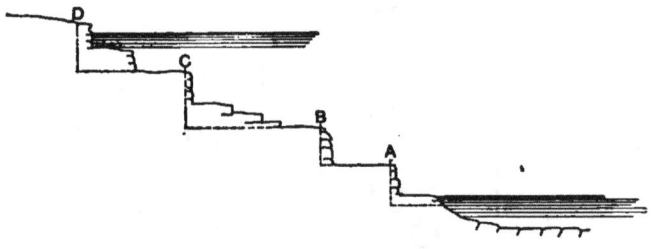

FIG. 238.

It is sometimes necessary to conduct temporary or even long-
continued observation of heights where no artificial structure
exists and where it is quite as important to record the high as
the low stages. A gauge-rod, set up on the bank of the
stream and braced to stakes and the like, is generally found
missing after any considerable rise of water, and never survives
the running of ice. Permanent arrangements for observing
the height of water in such situations are shown at Figs. 238
and 239. The first assumes a rock formation. Commencing
at the stream, a little below low-water level, cut a channel in
the rock back to *A*, where a vertical face of 5 feet or more can

be obtained. Cut such a face and bolt on the gauge. From
the level of the top of *A* cut back to *B*, and bolt on another.
From *B* go to *C*, thence to *D*, etc., till the highest possible
flood is covered. The graduation of *B* commences where *A*
terminates, *C* where *B* terminates, etc. Fig. 239 assumes a
gravel formation. Here we do not use a connected scale of
heights. We drive down the iron rods *A, B, C,* etc., each 12
or 15 feet long. The first is driven to a depth sufficient to
insure that it will never be above water, and the tops of the
others successively about 5 feet higher. The levels are carefully

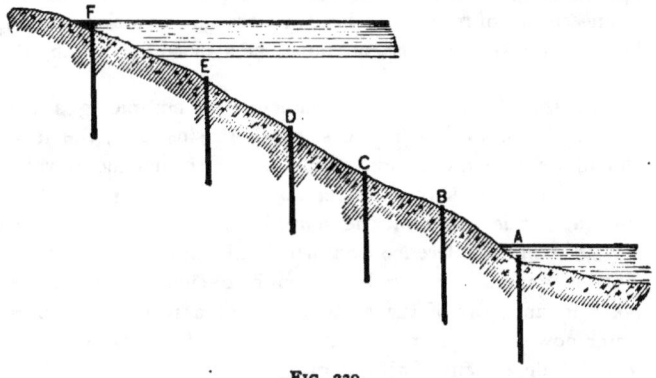

Fig. 239.

taken and recorded. The heights are taken by means of a
portable graduated rod placed on the head of the iron rod. A
discarded level-rod answers very well for this purpose. In
high water a skiff is used to observe the heights. To meet the
possibility of the malicious driving of the rods, the levels should
be repeated at intervals.

On great rivers, such as the Mississippi, Missouri, and Ohio,
floods of 40, 50, and 60 feet often occur. Excepting the con-
paratively rare case of a bridge, there are no vertical structures
to which a gauge can be attached, and the situation is usually
that of Fig. 239, which may be taken to represent the slope of a

natural bank or a levee, the latter being 2 or 3 base to 1 perpendicular. In this case the bank is dressed to a uniform slope,
sometimes paved, and the gauge-plank, some 9 inches wide and 2
or 3 inches thick, laid thereon, the face coming flush with the face
of the slope whether the latter is of earth or paved. It is confined
by stakes or long iron rods driven through it into the earth. The
several lengths of which it necessarily consists are spliced together into one continuous plank extending from low to high
water. The foot-marks of the graduation are determined by a
level and rod, and are subdivided into tenths. Figures are
applied only to the foot-marks. The graduation marks and
figures consist of rows of flat-headed brass nails or tacks. Caving
bends of the stream must, of course, be avoided in applying this
method.

Datum Plane.—In every water-power, embracing a connected system of canals, races, ponds, basins, etc., the levels
should refer to a common datum plane, so that the elevation
of any one part shows its relation to all other parts of the
system. It is common, therefore, in such works to assume a
certain plane of reference from which all heights are estimated.
This is generally taken low enough to underlie the lowest
point in any part of the system. Until a recent date every
water-power and every hydraulic work had its own system of
levels, independent of all others.

The important United States Government surveys,* the
Coast and Geodetic Survey, the Lake Survey, the United States
Geological Survey, now reckon their heights from the sea-level,
and bench-marks in this system of levels are now established
in all parts of the country, generally within reasonable distance
of every important hydraulic work. If it is a convenience to
know the relation of any one of a system of watercourses to all
other parts of the same system, it is a greater convenience to
know its relation to all other parts of the same river system and

---

* See Dictionary of Altitudes in the United States, Bulletin No. 160
of the U. S. Geological Survey. Washington, 1899.

all other parts of the country. The datum, therefore, should be determined with reference to the sea-level. At elevated points it is not necessary in every computation relative to levels to include the total height above the sea-level. It is sufficient to take the datum plane at some even hundred, or some even - fifty, or some even ten, feet above the world's bench-mark.

Precise Gauges.—When water stands in contact with a graduated gauge there is no distinct line of meeting between the water-surface and the gauge-surface. Through the action of capillary attraction these surfaces join in a curved line running up the face of the gauge. This curve obscures, in some measure, the height of the water so that the latter cannot be estimated within less than an eighth of an inch. For many purposes greater accuracy is required than is attainable by such means. In measuring the difference of level between the two columns in the Pitot tube, capillary attraction is taken advantage of. A ring-formed index, partly encircling the tube, moves in contact with a graduated scale. Capillary attraction within the tube gives a concave form to the surface of the water, and, owing to the reflection of light, the vertex of this cavity is very distinct, so that it is easy to adjust the index to the exact height of this vertex and read the height on the scale.

The Hook-gauge, Figs. 240 and 241, is an instrument invented by Mr. U. A. Boyden, and used by Mr. J. B. Francis in experiments and general water-measuring operations. A graduated rod carries a hook turning upward, with a sharp point. This hook being placed below the surface, and drawn up by a screw motion, when the point touches the surface it creates a slight distortion and a change in the direction of reflected light which is very perceptible. A vernier in contact with the graduated rod enables the height to be read to the $\frac{1}{1000}$ part of a foot. Neither this nor any other precise method of determining heights can be used upon the free surface of a stream or body of moving water, owing to continual oscillations and disturbances. Such methods must be applied to a limited space called a still-box, segregated from the main body and communicating therewith by a small opening through which abrupt oscillations cannot be transmitted, though the height of water is the same as the average height in the main body. The still-box must be high enough to include the highest and lowest levels of the water, and when, as usually happens, the level is liable to considerable fluctuation, it becomes, in the low stage, a source of inconvenience by obstructing the light. The hook-gauge, with its attachments, is bolted to a post immovably fixed in the ground or solidly attached to a permanent structure.

Fig. 242 is an instrument which the writer has found very convenient for precise measurement of the height of water. It is a portable rod accurately graduated, and is designed to measure the height of the surface above an immovable bench, generally consisting of an iron rod driven deep into the ground and surrounded by a still-box. The latter should be provided with a rest so that the rod may be plumb when placed in the water. A wooden slide clasps the rod by means of light springs, and is easily moved up and down. The slide carries a lip in contact with the rod, and an index-finger far enough from the rod to touch the water where it is unaffected by capillary attraction. In measuring, the rod is placed in the water resting on the bench; the index-finger is moistened and the slide is pushed down nearly to the water-surface, or till it almost joins its image seen in the water as coming up from below. Then the slide is tapped lightly with a pencil, each tap advancing it less than $\frac{1}{1000}$ of a foot. When the index touches the water, a distinct and unmistakable phenomenon occurs: Capillary attraction causes the water to spring upward and a tremor runs over the surface. The lip coincides exactly in height with the point of the index or with the true surface of the water. A vernier reading to thousandths of a foot can be applied to the slide if desired, but it is hardly necessary, as the thousandths can be estimated with sufficient accuracy.

Gauge to Show a Head or Difference of Level.—Fig. 244 shows the principle of a device whereby we may observe directly the difference in level between two bodies of water, as, for instance, the head acting upon a water-wheel.

On the light shaft 5 are mounted the disk 4 and the two bevel-gear wheels 1 and 2. These three pieces turn freely upon the shaft without being susceptible of any lateral movement. Their

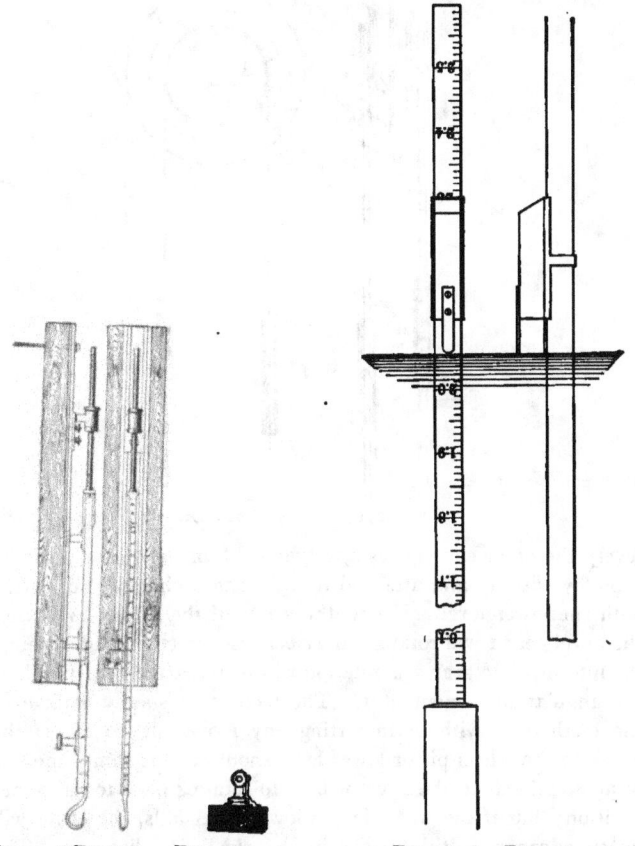

FIG. 240. FIG. 241.   FIG. 241a.   FIG. 242.   FIG. 243.

only connection is by means of the small bevel 3, turning on an axis attached to 4 and in mesh with the two gears. A pulley forms a part of the bevel-gear 1, and over this pulley passes a

light band of steel tape, also marked 1. One end of this band
is attached to a float in a chamber connected with the upper

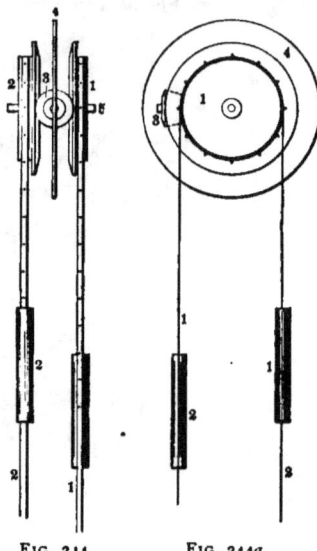

FIG. 244.   FIG. 244a.

level; the other end carries a weight 1. Similarly, gear 2 carries
a pulley with a band attached to a float in a chamber connected
with the lower level. If now the water of the upper level rises,
the bevel-gear 1 will rotate in the clockwise direction, and, through
the intermediate gear 3 a rotation will be imparted to 4, viz., half
the angular movement of 1. The teeth of 3 simply walk over
the teeth of 2 without imparting any movement to 2. If the
water in the wheel-pit or lower level should at the same time rise
to an equal extent, the effect will be to rotate 4 back to its former
position; but if the water in the lower level falls, the effect will
be to advance 4 still further in the clockwise direction. The
aggregate movement of 4 therefore is proportional to the change
of fall, and by suitable graduation the latter may be read from
the disk.

The Weir.—Although the word "weir" or "wear" has the general signification of "a dam," its meaning has latterly been restricted to a dam specially designed to measure the

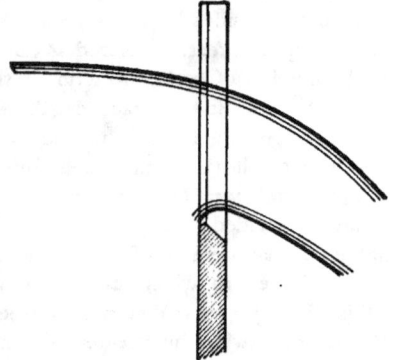

FIG. 245.

quantity of water passing over it. Equations (26) and (28) relate to the flow of water over dams of ordinary form; but for exact measurements a form of weir is presupposed conforming to that from which the formula was deduced. Facilities for making such experiments have multiplied enormously of late, especially in scientific schools, and there is hardly any form of weir or dam that has not been made the subject of exact determination of discharge. We will adhere, however, to the standard form of measuring-weir, Fig. 245, presenting a sharp edge on the up-stream side, both on horizontal crest and vertical sides, after passing which the water leaps freely into the air. The surface of the water approaching the weir takes a curved form, and the depth must be measured at a sufficient distance up-stream to avoid the effect of this curve. It is found that a perforated pipe resting on the bottom of the channel, close to the up-stream side of the weir and opening into a still-box, will give, in the latter, the true height of the surface. The formula determined by Mr. James B. Francis for the discharge of such a weir is

$$Q = 3.33(L - 0.1nH)H^{\frac{3}{2}}, \quad . \quad . \quad . \quad . \quad (71)$$

in which $Q =$ the discharge in cubic feet per second;

$\quad\quad L =$ the length of the horizontal crest of the weir in feet, the sides being supposed vertical;

$\quad\quad H =$ the depth in feet, measured as above;

$\quad\quad n =$ the number of end contractions; that is to say, if the weir is the exact length of the channel of approach, there is no end contraction and $n = 0$; if the weir meets one side of the channel, $n = 1$; if neither side, $n = 2$.

The weir may be divided into several parts by means of bulkheads, that is, it may consist of several weirs with their crests on the same level, in which case n may be greater than 2. In this case the least distance between two weirs should be not less than twice the greatest depth of water on the weirs.

This formula assumes the weir-opening so small compared with the cross-section of the channel that the velocity of approach is small, say not more than 0.5 foot per second. If the water approaching the weir has any velocity worth considering, Q will be a little greater than given by formula (71). We must then find the head due the velocity and add it to H. Then recompute Q with the new value of H. If we place the still-box in the channel of approach, and, omitting the pipe, let it communicate with the water of the channel by an orifice in the up-stream side, we shall not need, as a general thing, to make the above correction, as the water in the box will stand above the general level by a height nearly equal to that due the velocity of approach.

When a weir extends from side to side of the channel, the stream, being uncontracted at the ends, falls in contact with the sides, and there is no opportunity for the admission of air under the sheet. The action of the water speedily exhausts the air and diminishes the pressure which opposes the flow, thereby increasing the discharge above what would be indicated by the formula. In order that the formula may be applied with confidence, means must be adopted for giving the air access to the space under the sheet. The weir formula cannot be applied with full confidence to depths greater than 3 feet nor to depths more than one-third the length; neither does it apply where the water is so low that it clings to the weir on the down-stream side.

Where the water on the down-stream side rises above the crest of the weir, the latter becomes a submerged weir and the formula (71) is no longer applicable. For this case Mr. Francis [*] gives the following formula:

$$Q = 3.33L \sqrt{H - H'}(H + 0.381H'), \quad . \quad . \quad (72)$$

in which H = height of water on up-stream side, above crest of weir;

$\quad\quad\quad H'$ = height of water on down-stream side, above crest of weir.

This refers to a weir extending across the channel. Where there are end contractions, $L - 0.1nH$ must be used instead of L.

Systematic Measurement of the Flow of a Natural Stream.—A single measurement of the flow of a natural stream has very little value unless at the period of extreme low or extreme high water. To determine the average flow of water to be expected from the stream, the measurement must extend over a series of years. How long a period must be taken in order to obtain the true average cannot now be stated. Records of the flow of streams have not continued long enough to enable that question to be answered. From existing information it appears probable that the average flow for a period of twenty years would not differ materially from the average of one hundred, and that the average of ten years would not differ more than 20 per cent from the same. In connection with rain-gauge records of long duration, a single year's observation of flow has a considerable value, especially if it happens to be a year of nearly average rainfall. In what follows we are presupposing a series of daily measurements to

[*] Trans. Am. Soc. C. E., 1884, p. 312.

last at least a year. The smaller streams are most conveniently measured by weir. For the larger class the most
suitable instrument is the current-meter.

Measurement by Weir.—Figs. 18, 19, 20, already described (page 46), show, on a small scale, the form of weir
proper for a small stream. The sharp-crested weir-plank over
which the water flows is seen at *B*, Fig. 245. In preparing for
such measurements, the first question to be considered is as to
the length of the weir. To take the flow of 20 square miles
of drainage-ground for a series of years without rising more
than 3 feet on the crest would require a length of 400 feet;
whereas a length of 20 feet would probably take the entire flow
for 360 days out of 365. The case is not that of a permanent
dam, where an overflow of the shore connections would involve
loss of life and property. The worst that could happen from
overflow would be the partial loss of the weir. The apron and
sheet-piling would remain, and the weir could be put in order
again with no great expense on subsidence of the flood. The
practical method would appear to be this: Give the weir a
length of not less than 1 foot and not more than 2 for every
square mile of drainage-ground, and let it take its chance.
Establish ranges and take the cross-section of the channel, so
that, in the event of a failure, the flow during the continuance
of the flood can be measured by surface or mid-depth floats, till
the stream resumes normal dimensions, when the weir can be
very readily reestablished.

There would be no great chance of error if the vertical
sides of the weir should rise 5 feet above the crest, provided the
latter were so placed that the weir would be drowned when the
water reached a height of 3 feet. To find this elevation of the
crest we must find the depth of a stream which would carry the
discharge of the weir with the existing slope and cross-section.

$$v = \frac{Q}{A} = c\sqrt{\frac{A}{P}s},$$

in which Q = discharge of weir with 3 feet depth; A = corresponding area of cross-section; P = perimeter of same; and

$s =$ slope of surface. Q, c, and s are known. From the plotted cross-section of the stream-valley P is found, corresponding to any assumed area A of cross-section. The value of A can be found by trial subject to the condition $\dfrac{Q}{c\sqrt{s}} = \sqrt{\dfrac{A^3}{P}}$. Having found A from the plotted cross-section, the required height is apparent. This method requires a precise gauge on the up-stream side of the weir, and another on the down-stream side, the latter to be used only when the weir is submerged. Arrangements for measuring the height of water in time of flood are shown in Figs. 238 and 239.

Measurement by Current-meter.—The best arrangement for this case is to stretch a wire rope across the stream. One end of the rope can usually be attached to a tree, the other requires an artificial support. Fig. 246 shows such an anchorage, consisting of two stout posts set deep in the earth and supporting a drum made of the rough trunk of a tree. By cutting four mortises in this drum, for the insertion of handspikes, it becomes a winch whereby a sufficient strain can be brought on the rope, and the latter can be dropped into the river on passage of boats or rafts. Tags bearing numbers are affixed to the rope at uniform intervals, by means of the ordinary wire-rope clips. In observing the current velocities a boat is used. This is made fast to the rope at each of the several stations while the velocities are being observed, generally near the surface, at mid-depth, and near the bottom. The meter-rod is marked with a scale. The first reading is taken about a foot below the surface, the second as near the bottom as the meter will run, and in observing this the meter-rod rests on the bottom. In this position the depth is observed and the third reading is taken at mid-depth. Thus a complete cross-section of the stream is put on record at each measurement. In some recent operations of the Geological Survey, the observer is mounted on a car suspended from the wire rope and running thereon by means of grooved wheels. In winter, as soon as the stream freezes, the operation becomes much more simple, the observations being made through holes cut

in the ice.   If the water at the wire rope remains long open,
or partly open and partly closed, a new line is selected, where
the ice is intact and a new series of holes cut.   Sometimes a
covered sledge is provided mounted on runners to be shoved
from hole to hole.   The holes will usually be found freshly
frozen every cold morning, so that a complete outfit must

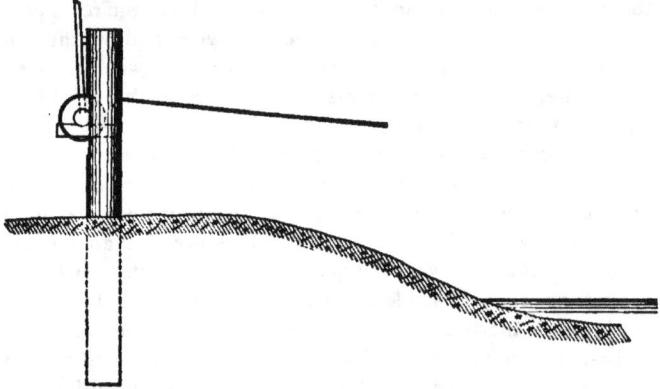

FIG. 246.

include an ice-chisel, ice-saw, and tongs.   A supply of hot
water is also essential, as ice is liable to form on the instrument
whenever it is withdrawn from the water, and obstruct its
working.   Ice does not form on the meter while it is immersed
in water.

Measurement by Means of Turbines.—At great water-
powers, where it is necessary, for the adjustment of rentals, to
keep an accurate account of the water drawn by different
lessees or users, the water-wheels are used to a considerable
extent as meters.   Each wheel is, preferably before its installation, subjected to experimental test, to determine the quantity
of water discharged under the varying conditions of use.   The
results so determined are arranged in tables which serve for
the finding of the discharge at any other time.   The discharge
of a turbine, at any given moment, is determined by three
things: 1. The opening of the gate which controls the dis-

charging orifices, called the speed-gate. 2. The head acting
on the wheel. 3. The velocity. Of course the discharge
cannot be determined experimentally for all possible variations
of these several elements. It is determined for heights of
speed-gate differing so little that the intermediate discharges
can be interpolated. These results are determined with reference to the ordinary head acting on the wheel, which we will
call the "standard head," and the ordinary velocity, which
we will call the "standard velocity." The discharge at each
height of gate is also observed for velocities varying from the
standard within the limits of practice. With these results we
are able to form a table giving the discharge of the wheel under
the standard head for all heights of speed-gate and all velocities within the range of practice. From this table the discharge under any conditions of working can be determined.
To fit the wheel for observing the several elements of the
discharge, it is provided with a gauge in the penstock and
another in the wheel-pit which serve to determine the fall or
head. A graduated scale is attached to the gate-rod by which
the opening of the wheel-passages may be observed. The
revolutions of the wheel-shaft per minute are counted with the
aid of a time-keeper. With these data, the discharge of the
wheel can be determined, from the table, for the observed gate-
opening and velocity and for the standard head. The discharge
under the observed head is obtained by the operation

Discharge under observed head

$$= \text{discharge under standard head} \sqrt{\frac{\text{observed head}}{\text{standard head}}}.$$

In making the experiments to determine the table of discharge,
the velocity of the wheel can be controlled by a brake or other
means, but the head cannot usually be closely controlled, and
will deviate more or less from the standard. In fact, when the
experiments are made subsequent to installation the erection
of a weir in the race necessarily diminishes the head largely.
The observed discharge must in this case be reduced to the
standard head by the above method, viz.,

Discharge under standard head

$$= \text{observed discharge} \sqrt{\frac{\text{standard head}}{\text{observed head}}}.$$

In its application to iron wheels this method is limited by the action of the water on the iron. Nodules or blisters form on the surface of the guides and floats, partly at the expense of the iron, partly from matters derived from the water, and in the course of a few years materially diminish the discharge. Little reliance can be placed upon this method without periodical cleaning of the surfaces, and even with this precaution the cleaning leaves the surfaces pitted and wasted in a manner to seriously affect the discharge.

The water-power interest at Lowell, Mass., was the first to adopt exact methods of water measurement, and the methods used at Lowell have been copied by other water-powers throughout the United States. In the method practised at Lowell the increment of discharge due to a change in the velocity of the wheel is regarded not as a function of the velocity, but as a function of another quantity, viz., the ratio of the velocity of the entrance to the wheel to the velocity due the head. So that if we know the discharge of a wheel acting under a given head and moving with a given velocity, then, if the head should vary without any change of the velocity, we should, after finding the increment of discharge due to a change of head, add another increment on account of the change in the above ratio, though no change has taken place in the velocity. The expression (47), page 306, for the centrifugal force in a Boyden turbine, shows that the introduction of the above ratio in computing the discharge is an unnecessary complication resting upon no rational basis whatever.

The Venturi Meter.—Water-meters for indicating the quantity of water delivered to consumers in municipal water-supply systems have been devised in innumerable forms, but with these devices this treatise has nothing to do. It would nevertheless be incomplete without some notice of a meter patented and introduced by Mr. Clemens Herschel, which is susceptible of use on a scale suited to the requirements of water-power. It is called the Venturi meter because it is founded upon facts in hydraulics to which attention was first called by Venturi, an Italian physicist, about 1796, viz. : Water moving in a pipe may pass from a condition of high pressure and low velocity to a condition of low pressure and high velocity, and *vice versa*, without great loss of energy.* In other words, pressure and velocity are mutually convertible.

Referring to Fig. 247, suppose *CB* to be part of a line of

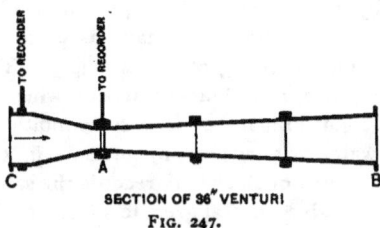

SECTION OF 36" VENTURI

FIG. 247.

pipe which has its normal diameter at *C* and *B*, but between *C* and *A* contracts to a small diameter, and between *A* and *B* expands to normal diameter. Suppose we insert a small vertical pipe at *A*, and another at *B*. The water while in motion will stand lower at *A* than *B* by a height which we will call *h*. Let the cross-section at *B* be *a* times that at *A*, and let *w* be the weight of any quantity of water that we choose to consider. Then *v* being the velocity of the water at *A*, its energy is $w\dfrac{v^2}{2g}$, at *B* it is $\dfrac{w}{a^2}\dfrac{v^2}{2g}$. That is, if $a = 4$, the energy at *B* is only $\frac{1}{16}$ that at *A*. What has become of the other $\frac{15}{16}w\dfrac{v^2}{2g}$? It has been converted into the head *h*. In other words, the slowing

---

* See a very interesting series of experiments by James B. Francis on the movement of water in expanding tubes. Lowell Hyd. Expts. 1868, p. 209. The same principles also apply to the Diffuser.

up of the water between A and B creates a difference of pressure represented by h. We have therefore $w\dfrac{a^2 - 1}{a^2}\dfrac{v^2}{2g} = wh$, whence $v = a\sqrt{\dfrac{2gh}{a^2 - 1}}$.

Of course the velocity v could not be computed from this formula, because it ignores the frictional head; but this result shows that the difference of level h is a correct measure of the velocity, and that the latter can be determined when the former is known. Practically the relation between h and v is to be determined by experiment. Fig. 247 shows a pipe from A, and another from C, though we have assumed a pipe from B for the better understanding of the principle. The pipes from A and C lead to a recording-instrument which by means of floats and mechanism may be made to exhibit upon a dial the quantity of water passing at any instant, in cubic feet per second, or by means of clockwork records the same in the form of a diagram, or gives the aggregate discharge for an hour, a day, or a week.

This meter can be inserted in a line of penstock to show the quantity of water drawn by a turbine wheel. It can be applied to an existing penstock by narrowing the cross-section. There is no limit to the size of pipes to which it can be applied. In its application to water-wheels there is not necessarily any serious loss of head. The size of the throat A with reference to the normal size of the pipe is governed by the minimum quantity of water that it is designed to measure. Where this is as low as the sixteenth or twentieth of the normal flow, the throat must be so small that a serious loss of head occurs in passing the normal flow. In applying the meter to water-wheels this objection has little force because it is not necessary, in that case, to measure quantities less than the fifth part of the normal flow. A water-wheel which gives its best efficiency on a discharge of 175 cubic feet per second would never need to draw less than 35 cubic feet. When this quantity cannot be obtained it is always better to stop the wheel and let the water accumulate. This method can be introduced into an ordinary manufacturing establishment with little disturbance of existing arrangements and at moderate expense. Where water is conveyed from the canal to the wheels in an iron pipe, a lining of timber can be introduced into the latter to contract the diameter and transform it into a Venturi meter. The same method can be adopted where the channel is a wooden penstock either round or square, and where exact methods are used in measuring the change of head the latter need not be great enough to occasion any appreciable loss.

CHAPTER XXIV.

STORAGE AND PONDAGE OF WATER.

THE total annual flow of a stream in Massachusetts is distributed throughout the several months, taking the average of a series of years, in about the manner stated on page 4.

This ratio of distribution is modified to some extent in other parts of New England, the time of extreme high and extreme low water being later in the northern part and earlier in the southern.

This statement exhibits in a very strong light the chief defect of water-power, viz., its extreme variability. An establishment depending wholly on water-power, which would run its machinery throughout the year without serious interruptions, cannot use more than 2 per cent of the annual flow of the stream in any one month, or more than 24 per cent in a year. Even with this use of water it would be exposed to more serious interruptions than appear above. In the months of August and September the flow would naturally be below the average half the time, and often below 2 per cent. In June and October also, though the flow is 4 or 5 upon an average, there would be many days on which it would be below 3 and even below 2. Moreover, the above statement represents the average of a long series of years. In such a series years will occur with a total flow of not more than half the average, and others with nearly or quite double. Several years will occur in succession falling 25 to 50 per cent below the average; others, as much above.

It results from these conditions that no rational use can be made of water-power, subject to the requirements of modern industry, without either steam-power to supply deficiencies in dry seasons, or storage-reservoirs to hold back the surplus of the high-water period for use in time of scarcity. With reservoirs it is possible, though hardly ever practicable, to make use of the entire flow of the stream, reducing it to a uniform average stage throughout the entire year or even for a series of years, or what may be called a cycle of years, that is, for a period long enough to embrace all possible fluctuations of the flow. With the aid of steam-power alone it is never possible to make use of the entire flow of the stream. This would imply the establishment of wheels and other appliances of water-power to the extent of the maximum flow of the stream, a proceeding obviously irrational. It is never worth while to install such plant to a greater extent than can be kept in operation for four months in an average year, and six months is more in accordance with common practice. This necessarily leaves large volumes of water to run to waste, so that the use of steam-power does not preclude the profitable use of reservoirs.

In tables which follow, the writer has endeavored to compute the values of reservoirs used in connection with steam, the several elements of the computation being:

First.—The quantity of water to be expected, month by month and year by year, from a given extent of drainage-ground.

Second.—The quantity of steam-power required to make up the deficiencies of the water-power and maintain a constant total of power.

Third.—The change in the relative proportions of water and steam consequent upon reservoirs of different sizes.

Fourth.—The cost of steam-power.

The saving in steam-power resulting from reservoirs determines the amount of money that can be judiciously applied to their construction.

The flow of water from a drainage-area of known extent has been observed and recorded regularly by the engineers of the Boston Water-works for nearly forty years past. Of these records, those of the last twenty-five years, relating to the Sudbury River drainage-area, a district of 75 square miles, are selected as being most reliable. It is assumed that twenty-five years covers all possible variations of flow, and that these results can be applied without serious risk of error to all parts of New England and, *mutatis mutandis*, to other regions of the United States.

Assume an establishment on a stream of 500 square miles drainage-area, with a fall of 15 feet, and a requirement of 1000 horse-power, being furnished to that extent, both with water-wheels and steam-engines. To avoid the complication of pondage, that is, the holding back of the flow during one part of the day for use during another, assume the use of water to be continuous during the 24 hours. On a fall of 15 feet 1000 h.p. requires about 800 cubic feet per second. In computing the expenditure of water we count only the working-days, deducting the Sundays, and a holiday each from the months of February, April, May, July, September, November, and December. So that we have months of

27 days requiring 1866 millions of cubic feet;
26 " " 1797 " " "
25 " " 1728 " " "
24 " " 1659 " " "
23 " " 1590 " " "

Table 17 exhibits the flow of water in the natural condition of the stream, being the monthly average in cubic feet per second for the entire period of twenty-five years.

Table 18 gives the aggregate volume of water furnished each month in millions of cubic feet.

These two tables show in a marked manner the enormous fluctuations of the stream. The month of March, 1877, furnished 9973 million cubic feet, September only 119 million; the former being 3723 cubic feet per second, the latter only 46. During March, assuming the flow to be uniform, thirty 48-inch wheels of the Rodney-Hunt pattern could have run at full gate. In September a single wheel could hardly have run at half gate.   Other equally striking variations appear.   Thus, in 1887, the highest monthly average exceeds the lowest twenty-six times; in 1889 only a little over four times.

Table 19 gives the aggregate steam-power required each month to supplement the water-power and make up a constant total of 1000 h.p.   The figures represent horse-powers 24 hours a day throughout the month.

Table 20 shows the effect of reservoirs of different sizes upon the quantity of water available for power.   It shows, in millions of cubic feet, the volumes of water drawn from the reservoir, used and wasted, during the month, and that remaining in the reservoir at the close of the month, for reservoirs of 3000, 5000, and 10 000 millions of cubic feet respectively. This table runs through the whole period of twenty-five years month by month, and is necessarily voluminous.

Tables 21, 22, and 23 show, month by month, the steam-power required with reservoirs of 3000, 5000, and 10 000 million cubic feet respectively, expressed in Table 19 for the case of no reservoirs.

Table 24 is a summary of the whole matter, being the cost of the steam-power each year.   The first column contains the year; the second, the cost of steam-power in the natural condition of the stream; the third, the cost with reservoirs of 3000 million cubic feet; the fourth, with 5000; the fifth, with 10 000.

Cost of Steam-power.—The leading item in the cost of steam-power is coal; the other items are attendance, lubricants, water for feed and condensation, repairs, waste, and miscellaneous small items.   It is an approximately correct statement that the cost of coal is about half the running cost of the power. Where the constant aim of the steam-user is to keep down the consumption of coal by maintaining the engine and appurtenances in the highest state of efficiency he will diminish the cost of coal and increase all the other items, and the latter will be much the larger half.   On the other hand, when little attention is given to economy of coal, the coal bill will be the larger half of the cost.   The consumption of coal therefore is a very variable factor. On trial runs of engines made by makers in fulfilment of guarantees, a rate as low as 1.5 pounds per h.p. hour and even less is often shown, but it is probable that the constant running of an engine at this rate would involve expense for attendance and refinements of practice more than offsetting the saving in coal.

In the trial of a case of a large number of mill-owners against the city of Worcester, Mass., in 1898, on account of diversions of water from Blackstone River, it was shown that the several engines in practical running consumed coal at the rate of 1.78 to 5.26 pounds per h.p. hour.

At the convention of the United States National Electric Light Association for 1895, the average consumption of ten stations, producing over 5000 kilowatts per diem, was stated at 5 pounds per indicated horse-power per hour. These engines run under conditions unfavorable to economy on account of the great variation of load from hour to hour.

A mill-engine running to supplement water-power works under conditions unfavorable to economy. It starts as soon as the water-power falls below the requirements of the mill, in which case a 500-h.p. engine may run to furnish 50 h.p., which would cost per h.p. twice or three times as much as when working at full load. The same condition recurs toward the end of the dry season, when the water-power is but little short of the mill's requirements. The cost of power in such engines is difficult to obtain, for, though the running expenses are shown by the mill's records, the power developed by the engine, varying from day to day, is not usually obtainable. No mill-engine works permanently under the most favorable conditions. It is the inevitable tendency in all mills to expand. When the engine is new it has usually a capacity considerably in excess of requirements, and works with too light a load for full economy. In its later history, requirements increase, the engine works under excessive load, with diminished expansion and poor economy.

Pumping-engines which work into a service-reservoir act

under conditions favorable to economy, usually working at full load while working at all. The records of such engines usually include the performance as well as the expenses.

Five cents per million gallons raised 1 foot is a rather low cost of pumping. This is at the rate of 1.2 cents per h.p. hour.

The average cost in Boston from 1884 to 1892 was at the rate of 6.06 cents per million gallons raised 1 foot. This, if we make the common allowance of 5 per cent for the "slip" of the pumps, would be at the rate of 1.51 cents per h.p. hour. At the Edison Electric-light Works, Boston, in 1895, the running cost of power was at the rate of 1.40 cents per h.p. hour.

TABLE 17.—FLOW IN CUBIC FEET PER SECOND FROM 500 SQUARE MILES OF NEW ENGLAND DRAINAGE-AREA. Average for Each Month.

Year.	Jan.	Feb.	March	April	May.	June.	July.	Aug.	Sept.	Oct.	Nov.	Dec.
1875	80	1175	1241	2358	919	672	248	305	160	499	1079	451
1876	497	1058	3431	2547	881	171	141	313	142	181	841	351
1877	508	734	3723	1851	1074	462	156	93	46	488	1096	997
1878	1400	1907	2713	1258	1078	391	99	367	124	399	1306	2458
1879	506	1323	1802	2410	861	319	122	305	109	54	159	358
1880	807	1382	1060	904	397	136	130	92	62	78	158	135
1881	320	1224	3097	1196	746	1035	214	114	152	143	305	600
1882	959	1859	2196	670	1000	409	67	43	237	231	167	243
1883	259	799	1246	1044	725	232	89	607	70	143	158	149
1884	770	2198	2928	2222	797	322	173	198	34	64	135	715
1885	955	1047	1216	1404	1033	329	48	186	93	259	911	908
1886	1131	3713	1592	1506	557	157	89	73	91	112	520	789
1887	2003	2187	2219	2026	789	320	88	165	85	147	285	497
1888	813	1509	2504	2046	1263	326	93	293	893	1546	2133	2354
1889	2152	925	1010	1090	680	505	490	1107	637	951	1501	1733
1890	970	1182	2818	1450	1057	438	83	102	354	1757	940	770
1891	2334	2696	3445	1816	440	320	115	125	157	162	235	421
1892	1446	716	1512	674	973	331	165	217	177	97	544	375
1893	335	1193	2510	1644	2230	340	122	140	83	171	246	616
1894	536	766	1731	1209	649	324	124	162	115	289	646	554
1895	800	418	1864	1946	492	134	178	177	69	1067	2143	1378
1896	838	2070	2966	1156	278	308	73	44	300	458	510	308
1897	653	825	1984	1172	708	744	509	457	141	72	703	1225
1898	1267	2338	2014	1415	963	410	178	856	285	897	1536	1391
1899	1720	1068	3253	1950	395	51	15	−27	72	89	237	170
Av.	965	1408	2243	1561	839	367	153	261	157	414	740	795

During April, 1896, the street-railway system of Boston was obtaining its power at a running cost of 1.20 cents per h.p. hour. These companies obtain coal at less than $3 per U. S. ton. Where coal is subject to a long railway haul and at points not accessible to railroads the cost must be considerably higher.

TABLE 18.—FLOW FROM 500 SQUARE MILES OF NEW ENGLAND DRAINAGE-AREA.

Aggregate for Each Month, in Millions of Cubic Feet.

Year.	Jan.	Feb.	March	April.	May.	June.	July.	Aug.	Sept.	Oct.	Nov.	Dec.
1875	213	2800	3324	6113	2461	1743	665	820	416	1338	2611	1209
1876	1332	2651	9189	6602	2359	445	378	840	369	484	2181	959
1877	1363	1776	9973	4799	2883	1197	418	251	119	1309	2842	2671
1878	3749	4614	7267	3260	2889	1014	266	985	322	1070	3394	6583
1879	1451	3201	4827	6248	2308	828	326	819	282	146	412	958
1880	2323	3464	2847	2343	1065	352	366	246	160	210	411	362
1881	859	2893	8296	3100	1999	2682	572	306	395	384	792	1606
1882	2570	4497	5882	1738	2676	1060	179	115	614	619	420	651
1883	693	1933	3337	2706	1943	601	239	1626	182	384	411	400
1884	2062	5508	7843	5721	2135	835	463	532	88	172	351	1916
1885	2559	2533	3258	3639	2768	854	129	498	243	696	2361	2342
1886	3027	8984	4265	3904	1492	406	239	195	236	302	1348	2113
1887	5365	5291	5942	5253	2089	829	237	443	222	393	939	1332
1888	2181	3781	6708	5304	3382	845	243	786	2316	4142	5530	6305
1889	5765	2237	2774	2827	1822	1310	1312	2966	1652	2548	3892	4643
1890	2598	2861	7548	3759	2831	1138	222	273	917	4708	2436	2063
1891	6253	6523	9227	4806	1207	829	309	337	406	435	611	1128
1892	3874	1793	4051	1747	2608	858	443	581	460	260	1398	1005
1893	897	2886	6723	4272	5973	891	327	375	215	458	648	1650
1894	1436	1853	4636	3289	1738	850	332	434	298	774	1674	1483
1895	2143	1011	4993	5044	1318	347	477	474	179	2858	5568	3691
1896	2245	5186	7933	2996	745	798	196	118	778	1228	1322	1361
1897	1749	1998	5314	3038	1896	1928	1363	1224	365	193	1822	3281
1898	3394	5656	5394	3668	2579	1063	477	2293	739	2403	3981	3724
1899	4607	2582	8713	5054	1058	132	40	—72	187	238	614	455
Av.	2588	3540	6011	4049	2249	953	409	699	486	1110	1919	2155

In the attempt to fix a price which will hold good for the future, it must be remembered that coal is probably as low now as it will ever be. The most accessible deposits of coal have been mined; the most favorable workings have been occupied, and in some degree exhausted. Future winnings must be made at a progressively increasing expense. This condition has been reached in England, where the price of coal has been advancing for several years past. It is fully realized there that the supply of coal is not inexhaustible, and the same fact is beginning to be understood in this country.

TABLE 19.—STEAM-POWER REQUIRED TO MAKE UP A UNI-FORM TOTAL OF 1000 H.P. 24 HOURS A DAY IN A MILL COMMANDING 500 SQUARE MILES OF NEW ENGLAND DRAINAGE-AREA, WITHOUT RESERVOIRS.

Head 15 ft. Power expressed in horse-powers, 24 hours a day. 12 cu. ft. per second on 1 ft. fall taken as a h.p.

Year.	Jan.	Feb.	March	April.	May.	June.	July.	Aug.	Sept.	Oct.	Nov.	Dec.
1875	900					160	690	619	800	376		436
1876	379					786	824	609	823	774		561
1877	365	83				423	805	884	943	390		
1878						511	876	521	845	501		
1879	368					601	848	619	864	933	801	553
1880					504	830	830	885	923	903	803	831
1881	600				68		733	858	810	821	619	250
1882				163		489	916	946	704	711	791	696
1883	676	1			94	710	889	241	913	821	803	814
1884	38				4	598	784	753	958	920	831	106
1885						589	940	768	884	676		
1886					304	804	889	909	886	860	350	14
1887					14	600	890	794	894	816	644	379
1888						593	884	634				
1889					150	369	388		204			
1890						453	856	873	558			38
1891					450	600	856	844	804	798	706	474
1892		105		158		586	794	729	779	879	320	531
1893	581					575	848	825	896	786	693	230
1894	330	43			189	595	845	798	856	639	193	308
1895		478			385	833	778	779	914			
1896					653	615	909	945	625	428	363	615
1897	184				115	70	364	429	824	910	121	
1898						488	778		644			
1899					506	936	981	1000	910	889	704	788

In these computations we do not deal with the absolute cost of steam-power, but rather with the expense that is saved when water is substituted for steam in an establishment provided with both steam- and water-motors; and, *vice versa*, the expense incurred when water is replaced by steam. This will of course vary with the locality, but the computation is made on the basis of 1 cent per hour per horse-power. When there

TABLE 20.—EFFECT OF RESERVOIRS OF DIFFERENT SIZES UPON THE FLOW OF 500 SQUARE MILES OF NEW ENGLAND DRAINAGE-AREA FROM 1875 TO 1899.

Quantities in millions of cubic feet.

Capacity of Reservoir		3000				5000				10 000			
Date	Working Days	Drawn from Reservoir	Used	Wasted	In Reser- voir at End of Month	Drawn from Reservoir	Used	Wasted	In Reser- voir at End of Month	Drawn from Reservoir	Used	Wasted	In Reser- voir at End of Month
1875. January	26		213		0		213		0		213		0
February	23		1590		1210		1590		1210		1590		1210
March	27		1866		2668		1866		2668		1866		2668
April	25		1728	4053	3000		1728	2053	5000		1728		7053
May	25	54	1728	733	3000	54	1728	733	5000	54	1728		7786
June	26		1797		2946		1707		4946		1707		7732
July	26	1132	1797		1814	1132	1707		3814	1132	1797		6600
August	26	977	1797		837	977	1728		2837	977	1797		5623
September	25	837	1253		0	1312	1797		1525	1312	1797		4311
October	26		1338		883	459	1797		1066	459	1728		3852
November	25		1728		295		1728		1949		1797		4735
December	26		1797		0	588	1797		1361		1728		4147
1876. January	26		1659		992		1797		896		1797		3682
February	24		1627		0	465	1659		1888	465	1659		4674
March	27		1866	5315	3000		1866	4211	5000		1866	1997	10000
April	24		1659	4943	3000		1659	4943	5000		1659	4943	10000
May	26	1352	1797	562	1648	1352	1797	562	5000	1352	1797	562	10000
June	26	1350	1797		298	1350	1728		3648	1350	1728		8648
July	25	298	1728		0	1026	1866		2298	1026	1866		7298
August	27		1138		0	1272	1641		1272	1359	1728		6272
September	25		369		0		484		0	1313	1797		4913
October	26		484		453		1728		0		1728		3600
November	25		1728		0		1728		453		1728		4053
December	25	453	1392		0	453	1392		0	789	1728		3264

Year	Month	Days												Total
1877.	January	27	....	....	1363	....	0	....	1363	....	0	503	1866	....	2 661
	February	23	....	....	1590	5293	186	....	1590	3293	186	....	1590	954	2 847
	March	27	....	....	1866	3140	3000	....	1866	3140	5000	....	1866	3140	10 000
	April	24	....	....	1659	1086	3000	....	1659	1086	5000	....	1659	1086	10 000
	May	26	600	1797	....	....	2400	600	1797	....	4400	600	1797	....	9 400
	June	26	1310	1797	....	....	1090	1310	1797	....	3090	1310	1728	....	8 090
	July	25	1090	1728	....	....	0	1615	1728	....	1475	1615	1866	....	6 475
	August	27	....	1341	....	....	0	1475	1594	....	0	1540	1659	....	4 935
	September	24	....	119	....	....	0	....	1309	....	0	557	1866	....	4 378
	October	27	....	1309	....	....	1114	....	1728	....	1114	....	1728	....	5 492
	November	25	....	1728	....	....	2057	....	1728	....	2057	....	1728	....	6 435
	December	25	....	1728	940	....	3000	....	1866	....	3940	....	1866	....	8 318
1878.	January	27	....	1866	3024	....	3000	....	1590	1964	5000	....	1590	1342	10 000
	February	23	....	1590	5470	....	3000	....	1797	5470	5000	....	1797	5470	10 000
	March	26	....	1797	1532	....	3000	....	1728	1532	5000	....	1728	1532	10 000
	April	25	....	1728	1092	....	3000	714	1728	1092	5000	714	1728	1092	9 286
	May	26	714	1707	....	....	2286	1531	1728	....	4286	1531	1797	....	7 755
	June	25	1531	1728	....	....	755	861	1797	....	2755	884	1866	....	6 874
	July	26	755	1740	....	....	0	537	1866	....	1874	1406	1728	....	5 468
	August	27	....	1070	....	....	0	1337	1659	....	537	....	1866	....	4 672
	September	24	....	322	....	....	0	537	1607	....	0	796	1728	....	6 338
	October	27	....	1728	....	....	1666	....	1728	....	1666	....	1728	1193	10 000
	November	25	....	1728	3521	....	3000	415	1728	1521	5000	900	1728	....	9 585
	December	25	....	1728	....	....	2585	415	1866	1196	4585	....	1866	1196	10 000
1879.	January	27	415	1866	1196	....	3000	....	1590	1196	5000	....	1590	1196	10 000
	February	23	....	1590	1196	....	3000	....	1797	3030	5000	1471	1797	3030	10 000
	March	26	....	1797	3030	....	3000	978	1728	4520	5000	978	1728	4520	10 000
	April	25	....	1797	4520	....	3000	....	1728	511	5000	1441	1728	511	10 000
	May	26	....	1797	511	....	3000	205	1797	....	5000	1720	1797	....	9 100
	June	25	900	1728	....	....	2100	900	1728	....	4100	900	1728	....	7 629
	July	26	1471	1797	....	....	629	1471	1797	....	2629	1471	1797	....	6 651
	August	26	629	1448	....	....	0	978	1728	....	1631	978	1797	....	5 205
	September	25	....	282	....	....	0	....	349	....	205	1441	1728	....	3 485
	October	27	....	146	....	....	0	205	412	....	0	1720	1866	....	2 238
	November	24	....	412	....	....	0	....	958	....	0	1247	1659	....	2 238
	December	26	....	958	....	....	0	....	958	....	0	839	1797	....	1 399

TABLE 20.—*Continued.*

Capacity of Reservoirs.		3000				5000				10 000			
Date.	Working Days.	Drawn from Reservoir.	Used.	Wasted.	In Reservoir at End of Month.	Drawn from Reservoir.	Used.	Wasted.	In Reservoir at End of Month.	Drawn from Reservoir.	Used.	Wasted.	In Reservoir at End of Month.
1880.													
January	27	....	1866	....	457	....	1866	....	457	....	1866	....	1 756
February	23	....	1590	....	2331	....	1590	....	2331	....	1590	....	3 630
March	27	....	1866	312	3000	....	1866	....	3312	....	1866	....	4 611
April	25	663	1728	615	3000	663	1728	....	3927	663	1728	....	5 226
May	25	1445	1728	....	2337	1445	1728	....	3264	1445	1728	....	4 563
June	26	892	1797	....	892	1431	1797	....	1819	1431	1797	....	3 118
July	26	....	1258	....	0	1431	1797	....	388	1551	1797	....	1 687
August	26	....	246	....	0	....	634	....	0	....	296	....	136
September	25	....	160	....	0	....	160	....	0	....	210	....	0
October	26	....	210	....	0	....	210	....	0	....	411	....	0
November	26	....	411	....	0	....	411	....	0	....	362	....	0
December	26	....	362	....	0	....	362	....	0	....	859	....	0
1881.													
January	23	....	859	....	1303	....	859	....	1303	....	1590	....	1 303
February	27	....	1590	4733	3000	....	1590	2733	5000	....	1866	....	7 733
March	25	....	1866	1372	3000	....	1866	1372	5000	....	1728	....	9 105
April	25	....	1728	271	3000	....	1728	271	5000	....	1728	....	9 376
May	26	....	1728	885	3000	....	1728	885	5000	....	1797	261	10 000
June	25	....	1797	....	1844	....	1797	....	3844	....	1728	....	8 844
July	27	1156	1728	....	284	1156	1728	....	2284	1156	1866	....	7 284
August	25	1500	1866	....	0	1500	1866	....	951	1560	1728	....	5 951
September	26	284	679	....	0	1333	1335	....	0	1333	1797	....	4 538
October	25	....	384	....	0	951	792	....	0	1413	1728	....	3 602
November	26	....	792	....	0	....	1606	....	0	936	1797	....	3 411
December	26	....	1606	....	0	....	1606	....	0	192	1797	....	....

1882.	January	26	......	1797	......	773	......	1797	......	773	......	1797	......	4 184
	February	23	......	1590	680	3000	......	1590	......	3680	......	1590	......	7 091
	March	27	......	1866	4010	3000	......	1866	2696	5000	......	1866	1107	10 000
	April	24	......	1659	79	3000	......	1659	79	5000	......	1659	79	10 000
	May	26	......	1797	879	3000	......	1797	879	5000	......	1797	879	10 000
	June	26	737	1797	......	2263	737	1797	......	4263	737	1797	......	9 263
	July	25	1549	1728	......	714	1549	1728	......	2714	1549	1728	......	7 714
	August	27	714	829	......	0	1751	1866	......	963	1751	1866	......	5 963
	September	25	......	614	......	0	963	1577	......	0	1114	1728	......	4 849
	October	26	......	619	......	0	......	619	......	0	1178	1797	......	3 671
	November	25	......	420	......	0	......	420	......	0	1308	1728	......	2 363
	December	25	......	651	......	0	......	651	......	0	1077	1728	......	1 286
1883.	January	27	......	693	......	0	......	693	......	0	......	693	......	0
	February	23	......	1590	......	343	......	1590	......	343	......	1590	......	343
	March	27	......	1866	......	1814	......	1866	......	1814	......	1866	......	1 814
	April	24	......	1659	......	2861	......	1659	......	2861	......	1659	......	2 861
	May	26	......	1797	7	3000	......	1797	......	3007	......	1797	......	3 007
	June	26	1196	1797	......	1804	1196	1797	......	1811	1196	1797	......	1 811
	July	25	1489	1728	......	315	1489	1728	......	322	1489	1728	......	322
	August	27	240	1866	......	75	240	1866	......	82	......	1866	......	82
	September	24	75	257	......	0	82	264	......	0	82	264	......	0
	October	27	......	384	......	0	......	384	......	0	......	384	......	0
	November	25	......	411	......	0	......	411	......	0	......	411	......	0
	December	25	......	400	......	0	......	400	......	0	......	400	......	0
1884.	January	27	......	1866	......	196	......	1866	......	196	......	1866	......	196
	February	24	......	1659	1045	3000	......	1659	......	4045	......	1659	......	4 045
	March	26	......	1797	6046	3000	......	1797	5091	5000	......	1797	91	10 000
	April	25	......	1728	3993	3000	......	1728	3993	5000	......	1728	3993	10 000
	May	26	......	1797	338	3000	......	1797	338	5000	......	1797	338	10 000
	June	25	893	1728	......	2107	893	1728	......	4107	893	1728	......	9 107
	July	26	1334	1797	......	773	1334	1797	......	2773	1334	1797	......	7 773
	August	26	773	1305	......	0	1365	1797	......	1508	1265	1797	......	6 508
	September	25	......	88	......	0	1508	1506	......	0	1640	1728	......	4 868
	October	27	......	172	......	0	......	172	......	0	1694	1866	......	3 174
	November	24	......	351	......	0	......	351	......	0	1308	1659	......	1 866
	December	26	......	1797	......	119	......	1797	......	119	......	1797	......	1 985

TABLE 20.—Continued.

Capacity of Reservoirs.		3000				5000				10 000			
Date.	Working Days.	Drawn from Reservoir.	Used.	Wasted.	In Reservoir at End of Month.	Drawn from Reservoir.	Used.	Wasted.	In Reservoir at End of Month.	Drawn from Reservoir.	Used.	Wasted.	In Reservoir at End of Month.
1885. January	27	....	1866	....	812	....	1866	....	812	....	1866	....	2 078
February	23	....	1590	....	1755	....	1590	....	1755	....	1590	....	3 621
March	26	....	1797	216	3000	....	1797	127	3216	....	1797	....	5 082
April	25	....	1728	1911	3000	....	1728	1040	5000	....	1728	....	6 993
May	25	....	1728	1040	3000	....	1728	....	5000	....	1728	....	8 033
June	26	943	1797	....	2057	943	1797	....	4057	943	1797	....	7 090
July	26	1668	1797	....	389	1668	1797	....	2389	1668	1797	....	5 422
August	26	389	887	....	0	1299	1797	....	1090	1299	1797	....	4 123
September	25	....	243	....	0	1090	1333	....	0	1485	1797	....	2 638
October	27	....	696	....	702	....	696	....	702	1170	1866	....	1 468
November	24	....	1659	....	1247	....	1659	....	702	....	1659	....	2 170
December	26	....	1797	6871	2477	....	1797	....	1247	....	1797	....	2 715
1886. January	26	....	1797	....	3000	....	1797	....	2477	....	1797	....	3 945
February	23	....	1590	2399	3000	....	1590	4871	5000	....	1590	1339	10 000
March	27	....	1866	2176	3000	....	1866	2399	5000	....	1866	2399	10 000
April	25	236	1728	....	2764	236	1728	2176	5000	236	1728	2176	10 000
May	26	1391	1797	....	1373	1301	1797	....	4764	1331	1797	....	9 764
June	26	1373	1612	....	0	1558	1707	....	3373	1558	1797	....	8 373
July	26	....	195	....	0	1602	1767	....	1815	1602	1797	....	6 815
August	25	....	236	....	0	213	459	....	213	1492	1728	....	5 213
September	26	....	302	....	0	....	302	....	0	....	1797	....	3 721
October	25	....	1348	....	0	....	1348	....	0	1495	1728	....	3 226
November	26	....	1348	....	0	....	1348	....	0	380	1728	....	1 846
December	26	....	1797	....	316	....	1797	....	316	....	1797	....	2 162

Year	Month													
1887.	January	26	.	1797	.	3000	.	1797	3884	.	1797	.	.	5 730
	February	23	.	1590	884	3000	.	1590	5000	.	1590	.	.	9 431
	March	27	.	1866	3701	3000	.	1866	5000	.	1866	.	.	10 000
	April	25	.	1728	4076	3000	.	1728	5000	.	1728	3507	.	10 000
	May	26	.	1728	3525	3000	.	1728	5000	.	1728	3525	.	10 000
	June	25	968	1797	361	2032	968	1797	4032	968	1797	361	.	9 032
	July	27	1491	1728	.	541	1491	1728	2541	1491	1728	.	.	7 541
	August	25	541	984	.	0	1423	1866	1118	1423	1866	.	.	6 118
	September	26	.	222	.	0	1118	1340	0	1506	1728	.	.	4 612
	October	25	.	393	.	0	.	393	0	.	1797	.	.	3 208
	November	26	.	939	.	0	.	939	0	789	1728	.	.	2 419
	December	26	.	1332	.	384	.	1332	0	465	1797	.	.	1 954
1888.	January	26	.	1797	.	384	.	1797	384	.	1797	.	.	2 338
	February	24	.	1659	4348	2506	.	1659	2506	.	1659	.	.	4 460
	March	27	.	1866	3645	3000	.	1866	5000	.	1866	2947	.	9 302
	April	24	.	1659	1585	3000	.	1659	5000	.	1797	1585	.	10 000
	May	26	.	1797	.	3000	.	1797	5000	.	1797	.	.	9 048
	June	25	952	1728	.	2048	952	1728	4048	952	1728	.	.	7 563
	July	27	1485	1349	.	563	1485	1728	2563	1485	1866	.	.	6 483
	August	24	563	1659	.	0	1080	1866	1483	1080	1659	.	.	7 140
	September	27	.	1866	.	657	.	1659	2140	.	1866	.	.	9 416
	October	25	.	1728	.	2933	.	1866	4416	.	1728	3218	.	10 000
	November	25	.	1728	3735	3000	.	1728	5000	.	1728	4577	.	10 000
	December	27	.	1866	4577	3000	.	1728	5000	.	1866	3899	.	10 000
1889.	January	23	.	1590	3899	3000	.	1866	5000	.	1590	647	.	10 000
	February	26	.	1797	647	3000	.	1590	5000	.	1597	977	.	10 000
	March	27	.	1728	977	3000	.	1797	5000	.	1728	1059	.	10 000
	April	24	.	1797	1099	3000	.	1728	5000	.	1797	25	.	9 582
	May	27	418	1723	25	2582	418	1797	-4582	418	1728	.	.	9 997
	June	26	485	1797	.	2097	485	1728	4097	.	1797	.	.	10 000
	July	35	.	1797	.	3000	.	1866	5000	.	1866	197	.	9 993
	August	27	7	1866	197	2993	7	1659	4993	7	1659	.	.	10 000
	September	24	.	1659	675	3000	.	1866	5000	.	1866	675	.	10 000
	October	27	.	1866	2164	3000	.	1728	5000	.	1728	2164	.	10 000
	November	25	.	1728	2915	3000	.	1728	5000	.	1728	2915	.	10 000
	December	25	.	1728	.	3000	.	1728	5000	.	1728	.	.	10 000

TABLE 20.—Continued.

Date	Working Days	3000 — Drawn from Reservoir	3000 — Used	3000 — Wasted	3000 — In Reservoir at End of Month	5000 — Drawn from Reservoir	5000 — Used	5000 — Wasted	5000 — In Reservoir at End of Month	10 000 — Drawn from Reservoir	10 000 — Used	10 000 — Wasted	10 000 — In Reservoir at End of Month
1890. January	27		1866	732	3000		1866	732	5000		1866	732	10 000
February	23		1590	1271	3000		1590	1271	5000		1590	1271	10 000
March	26		1797	5751	3000		1797	5751	5000		1797	5751	10 000
April	25		1728	2031	3000		1728	2031	5000		1728	2031	10 000
May	26		1797	1034	3000		1797	1034	5000		1797	1034	10 000
June	25	590	1728		2410	590	1728		4410	590	1728		9 410
July	26	1575	1797		835	1575	1797		2835	1575	1797		7 835
August	26	835	1108		0	1524	1797		1311	1524	1797		6 311
September	25		917			811	1728		500	811	1728		5 500
October	27		1866	619	2842		1866		3342		1866		8 342
November	24		1659	266	3000		1659		4119		1659		9 119
December	26		1797		3000		1797		4385		1797		9 385
1891. January	27		1866	4387	3000		1866	3772	5000		1866	3772	10 000
February	23		1590	4933	3000		1590	4933	5000		1590	4933	10 000
March	26		1797	7430	3000		1797	7430	5000		1797	7430	10 000
April	25		1728	3078	3000		1728	3078	5000		1728	3078	10 000
May	26	521	1797		2479	521	1797		4479	521	1797		9 479
June	26	968	1797		1511	968	1797		3511	968	1797		8 511
July	25	1488	1728		23	1488	1728		2023	1488	1728		7 023
August	26		366		0	1460	1797		503	1460	1797		5 563
September	25		406		0	563	969		0	1322	1797		4 241
October	27		435		0		435		0	1431	1866		2 810
November	24		611		0		611		0	1048	1659		1 762
December	26		1128		0		1128		0	669	1797		1 093

1892. January	27	....	1797	....	2077	....	1797	....	2077	....	1797	....	3 170
February	24	....	1659	....	2211	....	1659	....	2211	....	1659	....	3 304
March	27	....	1866	1396	3000	....	1866	....	4396	....	1866	....	5 480
April	25	....	1728	19	3000	....	1728	295	4415	....	1728	....	5 508
May	25	939	1728	880	3000	....	1728	....	5000	939	1797	....	6 388
June	26	1285	1797	....	2061	939	1797	....	4061	1285	1728	....	5 449
July	25	776	1728	....	776	1285	1728	....	2776	1285	1866	....	4 164
August	27	....	1357	....	0	1268	1866	....	1491	1268	1728	....	2 879
September	26	....	460	....	0	223	1728	....	223	1537	1797	....	1 611
October	25	....	260	....	0	....	483	....	0	74	1472	....	74
November	26	....	1398	....	0	....	1398	....	0	....	1005	....	0
December	26	....	1005	....	0	....	1005	....	0	....	897	....	0
1893. January	23	....	897	....	1296	....	897	....	1296	....	1590	....	1 296
February	27	....	1590	3153	3000	....	1590	1153	5000	....	1866	....	6 153
March	24	....	1866	2613	3000	....	1866	2613	5000	....	1659	....	8 766
April	26	....	1659	4176	3000	....	1659	4176	5000	....	1797	2942	10 000
May	26	906	1797	....	2094	906	1797	....	4094	906	1797	....	9 094
June	25	947	1797	....	693	1401	1728	....	2693	1401	1728	....	7 693
July	27	....	1068	....	0	1491	1866	....	1202	1491	1866	....	6 202
August	25	....	215	....	0	1202	1417	....	0	1513	1728	....	4 689
September	26	....	458	....	0	....	458	....	0	1339	1797	....	3 350
October	25	....	648	....	0	....	648	....	0	1080	1728	....	2 270
November	35	....	1650	....	0	....	1650	....	0	78	1728	....	2 192
December	27	....	1436	....	0	....	1436	....	0	430	1866	....	1 762
1894. January	23	....	1590	....	263	....	1590	....	263	....	1590	....	2 025
February	27	....	1866	33	3000	....	1866	....	3033	....	1866	....	4 795
March	24	....	1659	1630	3000	....	1659	....	4463	....	1659	....	6 425
April	26	59	1797	....	2941	59	1797	....	4604	59	1797	....	6 366
May	26	947	1797	....	1994	947	1797	....	3657	947	1797	....	5 419
June	25	....	1728	....	598	1396	1728	....	2261	1396	1728	....	4 023
July	27	....	1032	....	0	1432	1866	....	829	1432	1866	....	2 591
August	24	....	298	....	0	829	1127	....	0	1361	1659	....	1 230
September	27	....	774	....	0	....	774	....	0	1092	1866	....	138
October	25	....	1674	....	0	....	1674	....	0	54	1728	....	84
November	25	....	1483	....	0	....	1483	....	0	....	1567	....	0

TABLE 20.—Continued.

Capacity of Reservoirs → Date	Working Days	3000 — Drawn from Reservoir	3000 — Used	3000 — Wasted	3000 — In Reservoir at End of Month	5000 — Drawn from Reservoir	5000 — Used	5000 — Wasted	5000 — In Reservoir at End of Month	10 000 — Drawn from Reservoir	10 000 — Used	10 000 — Wasted	10 000 — In Reservoir at End of Month
1895.													
January	27		1866		277		1866		277		1866		277
February	23	277	1288		0	277	1288		0	277	1288		0
March	26		1797	196	3000		1797		3196		1797		3196
April	25		1728	316	3000		1728	1512	5000		1728		6612
May	26	479	1797		2521	479	1797		4521	479	1797		6033
June	25	1381	1728		1140	1381	1728		3140	1381	1728		4652
July	26	1140	1617		0	1320	1797		1820	1320	1797		3332
August	27		474		0	1392	1866		428	1392	1866		1940
September	24		179		0	428	667		0	1480	1728		460
October	27		1866		992		1866		992		1866		1452
November	25		1728	1832	3000		1728		4832		1728		5292
December	25		1728	1963	3000		1728	1795	5000		1728		7255
1896.													
January	27		1866	379	3000		1866	379	5000		1866		7634
February	24		1659	3527	3000		1659	3527	5000		1659	1161	10 000
March	26		1797	6136	3000		1797	6136	5000		1797	6136	10 000
April	25		1728	1268	3000	983	1728	1268	4017		1728	1268	10 000
May	25	983	1728		2017	999	1728		3018	983	1728		9017
June	26	999	1797		1018	1601	1797		1417	999	1797		8018
July	26	1018	1214		0	1417	1797		0	1601	1797		6417
August	26		118		0		1535		0	1679	1797		4738
September	25		778		0		778		0	950	1728		3788
October	27		1228		0		1228		0	638	1866		3150
November	24		1322		0		1322		0	337	1659		2813
December	26		1361		0		1361		0	436	1797		2377

1897. January	26	....	1749	....	0	....	1749	....	0	408	1797	48	....	2 339
February	23	....	1590	856	408	....	1590	856	408	3856	1590	....	....	2 737
March	27	....	1866	1310	3000	....	1866	....	5000	1866	....	....	....	6 185
April	25	....	1728	169	3000	166	1728	....	5000	1728	....	....	....	7 495
May	25	....	1728	131	3000	168	1728	....	5000	1728	....	....	....	7 663
June	26	....	1797	....	3000	131	1797	....	5000	1797	....	....	....	7 794
July	26	434	1797	....	2566	....	1797	434	4566	1797	434	....	....	7 360
August	26	573	1797	....	1993	....	1797	573	3993	1797	573	....	....	6 787
September	25	1363	1728	....	630	....	1728	1363	2630	1728	1363	....	....	5 424
October	26	630	823	....	0	....	1728	1604	1026	1728	1604	....	....	3 820
November	25	....	1728	....	94	....	1797	....	1120	1797	....	....	....	3 914
December	26	....	1797	....	1578	....	1728	....	2604	1728	....	....	....	5 398
1898. January	26	....	1797	175	3000	....	1797	....	4201	1797	....	....	....	6 995
February	23	....	1590	4066	3000	3257	1590	....	5000	1590	1061	....	....	10 000
March	27	....	1866	3528	3000	3528	1866	....	5000	1866	3528	....	....	10 000
April	25	....	1728	1940	3000	1940	1728	....	5000	1728	1940	....	....	10 000
May	25	....	1728	851	3000	851	1728	....	5000	1728	851	....	....	10 000
June	26	734	1797	....	2266	....	1797	734	4266	1797	....	734	....	9 266
July	25	1251	1728	....	1015	....	1728	1251	3015	1728	....	1251	....	8 015
August	27	....	1866	....	1442	....	1866	....	3442	1866	....	....	....	8 442
September	25	989	1728	....	453	989	1728	....	2453	1728	....	989	....	7 453
October	26	....	1797	....	1050	....	1797	....	3059	1797	....	....	....	8 059
November	25	....	1728	....	3000	....	1728	....	5000	1728	....	....	....	10 000
December	26	....	1797	312	3000	312	1797	....	5000	1797	312	....	....	10 000
1899. January	26	....	1797	1927	3000	1927	1797	....	5000	1797	1927	....	....	10 000
February	23	....	1797	2810	3000	2810	1797	....	5000	1590	2810	....	....	10 000
March	27	....	1866	992	3000	992	1590	....	5000	1866	992	....	....	10 000
April	24	....	1659	6847	3000	6847	1866	....	5000	1659	6847	....	....	10 000
May	26	....	1797	3395	2261	3395	1659	....	5000	1659	3395	....	....	9 261
June	26	739	1797	....	596	....	1797	739	4261	1797	....	739	....	7 596
July	25	1665	636	....	0	....	1728	1665	2596	1797	....	1665	....	5 908
August	27	596	0	....	0	....	836	836	908	1728	....	1688	....	3 970
September	25	....	187	....	0	....	187	....	0	1866	....	1938	....	2 429
October	26	....	238	....	0	....	238	836	0	1728	....	1541	....	870
November	25	....	614	....	0	....	614	....	0	1797	....	1559	....	0
December	25	....	455	....	0	....	455	....	0	455	455	870	....	0

is reason to think that the cost should vary from this figure the results can be modified accordingly. We do not here consider the cost of installation either of the steam-engines or the wheels.

The area occupied by the reservoirs must be regarded as ineffective, since the evaporation therefrom will be about equal to the rainfall. It is impossible to form any general estimate of this area. We must therefore regard the area of the drainage-ground as 500 square miles in excess of that occupied by the reservoirs.

TABLE 21.—STEAM-POWER REQUIRED TO MAKE UP A UNIFORM TOTAL OF 1000 H.P. 24 HOURS A DAY IN A MILL COMMANDING 500 SQUARE MILES OF NEW ENGLAND DRAINAGE-AREA.

With reservoirs to the extent of 3000 millions of cubic feet. Head 15 feet. Power expressed in h.p. 24 hours a day.

Year.	Jan.	Feb.	March	April.	May.	June.	July.	Aug.	Sept.	Oct.	Nov.	Dec.
1875	881								275	255		
1876	95							390	786	731		194
1877	271							281	925	299		
1878							68	806		427		
1879								194	837	922	752	467
1880							300	863	907	882	762	798
1881	522								607	786	542	106
1882								556	645	655	758	623
1883	629								845	794	762	769
1884								273	949	908	788	
1885								507	859	628		
1886							103	891	863	832	222	
1887								473	872	782	457	258
1888								277				
1889												
1890								383	469			
1891								800	765	767	632	373
1892								273	734	856	191	442
1893	501							428	876	745	625	45
1894	230							447	820	585	31	142
1895		190					100	746	892			
1896							324	934	550	342	203	243
1897										542		
1898												
1899							632	1000	892	868	645	737

The anomaly of August, 1899, actually occurred on the Sudbury drainage-basin. The flow fell short of the evaporation. Such a result probably could not occur on a drainage-area of 500 square miles, as such an area could hardly fail to embrace streams furnishing considerable water in the driest time.

In Table 24, subtracting the average of the third, fourth, and fifth columns, respectively, from that of the second, we find:

The annual saving consequent upon a reservoir
capacity of 3000 millions.................. $13 415.92
The annual saving consequent upon a reservoir
capacity of 5000 millions.................. 18 645.41
The annual saving consequent upon a reservoir
capacity of 10 000 millions................ 24 944.96

TABLE 22.—STEAM-POWER REQUIRED TO MAKE UP A
UNIFORM TOTAL, ETC.

With reservoirs to the extent of 5000 millions of cubic feet.

Year.	Jan.	Feb.	March	April.	May.	June.	July.	Aug.	Sept.	Oct.	Nov.	Dec.
1875	881											
1876									950	731		194
1877	271								39	299		
1878										139		
1879										813	752	467
1880								647	907	883	762	798
1881	522								257	542	106
1882									87	656	757	623
1883	629								841	794	762	769
1884									76	907	788	
1885									229	627		
1886									734	832	220	
1887									225	781	457	259
1888												
1889												
1890												
1891									439	767	632	372
1892										731	191	441
1893	501								180	745	625	45
1894	230								321	585	31	142
1895		190							598			
1896								146	550	342	203	243
1897												
1898												
1899								552	892	867	645	737

These figures show what we knew perfectly well before, viz., that the benefit of reservoirs to one single establishment, or one mill privilege, is hardly ever sufficient to warrant their construction. The total head to be served must usually be as much as 100 feet to bring such projects within the range of feasibility. Exceptions occur where the level of natural lakes can be raised at slight expense, but with the exception of the northern part of Maine, such opportunities in New England are nearly all appropriated.

Suppose six such establishments lying consecutively on the same stream with an aggregate fall of 90 feet. In this case the aggregate saving would be:

For a capacity of 3000 millions........ $80 495.52
       "   "   5000   " 111 872.46
       "   "   10 000   " 149 669.76

TABLE 23.—STEAM-POWER REQUIRED TO MAKE UP A UNIFORM TOTAL, ETC.

With reservoirs to the extent of 10 000 millions of cubic feet.

Year.	Jan.	Feb.	March	April.	May.	June.	July.	Aug.	Sept.	Oct.	Nov.	Dec.
1875	881											
1876												
1877												
1878												
1879					.							
1880	....	....	....	....	....	....	....	....	829	883	762	798
1881	522											
1882												
1883	629	....	....	....	....	....	....	....	841	794	762	769
1884												
1885												
1886												
1887												
1888												
1889												
1890												
1891												
1892	....	....	....	....	....	....	....	....	....	....	148	441
1893	501											
1894	....	....	....	....	....	....	....	....	....	....	....	93
1895	....	190										
1896												
1897												
1898												
1899	....	....	....	....	....	....	....	....	....	....	141	737

And the cost at which such reservoirs would be a 10 per cent investment would be:

For 3000 millions $804 955.20, or $268.32 per million cu. ft.
"   5000   "   1 118 724.46, "   223.74 "   "   "
" 10 000   "   1 496 697.60, "   149.67 "   "   "

TABLE 24.—COST OF STEAM-POWER TO MAINTAIN 1000 H.P
24 HOURS A DAY IN A MILL COMMANDING 500 SQUARE
MILES OF NEW ENGLAND DRAINAGE-AREA.

With reservoirs of different sizes. Head 15 ft. Power reckoned at 1 cent
per hour per h.p.

Year.	No Reservoirs.	3000 Million Cubic Feet.	5000 Million Cubic Feet.	10 000 Million Cubic Feet.
1875	24 841.44	8 804.64	5 497.44	5 497.44
1876	29 677.44	13 703.04	11 700.00	
1877	24 292.32	11 082.24	3 800.16	
1878	20 304.96	8 118.24	867.36	
1879	34 862.88	19 793.28	12 679.68	
1880	40 616.16	28 154.88	24 941.28	20 417.28
1881	29 697.16	15 993.12	8 904.48	3 257.28
1882	33 795.84	20 198.88	13 247.52	
1883	37 202.88	23 705.76	23 680.80	23 680.80
1884	31 150.08	18 208.32	11 051.04	
1885	24 067.68	12 442.56	5 341.44	
1886	31 299.84	18 164.64	11 144.64	
1887	31 393.44	17 734.68	10 745.28	
1888	13 172.64	1 728.48		
1889	6 932.64			
1890	17 584.32	5 316.48		
1891	34 519.68	20 822.88	13 790.40	
1892	30 457.44	15 575.04	8 505.12	3 675.36
1893	33 908.16	20 092.80	13 079.04	3 126.24
1894	29 927.04	14 071.20	8 168.16	558.00
1895	26 002.08	12 030.72	4 917.12	1 185.60
1896	32 154.72	16 199.04	9 260.16	
1897	18 826.08	3 382.08		
1898	11 918.40			
1899	41 895.36	29 779 76	23 044.32	5 478.72
Totals......	690 500.68	355 102.76	224 365.44	66 876.72
Averages...	27 620.03	14 204.11	8 974.62	2 675.07

At the highest of these figures, opportunities for storage can usually be found in most New England river valleys; at the lowest, their occurrence cannot usually be counted on. If we suppose a series of establishments on the same stream, circumstanced as above with regard to power, having an aggregate fall of 150 feet—a case by no means unusual,—the cost at which reservoirs would be a 10 per cent investment would be $444 per million cubic feet capacity for 3000 millions, $373 for 5000, and $249 for 10 000. It would be rare indeed if opportunities for storage could not be obtained at these figures. Of course any capacity under 3000 millions could be obtained upon equally or more advantageous terms. Many situations exist in which reservoirs command a much greater fall than here assumed. Certain reservoirs on the headwaters of the Blackstone River, appropriated for domestic water-supply by the city of Worcester, Mass., in 1895, were available on a total head of more than 1000 feet, nearly all of which is utilized for power.

Value of Reservoirs.—The saving consequent upon a capacity of 3000 millions, which is at the rate of 6 million per square mile of drainage-area, is $4.47 per annum per million cubic feet of capacity, and this is at the rate of about 30 cents per foot of fall. The following statement is therefore approximately true: The value of reservoirs, to the extent of 6 million cubic feet capacity per square mile of drainage-ground, is 30 cents per annum per million cubic feet of capacity on each foot of fall. The preceding tables show that the value per unit of capacity is greater as the relative capacity is less. Where the capacity is merely sufficient to avoid working the engines at a very uneconomical load (see page 567), the value may be twice or three times the above figure.

These tables throw light upon some controverted questions in the subject of power. How often, for instance, do we hear it said that water-power has had its day; is going out of use; is being superseded by steam, etc.!

The total volume of water discharged by the 500 square miles of drainage-ground in the assumed period of twenty-five years is 654 200 millions of cubic feet. The quantity used for power in connection with steam is 329,904 million in the natural condition of the stream. A mill using 200 cubic feet per second, which is about 450 millions per month, would, as appears from Table 18, have been short of water sixty or seventy times during the twenty-five-year period, and this quantity is the utmost that could have been used by such a mill. The entire volume used by such a mill would have been 135 000 million cubic feet, against 329 904 million available when used in connection with steam. We are entitled to say, therefore, that, of the total flow of any stream, two or three times as much water can be used in connection with steam as can be used to any rational purpose without it. So far from water being superseded or rivalled by steam, it is only in connection with steam that it attains its full development.

Another point may be adverted to in this connection. It is sometimes asserted that it is cheaper to use steam alone than to run steam in connection with water-power. Compare the cost in a mill running as assumed in these tables with that in a mill using the same power obtained from steam alone. We will roughly estimate the cost of water-power plant at $100 per horse-power. We have taken 1 cent an hour, not as the absolute cost of steam, but as the excess of cost of steam over water, i.e., as the additional expense incident to the substitution of steam for water in a mill provided with both steam- and water-plant. We will suppose the absolute running cost of steam to be 1.2 cents per h.p. hour, and that of water 0.2 cent. We will concede that in a mill running wholly by steam 1 cent per hour may be taken as the absolute running cost of the power. For a mill using water and steam the account stands thus:

Cost of steam, 20 per cent more than in Table 24 . $33 144.16
Interest on cost of water-power plant, at 5 per cent   5 000.00
Sinking-fund for its entire renewal in thirty years .   1 413.00
Running cost of water, 4 587 997 h.p. hours at
    0.002 = ...   9 175.99
       Total........................ ...... $48 733.15

The item of running cost appears excessive, though consistent with our preceding calculations. If it is so, it shows that we have done less than justice to the value of reservoirs.

For a mill using steam alone, 1000 h.p., 306 days, 24 hours a day at 1 cent per hour........ $73 440.00

The cost for steam alone is therefore about 50 per cent in excess of that for steam and water.

Value of Water-power.—These tables also throw light on the question of the absolute value of water-power, a subject on which widely divergent opinions are expressed. If we can correctly determine the cost of power for an establishment such as is contemplated when run wholly by steam and again when run in connection with water-power, the difference, as it appears to the writer, may be accepted as the accurate value of the water-power.

We have, as above, cost by steam alone........... $73 440
Cost by steam and water 48 733
Value of water per annum..................... $24 707

This sum is what the mill could afford to pay annually for the water used for power, and still stand upon an equal footing with an establishment run wholly by steam at the same place.

The foregoing figures do not apply to a mill engaged in bleaching, dyeing, and similar operations requiring great amounts of heat.

General Considerations touching Reservoirs.—The natural conformation of the earth was not adopted with any special reference to the construction of reservoirs. The sites most feasible for holding-grounds are most commonly near the headwaters of the stream, with small area of drainage-ground appertinent. Sites commanding large drainage-grounds are generally situated in broad river valleys involving high and expensive dams, covering valuable lands, flooding centres of population, necessitating changes of location in railroads and highways. The damages incident to flowage are generally the controlling factor in the latter case. So far as the cost of the dam is concerned, though absolutely great, it is generally less per unit of storage than in the former case. Occasionally there are exceptions to this conformation, as is strikingly seen in northern Maine, on the upper part of the Penobscot and St. Croix, where numerous lakes occur in the main line of the river. The west branch of the Penobscot, at a point 2 or 3 hundred feet above the sea-level, with a drainage-area of some 1800 square miles, commands a series of lakes susceptible at moderate cost of being converted into reservoirs capable of reducing the flow of the stream to uniformity, not only for a year, but for a cycle of years.

In this connection a very interesting question is, What reservoir capacity would we require on our assumed drainage ground to reduce the flow of the stream to uniformity for the entire period of 25 years? Dividing the total volume of water discharged by the stream, in the period considered, by the number of working seconds in that period, taken as $\frac{4}{7}$ of the total number of seconds, we find 967 cubic feet per second as the average flow of the stream during the working hours, being 1922 millions of cubic feet for a working month of 23 days, 2006 for 24, 2090 for 25, 2173 for 26, and 2257 for 27. An inspection of Table 20 shows that the reservoir should have been practically empty on January 1, 1884. Starting at that date, adding in the monthly volume furnished by the stream, and deducting the monthly draft on the supposition that no water is wasted, we shall find the maximum volume in store May 1, 1891, viz., 51 570 millions of cubic feet. That is to say, to utilize all the water furnished by the stream would require reservoir capacity equal to nearly two years' average flow.

The chief obstacle to the construction of reservoirs for water-power lies in the fact that the benefit to any single user of water is rarely sufficient to warrant the investment. Their construction presupposes a concert of action between a number, sometimes a large number, of separate concerns, and the adjustment of conditions and distribution of expense is generally a fatal difficulty. There will ordinarily be some willing to accept the benefits of the undertaking but unwilling to share the expense. This matter might well be controlled by law, as in projects of drainage, irrigation, maintenance of levees, and other cases in which many people share in the control of the same natural agency. As the law now stands in Massachusetts, cases of this kind have occurred: A few enterprising manufacturers have constructed reservoirs on the headwaters of the stream for their own benefit. Manufacturers further down the stream, though freely accepting the benefits of the reservoirs, have refused to assume any share of the expense. Later the reservoirs have been appropriated by a city for purposes of domestic water-supply, with ample compensation to the owners. The manufacturers further down have sued and recovered damages from the city for the loss of these benefits which they had refused to pay for.

The writer has long been of the opinion that the following simple legal provision would go far toward remedying existing difficulties in the organization of reservoir projects. Let it be enacted that whenever a reservoir is constructed for the purpose of holding flood-waters which are of no value for power, and discharging them in form and manner to be available for power; then whenever water is discharged from said reservoir, every user of water on the stream below shall either pay for the same or permit an equal quantity to run to waste. Noncompliance by the manufacturer would be readily detected, and no one could deny the justice of the law, which excludes him from benefits that he declines to pay for. The variations of level in the mill-pond incident to ponding the water need not impair the practicability of discharging a fixed quantity of water, since a device for so controlling the gate of an orifice as to discharge a fixed quantity at all stages of the head is wholly within the resources of mechanical skill.

Cost of Reservoirs.—The following figures are given on the authority of Mr. Geo. W. Rafter.*

---

* Report of the State Engineer and Surveyor of New York, 1895, p. 177.

A system of reservoirs on the headwaters of the Hudson, contemplated by the State of New York for compensation to water-power on account of water diverted for the use of canals, surveyed and examined by Mr. Rafter, with a total capacity of 41 593 million cubic feet, is expected, according to that gentleman's estimate, to cost $62.69 per million cubic feet of storage.

The Bear Valley Reservoir in California, with a capacity of 1766 million, cost $121.67 per million.

The Bhatgur Reservoir in India, 5510 million capacity, cost $73.46 per million.

The Long Valley Reservoir in California, 1436 million, cost $50.73 per million.

The Erkuk Reservoir in India, 3315 million, cost $91.83 per million.

The Sweetwater Reservoir in California, holding 784 million, cost $938.70 per million.

The Hemet Valley Reservoir in California, holding 6111 million, cost $229.16 per million.

The Periar Reservoir in India, holding 6970 million, cost $106.75 per million.

The Betwa Reservoir in India, holding 1603 million, cost $204.25 per million.

The proposed Genesee Reservoir, holding 7700 million, is estimated to cost $311.69 per million.

The system of reservoirs on the headwaters of the Mississippi, constructed by the United States for the improvement of navigation, has a total capacity of over 90 000 millions and cost less than $9 per million. This was a case where the level of natural lakes was raised by cheap wooden dams and the land damages were very slight.

Reservoirs surveyed and planned by the writer in Massachusetts have usually shown a cost of about $200 per million.

Relation of Reservoirs to Coal.—The extreme cheapness of coal, during the last quarter of a century, has predisposed manufacturers in New England to look with slight favor upon reservoir projects as adjuncts of water-power. To the writer's mind these works are destined hereafter to play a more important part in manufacturing than hitherto. There are two agencies at work tending in that direction: 1. The diminishing rate of interest, which makes large permanent investments preferable to high annual charges. 2. The inevitable exhaustion, in the course of time, of our supplies of coal. The consumption of coal in the United States doubled from 1862 to 1871, again from 1871 to 1881, and again from 1881 to 1893, amounting now (1900) to something over 200 million U. S. tons per annum. With these facts to guide us we might look for a consumption of 400 million about 1915, 800 million in 1933, 1600 million in 1953, and by 1960 a consumption which would exhaust 500 square miles of average workable coal land in a single year. There are probably some 300 000 square miles of land in the United States underlaid with coal, not more than 50 000 of which is workable at anything like the existing price. Premonitory symptoms of exhaustion already appear in the coal-supply of European countries. It is probable that the price of coal in England has passed its culminating point, and will never again be so low as it has been. Without looking forward to the absolute exhaustion of the coal-supply, these figures make it evident that materially higher prices of coal may be expected within a period not too long to enter into the calculations of a corporation endowed with perpetual succession, when, as often happens, the ownership remains in the same family and it is managed with some reference to the interests of posterity. Moreover, in these days of gigantic combinations, the monopolization of all existing workable coal lands by a single concern is an event entirely within the range of possibility. However much such an event would be against the immediate interests of consumers, it would, on a broader view, be in harmony with the general interests of mankind, tending to preserve for the use of future generations some part of the bounties of nature so lavishly wasted by the present.

It may be interesting to note in this connection that the power derivable from the St. Lawrence River and its tributaries, from their sources to the sea, is not far short of that obtainable from all the coal mined in the United States.

Reservoirs on a Grand Scale.—It is probable that New England will remain, in the future as in the past, a manufacturing country. With her rugged and sterile soil and rigorous climate she will ever find it more advantageous to procure her supplies of food and materials mainly by producing manufactured goods to exchange for them by direct production. This career is forced upon New England, not so much from the possession of peculiar advantages in the pursuit of manufacturing as from the lack of advantages in the pursuit of agriculture and other industries. This, in common with other sections of the country, is well situated for commerce and navigation. In common with other sections it has an active and energetic population. The remoteness from the coal regions puts it at some disadvantage as regards the use of steam-power. It is at a great disadvantage as regards agriculture and general mining. If New England has any positive advantage in respect to manufacturing industry it is this: The rugged and broken outline of the country which put it at a great disadvantage in respect to agriculture is very favorable to the existence and development of water-power. Inasmuch as not more than half the ultimate capacity of any water-power can ordinarily be realized without the aid of storage reservoirs it would seem that works of this character merit increased attention.

An acre of New England land used for farming purposes, does not yield on an average an income of more than $10 per annum. An acre lying 100 feet above the sea-level and capable of being submerged to a depth of 25 feet, represents a storage capacity (see p. 586) worth $20 or more per annum.

Such projects in New England have hitherto been undertaken in a faltering and tentative spirit, and have been confined to the most obviously favorable localities, chiefly to natural lakes whose surface could be raised a few feet by a cheap dam without entailing great expense for flowage. No organization, however, has yet had the courage to attempt any of the great projects which the topography of the country suggests, requiring high dams on large streams, subject to all the obstacles incident to the preoccupation of the country by lines of communication and aggregations of population involving startling changes in the landscape, migrations of people, and relocations of highways and railroads. The mind of the New England manufacturer does not grasp projects of this kind. They appear to him as amusing extravaganzas outside the range of practical consideration. Nevertheless if such projects abide the test of figures, New England in putting them in execution is merely adjusting herself to her natural destiny and strengthening the foundations of her industrial prosperity.

General reasoning and computation cannot be entered into in such a matter, because nature never presents two identical cases. We will consider an actual situation where a great reservoir might be created. We select for this purpose the situation presented by the Housatonic River, which admits at North Canaan, Conn., of the construction of a dam to flow some 40 square miles of ground, and capacious enough to reduce the flow of the stream to uniformity for a long series of years or for an unlimited period. With no personal knowledge of this situation we will examine the project of a great reservoir at this site upon the information given in the atlas of the geological survey, census returns, and similar data. Of course such data would be entirely insufficient for a practical project, but it will suffice for our present purpose.

The drainage area tributary to this dam is 523 square miles, including the water surface. The latter would fall sometimes below 15 square miles and rise at rare intervals, once in 25 or 30 years, to 40. Its average extent would be 25 or 30 square miles. Though the rainfall upon a water surface is ordinarily taken equal to the evaporation, it has been found by careful experiment that in New England the latter slightly exceeds the former. We may, therefore, consider the effective drainage area as 500 square miles, thus making available for this case some of the computations that we have made.

The reservoir would lie in the towns of Egremont, Great Barrington, and Sheffield, Mass., mainly in the two latter, with perhaps a couple of square miles in North Canaan, Conn. The two towns of Sheffield and Great Barrington contain, according to the Mass. State Census of 1895,

Cultivated land .... 18 442 acres valued at near $50 per acre
Uncultivated, pasture, etc. 18 700 " " " " 12 " "
Forest .... 16 406 " " " " 15 " "

Total 53 748

The reservoir would require about half as much land as there is in the two towns, viz., 25 600 acres, but the flowage would probably consist in larger measure of cultivated ground than of pasture or forest. We will therefore assume 15 000 acres at $50 and 10 600 at $15. A large part of this land would be susceptible of profitable use 9 years out of 10 after the construction of the reservoir, but we will assume that it will all have to be paid for.

There are in the two towns 487 farm dwellings valued at $926 each and 668 barns valued at $426 each. We will assume the number of buildings taken to be in proportion to the land taken, say 250 dwellings at $1000 and 340 barns at $500. The greater part of these buildings would simply be moved to other localities at a great deal less expense than here allowed, but we will assume that they are all paid for. Many of the buildings taken will be outside of the two towns named, but this does not matter so long as we allow for a sufficient number.

There are aggregated in villages on the reservoir site, as near as can be ascertained, 7500 people. The larger half of these are in the village of Great Barrington, which carries on the industries of carriage-making, clocks, watches, and jewelry, cotton goods, food preparations, drugs, medicines, etc. In this case the buildings below the flow-line would simply be removed to a higher level without much disturbance of the industries, and the expense would be much less than here allowed for. The other villages would be blotted out. We will assume payment for 1500 dwellings at $2000 each. The capital stock of the manufactories in these villages is reported as near $850 000.

Railroads.—This site is crossed from east to west by the Connecticut Western, and from north to south by the Housatonic Railroad, with an aggregate of 22 miles of track to be relocated. This work would have to be entirely completed previous to any disturbance of existing tracks, but the rails, ties, frogs, switches, etc., would be available to the companies for renewals on other parts of their lines. We will allow for this work at the rate of $25 000 per mile.

Cemeteries.—It is an unfortunate feature of most public improvements, and especially of reservoirs, that they must in some cases disturb the resting places of the dead. The place of interment of a dead body, though probably a matter of entire indifference to the original owner thereof, is an object of keen solicitude to his surviving relatives, and its submergence by water would be a source of great grief. Provision must therefore be made for the removal and reinterment of the bodies of all persons having surviving relatives. New cemeteries above the flow-line must be laid out and appropriately embellished in keeping with those that are submerged, tombs rebuilt, all monuments, tablets, and ornamental fences removed and re-erected, whether the appertaining bodies can be distinctly traced or not. Of course bodies which have lain in the earth twenty-five years, except in tombs or metallic cases, are mixed with the indiscriminate clay beyond hope of separation. In a transient and migratory manufacturing population it is not probable that one in five of the bodies in a state of preservation would have any living representative. It appears to me that this item of expense would be abundantly covered by an allowance of $100 000.

Dam and Sluices.—No data exist for an exact estimate of this item. The dam would be 90 or 100 feet above low water. The contoured State map of Massachusetts shows a fair situation for such a work. Analogy with similar works would indicate a cost something under 2 millions. To insure unquestioned

permanency I allow $2 500 000. We have then for the total cost
of the work:

Dam and sluices	$2 500 000
15 000 acres land @ $50...................	750 000
10 600 " " @ $15...................	159 000
Farm buildings...........................	420 000
1500 village tenements @ $2000...............	3 000 000
Capital stock of manufactories...............	850 000
Railroads, 22 miles relocation @ $25 000	550 000
Cemeteries	100 000
Add for legislation and legal expenses, engineer-ing and contingencies...................	1 671 000
Total cost of reservoir	$10 000 000

This sum would undoubtedly build the work. Let us now
inquire whether the income derivable from this enterprise would
be a satisfactory return on the investment. We cannot here
adopt the simple rule of 30 cents per annum per million cubic
feet capacity and per foot of fall. We will take the value of the
water as simply equal to that of the equivalent coal, calling the
latter 2 mills a pound and taking 3 pounds per hour as a practical
rate of consumption per horse-power. It may be presumed that
the opportunity of getting power on such terms—about half the
cost of steam—combined with perfect steadiness of flow from
month to month and from year to year, would lead to a rapid
development of water-power on the stream.

The water would be delivered into the stream at an elevation
of 660 feet above tide-water and 120 miles distant therefrom.
We will allow 60 feet for maintaining the flow in the stream,
which would be abundant in case the entire river were reduced to
slack water by dams. The remaining 600 feet is the fall that
could be used at mills, where we may assume that 1 cubic foot
of water per second on 12-foot head, or its equivalent, represents
a horse-power.

Table 19 gives the steam-power required for a reasonable use
of the flow from 500 square miles on a head of 15 feet. Adding

up the several columns and dividing the grand total by 25, we find the average steam-power equal to 4426.   That is to say, the average annual cost of steam is represented by the cost of maintaining 4426 horse-power for one month.   Counting 25 working days to a month, and remembering that our head is 40 times that of Table 19, we have for the total cost of supplementary steam-power $4426 \times 24 \times .002 \times 3 \times 25 \times 40 = \$637\ 360$.   Were the water-power on the Housatonic fully utilized by the aid of steam, this item could be counted on as certain, since no manufacturer would hesitate to turn over the money which he actually pays for coal to secure equivalent water-power.

We have seen (p. 586) that the annual volume of water to be expected from 500 square miles is $\dfrac{654\ 200}{25} = 26\ 168$ millions of cubic feet.   The total volume usable without reservoirs is $\dfrac{329\ 904}{25} =$ 13 196 millions.   Subtracting the latter from the total volume, the remainder 12 972 millions represents the addition to the usable volume made by the reservoir.   Of this amount $4426 \times \dfrac{12}{600} \times 25 \times 86\ 400 \times \dfrac{600}{15} = 7648$ millions is the equivalent of the supplementary steam.   The remainder, 5324 millions, is the volume rendered usable by the aid of the reservoirs, over and above the greatest volume susceptible of use by the aid of steam, in the natural condition of the stream, together with water equivalent to the supplementary steam.   The value of this volume of water is determined as follows: A cubic foot of water discharged from the reservoir represents a horse-power for 50 seconds; 72 cubic feet is a horse-power for an hour, representing 3 pounds of coal worth 6 mills; 12 cubic feet, or say for convenience 12.5, is worth 1 mill; 1000 cubic feet is worth 80 mills = 8 cents; 1 000 000 is worth 80 dollars.

The volume of 5324 millions is worth. $425 920
Add coal for supplementary steam 637 360
                                                          ─────────
Total annual value of reservoir. $1 063 280

Making due deduction for taxes, maintenance, and adminis-
trative expense this would be an extremely satisfactory income
on the investment. This result assumes, however, that the
power of the stream is utilized to the greatest advisable extent
by the aid of steam, which, for the case under consideration, is
far from being the fact. If half the fall were utilized it would be
a feasible enterprise, showing a fair rate of interest at the start
and promise of increase from year to year. There are many
great reservoir sites in New England, however, lying at equal or
greater elevations, where the fall of the stream is more than half
utilized and some where it is practically all utilized.

The dazzling projects of water-power development recently
carried out on the outlets of the great lakes, some of which have
signally failed to meet the expectations of their promoters, where
vast quantities of power are obtainable at low cost, but where
useful application and all co-ordinate elements of value are
wanting, have diverted attention from many less striking but
intrinsically far safer and more promising opportunities for the
extension of existing water-power, in situations where remunerative application in established industries awaits every horse-
power that can be added.

Pondage.—Hardly less important than storage, or the
retention of flood-water for use during times of scarcity, is
pondage, or the retention of the flow of a stream during non-
working hours for use during working hours.

To have assumed the water used during the working hours
of the secular day, in the preceding computations, would have
introduced too large an element of hypothesis and made the
matter too complex. This embarrassment we have avoided
by assuming the mill to run throughout the 24 hours. Flour-,
lumber-, and paper-mills usually run in this manner, and require
no pondage unless it be to hold the Sunday flow. Textile
mills more commonly run in the daylight and during the working-hours established by law or custom.

We have taken the maximum limit of the use of water at
800 cubic feet per second, which is very nearly the average flow

of the stream for the entire period considered. This is as large a use as is ordinarily made by an establishment running 24 hours a day, though no doubt a still larger use might in some cases be made with advantage.   An establishment running 10 hours a day and designed to make full use of this volume of water must be organized upon an entirely different basis.   All its buildings, machinery, boilers, engines, water-wheels, race-ways, etc., would require to be on a scale of magnitude about $2\frac{1}{2}$ times as great, and it would require pondage to hold 800 cubic feet per second during the time that the wheels are not running, viz., 50 400 seconds each day; that is to say, something over 40 million cubic feet of pondage.   The maximum limit of the use of water would be $800 \times 24 \div 10 = 1920$ cubic feet per second.

In the actual running of mills it is very seldom practicable to use the water to this extent. The pondage appertinent to a mill privilege depends usually upon the height of its dam and the declivity of the bed of the stream above the dam, and has no relation to the requirements for power.   Practically the mill makes use of what pondage it has, and beyond this allows the night flow to run to waste.   The flow of Sundays and holidays would largely run to waste even with the above pondage of 40 millions.   To avoid any waste of this flow, we should have to hold the flow for about 40 hours, requiring a pondage of $800 \times 40 \times 3600 = 115\ 200\ 000$ cubic feet.

To exhibit the value of pondage by the same rigorously exact methods as have been adopted with reference to storage, Tables 25 and 26 have been computed, Table 25 being merely auxiliary to the computation of 26.   These tables assume an establishment with the same drainage-area and the same fall as in the former case, running 10 hours a day and stopping 1 hour at noon, its maximum use of water being 1000 cubic feet per second.   In Table 25 the first column is the natural flow of the stream in cubic feet per second, running from 50 to 1000 with intervals of 25.   The second column gives the quantity of water which the mill is able to use with any given

TABLE 25.—VALUE OF PONDAGE TO A MILL COMMANDING 500 SQUARE MILES OF NEW ENGLAND DRAINAGE-AREA.
Running 10 hours a day and capable of using 1000 cubic feet of water per second. Head 15 feet.

1	2	3	4	5	6	7	8	9
	10 Million Cubic Feet Pondage.				20 Million Cubic Feet Pondage.			
Flow of Stream in Cu. Feet per Second.	Draft of Mill in Cubic Ft. per Sec.	Gain from Pondage.		Cost per Diem of Steam-power Equivalent to Gain from Pondage.	Draft of Mill in Cubic Ft. per Sec.	Gain from Pondage.		Cost per Diem of Steam-power Equivalent to Gain from Pondage.
		Cubic Ft. per Sec.	Horse-power.			Cubic Ft. per Sec.	Horse-power.	
50	140	90	112	$11.20	140	90	112	$11.20
75	210	135	169	16.90	210	135	169	16.90
100	265	165	206	20.60	280	180	225	22.50
125	319	144	242	24.20	350	225	281	28.10
150	372	222	277	27.70	420	270	337	33.70
175	428	253	316	31.60	475	300	375	37.50
200	483	283	354	35.40	530	330	412	41.20
225	525	300	375	37.50	584	359	449	44.90
250	553	303	379	37.90	639	389	486	48.60
275	580	305	380	38.00	693	418	522	52.20
300	608	308	385	38.50	748	448	560	56.00
325	635	310	387	38.70	802	477	595	59.50
350	663	313	391	39.10	857	507	634	63.40
375	690	315	394	39.40	911	536	670	67.00
400	718	318	397	39.70	960	560	700	70.00
425	745	320	400	40.00	995	570	712	71.20
450	773	323	404	40.40	1000	550	687	68.70
475	800	325	406	40.60	1000	625	656	65.60
500	828	328	410	41.00	1000	500	625	62.50
525	856	331	414	41.40	1000	475	594	59.40
550	883	333	416	41.60	1000	450	562	56.20
575	911	336	420	42.00	1000	425	531	53.10
600	938	338	422	42.20	1000	400	500	50.00
625	965	340	425	42.50	1000	375	469	46.90
650	993	343	429	42.90	1000	350	437	43.70
675	1000	325	406	40.60	1000	325	406	40.60
700	1000	300	375	37.50	1000	300	375	37.50
725	1000	275	344	34.40	1000	275	344	34.40
750	1000	250	312	31.20	1000	250	312	31.20
775	1000	225	281	28.10	1000	225	281	28.10
800	1000	200	250	25.00	1000	200	250	25.00
825	1000	175	219	21.90	1000	175	219	21.90
850	1000	150	187	18.70	1000	150	187	18.70
875	1000	125	156	15.60	1000	125	156	15.60
900	1000	100	125	12.50	1000	100	125	12.50
925	1000	75	94	9.40	1000	75	94	9.40
950	1000	50	62	6.20	1000	50	62	6.20
975	1000	25	31	3.10	1000	25	31	3.10
1000	1000	0	0	0	1000	0	0	0

flow of the stream and 10 million cubic feet of pondage. Thus, with 50 cubic feet per second, the entire flow of 168 hours can be saved and drawn in the 60 working hours, giving a draft of 140 cubic feet per second. The same with a flow of 75 cubic feet per second. With a flow of 100 cubic feet per second there is a little wastage of the Sunday flow. Above 214 cubic feet per second the Sunday flow is wholly wasted. Above this stage the computation is as follows, say for a flow of 225 cubic feet per second:

In pond at starting of mill................ 10 000 000 cu. ft.
1 hour flow held at noon, 225 × 3600....... 810 000 "
Pondage drawn in 10 hours or ————
         36 000 seconds = 300 cu. ft. per sec.. 10 810 000 cu. ft.
Add natural flow of stream, 225 " "
         ————
         Total draft of mill 525 " "

TABLE 26.—VALUE OF PONDAGE TO A MILL COMMANDING 500 SQUARE MILES OF NEW ENGLAND DRAINAGE-AREA.

Running 10 hours a day and capable of using 1000 cubic feet of water per second on a head of 15 feet.

Year.	Saving in Steam-power with		Year.	Saving in Steam-power with	
	10 Mil. Cu. Ft. of Pondage.	20 Mil. Cu. Ft. of Pondage.		10 Mil. Cu. Ft. of Pondage.	20 Mil. Cu. Ft. of Pondage.
1875	$6619.20	$8776.70	1888	$3459.94	$4 486.81
1876	6945.00	9214.56	1889	4535.54	5 715.74
1877	5530.28	7580.67	1890	4024.94	5 451.19
1878	4230.90	6636.11	1891	6693.30	9 451.68
1879	6741.05	9231.94	1892	8091.25	10 290.28
1880	6432.78	7346.67	1893	6590.13	8 557.46
1881	6748.62	8489.39	1894	9003.68	11 197.83
1882	6615.00	7992.54	1895	5433.84	7 084.76
1883	7580.56	8739.64	1896	7225.75	10 212.52
1884	6244.65	7143.49	1897	7494.04	8 876.20
1885	4374.88	5447.31	1898	3699.62	5 061.02
1886	5489.30	6614.11	1899	4053.07	5 197.05
1887	6132.66	7950.17			
Averages.				5999.60	7 709.83

The third column gives the gain from pondage in cubic feet per second; the fourth, the same in h.p.; the fifth, the value of the equivalent steam-power, or, rather, the expense saved where water is substituted for steam, which we take at 1 cent per hour or 10 cents per diem per horse-power. When the draft reaches 1000 cubic feet per second the gain from pondage begins to diminish; and when the natural flow of the stream reaches 1000 cubic feet per second there is no further gain from pondage. The sixth, seventh, eighth, and ninth columns are the same as the second, third, fourth, and fifth, except that the former refer to a pondage of 20 million, the latter to one of 10.

Table 26 gives the aggregate value of pondage for each year of the series for 10 and 20 millions respectively. It is computed from Tables 25 and 17 with the aid of the calendar for the several years. Thus for the year 1875 with a pondage of 10 millions:

Value for Mo.

January. Average flow 80 cu. ft. per sec. Value of pondage
per diem, Table 25, $18.00. No. of working days 26.
Value for the month 18 × 26 =........................ $468.00
May. Flow 919. Val. per diem $10.14. Working days 25....... 253.50
June. " 672. " " " 40.88. " " 26....... 1062.88
July. " 248. " " " 37.89. " " 26....... 985.14
Aug. " 305. " " " 38.54. " " 26....... 1002.04
Sept. " 160. " " " 29.26. " " 25....... 731 50
Oct. " 497. " " " 40.98. " " 26....... 1065.48
Dec. " 451. " " " 40.41. " " 26....... 1050.66

Value of pondage for the year at 1 cent per h. p. hour $6619.20

If for any reason it should seem proper to take the value of power at a higher or lower rate, this result can be changed accordingly.

As shown by Table 26, the average result for the entire period of twenty-five years is $6000 for a capacity of 10 millions, and $7710 for 20 millions. These sums represent the average annual benefit of the respective amounts of pondage to an establishment running under the conditions supposed, as compared with one running under the same conditions without pondage. These results would appear larger if we should assume a larger maximum use of water; that is, if we assume the establishment capable of using 1200 or 1500 cubic feet of water per second instead of 1000.

In the case of a series of mills running 10 hours a day and supplied by storage-reservoirs, it often happens that the daily discharge from the reservoirs does not reach the mills within the working hours, and more or less pondage is necessary in order to make the discharge of the reservoirs available and avoid waste. No greater amount of pondage, however, is necessary for this purpose than would be required to make an equally complete use of the flow of the stream in its natural condition; generally less.

The fall of a stream is hardly ever uniform throughout its course, but is often concentrated in a few miles of rapids with long reaches of quiet water above and below. These rapids are occupied by a succession of dams with little opportunity for the creation of pondage. The common interest, in that case, calls for a large pond at the head of the line, and each mill, on starting, draws from its own pondage only till it receives water from the next pond above, so that the single pond serves the whole series. This pond may have a value many times as great as computed above. Such a case exists at Gardiner, Maine, where there is a series of dams supplied by the Cobbosseecontee River, which stream, draining about 215 square miles, has some 2500 or 3000 million cubic feet of storage capacity, and has a pond at the head of the series of dams, capable of holding the discharge of the reservoirs while the mills are stopped.

The most fortunate condition is where the same basin serves both for storage and pondage, as occurs at Rockville in Connecticut. Here Lake Shenipset, near the head of the Hockanum River, has a drainage-area of some 15 square miles, and is controlled by a dam to the extent of some 24 feet, holding practically the entire flow of the drainage-ground. Below the dam is a series of mill privileges with an aggregate fall of more than 250 feet. The declivity is so great that the pondage appurtenant to each dam is very small. These are mostly textile mills and run but 10 hours a day. The reservoir-gates are opened only during working-hours. As soon as any mill commences to draw from its pond, the mill above commences to discharge into the same, and very little water runs to waste. Very rarely does so small an extent of drainage-ground furnish so large an amount of power.

Common Interest of Mill-owners on the Same Stream.— The water-powers of a stream have many interests in common. The advantage of increasing the power by means of storage-reservoirs is common to all the sites below the reservoir and cannot be made available without mutual co-operation. The injuries to water-power incident to the discharge of sawmill refuse into the stream and to the running of logs are of common interest. The difficulties incident to sawmill refuse, leaves, and other trash are sometimes met by devices of the nature of revolving screens, which withdraw the stuff from the water and usually return the same to the tail-races for the annoyance of mills below, whereas a regard for the common interest would dictate its being dried and burnt. Diversion of water for the supply of towns and legislation adverse to the interests of mill-owners are matters of common interest. So to a greater or less extent are the subjects of pondage and back-water. Cases frequently occur where one dam could secure an addition of two feet or more to its head, with great improvement of its pondage, by taking as many inches from the head appurtenant to the next dam above. For these reasons it is common for mill-owners on the same stream to maintain a mutual organization and an agency for the control of affairs common to the entire stream.

CHAPTER XXV.

COMPUTATION OF DAMAGES TO MILL-OWNERS RESULTING FROM THE DIVERSION OF WATER.

WATER applied to the generation of power is a utility which can readily be replaced by steam. Water for drinking, cleanliness, sanitary and domestic uses is a necessity for which no substitute can be found. In the growth of communities and the aggregation of population in great masses, the necessity is constantly arising for the diversion of streams from their natural channels for purposes of domestic water-supply. The State readily grants such rights of diversion on the ground that the new application of the water is more beneficial to mankind than the old; but such rights, whether so stipulated in the grant or not, are always coupled with the obligation to pay the damages incident to the diversion. The estimation of these damages is therefore a duty which frequently devolves upon the hydraulic engineer.

The nature of the mill-owner's right in the water is not that of ownership. He does not own the water. The only thing he has a right to is the effect of the water while passing from the upper to the lower level. He has the right to have the water flow past his mill and through his wheel, and after it has made this journey he has no further control over it. His loss is fully made good by a sum of money sufficient, under existing conditions, to supply in perpetuity, the power lost by his mill. The theory of the law is that the loss is represented by the diminution of market value, i.e., by the difference in the market value of the property before and after the diversion, technically

606

called the "taking," of the water; but the extreme clumsiness and inapplicability of this rule has led courts to accept the cost of replacing the lost power as the diminution in market value.

The legislative grant of authority to divert or "take" the water takes different forms. Sometimes it is a certain number of millions of gallons daily. Sometimes it is the right to divert the entire flow of a stream, that is, to abstract a definite number of square miles of drainage-ground. Sometimes it is the right to divert the flood-waters or a certain portion thereof, in which case little injury is done to the water-power unless favorable opportunities exist for the storage of water.

In the trial of such cases many false and untenable claims are set up on either side. The mill-owner usually figures up his claim in this way: "You take from me so many, say 25, cubic feet of water per second. This, on the existing fall of 15 feet, and assuming an efficiency of 80 per cent in my wheels, amounts to some 34 h.p. Power furnished to small consumers commands, say, $75 per annum per h.p., which would amount to $2550 per annum. The sum which would yield this annuity at the highest attainable secure and permanent rate of interest, say 3 per cent, is $85 000." In addition to this the loser usually sets up a claim for the installation of additional steam-power to the extent of the diversion. The taker reckons in this manner: "We concede that we take 25 cubic feet per second usable on a fall of 15 feet. It is only for six months in the average year that this diversion does you any injury, the remainder of the year you have an abundance without it. The efficiency of your wheels, in their ordinary running, cannot be placed higher than 75 per cent, so that your loss is but 31 h.p. This loss subjects you to no expense that you can clearly define and specify other than the burning of some 93 pounds of additional coal per hour, worth, say, 2 mills a pound or $1.96 per diem. This for half a year would amount to $300. This annuity used in your business would be worth 6 per cent to you and should be capitalized on that basis. Therefore $5000 is all you are entitled to." With such divergence of views it is not strange that a mutual agreement is difficult.

The writer regards the following as the equitable principles governing such estimates. The loser can, in general, make no claim for additional steam-plant necessitated by the diversion. It is only in virtue of his ability to make a rational use of the water that he suffers loss by the taking of it. Such rational use implies steam-power sufficient to supplement the water-power at all stages of the stream. Consider a mill using 1000 h.p. on the stream of 500 square miles, as already assumed, and suppose 50 square miles to be diverted. The total flow, shown by Table 17, falls many times below 100, often below 50, cubic feet per second. The engine should be capable of working with fair economy when the stream is as low as 100 cubic feet per second. That is, it should furnish 875 h.p. with fair economy. Such an engine would yield 1000 h.p. with slightly diminished economy or slightly increased cost per h.p. After the diversion the stream would yield 90 cubic feet per second where it now yields 100, 45 where it now yields 50, etc. The only effect of the diversion is to prolong the time during which the engine works with diminished economy. The diversion unquestionably necessitates increased steam-power, but nothing could be more fanciful than the claim that it necessitates increased steam-plant. It is to be remembered that there will be years after the diversion with greater flow than others before, and years before the diversion with less flow than others afterwards. The existence of storage-reservoirs does not alter this conclusion, because the mill-owner, in fixing the capacity of his engine without any reference to the diversion, is bound to take notice that the reservoirs are liable to be exhausted while the stream is in its lowest stage.

Still more irrational is the claim that the cost of supplying the lost power should be computed upon the basis of an extra steam-plant of the precise capacity required to supply the same, which would be vastly more expensive in operation. The owner is entitled to reimbursement of the expense of repairing the loss incident to the diversion. But this loss must be repaired in a reasonable manner. Now, the reasonable mode of meeting fluctuations in the flow of a stream is the one universally employed, viz., an engine of sufficient capacity, and whether these fluctuations arise from vicissitudes of season or from human agency is immaterial in this connection.

The mill-owner cannot claim the market value of the power producible from the diverted water, that is, of power transmitted electrically or otherwise, and placed at the disposal of the consumer. The loser is entitled to the market value of the thing lost. The loser of a basket of eggs is entitled to the market value of the eggs, not to the market value of the chickens that might, with trouble and labor, have been produced from them. The loser of water is entitled to the market value of the water, in the market that is open to it. This market is the water-wheel, and the only value that can be obtained for the water is a certain saving in the cost of the power. The value of this saving is what the user is entitled to.

Wildly extravagant claims are often set up when storage-reservoirs are taken by municipalities. It is contended that these have a great value by reason of their adaptability to purposes of water-supply; that they are indispensable and would command a high price for that purpose, etc. In the taking of Whitehall Pond by the city of Boston, several years ago, an unheard-of award was made on this ground. Municipalities are entrusted by the State with the power of seizure and condemnation. Without this power municipal development might be wholly arrested by the cupidity of individual landowners. The price paid for property taken is what the property is worth to the owner, not what it is worth to the taker. The value of a cup of water to the owner of a spring is trifling, not worth taking account of. To the wayfarer perishing with thirst it may be of such value that he would give all his wealth rather than not have it. To say that a town having the right of seizure and condemnation ought to pay for a reservoir or other adjunct of water-power the same price that it might be compelled to pay if it did not have that power, is simply an attempt to evade the legal maxim that the value of the thing taken cannot be regarded as enhanced by the taking. The value of the reservoir must be determined by the use which the owner can make of it for his own purposes.

On the other hand, the theory sometimes set up by municipalities that the value of the power diverted is measured by the cost of coal required to replace it is untenable. It is true that it is difficult precisely to define and specify the other elements of cost. Nevertheless we know that the entire cost of steam-power is ordinarily something like double the cost of the coal consumed, and that if the entire power were taken it would have to be paid for upon that basis or something like it. To say that a mill-owner ought to dispose of a small part of his power at a lower rate than he would be entitled to if he sold the whole of it is certainly an unfair proposition. The reverse statement would be more equitable, viz., a small part of the power should command a higher rate per h.p. than the whole.

In such cases some application can be given to the proposition that doubtful points should be construed in the claimant's favor. He does not seek to sell his power. It is taken from him, presumably without his consent. He must be fully paid for it. The sufficiency of the payment must not be open to doubt. He must not be subjected to loss or to serious risk of loss in the transaction. This requires that points of uncertainty and doubt should be construed in his favor. A different principle must be adopted in estimating the value of a water-power for the guidance of an investor. Here, the estimate must not leave any doubt as to the value assigned the property. It must not subject the investor to loss or serious risk of loss. Doubtful elements of value must be excluded, and doubts must be construed against the property. For an engineer, therefore, to estimate the value of water rights at a higher figure when acting on behalf of a claimant than on behalf of an investor is not necessarily inconsistent with good conscience.

The tables, under the head of Storage and Pondage, suggest the method to be adopted for computing the value of a given water-power, but they can be more specific when applied to a specific case. The elements of the computation are: (1) The drainage, and (2) the flow per square mile to be expected for a series of years. From these we find (3) a table of the quantity of water to be expected from the entire drainage-ground for a series of years (table 17). (4) The fall, which we have assumed at 15 feet, but in fact it will vary from day to day according to the stage of the stream. (5) The pondage, whether directly appertinent to the dam or derived from ponds higher up the stream. (6) The largest flow of the stream that it would be judicious to provide for the use of. This is usually taken as about the average of the fourth highest month, being 965 cubic feet per second in Table 17, excluding pondage.

We would now proceed to form a table of averages running through the entire series of years, generally month by month. This would contain:

1. The flow of the stream.
2. The quantity of water added to the natural flow, during working-hours, by pondage. This may require more than one column, as there may be more than one pond.
3. The quantity of water susceptible of use by the mill.
4. The head acting on the wheels. This may require several columns, as there are several sources of loss of head, viz., in head-race; in tail-race; in wheel-pits and penstocks. These losses must be determined by observation coupled with the mechanical principle that either of these losses is proportional to the square of the velocity.
5. The power. This is most conveniently expressed in horse-power hours, the unit being a horse-power for one hour.

6. The value of the power per diem.

7. The number of working-days in the month.

8. The value of the power for the month.

9. The value of the power for the year.

The average of the quantities in the ninth column, for the series of years, must be taken as the true value of the power which the entire drainage-ground is capable of yielding estimated with reference to the given fall.

Assuming a case of diversion of a certain part of the drainage-area, 50, 60, etc., square miles, we form a new table embracing the remaining area, and compute the value of power on the diminished area in the same manner. The difference between these two results would be the value of the power lost if the mill were in a position to use the assumed quantity of water. Generally, however, the existing use of water is much less than the maximum on which these results are based. The power available but not in use cannot be accounted valueless, but it is of less value than that already in use. We now find the maximum power in present use by the mill, and go over the tables and separate the power into two parts: (1) the power actually in use; (2) the power not now in use but available at a later date. The history of the mill and the growth and expansion of the manufacuture will show the average date at which it will have use for the entire power. The value of the power in the second category may be regarded as a sum of money due at that later date, and subject to a corresponding discount. Making this discount and adding the sums, we have the true value of the water-power before and after diversion. The difference is the annual value of the loss sustained by the owner, and he should receive a sum of money sufficient to yield this annuity at such rate of interest as can be obtained on safe and permanent investment.

It would appear at first view that instead of forming a table running through a series of twenty-five years we might take the average of the several months in Table 17 and thus form an ideal year to serve as the basis of our computations. The objection to this method is that it would include a large quantity of water which actually runs to waste. Thus the average of all the Decembers in Table 17 is 798 cubic feet per second, which, being below the assumed maximum use, would all be included as usable water, whereas in point of fact there are no less than six months of the series in which large quantities of water ran to waste.

It would be equally fallacious to compute the loss on the assumption that the power lost is in proportion to the area diverted. In the case we have supposed of 50 square miles diverted from a water-power commanding an area of 500, suppose the mill to have use for water up to the average of the seventh month in the order of flow, i.e., to waste water for five months. There are five months of the year that the diversion does the mill no injury, and during four of the remaining months the flow is very small. So that the quantity of power diverted from the mill is in no sense proportional to the drainage-area diverted. If we suppose a number of successive diversions of 50 square miles each, the second would reduce the period of wasting water. Each successive diversion would cut more and more into the months of abundant flow, and each would do a greater injury to the water-power than the one preceding. If the drainage-area were provided with storage-reservoirs to the extent of impounding all the water, then the injury consequent upon any specific diversion would be proportional to the area diverted.

Industries Requiring Heat.—A case which occasions much perplexity is that of mills carrying on the operations of bleaching, dyeing, printing, etc., requiring great quantities of heat. In this case, the steam from the engines, instead of passing to the condensers and so increasing the available power, is led, under a pressure of 5 or 6 pounds, to the heating-pipes. By sacrificing the power due to condensation, more than nine-tenths of all the heat imparted to the steam may be made available for industrial purposes. Establishments of this kind exist in which it makes but slight difference in the consumption of coal whether the water-wheels run or not, and where a question of the value of water-power so applied arises, it is contended, with a show of reason, that it is of no value, since it could be dispensed with with but slight increase of expense—an increase not exceeding the expense of maintaining wheels.

The true view of this case appears to me this: Water-power is perpetual; the applications of water-power are temporary and transient. The running by water-power of an establishment which is susceptible of being operated by a small fraction of the heat required in the industry is not a judicious use of water-power. But, like all other property, water-power cannot be accounted valueless because owners are not making a profitable use of it. Its value is to be determined from the uses of which it is susceptible, not the use to which it is, for the time being, applied.

Nevertheless water-power so applied cannot be accounted of the full value of water-power in profitable application, because, although it is susceptible of a profitable use, a certain time must necessarily elapse before it can be converted to that use, and this time during which it must continue in unprofitable use may properly be considered in abatement of its value.

Water returned to the River through Sewers.—Cities on tide-water are usually supplied by aqueducts from distant points, and the water, after passing the sewers, is discharged into the sea when it has no further value for water-power. In this case all water-powers on the supplying stream, from the point of diversion to the sea, have equitable claims on account of the diversion. In inland cities, the water, after passing the sewers, usually falls into the same stream in a condition quite as available as before for water-power to mills farther down. In fact the operations of the city often augment, in this respect, the value of the stream which it takes. By impounding the flood-waters of the stream and drawing them at a uniform rate it renders much of the water available for power which would otherwise run to waste. Hence, in the earlier cases of diversion, no claim was set up on behalf of water-powers lying below the sewer-outfall of the diverting municipality.

In later cases, however, claims have been allowed on be-half of all water-powers lying below the point of diversion. This change is due to the rapidly changing conditions of municipal growth. The demands of public health are becoming more and more imperative. It is within the power of the state, and this power is often exercised, to require municipalities to divest their sewage of all noxious qualities before turning it into a running stream. This is often done by spreading it upon land after the manner of irrigation, in which case a large part of it is taken up by vegetation never to reenter the stream. They may be required to turn the sewage into a channel conducting it to the sea separate from the stream, thus excluding its further use for water-power. The court is bound to adjust the claims of mill-owners once for all. It can leave nothing contingent or undetermined. It therefore treats the water as permanently withdrawn from the stream to the injury of all mill-owners farther down. Of course it is presumable that the legislative act authorizing the diversion might be so framed as to preclude claims of this kind by authorizing or requiring the municipality to return the diverted water in un-impaired volume to the stream.

Hydraulic Experts.—From time immemorial it has been the practice of judicial tribunals, in cases involving scientific and mechanical principles or special technical knowledge, to avail themselves of the testimony and opinions of persons skilled in the art. Such witnesses are usually a conspicuous figure in trials for the diversion of water. Witnesses of this character are not subject to the same rules as ordinary witnesses, who are supposed to confine their testimony to facts of their own knowledge, abstaining from comments, inferences, or deductions. The expert witness is expected to give opinions and inferences founded upon facts stated by himself or others. This practice is admitted as the shortest way of arriving at conclusions. Many questions coming before a judicial tribunal are of that character that would require for their solution an amount of instruction and an expenditure of time entirely inconsistent with the transaction of legal business. All this is avoided by referring the question to the decision of one who has made a special study of the subject. Nevertheless the court is in no way bound to accept the opinion of the expert. It may even exclude the testimony of experts without subjecting itself to the imputation of error. The expert can only influence the decision of the court by his reputation for attainments and honesty, and by the soundness of his reasons. His position before the court partakes both of the character of a witness and that of a scientific or technical adviser.

Experts in this country and in England are employed and paid by the litigants, and, naturally, preference is given to those most ready to make statements favorable to their employers. The common practice is to engage the expert to make an examination and report upon the question at issue. If the report is sufficiently favorable, he is placed upon the witness-stand; if otherwise, he is paid for his report and his connection with the case terminates. In important cases several experts are consulted, and those who make the most favorable reports are chosen. Often the report and testimony of an expert are so manipulated by sharp lawyers as wholly to misrepresent his real views. By adroit questioning opinions favorable to the contention are elicited, those unfavorable suppressed. From a number of expert witnesses a chain of propositions can thus be evolved going to support contentions which neither of them would have maintained separately.

These causes—the desire of the expert for profitable employment, and the furious zeal of lawyers in pursuit of legal advantage—have led to deplorable abuses in the use of expert testimony, and the evidence of the professional expert is always looked upon with distrust by the courts. "They come," says Lord Campbell, " with a bias in their minds to support the cause in which they have embarked, and hardly any weight should be given to their evidence." Many legal opinions might be quoted in which similar views are expressed.

It must be conceded that trials of water cases, in our own country, furnish too much verification of these ill opinions. Hydraulic engineers as a class are probably as truthful and scrupulous in their statements as any other class of men, and if all litigants were in court in pursuit of exact justice there would be no difficulty. But the claimant often regards the diversion of his water as the opportunity of a lifetime, which he is bound to make the most of, and the respondent regards any claim on that account as an attempt at robbery; so that, each having an unfair and unjust proposition to maintain, they are compelled to have recourse to the more reckless and unscrupulous members of the profession.

In a recent Massachusetts case of claims by mill-owners against a municipality for diversion of water some twelve or more experts were employed on each side. Those acting for the city estimated the damages, upon an average, at about $125 000. Those acting on behalf of the mill-owners swore to estimates aggregating, as an average, nearly $1 500 000. The award of the tribunal was about $500 000. This result would go far to justify the learned Campbell's opinion, and the judicial commission before whom this case was tried were compelled of necessity to act upon the same, for certainly two estimates which differ by twelve-fold can afford but little guidance to the impartial judge.

Nevertheless this dictum of Lord Campbell is too harsh and sweeping. The expert's claim to credence does not rest solely upon his word or his oath. It rests mainly upon the reasons that he is able to give for his statements; and if he thoroughly understands the subject, his views and elucidations cannot fail to aid the court in the pursuit of its own conclusions. It is, moreover, always open to the litigant to employ men whose character, standing, and habitual fairness preclude any suspicion of bias, and this is the usual course when the contention is just and fair. It is the unjust contention which calls for the employment of the unscrupulous witness.

It is often suggested to the expert that an extravagant estimate on his part is the only mode of securing a fair award, seeing that the opposing party will present extravagant estimates and the court is naturally disposed to adopt an average between the two. Not to urge the antiquated maxim which forbids us to do evil in order that good may come of it, it is sufficient to observe that this method is grievously unfair to the respondent, since his expert, in his estimate of damages, cannot put them below zero, whereas the claimant having at his command the nine digits of the numerical system together with unlimited ciphers, can present figures which averaged against zero would still be extravagant.

The method of the French courts appears well calculated to avoid the abuses incident to the employment of expert testimony which have become so flagrant in England and America. In certain cases experts are required by law, in others their employment is optional with the litigants. In every case they are appointed by the court. By the Code of Civil Procedure, Art. 302, " L'expertise ne pourrera se faire que par trois experts, à moins que les parties ne consentent qu'il soit procédé par un seul," i.e., three experts are required unless the litigants consent to trust the matter to one. Even where experts are neither called for by law nor demanded by the litigants it is in the discretion of the court to employ such, and in that case the number is limited to one. The expert is not permitted to regard himself as acting on behalf of either of the litigants. His compensation is fixed by the court, indorsed on his report, and payable like other costs. This method makes the expert what he ought to be, viz., the unbiassed adviser of the court.

CHAPTER XXVI.

LEASES AND RENTALS OF WATER-POWER.

SMALL water-powers, namely, those not capable of supplying more than one establishment, company, or corporation, are generally owned and controlled by the user; but large powers, capable of supplying a number of concerns, are usually the subject of separate ownership. The power requires to be divided and distributed and is furnished to the users on payment of specified rentals.

At old water-powers of minor importance, water was formerly furnished to more than one user on a system of this kind: The first grant would be for water sufficient to drive a specified number of mill-stones, a specified number of sets of woollen machinery, as much water as would be necessary for a wheel of specified dimensions, etc., etc. At a later date a second grant would be made of similar tenor, and still later a third, etc. Under the law governing such transactions, the first grantor is entitled to draw the specified quantity at all times that the stream will furnish the same. The second can only draw from the excess remaining after fulfilling the first. The third after the first and second are satisfied, etc. These privileges were called first rights, second rights, third rights, etc. In the course of time it becomes desirable to introduce more efficient motors, a different kind of machines, etc., and then literal compliance with the terms of the grant becomes impossible. Suits at law are often instituted in such cases for the purpose of obtaining a legal definition of the quantity of water to which the several lessees are entitled.

Courts usually hold that the expressions "run of stones," "set of woollen machinery," "breast-wheel," etc., are introduced merely as convenient measures of the quantity of water intended to be granted, and that the quantity so intended may be used for any purpose whatever at the discretion of the grantee. It is also the legal presumption that any improvement made by the grantee subsequent to the date of the grant, whereby the power of the water is increased, such as an increase of fall, the adoption of more efficient motors, etc., inures to the advantage of the grantee, and not to that of the grantor. The court inquires what quantity of water would be necessary to fulfil the purpose of the grant, on the head, with the machines, motors, and appliances, and in accordance with the methods in use at the date of the grant.

Many old leases express the quantity of water in inches, often with apparently no clear conception of what is meant by an inch of water. This term originated in mining, where "a miner's inch" means the quantity of water discharged through an orifice 1 square inch in cross-section under a specified or implied head, usually 6 inches. This is about 1 cubic foot in 41 seconds or about 2100 cubic feet in 24 hours. In American practice an inch of water more commonly signifies a stream of water 1 square inch in cross-section moving with the velocity due to the existing head. It is often used in this sense by makers of water-wheels, and in this sense is sometimes employed in leases.

At Lowell the water is held and controlled by a corporation entitled The Proprietors of the Locks and Canals on Merrimack River, that being the name of the old navigation company which owned the rights under which the development of water-power was accomplished. In point of fact, however, the water is here owned in common by the users, the several manufacturing companies being owners of the stock of the Proprietors of Locks and Canals, and each company being entitled to water-power in proportion to its stock. The Proprietors of Locks and Canals is thus a mere agency for the management of the water-power.

In the indentures of lease between this corporation and the

several users, the practice originated of defining water-power in a
scientific manner by the adoption of a unit called a mill power,
involving both elements of water-power, viz., quantity of water
and head, and which is thus defined:

"The right to draw from the nearest canal of the said proprietors so
much water as during fifteen hours in every day of twenty-four hours shall
give a power equal to 25 cubic feet per second at the great fall, when the
head and fall there is 30 feet; to 45.5 cubic feet per second at the lower
fall, when the head and fall there is 17 feet; and to 60.5 cubic feet per second
at the middle fall, when the head and fall there is 13 feet. And in order to
prevent disputes as to the power of each mill privilege in the variations of
the height of water from changes of the season or other causes, it is understood
and declared that the quantity of water shall be increased in proportion to
the reduction of the height, . . . one foot being allowed and deducted from
the height of the actual head and fall and also from that with which it is
compared before computing the proportion between them."

That is to say: If the fall were reduced to 29 feet when the
standard fall is 30, the quantity of water would be, not $\frac{29}{30}$, but $\frac{31}{30}$
of 25, viz., 25.89 cubic feet per second. The deduction of 1 foot
is intended to allow for the loss of head in penstocks and race-
ways. It is a very crude correction, because this loss is proportional to the square of the velocity and varies greatly with the quantity of water drawn.

The head is diminished in time of high water by the rising
of the river below the mills more than it rises above, and in time
of low water by the lowering of the mill-pond in the afternoon, and
the consequent increased velocity in the canals, consuming a large
amount of head in giving motion to the water. This loss of head
is an evil that aggravates itself. The more the mills draw the
greater is the velocity in the canals, and the greater the resulting
loss of head. The result is that, unless the draft of water is under
constant and firm control, the lessees may so diminish the head
as to inflict great mutual injury.

The lessees at Lowell pay to the central organization $300 per
annum per mill-power as a fund for maintenance and adminis-
tration. This applies to the "permanent power," that which
can be guaranteed at all times. The great and variable volume

of water which the river furnishes in excess of the permanent power, called the "surplus power," is leased subject to interruption at so much per mill-power per diem. When this system was introduced, about 1859, the prices were graduated with the view of discouraging the excessive use of surplus power; increasing as the surplus used bore a larger proportion to the permanent power belonging to the user. They ranged from $3.50 to $14 per diem per mill-power. Later, since the great development of steam-power, the prices have been graduated with the view of encouraging the use of surplus power, diminishing as the surplus used bears a larger proportion to the permanent power. These rates, being merely arrangements for adjusting conflicting interests between common owners, are of no importance as indicating the value of water-power.

This company has of late been making improvements in its canals, designed to increase their carrying capacity, whereby a larger income may be derived from the surplus water. These improvements consist, to some extent, in enlarging and deepening the canals, but mainly in diminishing the resistance to the movement of the water, by lining the beds and slopes with timber and planking as pointed out under the head of " Canals," and by pointing and otherwise smoothing the masonry side walls. The velocity on a rough and irregular canal may often be increased from two- to three-fold, without change of cross-section or slope, by reducing it to smoothness and regularity.

At Lawrence, Holyoke, Turner's Falls, and other places in New England the water-power is owned by a separate corporation having its own stockholders and paying its own dividends. At Lawrence the permanent powers pay an annual rental of $1200 per mill-power, which is here 25 cubic feet per second on 30 feet head, or its equivalent. This, however, does not measure the value of the power, as the sale of water-power usually has carried with it the sale of land, which owes its value chiefly to the water-power. The owners of the water-power (the Essex Company) also derive a large revenue from the rental of surplus power, and the existing rates for the latter may be assumed to represent in some measure the market value of water-power at Lawrence.

The present rate for amounts not exceeding 50 per cent of the permanent power to which the user is entitled is $4 per mill-power per diem, $3 for anything in excess of 50 per cent. These rates are for water used between 6 A.M. and 6 P.M. For water used between 6 P.M and 6 A.M., which would otherwise run to waste, the rates are merely nominal: about $1.20 per mill-power. In times of low water the Essex Company limits the quantity of water that may be drawn, and for water drawn in excess of such limitation the charge is $20 per diem per mill-power. This rate is not fixed as a penalty, but as the extreme limit which users will pay for water available in time of scarcity.

At some great water-powers where large tracts of land have been acquired by the promoters, the main reliance for profit is placed upon the sales of the latter, and the water is leased at nominal rates in the expectation that this method will lead to a more rapid development of industry and a better market for the land. This was notably the case at Turner's Falls on the Connecticut, where the water was largely disposed of in perpetual leases at $7.50 per annum per horse-power.

This company fixes the price of surplus power at $10 per annum per horse-power. The horse-power contemplated in these rentals is a theoretical horse-power. That is to say, 550 pounds of water per second is accounted a horse-power for each foot of available fall. Any other definition of "horse-power" in leases of water leaves a question open for dispute and litigation. This company's experience has not demonstrated the soundness of its methods. There was at first a large sale of land and the place had a period of rapid growth. Later, however, sales have fallen off and the population has remained nearly stationary for a number of years.

In any project for the development of water-power for manufacturing industry, the increase in the value of land consequent upon the improvement is a legitimate source of profit and should be made available by securing all the land within the direct sphere of influence of the improvement. It does not follow that the capital representing this property must lie idle and unproductive till the land becomes salable for building purposes. Farm property may usually be leased at rentals which will pay interest on the purchase-money, and these rentals may be expected to increase from year to year as the enterprise affords a better market for produce. Instead of direct purchase, a system of bonding the land may be adopted, each owner being induced for a comparatively trifling consideration to encumber his estate with an agreement to sell the same at a stipulated price if tendered within a stipulated term of years, the price being fixed distinctly above the existing market value of the land and subject to compound interest at a moderate rate from the date of the bond to the date of the alienation. This makes the landowner a sharer in the profits and leaves it optional with the promoter to decline the purchase should the success of the enterprise not warrant it. The acquisition of the land is the most delicate part of the undertaking and may occupy years in its consummation. It should be completed, so far as the more desirable tracts are concerned, before any work of construction appears, to give an exaggerated idea of prospective enhancement in value of land.

The great water-power companies of New England made two mistakes in reference to land: 1. They did not secure a sufficient quantity of land. 2. They were in too great haste to dispose of it. The land has rapidly passed out of the hands of the companies and become the property of the general public. A class of population has arisen having aims and purposes entirely different from the manufacturers. The latter generally hold the greater part of the property of the municipality, but the power of taxation is with the former. Taxes are levied and expended with usual municipal extravagance, for purposes in which the manufacturers have little direct concern. The manufacturing interest finds itself compelled to pay the larger half of the taxes, with no direct voice in the assessment of the same. Fierce litigation is continually occurring over taxes and assessments. Such a state of affairs could not arise if the company retained control of its land, allowing it to be held on leasehold tenure, but refusing to alienate. By retaining the fee of the land other than that required for manufacturing, not only could the company control the assessment of taxes, but it could control the liquor traffic, vicious resorts, gambling places, and other matters of vital importance to its working people.

Neither is it to be apprehended that the company would neglect improvements essential to the health, comfort, and welfare of the people; for, being the owner of all real estate, it would have the strongest interest in all measures tending to the prosperity and increase of the population and the consequent advance and maintenance of rental values. It could, moreover, probably execute all improvements of this kind for less than half what they ordinarily cost under municipal control.

Electrical Power.—The companies who have developed great water-powers for transmission to distant points do not count upon the enhancement of land values as a part of their profits. In many of these enterprises, however, the power is consumed in the immediate vicinity of the power-house or on land secured for that purpose, in which case this element comes legitimately into consideration.

The Niagara Falls Power Company, as one of the preliminaries of its enterprise, secured 1200 acres of land in the vicinity of its power-house, which it leases to users of its power. This company furnishes power in blocks of 1000 horse-power or more at the rate of $20 per annum per horse-power. This is understood to be 24-hours power, delivered on premises of tenants in the form of electric current. It is an alternating current of 2200 volts with a frequency of 25 cycles per second. To a small user of 10-hour power the rental runs from 1 cent to 5 mills per kilowatt hour, depending on the amount taken, in addition to which there is a charge of 75 cents per month per kilowatt reckoned on the maximum consumption of power. Thus: To a party using an average of 8 kilowatts, and a maximum of

12, per month of 25 working days, which, holidays excluded, is about the average:

For service, $12 \times \$0.75$	$ 9.00
$8 \times 25 \times 10 = 2000$ kilowatt hours @ $0.01	20.00
Total per month........................	$29.00

Being at the rate of $3.625 per kilowatt per month, and for a horse-power $2.72 per month, =$32.64 per annum.

For a 60-kilowatt motor running full capacity at times, but averaging 50 kw. = $50 \times 25 \times 10 = 12\,500$ kilowatt hours per month, the charge would be:

For service, $60 \times \$0.75$	$45.00
10 000 kw. h. @ 0.64 cents.	64.00
2 500 kw. h. @ 0.60 "	15.00
Total per month........................	$124.00

Dividing by 50, the average use of power, this is at the rate of $2.48 per month per kw., or for a horse-power $1.86 per month =$22.32 per annum.

To a party using an average of 240 kw. = 60 000 kw. h. per month and a maximum of 300 kw. the charge would be:

For service, 300 kw. @ $0.75	$225.00
40 000 kw. h. @ 0.55 cents.	220.00
20 000 kw. h. @ 0.50 "	100.00
Total per month........................	$545.00

=$2.27 per month per kw.=$1.70 per month=$20.40 per annum per horse-power.

For power transmitted from Niagara Falls to Buffalo the rental is somewhat higher. The charge for service is one dollar per kilowatt, reckoned on the maximum use. The additional charge by meter is:

For 1 000 units or less per month.... 2 cents per unit
Excess over 1 000 up to 2 000 units 1.5 " " "
" " 2 000 " " 3 000 " 1.2 " " "

Excess over 3 000 up to 5 000 units 1   cent per unit
"   "   5 000 " " 10 000 " 0.8   "   "   "
"   "   10 000 " " 20 000 " 0.75   "   "   "
"   "   20 000 " " 40 000 " 0.70   "   "   "
"   "   40 000 " " 80 000 " 0.66   "   "   "
"   "   80 000. . . . . 0.64   "   "   "

For a small motor, therefore, using an average of 5 h.p. = 3.75 kw. and a maximum of 6 h.p. = 4.50 kw., the charge per month would be:

For service, 4.5 kw. @ $1.00. . . . . $ 4.50
                $10 \times 25 \times 3.75 = 937.5$ units @ $0.02 . . . 18.75

Total per month. $23.25

Being at the rate of $4.65 per month = $55.80 per annum for each horse-power actually used.

For an average use of 450 h.p., which is 360 kilowatts or 90 000 kilowatt hours per month (on a 10-hour basis), but running at times to 450 kilowatts or 600 h.p., the charge, according to the letter of the above table, would be:

For service, 450 kw. @ $1.00 . . . . $450
                1 000 units @ 2   cents   20
                1 000   " "   1.5   "   15
                1 000   " "   1.2   "   12
                2 000   " "   1   "   20
                5 000   " "   0.8   "   40
               10 000   " "   0.75 "   75
               20 000   " "   0.7   "   140
               40 000   " "   0.66 "   264
               10 000   " "   0.64 "   64

Total per month. $1100

Being $\frac{1100}{450}$ = $2.444 per month = $29.33 per annum per horse-power actually used.

The St. Lawrence Power Company, as a preliminary to its development of power at Messina, N. Y., acquired 1600 acres of land, which it holds for sale or rent.   It distributes a current of 2000 volts, an alternating 3-phase current of 3000 alternations. It offers to lease 24-hours power, in units of 1000 h.p. or more, delivered within one-half mile of its power-house, at $15 per h.p. per annum.*

---

* At the date of this third edition we are obliged to chronicle the foreclosure of the bonds of this company, and the sale of its property and franchise at a fraction of their original cost.

CHAPTER XXVII.

TESTING WATER-WHEELS.

Testing-flumes.—The water-wheels now in possession of the American market have been developed by a tentative process,—by making a wheel and observing its action, altering it in some point and re-observing till a tolerably good result is obtained. A good result in one wheel is no guaranty that another wheel bearing the same name will perform equally well. Neither is a good efficiency on a low head a guaranty of equally good performance on a high head. Manufacturers cannot afford to use wheels of low efficiency at any price. Where the water for power costs at the rate of $20 per annum per horse-power, a difference of 5 per cent in the performance of two wheels of 500 nominal horse-power would amount to $500 per annum, justifying an extra expenditure of at least $5000 for the purpose of securing the higher result. In fact the figure of $20 per annum is too low, because the better wheel would furnish 25 horse-power on the shaft more than the other and we should use the price of applied power, not the price of water. This should be taken as high as $40 per annum per horse-power, making a difference in value of $10 000 at least, and more probably of $20 000.

New wheels are constantly being offered to the trade, and assertions as to their efficiency command no attention whatever unless backed by a responsibly certified test. Moreover, in adjusting the rentals of water used for power, records of the head, gate-opening, and velocity of the wheels are largely depended on for determining the quantity of water consumed, a method obviously valueless unless founded upon accurate experimental determination of the discharge under conditions of practice.

629

For these reasons the testing of wheels has become an operation of very frequent occurrence, and special establishments, called testing-flumes, are provided for that purpose. In the early history of the turbine such tests were often called for; but as all apparatus and arrangements had to be specially made and prepared, they were very expensive, the cost usually running up into the thousands of dollars, so that such tests were beyond the means of inventors who had new ideas to offer to the trade. Mr. James Emerson, then of Lowell, a man of much natural sagacity but limited education, first carried out the idea of setting up a permanent establishment for the accommodation of such parties. He set up at Lowell, early in the seventies, a flume where wheels of moderate size could be tested on heads of 5 to 18 feet. This idea commended itself to Mr. Chase, the manager of the Holyoke Water-power Company, so strongly that he offered Mr. Emerson some special inducements to transfer the business to Holyoke. He there set up a timber flume, which subsequently was replaced by a more elaborate establishment constructed by the Holyoke Company, about 1880. This was built of masonry, in a permanent manner, and was suited to the test of wheels of standard size, under heads up to 18 feet. In such an establishment the test of a wheel cost the owner about as many hundreds as it formerly cost thousands. The company has continued, since that time, to test all wheels offered for that purpose, within the capacity of the flume; but the main purpose of the establishment has been to determine the discharge of wheels installed by the company's tenants, as a basis for the assessment of rentals. Wheels have now attained dimensions beyond the capacity of this flume. It is no uncommon thing to find wheels in cases 15 feet diameter, with supply-pipes of 10 feet, requiring with draft-tube and quarter-turn a head of full 30 feet.

Figs. 248 and 249 show a testing-flume suited to the largest modern wheels, while capable of dealing with the smallest. The structure is supposed to rest upon a formation

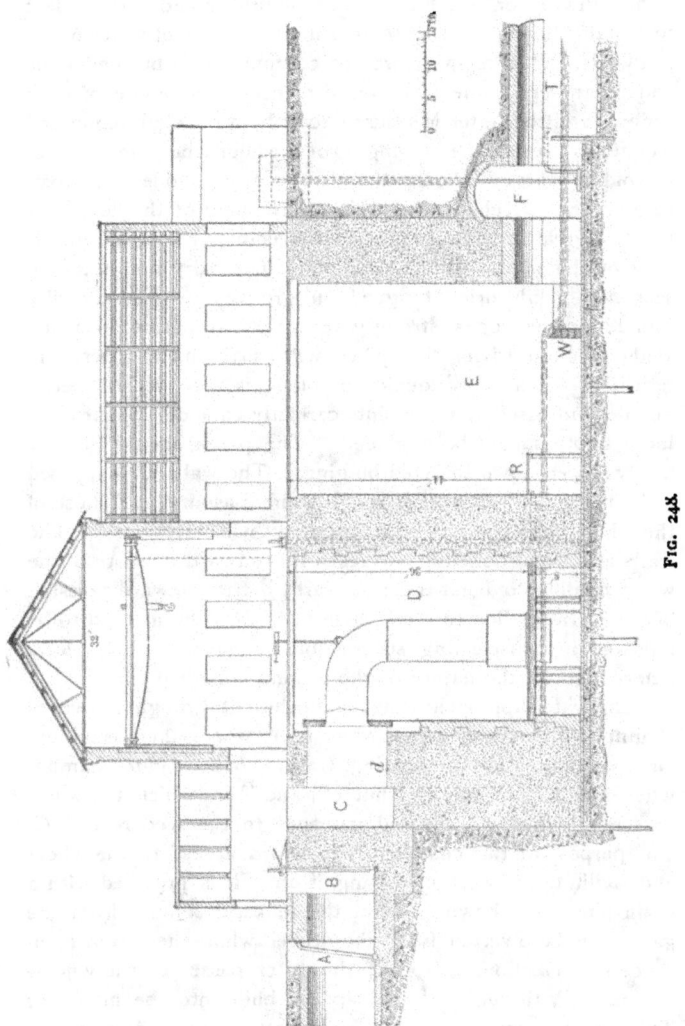

Fig. 248

of firm gravel, on which a grillage of timber and plank is laid
to sustain the masonry and prevent the escape of water under
the walls.   Bearing-piles are not contemplated, but under the
wide chambers a pile or two is driven at the middle of each
timber, and the latter is secured to it by a dovetail tenon and
wedge, to prevent the bulging up of the floor when the chamber
is void of water.   Should the walls settle a trifle, the worst
result to be apprehended is a slight crowning of the floor.   A
point of great importance in such a structure is to secure tightness of the floor.   White-pine plank fairly dry but not long
seasoned or kiln-dried, tongued and grooved, or even carefully
jointed on the edges, strongly set up by means of dogs and
wedges before driving the spikes, will usually make a perfectly
tight job.   If a poorer quality of plank is used, it had better
be laid undressed and the joints carefully calked.   Shakes and
loose knots should be excluded.   The plank should be very
firmly spiked down to avoid bulging.   The walls are supposed
to be formed of concrete or rough stone masonry, and those of
the wheel-chamber are lined internally with brickwork.   The
walls are proportioned not to resist the outward pressure of the
water tending to burst them outward, but the inward pressure
of the earth tending to crush them inward.   The method to be
followed in constructing such a foundation can not be fully
determined till the nature of the ground is known.

The water approaches the establishment through a covered
channel which comes first to view at *A*, where there is a rack
for the elimination of floating trash.   *B* is a gate-chamber
where the supply may be shut off, and from which the water
passes through two covered passages to the feeder-head *C*.
The purpose of this chamber is to afford access to the wheel
and facilitate the work of setting it up.   It is provided with a
drain-pipe, not shown, so that the leakage coming from the
gates can be diverted from the wheel while the latter is in
process of erection.   From *C* the water passes to the wheel-
chamber *D*, through a large pipe, *d*, built into the masonry.
This pipe is large enough for the largest wheel, and ordinarily

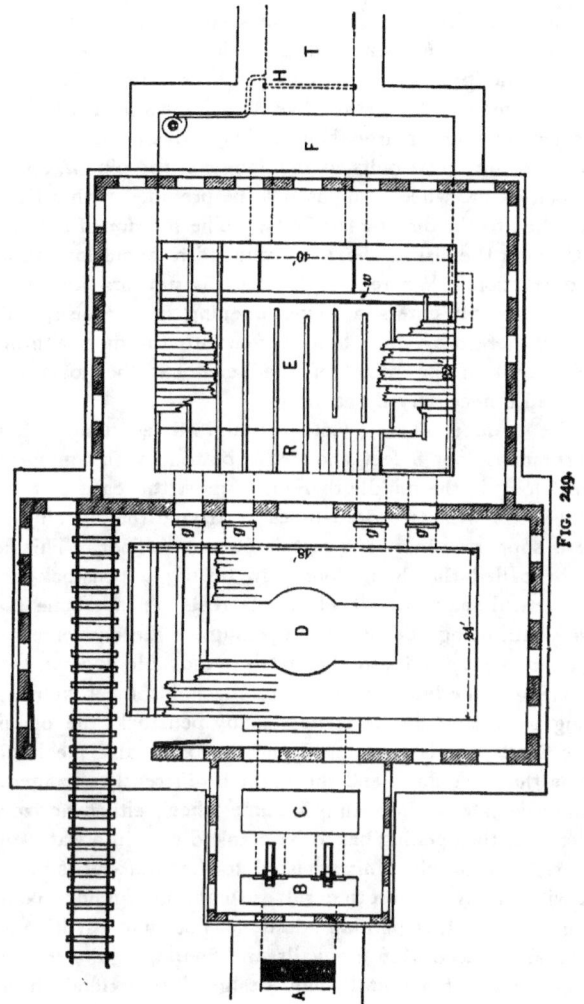

FIG. 249.

its inner end must be partly closed by timber to admit the attachment of the penstock. The section, Fig. 248, shows a wheel in position for test, and the mode of attaching its penstock to the pipe *a*. A circular disk is formed of 12-inch blocks of wood, put together like the fellows of a wheel, and encircled by a strong iron band tightly driven on. This disk is attached by screw-bolts to the flange of the pipe *d*, and the penstock of the wheel, supposed to be provided with a flange, is attached to the disk by lag-bolts. The interior of the opening through the disk is shown as rounded to avoid loss of head by contraction. Where the penstock is not provided with a flange, it simply enters a round opening in the disk, and is confined thereto by short bolts driven into the timber through holes bored in the metal of the penstock, the joint being tightened if necessary by calking.

The chamber *D* is 24 feet wide and 48 feet long. It has a permanent floor 8 feet above the bottom, with an opening 26 feet long in the middle thereof, to meet the case of a horizontal wheel with two draft-tubes. Around this opening, the floor is supported by pillars resting on the pit floor. This floor may be called the flume floor. In testing a penstock-wheel with vertical shaft, the wheel simply rests on the flume floor, loose plank being laid over the opening for the convenience of workmen. At the top of the pit is a movable floor on which are arranged the brake and testing mechanism. Likewise, in testing a horizontal wheel supplied by penstock, the opening in the flume floor is not closed, but the brake and mechanism rest on the same floor, and the brake is differently arranged, as will appear later. In testing a flume-wheel, either horizontal or vertical, the opening has to be planked over in a water-tight manner, and the planking adjusted to the draft-tubes in ways that will readily suggest themselves to an intelligent carpenter. Some wheels will require an additional floor and partition supported and braced from the walls and flooring. For instance, a wheel on a horizontal shaft designed to extend through stuffing-boxes in the flume will require a partition reaching from the flume floor to the water-surface, braced from the walls and made tight by calking against the brickwork. In this case, however, such a disposition can usually be avoided by arrangements which permit the effort of the brake and the velocity of the wheel to be observed above water, the brake itself running under water. Between the chambers D and E are the gates g g g g for the purpose of controlling the head acting on the wheel, when required. By partly closing these gates, in the case of a penstock-wheel, the water in D can be raised to any desired level, submerging the wheel and diminishing the head. In the case of a flume-wheel the closing of the gates diminishes the head by increasing the pressure acting against the discharge. The total head available is supposed to be 32 feet, and the surface of the ground at the building is above the level of the canal from which the feeder draws.

The chamber E, called the weir-chamber, is 40 feet by 32. It contains the weir for measuring the water discharged by the wheels, and the gauges for observing the depth of water on the weir, and the height of water in the wheel-pit. When the water is backed up in the wheel-pit, it will issue through the gates g with great violence, and would be in such a state of commotion in E as to make the depth on the weir uncertain. The rack R, extending entirely across the chamber, is introduced for the purpose of suppressing this disturbance in the water. The rack is surmounted by a platform, above the water-level, which is convenient for observers to stand on. The height corresponding to the pressure in the wheel-pit is shown by a glass pipe affixed to a graduated scale extending up the wall. An iron ladder is permanently fixed to the wall for the convenience of the observer. In testing flume-wheels the height above the wheel is observed in the chamber D, and in penstock-wheels in the chamber C. The water-level in a still-box in the chamber E is observed with a precise gauge to determine the depth of water on the weir, electric lights being introduced to facilitate these observations. A floor of planking resting on iron beams is laid over the weir-chamber, and the room above serves as a shop and storeroom. A removable floor is laid over the chamber *D*, consisting of wooden beams resting in sockets in the masonry not shown in plan, and covered with 2- or 3-inch planking. The beams which sustain the bridge-trees of wheels under experiment are fixed immovably in their sockets by wedges. The room above the last-mentioned floor is commanded by a travelling-crane, all heavy weights being brought within its reach on cars running on a track as shown. In order to give room for this track the building extends somewhat beyond the necessary thickness of the wheel-pit wall. It would not be good engineering to rest such a building partly on masonry and partly on earth. The projecting part is sustained by piers built up from the wooden grillage and covered with arches.

In setting up wheels without draft-tubes, which rest on the bottom of the pit, the latter must be freed from water. For this purpose a barrier is erected at *H* consisting of thick stop-plank set into grooves in the side walls, and the water is removed by a centrifugal pump discharging on the down-stream side of the barrier. The shaft of this pump passes through an opening in the arch over the chambers *F*, ascends through a vertical pipe set in the ground, and is driven by an electric motor in a little shed adjoining the main building. The above are the main features of the establishment. Many modifications would suggest themselves on a careful study such as would devolve upon an engineer charged with its construction.

Testing Mechanism.—The power developed by a wheel undergoing a test is consumed in friction. A stout pulley *B* is attached to the shaft *S*, and its velocity is controlled by a brake, a convenient form of which is shown at Fig. 250. This consists of a hollow strap encircling the pulley and prevented from rotating by the arm l_s. The pressure necessary to be applied to this arm to prevent the brake from rotating is a correct measure of the effort exerted by the wheel, and this pressure combined with the velocity of the wheel correctly measures the power developed by the latter. The pressure of the strap on the wheel is controlled by means of the hand-wheel w.

This application of the power creates great heat. In fact, the whole energy of the wheel is developed in the form of heat,

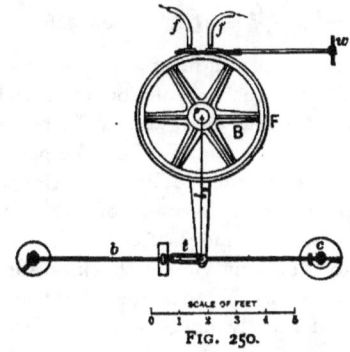

FIG. 250.

at the rate of 1 unit of heat for every 772 foot-pounds of energy, so that a wheel of 500 horse-power would, in a run of an hour, develop heat enough to raise 3 tons of iron to a white heat. To remove the heat as fast as it is generated and prevent the destruction of the apparatus, the hollow strap is traversed by a current of water entering through the flexible pipe f and discharged through f_1. Figs. 250 and 251 relate to a vertical shaft in which the pressure exerted by the

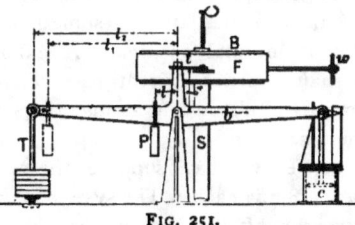

FIG. 251.

arm l_3 must be ascertained by weights acting through a combination of levers. On a horizontal shaft the operation is simpler, the weights being applied directly to the arm l_3. In the present case, the pressure exerted by the arm l_3 acts, through the link t, on the arm l_4 of the lever b, and is balanced by the weight P and the weights on the hanger T. An arrangement of this system of levers convenient for computation is to make $2\pi l_3 \dfrac{l_2}{l_4} = 100$. Then, the expenditure of energy represented by a weight of m pounds on the hanger, is $100m$ foot-pounds for each revolution of the wheel. P is a weight or poise, movable along the scale-beam, its extreme positions being at distances l and l_1 from the point of suspension of the latter. c is a cylinder partly filled with water in which a piston attached to the scale-beam, of a diameter a little less than that of the cylinder, moves freely. The purpose of this attachment is to avoid any violent shock on the beam, as the piston will move slowly, under the slightest pressure, but cannot be moved rapidly under the most violent impulse. The writer does not regard this attachment as essential, but it has usually been applied under that impression, and as it serves the necessary purpose of a counterweight on the scale-beam, it has been adopted in all such mechanism without close consideration of its necessity for the ostensible purpose of avoiding shocks. An upright on the cylinder c carries a short scale showing, by means of a finger on the scale-beam, how much the latter deviates from its horizontal position. Another upright has a slot through which the beam passes, and which limits the movement of the latter to something like an inch above and below its normal position. A rod attached to the free end of the shaft s carries a worm-gear which turns a small wheel and by means of a stud and springs rings a bell at every 100 turns of the wheel.

For convenience we may suppose the weights on the hanger to be 100 pounds each. The system is so proportioned that, with the weights off the hanger, the piston attached and immersed in water, and the poise P in its indicated position, the beam is in exact equilibrium. The poise P has a weight such that at a distance l_1 it is equivalent to a 100-pound weight on the hanger. The value of P is therefore determined from this consideration: With P at the distance l, the beam and attachments are in exact balance. Removing P leaves an unbalanced moment of Pl tending to raise the left end of the beam. Applying P at l_1 calls for an additional moment of $100 l_2$. Therefore $Pl_1 - Pl = 100 l_2$ or $P = 100 \dfrac{l_2}{l_1 - l}$. The distance $l_1 - l$ is divided into 100 equal parts which are numbered toward the left and marked by notches. The advancing of the poise by one of these notches counts as a pound added at T. With five 100-pound weights on the hanger and the poise at the sixty-fifth notch the weight is written 565. If, in this condition, the scale-beam is between its limits and the wheel is making n turns per second, the power developed is $100 n \times 565$ foot-pounds or

$$\frac{100 \times 565 n}{550} \text{ horse-power.}$$

In testing a wheel, one assistant controls the tension on the brake, and aims to so control it as to give the wheel a uniform velocity. Another controls the weights on the scale-beam, aiming to maintain the latter in a horizontal position. Another records the weight on the scale-beam and the time occupied by the wheel in making 100 revolutions. Other assistants note the depth on the weir, the opening of the speed-gate, and the height of water above and below the wheel. From the record so obtained we may compute the power developed by the wheel, the quantity of water discharged, and the head acting thereon. In the case above supposed, if q represent the discharge of the wheel in cubic feet per second, h the head in feet acting on the wheel, y the weight of a cubic foot of water, which, for exact purposes, varies slightly with the temperature, then

$$e = \text{efficiency of wheel} = \frac{100 \times 565 n}{yqh}.$$

After the test is concluded and the water has been shut off
and has dropped in the weir-chamber below the crest of the
weir, its rate of settlement should be carefully noted in order to
determine the rate of leakage and thus correct the quantity q.

INDEX.

WATER-POWER.

CHAPTER I.

NATURAL WATERCOURSES.

PHYSICISTS are wont to assert that the sun is the source of all power upon the earth. It creates fuel by separating carbon from the carbonic acid of the atmosphere and storing it in the substance of plants. In former times, when this constituent of the atmosphere was much more abundant than at present, it thus originated the vast accumulations of vegetable growth which now constitute the coal-mines. By rarefying parts of the atmosphere and disturbing the barometric equilibrium, it creates the winds, which involve enormous power. A current of air one mile wide, 100 feet high, and moving with the moderate velocity of 30 miles an hour represents an expenditure of more than 100 000 horse-power. The sun creates the food which nourishes animals and gives them the strength to perform labor. It also sets in motion the agencies which give rise to water-power.

Watercourses.—About three-fourths of the earth's surface is water, from which evaporation is constantly going on. Water in the form of vapor is constantly rising from the seas. It is distributed by the winds in very unequal measure through-out the land, falls in the form of rain or of snow which resumes the liquid form, and, gathering into streams, pursues its course

www.ingramcontent.com/pod-product-compliance
Lightning Source LLC
Chambersburg PA
CBHW081101170526
45165CB00008B/2285